TAURINE
**Nutritional Value and
Mechanisms of Action**

ADVANCES IN EXPERIMENTAL MEDICINE AND BIOLOGY

Recent Volumes in this Series

TAURINE
Nutritional Value and Mechanisms of Action

Edited by

John B. Lombardini

Texas Tech University Health Sciences Center
Lubbock, Texas

Stephen W. Schaffer

University of South Alabama School of Medicine
Mobile, Alabama

and

Junichi Azuma

Osaka University Medical School
Osaka, Japan

PLENUM PRESS • NEW YORK AND LONDON

Library of Congress Cataloging-in-Publication Data

Taurine : nutritional value and mechanisms of action / edited by John
  B. Lombardini, Stephen W. Schaffer, and Junichi Azuma.
        p.   cm. -- (Advances in experimental medicine and biology ; v.
  315)
     "Proceedings of the Waltham Symposium on Taurine and Cat
  Nutrition, held October 8, 1991; and of the International Taurine
  Symposium: New Dimensions on its Mechanisms of Action, held October
  9-10, 1991, in Orange Beach, Alabama"--T.p. verso.
     Includes bibliographical references and index.
     ISBN 0-306-44224-8
     1. Taurine--Physiological effect--Congresses.   2. Cats--Nutrition-
  -Congresses.   I. Lombardini, J. Barry.   II. Schaffer, S. W.
  III. Azuma, Junichi.   IV. Waltham Symposium on Taurine and Cat
  Nutrition (1991 : Orange Beach, Ala.) V. International Taurine
  Symposium: New Dimensions on its Mechanisms of Action (1991 : Orange
  Beach, Ala.) VI. Series.
  QP801.T3T39 1992
  599'.019245--dc20   92-14174
                                                                  CIP

Proceedings of the Waltham Symposium on Taurine and Cat Nutrition,
held October 8, 1991; and of the International Taurine Symposium:
New Dimensions on Its Mechanisms of Action, held October 9-10, 1991,
in Orange Beach, Alabama

ISBN 0-306-44224-8

© 1992 Plenum Press, New York
A Division of Plenum Publishing Corporation
233 Spring Street, New York, N.Y. 10013

Printed in the United States of America

PREFACE

The underlying philosophy of these two symposia on taurine remains the same as all those that have been held previously: the best way to remain current in the subject matter is to talk directly with the investigators at the forefront of the field. Thus, we brought together some 50 individuals from 11 different countries who have keen interests and active research programs in the many-faceted areas of taurine research. The meetings were held on October 8-10, 1991, in an elegant setting in a resort area at Orange Beach, Alabama, approximately 50 miles outside of Mobile on the Gulf Coast. The meetings were programmed as two separate Symposia held sequentially. The first symposia on October 8 was devoted exclusively to taurine research in the cat, primarily in the area of nutrition, and entitled "The Waltham Symposium on Taurine and Cat Nutrition". The second symposia on October 9 and 10 was open to all fields of taurine research and was entitled "International Taurine Symposium: New Dimensions on its Mechanisms of Actions".

If the philosophy of these meetings was to bring both experts and novices together in a discussion and presentation of current taurine research, then the major purpose of the Proceedings is to document the current research efforts and to present an objective summary of where the taurine field stands today and where the focus will be in the future. By providing this basic and critical information, we hope we have made the job easier and perhaps provided some future direction in the expanding areas of taurine research.

Finally, we would like to thank a number of people and organizations who helped make these two symposia possible. First, the contributors of this proceeding who provided the scientific and intellectual curiosity to generate the data which were vigorously discussed. Also thanked are all other individuals who attended the meetings and shared in the discussions. Second, the organizing committee composed of Akemichi Baba, Flavia Franconi, Simo Oja, Herminia Pasantes-Morales, and Jang-Yen Wu. Third, we thank the following companies and foundations for their most generous financial support: Taisho Pharmaceutical Co., Ltd., Waltham Centre for Pet Nutrition, Uehara Memorial Foundation, and Mead Johnson Nutritional Group. And last, we acknowledge the following individuals Ms. Christine Eaton who helped with transportation of the guests, Mr. J. Duan the projectionist, and Mrs. Kim Schaffer who gave excellent advice in moments of crises before and during the meetings. A special thanks must be given to our two secretaries Mrs. Mary Alba and Mrs. Josie Aleman who had the arduous and heroic task of retyping most, if not all, the manuscripts and entering them into a central computer.

John B. Lombardini
Stephen W. Schaffer
Junichi Azuma

WALTHAM: ITS AIMS AND ACHIEVEMENTS

G.D. Wills and I.H. Burger

Waltham Centre for Pet Nutrition
Waltham-on-the-Wolds
Melton Mowbray, Leics, LE14 4RT
United Kingdom

WALTHAM has been closely associated with taurine since the late 1970's when we conducted studies on the taurine requirement of the cat in collaboration with Dr. Keith Barnett of the UK Animal Health Trust. We are, therefore, particularly pleased and proud to be co-sponsors of this International Taurine Symposium. However, our involvement with taurine research is only one area in which WALTHAM has made significant contributions to the global pool of knowledge of companion animal nutrition.

The name WALTHAM signifies the internationally recognised research authority which provides the science behind the brands of the Mars Pet Care companies worldwide, such as Pedigree Petfoods in the UK and Kal Kan Foods Inc. in the USA. At the heart of WALTHAM research is the WALTHAM Centre for Pet Nutrition (WCPN), located near Melton Mowbray in England, which represents our focus of knowledge in nutrition, behaviour and husbandry of cats, dogs and both cage and wild birds.

Mars Pedigree Petfoods founded its first Nutrition Research facility in the UK in 1965, and in 1973 the Centre was moved to its present site at Waltham-on-the-Wolds. At this time, a Fish Research Centre was also established. In 1989, the animal facilities at the WALTHAM Centre were subjected to a state-of-the-art expansion scheme and we have successfully created conditions for our resident animals which approximate those of a domestic environment.

In 1990, the WALTHAM Centre for Equine Nutrition and Care was established in Verden, Germany, where maintenance, growth and performance studies in a number of different breeds of horse are being undertaken.

The Fish Research Facility has been updated, and in 1991, the WALTHAM Aquacentre was set up in Leeds, England. In addition to providing an advisory service for the public, the Aquacentre conducts research into the nutritional and energy needs for ornamental fish; methods of quantifying palatability; and investigating optimum conditions for aquaria.

WALTHAM research is aimed at determining the factors that will enhance the health and quality of life of companion animals and its scope embraces a wide range of subjects. Basic nutrition is not necessarily optimum nutrition, and in the pursuit of the elucidation of this concept, studies are carried out into palatability; dietary management; feeding behaviour; husbandry and care; the role of pets in society; the human-pet relationship; owner/animal expectations; general behaviour; and food science and technology. To this end, WALTHAM associates are drawn from a variety of disciplines including veterinary surgeons, nutritionists, biochemists, behaviourists, animal husbandry specialists and food technologists, all of whom have specific expertise and experience and are internationally recognised for their contribution to the field of companion animal nutrition and welfare.

WALTHAM research is widely documented and over 200 WALTHAM publications are currently cited in the literature. In addition to its own research projects, WALTHAM also supports collaborative studies with universities and research institutes worldwide. Our work is widely acknowledged and, in several instances, has been decisive in helping establish dietary recommendations by official bodies such as the National Research Council (NRC) of the United States Academy of Sciences.

WALTHAM encourages the dissemination of its body of knowledge by a variety of means. As well as publishing articles in the scientific and the lay press, WALTHAM has recently introduced its own clinical veterinary journal called "WALTHAM International Focus" which is translated into 8 languages and distributed to companion animal veterinary surgeons in 25 countries. WALTHAM associates give lectures and practical instruction at conferences, colleges and other institutes around the world, and organise the internationally acclaimed "WALTHAM Symposia" of which the first was held in 1977. These symposia bring together experts in a particular field and provide the opportunity for a thorough discourse on the chosen subject or subjects. They have become an important event in the veterinary calendar and the proceedings are widely distributed. The WALTHAM Symposium on Taurine and Cat Nutrition brings the total in this series to well over 20.

Our early work on taurine confirmed that the cat requires a dietary source of taurine to maintain the structural integrity of the retina, and subsequent investigations established the level that was required to maintain the adult cat in adequate taurine status[1]. Our interest in this enigmatic nutrient has been unremitting, and our most recent research (reported at this symposium) suggests that, in some cats, cysteic acid may be used as a precursor in the biosynthesis of taurine.

Over the year, WALTHAM has built up a formidable reputation for its research into the energy requirements of companion animals. The complex nutritional profile of the cat represents a particular challenge in the field of pet nutrition. One area that WALTHAM has studied extensively is the energy intake and bodyweight changes that occur in pregnant and lactating queens. Our results have provided the most definitive account of the nutritional aspects of the reproductive cycle. We demonstrated that, unlike most other mammals, the cat gains weight at a steady rate throughout pregnancy, and this is thought to be due to the deposition of extra-uterine tissue which can be mobilised when demands are high in late pregnancy and lactation[2]. Scientists at the WCPN have also designed unique "kittening cabins" which allow kittens to feed separately from the queen whilst still remaining with their mother, thus permitting a more accurate assessment of the queen's requirements prior to weaning. Our own data on the energy requirements of cats[3,4] and the energy content of cat foods[5] were incorporated into the NRC's 1986 recommendations in "Nutrient Requirements of Cats".

In the dog, adult bodyweights can vary between 1 kg and 115 kg, the widest range within any single species, and this parameter is the single most important factor in determining the dog's energy requirements. We have been investigating the effect of breed differences on predicted energy requirements and have found that large breeds do not always follow the general pattern, for example Newfoundlands require less and Great Danes require more energy than the predicted amounts[6]. This research has been further developed through the use of a direct whole body calorimeter in which energy production can be assessed with a high degree of accuracy. We have recently published data on the allometry of energy requirements in the dog using this calorimetric technique[7].

In both dogs and cats, glucose is required as an energy source, but it has been shown that, in these species, this can be provided by the breakdown of protein. Research at WCPN has shown that assuming an adequate supply of dietary protein is available, carbohydrate is not an essential component of the diet[8].

Cats are traditionally thought to have a higher protein requirement than dogs and this has been confirmed by studies carried out at the WCPN[9]. These also provided evidence of practical importance that cats are unable to adapt to diets that have a very low protein content. The NRC also employed our data in their recommendations for the minimum protein requirements of adult dogs[10], and of dietary levels of methionine required by immature dogs[11].

The nutritional needs of an animal are inter-related with its behavioural needs. Although we conduct studies on many aspects of companion animal behaviour, it is their feeding behaviour that is of particular interest to us. We have found that food preferences are affected mainly by odor and taste, but that other characteristics such as texture and temperature also play an important role. By understanding these factors, we can not only improve the palatability of our products, but we can also utilize these factors to encourage inappetant, ill or simply fussy individuals to eat.

At WALTHAM, our concern is not restricted to healthy pets, and a significant part of our research deals with the ways in which many disease conditions may be managed by dietary means. We have formulated a range of special diets that may be dispensed by the veterinary profession to assist in the dietary management of certain clinical conditions as obesity, feline lower urinary tract disease, food allergy and renal disease. The accumulation of plaque is a major cause of periodontitis in the dog and the cat. WALTHAM has embarked on an extensive programme of investigation into the causes of this disease and the preventative measures that may be taken, such as adjusting the texture of the food. Early indications are that progress in this area can be made.

The bond that exists between a pet and its owner is not only special, but of great interest and importance. We are currently supporting a number of research projects in this field and a recent "WALTHAM Symposium" (No 20)[12] was devoted entirely to this subject. There is now growing evidence that pet ownership can exert some very positive effects on human health, behaviour and well-being.

The achievements of WALTHAM research are considerable, and within the petfood industry as well as the veterinary and nutrition communities, WALTHAM continues to act as an international catalyst for the development and improvement of nutritionally balanced, prepared petfoods. Our work on taurine and involvement in this Taurine Symposium are some examples of this commitment to pet care and nutrition. We are confident that companion animals will continue to benefit from WALTHAM research for many years to come.

REFERENCES

1. I.H. Burger and K.C. Barnett, The taurine requirement of the adult cat, *J.S.A.P.* 23:533-537 (1982).
2. G.G. Loveridge and J.P.W. Rivers, Bodyweight changes and energy intakes of cats during pregnancy and lactation, *in*: "Nutrition of the Dog and Cat," I.H. Burger and J.P.W. Rivers, eds, pp. 113-132, Cambridge University Press, Cambridge, U.K. (1989).
3. G.G. Loveridge, Bodyweight changes and energy intakes of cats during gestation and lactation, *Anim. Technol.* 37:7-15 (1986).
4. P.T. Kendall, S.E. Blaza, and P.M. Smith, Comparative digestible energy requirements of adult Beagles and domestic cats for bodyweight maintenance, *J. Nutr.* 113:1946-1955 (1983).
5. P.T. Kendall, I.H. Burger, and P.M. Smith, Methods of estimation of the metabolizable energy content of cat foods, *Fel. Pract.* 15(2):38-44 (1985).
6. A. Rainbird and E. Kienzle, Studies on the energy requirement of dogs depending on breed and age, *Kleintierpraxis* 35:149-158 (1990).

7. I.H. Burger and J.V. Johnson, Dogs large and small: the allometry of energy requirements within a single species, *J. Nutr.* 121:S18-S21 (1991).

8. S.E. Blaza, D. Booles, I.H. Burger, Is carbohydrate essential for pregnancy and lactation in dogs?, *in*: "Nutrition of the Dog and Cat," I.H. Burger and J.P.W. Rivers, eds, pp. 229-242, Cambridge University Press, Cambridge, U.K. (1989).

9. I.H. Burger, S.E. Blaza, P.T. Kendall, and P.M. Smith, The protein requirement of adult cats for maintenance, *Fel. Pract.* 14(2):8-14 (1984).

10. P.T. Kendall, S.E. Blaza, and D.W. Holme, Assessment of endogenous nitrogen output in adult dogs of contrasting size using a protein-free diet, *J. Nutr.* 112:1281-1286 (1982).

11. S.E. Blaza, I.H. Burger, D.W. Holme, and P.T. Kendall, Sulphur-containing amino acid requirements of growing dogs, *J. Nutr.* 112:2033-2042 (1982).

12. I.H. Burger, "Pets, Benefits and Practice," British Veterinary Association Publications, London (1990).

TAURINE AND TAISHO PHARMACEUTICAL CO., LTD.

Yasuhide Tachi

Research and Development Planning and Coordination
    Division
Taisho Pharmaceutical Co., Ltd.
Toshimaku, Tokyo, 171
Japan

Taisho Pharmaceutical Co., Ltd., which was founded in Japan in 1912, has had a long term interest in taurine. Initial interest in taurine was generated when Taisho Pharmaceutical Co., Ltd. was informed that during World Wars I and II naval doctors administered taurine to soldiers to enhance their scotopic vision during night in order to increase their ability to detect enemy ships and also to recover from fatigue. In 1941 the company conducted product planning and in 1949 began selling "Taurine Extract", the first taurine product.

When the taurine extract was first sold, taurine was thought to be an all-round medicine. It was used for patients with acute diseases, such as pertussis and pneumonia and for bedsores. The extract was used most frequently for patients with diseases related to the heart and the liver, and it was believed to be effective in detoxification. The raw material for the extract (not to be confused with the synthetic chemical product) was extracted from octopus.

The main drugs in Japan at the time were those which promoted nutrition, such as vitamins and amino acid preparations, and maintained the homeostatic function of the body.

Under these conditions, it was thought that taurine, which has effects similar to those of some vitamins and is an amino acid, was necessary for maintaining health. In fact, this concept remains unchanged even after many studies have been reported.

Lipovitan D, a product containing taurine which still dominates the market, was sold for the first time in 1962. Since then, based on its physiological effects, taurine has been used widely in such products as tonics, drugs for the circulatory, nervous, and digestive systems and eyedrops. Subsequently, the company has produced more than 100 taurine-containing products although some have been changed or discontinued. At present, taurine is included in 20 products, such as health agents that contain vitamins, tonics and eyedrops.

Lipovitan D has retained its position as a top selling product for 30 years since it was introduced into the market in 1962. This product, which contains vitamin B_1, B_2 and B_6 and taurine (1,000 mg), is used as a nutrient in cases of physical fatigue or pyrogenic infectious diseases, and as a tonic. It is a unique liquid agent born in the Japanese culture. Since Lipovitan D was first exported to Taiwan in 1963, it has been exported or produced in Southeast Asia, the Middle East, Europe and other areas.

Among the various effects of taurine, the usefulness in ameliorating the sequalae of congestive heart failure and hyperbilirubinemia was recognized in a double blind study conducted in 1984. Accordingly, taurine powder (Taisho) was approved. Its main effects on the heart are protection of the myocardial cells against Ca^{2+} overload and the improvement in cardiac function. The main effects of taurine on the liver are the acceleration of bile acid secretion, stimulation of hepatic cells and amelioration of certain abnormalities in hepatic function. Therefore, agents such as taurine which have been used for many years as health agents and tonics are now also being used widely as a state of the art treatment in Japan.

Taurine, the main ingredient of the internationally marketed product Lipovitan D, is produced by chemical synthesis at Taisho's factory. This material is exported in its original form, as well as Lipovitan D, to various countries. It has found wide use as an additive for infant formula and for pet foods.

In accordance with the development of taurine products, Taisho Pharmaceutical Co., Ltd has carried out academic studies on taurine both inside and outside the company. In Japan, the Japanese Research Society on Sulfur Amino Acids was established in 1978. Taisho Pharmaceutical Co., Ltd. supported their annual research meetings and the publication of their academic journal for 10 years.

Taurine has been reported to have various physiological effects such as a nerve modulator and a regulator of cellular membrane function. Its pharmacological effects when administered externally include a depressor effect and a cholesterol-reducing effect. As for the effects of taurine as a nutrient, it has been shown that a retinopathy and cardiomyopathy may develop in taurine-deficient kittens and in newborns. However, there are still many phenomena regarding the action of taurine, including its mechanism of action, which remain to be clarified. Scientific studies of the physiological, pharmacological and nutritive effects of taurine may well be in their infancy.

Taisho Pharmaceutical Co., Ltd. supports studies of taurine throughout the world. The company is committed to disseminating information, manufacturing useful products and improving the health of people throughout the world.

CONTENTS

THE WALTHAM SYMPOSIUM ON TAURINE AND CAT NUTRITION

INTERNATIONAL TAURINE SYMPOSIUM:

NEW DIMENSIONS ON ITS MECHANISMS OF ACTION

SECTION I: TAURINE AND THE HEART

SECTION II: TAURINE: DEVELOPMENTAL ASPECTS

SECTION III: TAURINE AND THE CENTRAL NERVOUS SYSTEM

SECTION IV: TAURINE AND ITS ACTION AS
AN ANTI-OXIDANT IN LUNG

SECTION V: TAURINE AND OSMOREGULATION

SECTION VI: TRANSPORT AND METABOLISM OF TAURINE AND RELATED COMPOUNDS

REVIEW: TAURINE DEFICIENCY AND THE CAT

John Sturman

New York State Institute for Basic Research
   in Developmental Disabilities
Staten Island, NY   10314

INTRODUCTION

Although taurine was discovered more than 160 years ago, and much important research concerning its chemistry and biochemistry has been reported since that time, the real importance of taurine in biology has only been recognized within the past two decades. It is fair to say that the impetus for this later surge of interest in taurine was started by the first International Symposium on taurine held in March 1975, which was followed on a regular basis by nine others held around the world, including this one in October 1991 in Mobile, Alabama. These meetings have contributed the definitive literature on the subject in the form of proceedings and have inspired countless smaller workshops and symposia devoted to taurine under the auspices of a variety of scientific societies. It is not unreasonable to state that the single biggest factor to elucidating the many functions of taurine has been the special importance of this compound for the cat, and the recognition of the importance of this animal in taurine research. It is ironic that the landmark paper of Hayes and colleagues published in May 1975, which first associated a deficiency of dietary taurine with feline central retinal degeneration, had been submitted two months prior to the first taurine meeting, and yet none of us at the meeting were aware of this seminal observation. Subsequent meetings, of course, recognized this observation, which was soon followed by others relating taurine to such diverse actions in the cat as reproduction, development, cardiac function and immune function. This most recent meeting has seen the final recognition of the actions of taurine in the cat by the designation of a separate day's symposium devoted solely to this important species - The Waltham Symposium on Taurine and Cat Nutrition sponsored by the Waltham Centre for Pet Nutrition, England. This separation proved not only practical, since there were 11 separate presentations on the role of taurine in the cat, but also successful as judged by the numerous probing questions and intense discussion. The consensus of this meeting's participants was that this format should continue at future International Taurine meetings.

Taurine, Edited by J.B. Lombardini *et al.*
Plenum Press, New York, 1992

International Meetings Devoted Totally or Predominantly to Taurine Research

1975   International Symposium on **Taurine**, March 24-26, Tucson, Arizona.
Proceedings: *Taurine* (eds. R. Huxtable & A. Barbeau) Raven Press, 1976.

1977   2nd International Symposium on **Taurine in Neurological Disorders**, March 28-30. Tucson, Arizona.
Proceedings: *Taurine and Neurological Disorder* (eds. A. Barbeau & R.J. Huxtable) Raven Press, 1978.

1979   21st Annual A.N. Richards Symposium of the Physiological Society of Philadelphia - **The Effects of Taurine on Excitable Tissues**, April 23-24, Valley Forge, Pennsylvania.
Proceedings: *The Effects of Taurine on Excitable Tissues* (eds. S.W. Schaffer, S.I. Baskin, & J.J. Kocsis) Spectrum Publications, 1981.

1979   3rd International Meeting on **Low Molecular Weight Sulfur Containing Natural Products**, June 18-21, Rome, Italy.
Proceedings: *Natural Sulfur Compounds - Novel Biochemical and Structural Aspects* (eds. D. Cavallini, G.E. Gaull, & V. Zappia) Plenum Press, 1980.

1980   International Symposium - **Taurine: Questions and Answers**, November 16-18, Mexico City, Mexico.
Proceedings: *Taurine in Nutrition and Neurology* (eds. R.J. Huxtable & H. Pasantes-Morales) Adv. Exptl. Med. Biol. vol. 139, Plenum Press, 1982.

1982   International Symposium and Fifth Annual Meeting of the Japanese Research Society on Sulfur Amino Acids, August 7-10, Tokyo, Japan.
Proceedings: *Sulfur Amino Acids: Biochemical and Clinical Aspects* (eds. K. Kuriyama, R.J. Huxtable, & H. Iwata) Prog. Clin. Biol. Res. vol. 125, Alan Liss, 1983.

1984   Satellite Symposium of the 9th IUPHAR Congress of Pharmacology - **Taurine: Biological Actions and Clinical Perspectives**, August 6-8, Hanasaari, Finland.
Proceedings: *Taurine: Biological Actions and Clinical Perspectives* (eds. S.S. Oja, L. Ahtee, P. Kontro, & M.K. Paasonen) Prog. Clin. Biol. Res. vol. 179, Alan Liss, 1985.

1986   Symposium on **Sulfur Amino Acids, Peptides, and Related Compounds**, October 6-8, San Miniato, Italy.
Proceedings: *The Biology of Taurine: Methods and Mechanisms* (eds. R.J. Huxtable, R. Franconi, & A. Giotti) Adv. Exptl. Med. Biol. vol. 217, Plenum Press, 1987.

1989   Satellite Symposium of the 12th ISN Meeting - **Functional Neurochemistry of Taurine**, April 19-22, Moguer, Spain.
Proceedings: *Taurine: Functional Neurochemistry, Physiology, and Cardiology* (eds. H. Pasantes-Morales, D.L. Martin, W. Shain, & R. Martín del Río) Prog. Clin. Biol. Res. vol. 351, Wiley-Liss, 1990.

1991   **The Waltham Symposium on Taurine and Cat Nutrition**, October 8, Mobile, Alabama. **International Taurine Symposium: New Dimensions on Its Mechanisms of Action**, October 9-10, Mobile, Alabama.

Proceedings: *Taurine: Nutritional Value and Mechanisms of Action* (eds. J.B. Lombardini, S.W. Schaffer and J. Azuma) Adv. Exptl. Med. Biol., Plenum Press, 1992.

1993   Proposed: Cologne, Germany.

1995   Proposed: Kyoto, Japan.

Summary of Taurine Deficiency Effects on the Cat.

The dietary essentiality of taurine for the cat is now well established. The activity of cysteinesulfinic acid decarboxylase, the limiting enzyme on the pathway of taurine biosynthesis, is two orders of magnitude lower in the cat than in rodents. Although the cat can synthesize limited amounts of taurine endogenously, unless it receives adequate taurine in the diet, it becomes taurine-depleted. All tissues and organs have decreased taurine concentrations under such circumstances, although the rate of loss of taurine varies considerably. It is also clear that the availability of taurine and its precursors depends to some extent on the composition of the diet, a subject of great interest to nutritionists and pet food manufacturers. For example, a dry purified diet containing 0.05% taurine is sufficient to maintain the body taurine pools and prevent the abnormalities associated with taurine depletion, whereas the same amount of taurine in a wet commercial meat-byproduct diet can result in taurine depletion. Although it has been known for many years that cats are susceptible to central retinal degeneration, the cause was only recognized to be taurine deficiency in 1975. These changes, which are bilaterally symmetrical, are characterized by degeneration and disintegration of the photoreceptor cell outer segments, beginning in the area centralis and gradually extending across the entire retina. Ultrastructurally these changes appear as vesiculation and disorientation of the regular stacks of outer segment membranes followed by increasing disintegration. Retinal degeneration is accompanied by impaired vision as measured by reduced electroretinogram responses, reduced visual evoked potentials, reduced visual acuity, and eventually complete blindness. Taurine depletion also results in degeneration of the tapetum lucidum, a layer of cells behind the retina which acts as a biological mirror and reflects unabsorbed light back through the retina, maximizing retinal sensitivity. These are unique cells, derived from modified choroidal cells, which contain a large nucleus and regular parallel arrays of tapetal rods, each of which is surrounded by a membrane. Taurine deficiency results in disintegration of this membrane followed by collapse of the rods into droplets and eventually by destruction and loss of the entire cell. Such changes further compromise vision in taurine-depleted cats.

Over the last several years, myocardial failure based on echocardiographic evidence of dilated hearts in domestic cats has been associated with low plasma taurine concentrations. This condition was reversible by nutritional taurine therapy if treated in time, and led to the fortification of commercial cat foods, which already contained taurine, with additional taurine. In the laboratory, cats fed purified taurine-free diets develop echocardiographically-defined cardiomyopathy, and sometimes die suddenly, but do not have the extended clinical manifestations demonstrated by domestic cats. Furthermore, the cardiac taurine concentration in apparently healthy cats fed a purified taurine-free diet for many years is substantially lower than in domestic cats which died from myocardial failure. These observations have led to the speculation that some other component of commercial cats diets is involved in addition to taurine deficiency. The nature of the changes which occur in taurine-depleted cardiac muscle are unknown at present, although no obvious ultrastructural pathology has been noted.

A number of studies have shown that immune responsiveness can be adversely affected by inappropriate nutrition. While the majority of these studies have focused on changes in the immune response as a result of protein nutrition, immune competence is also affected by alterations in dietary trace elements (zinc, copper, selenium) and vitamins (A,E, B-6, and folate). Studies reported at this meeting and at the previous meeting show that there are changes in the host defense mechanisms in cats fed diets deficient in the amino acid taurine. These host-defense mechanisms are present to protect (animals) individuals against infections with pathogenic organisms (bacteria, viruses, parasites) and against the development and spread of malignant tumors. Studies reported previously showed quantitative and functional changes in the cells that comprise the immune system (B and T cells, monocytes and neutrophils). Histological changes in the spleens of taurine-depleted cats were consistent with depletion of B-cells and T-cells and follicular center reticular cells. The spleen, a lymphoid organ which is essentially a compartmentalized collection of lymphocytes and macrophages has an important function in clearing the blood of infectious organisms. Lung lavage fluid from cats fed taurine-deficient diets also contained a reduced proportion of neutrophils and macrophages which produced increased quantities of reactive oxygen intermediates associated with lower taurine concentrations. This observation is of great interest and importance since in other species supplemental taurine has been demonstrated to protect against pulmonary damage caused by ozone, nitric oxide, bleomycin and paraquat. Thus, histological changes in the spleens, as well as quantitative and functional changes in the peripheral blood and alveolar lavage leukocytes of cats fed a taurine-deficient diet signals potential alterations in host-defense mechanisms. These recent discoveries promise to provide exciting updates at future meetings.

Taurine deficiency has a profound adverse effect on feline pregnancy and outcome of the progeny. Such females have difficulty in maintaining their pregnancies, frequently resorbing or aborting their fetuses. Aborted fetuses have an abnormally wide spread in size and degree of development, and some with hydrocephalus have been noted. Abnormal relaxin and progesterone values have been noted in such pregnant cats. Pregnancies which do reach term frequently result in stillborn and low-birth-weight kittens, survivors have abnormally slow growth rates and exhibit a variety of neurological symptoms. Morphological examination of such kittens has revealed widespread abnormalities, especially in the nervous system, including retina, visual cortex, cerebellum and dorsal roots. A number of the ultrastructural changes described are consistent with damaged or weakened membranes. More extensive and detailed studies continue and also promise to figure prominently at future meetings. One extension of research deriving from these observations is the use of kitten cerebellar cells in culture, which allows many interesting manipulations to be applied, and detailed results have appeared at the last two meetings. Perhaps the single most stimulating and puzzling observation is that the role of taurine and analogues such as ß-alanine and guanidinoethanesulfonic acid which compete with taurine for the ß-amino acid uptake system, appears quite different for cat cells than for mouse cells. Undoubtedly, the explanation for this difference is of great importance, but as yet, is unknown.

Other differences between cats and rodents are emerging from the use of antibodies to taurine and other amino acids, prepared by using the free amino acid conjugated to bovine serum albumin with glutaraldehyde. Rodent cerebellum shows strong taurine-like immunoreactivity in Purkinje cells and their dendrites and in some granule cells. Adult cat cerebellar Purkinje cells, in contrast, are virtually devoid of taurine, and contain instead GABA and glutamate. In newborn kittens, cerebellar Purkinje cells appear to contain only glutamate. By weaning at eight weeks after birth, cat cerebellum is strongly positive for taurine, especially in Purkinje cells and their dendrites, and in granule cells, which both contain glutamate as well, but not GABA. The explanation for this intriguing developmental sequence in cat neurons awaits further study. Weaning kittens from taurine-deficient mothers are virtually devoid of taurine in their cerebellar Purkinje cells, dendrites, and granule cells, which show up as "ghosts". Such kittens also contain large numbers of reactive astrocytes which are rarely found in control kittens, perhaps the result of increased permeability of the blood-brain barrier.

Another unexpected phenomenon reported at this meeting is the distribution of taurine within cat muscles. In skeletal muscle there is strong taurine-like im-munoreactivity in some fibers while others nearby contain virtually none. The proportion of taurine-positive fibers is increased by denervation and decreased by electrical stimulation. Cardiac muscle and smooth muscle of the small intestinal wall, in contrast, contain taurine uniformly in all fibers. This distribution in skeletal muscle is also shared by rodents, and promises to shed further light on the role of taurine in muscle function.

This summary of abnormalities in cats attributable solely to dietary taurine deficiency is impressive and likely to be extended by ongoing and future research. The association of taurine deficiency with damaged membranes and/or altered membrane function in most, if not all, of these abnormalities is striking. It should be noted that all felines, including lions and tigers, have this exquisite dietary dependence on taurine. The opposite dietary situation, namely an excess of dietary taurine, was presented for the first time at this meeting, and found no obvious abnormalities in adult cats or their offspring resulting from prolonged feeding of such diets, probably to the relief of the sponsor!

TAURINE SYNTHESIS IN CAT AND MOUSE IN VIVO AND IN VITRO

Ekkhart Trenkner, Alice Gargano, Philip Scala, and John Sturman

Department of Developmental Biochemistry
Institute for Basic Research in Developmental Disabilities
Staten Island, NY   10314

Taurine, 2-aminoethanesulfonic acid, is essential for the development and survival of mammalian cells, in particular cells of the cerebellum and retina (Sturman, 1988). Taurine is the second most abundant free amino acid in mammals; its concentration in the CNS reaches the millimolar range. The highest concentration of taurine occurs in newborn and early postnatal brain, suggesting a role during development. In fact, two lines of evidence strongly suggest the importance of taurine in early postnatal cerebellar development of cat and mouse. Female cats raised on a taurine-deficient diet produce offspring with severe brain abnormalities (Sturman et al., 1985). Morphological analysis of cerebella of 8-week-old taurine-deficient kittens revealed the presence of an external granule cell layer and many granule cells still in the process of migrating to the internal granule cell layer. Furthermore, some granule cells in the external layer were still dividing. These observations of perturbed granule cell migration were accompanied by greatly reduced cerebellar taurine concentrations.

In this respect the taurine-deficient kittens resemble the neurological mutant mouse, weaver (wv/wv). Wv/wv is viewed as an autosomal recessive, single-gene mutation that affects various cell types of the cerebellum (Rakic and Sidman, 1973; Sotelo, 1975) and the striatum (Roffler-Tarlov and Graybiel, 1984). The wv/wv defect appears to involve both taurine and glutamate. The results of Roffler-Tarlov and Turey (1982) showed that concentrations of glutamate and taurine in the cerebellum of early postnatal wv/wv were reduced, and suggested that these changes might be caused by a reduction in the number of taurine- and glutamate-containing granule cells. Our results show that the addition of extracellular taurine counteracts the wv/wv defect *in vitro*, suggesting that the reduction of taurine or the distortion of the stoichiometric relationship between glutamate and taurine is functionally related to taurine and not only to cell loss (Trenkner, 1990). Thus, the weaver mutant mouse provides a suitable system to better understand the function of taurine.

During adult life taurine appears to play the role of a modulator of Ca^{2+} and Cl^- influx, acting directly on Ca^{2+} or Cl^- channels (for review Huxtable, 1992). It also acts as an osmoregulator, removing water from neuronal cells after depolarization (Pasantes-Morales and Schousboe, 1988; 1989), thereby playing a putative role in excitatory events. We have proposed that taurine acts as a modulator during excitotoxicity because it can protect neurons from excitotoxic death in the presence of glutamate and kainate but not in the presence of 3-hydroxyanthranilic acid, an

Taurine, Edited by J.B. Lombardini *et al.*
Plenum Press, New York, 1992

oxidative reagent which irreversibly alters cell membrane permeability (Trenkner, 1990). We conclude, therefore, that taurine is required not only in the developmental process but also in the prevention of excitotoxicity-inflicted injuries. Our preliminary results suggest that a stoichiometric relationship between glutamate- and taurine-concentrations has to be maintained in order to fulfill this suggested function. If this protective mechanism is physiologically important in the brain, it is conceivable that taurine synthesis may be inducible in order to provide sufficient taurine concentration after injury.

It has been generally accepted that the major pathway of taurine synthesis in the brain is through the decarboxylation of cysteinesulfinic acid to hypotaurine by cysteinesulfinic acid decarboxylase (CSAD) and the subsequent oxidation of hypotaurine to taurine. CSAD is considered the rate limiting step in taurine synthesis (Wu, 1982; Griffith, 1983; De La Rosa and Stipanuk, 1985). Divergent results were obtained pertaining to the specificity of brain CSAD. While Wu (1982) reported a brain specific CSAD activity which is different from the CSAD in liver, Tappaz's group described isoforms of CSAD in brain, one of which copurifies with GAD. Both CSAD (I and II) activities were found in brain and liver (Remy et al., 1990).

There is agreement, however, that species-specific differences exist in CSAD activity. While sufficient taurine is synthesized in rodents and many other mammals to maintain normal levels, in cats, monkeys and man, the required taurine concentrations are provided through dietary intake. This suggests to us that cat, with respect to taurine-utilization, is more closely related to man than mouse. Therefore, kittens appear to be an ideal system in which to study the role of taurine and its synthesis under controlled conditions. This was further substantiated *in vitro*. We developed a cell culture system for kitten cerebellar cells (Trenkner and Sturman, 1991) similar to that developed for early postnatal mouse cerebellum (Trenkner and Sidman, 1977) and analyzed cerebellar cell survival and function from kittens of taurine-supplemented and taurine-depleted mothers. Unexpectedly, kitten cerebellar cells died in the presence of taurine and survived in the presence of the taurine-uptake competitors β-alanine and GES, in contrast to cerebellar cultures of mice and particularly of the granule cell-deficient mutant weaver, in which taurine prevented granule cell death and restored granule cell function (Trenkner, 1990). This suggests a different role of taurine in cat and mouse.

This study compares taurine synthesis in the developing mouse and kitten cerebellum *in vivo* and *in vitro* by measuring CSAD activity. Since CSAD activity is significantly different in mouse and cat brain tissue such a comparison might explain the discrepancies between these two species and thus provides a better understanding of taurine's general role(s) in the development and maintenance of brain tissue. Our model (Figure 1) predicts a stoichiometric relationship between glutamate and taurine which has to be maintained for normal development and neuron survival. If this relationship is disturbed, e.g. through injury, taurine concentrations could be adjusted locally and transiently through regulation of CSAD activity. We report here the induction of CSAD activity after injury in three different systems: after cell death in the wv/wv cerebellum, after disruption of tissue integrity (tissue culture) and after excitotoxic injury.

METHODS

Cysteinesulfinic acid decarboxylase (CSAD) [EC 4.1.1.29] was measured by collecting the $^{14}CO_2$ produced from [1-^{14}C] cysteinesulfinic acid (Research Products International, IL) under conditions described by Jacobsen et al. 1964. Although it has been demonstrated that in brain CSAD is a different protein than glutamic acid

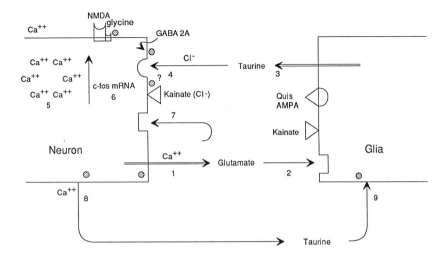

○ Site of Taurine action

Figure 1. The relationship between taurine and glutamate during neuron glia interaction; A Model. Glutamate is released from granule cells (1) and stimulates astroglial cells (2) to release taurine (3). Taurine then stimulates granule cells independent of external Ca^{2+} (4) to release Ca^{2+} from internal pools (5) which leads to a significant stimulation of c-fos mRNA (6). The induction of c-fosmRNA by glutamate at P1-P4 was insignificant. Glutamate (7) or taurine (4) stimulate neuronal release of taurine (8), which is taken up into glia (9). Possible sites of taurine action are indicated. However, which site(s) control or regulate the induction of CSAD is yet unknown.

decarboxylase (GAD); the latter enzyme can utilize cysteinesulfinic acid as substrate. Therefore, decarboxylation of cysteinesulfinic acid by GAD was prevented by saturating the enzyme with unlabeled substrate (100 mM L-glutamate), which will not contribute to the labeled CO_2 produced.

Tissue Culture

Cell cultures of early postnatal and mouse mutations and cat have been described in detail elsewhere (Trenkner and Sidman, 1977; Trenkner et al., 1978; Trenkner 1991; Trenkner and Sturman, 1991). In brief, cells from trypsinized cerebellum reaggregate in BME and 10% HS within hours into clusters that later develop interconnections consisting of fiber bundles of neurites, fascicles of granule cell axons, and astroglial processes, with cells migrating along their surfaces. This micro-tissue culture system allows the analysis of the dynamics of cell behavior important to histogenesis. Granule cells in several stages of differentiation, some larger neurons and two types of glial cells form reproducible non-random patterns. Axonal and dendritic processes and synapses are generated. This system is based on cell-cell recognition rather than cell-substrate interaction. In this study cerebellar cells from 7-day-old kittens from mothers raised on a diet of 0.05% taurine, or 7-day-old mice were cultured in BME supplemented with 10% horse serum. The horse serum batch used in these experiments contained 5×10^{4}M taurine.

RESULTS AND DISCUSSION

It is well established that rodents can synthesize ample taurine throughout their lifetime while cats, monkeys and humans acquire taurine mainly through dietary intake. Brain taurine concentration remains nearly constant after the blood brain barrier has been in place and taurine cannot readily penetrate the blood-brain-barrier. Although the taurine concentration varies from species to species it appears characteristic for taurine that in all species the blood brain barrier tightly controls taurine levels in the CNS, (Sturman, 1979). This is independent of taurine concentration in the blood-stream since high blood-taurine levels do not change brain-taurine levels (Sturman and Messing, 1992; these proceedings). Furthermore it is difficult to reduce brain taurine in most species, the cat being an exception. In cat, brain taurine levels rapidly decline (6 to 10 times) when fed a taurine-deficient diet (for review Sturman and Hayes, 1980). Thus the blood-brain-barrier controls taurine uptake which, if our hypothesis is valid (Figure 1), raises the question of how the balance between glutamate and taurine can be maintained, after injury. Does the barrier open transiently or is taurine synthesis enhanced? In adult rat cerebellum CSAD has been localized immunocytochemically (Chan-Palay et al., 1982; Almarghini et al., 1991). Although it remains controversial from these studies whether only neurons or both neurons and glial cells contain the enzyme, it is certain that the enzyme is present in rat brain cells after the blood-brain-barrier is in place. Unfortunately, when applied to cat and human brain, the antibodies available did not result in any staining (Trenkner, unpublished results). Since CSAD activity in cat brain is low (Sturman, 1988; Table 1), the amount of enzyme protein might be insufficient or the structure of the cat enzyme may be different from that of rodents.

This study measured CSAD activity in mouse brain throughout development *in vivo* and in mouse and cat cerebellar cell cultures. If our hypothesis is valid that taurine modulates excitatory and excitotoxic events (Trenkner, 1990), it is conceivable that in such events the established taurine concentrations (intracellular and extracellular) are not sufficient to fulfill this function. Thus, taurine concentrations might be regulated by inducing taurine synthesis locally and transiently, and/or by changing the high-affinity uptake system.

We found in mouse cerebellum that CSAD activity is developmentally regulated, reaching the lowest activity around postnatal day 14 (Figure 2), the approximate time of the blood- brain-barrier formation. Subsequently, the activity increases to mature levels.

Table 1. CSAD activity in cerebrum and liver of developing wild-type and weaver mice

Age	Cerebrum		Liver	
	+/+	wv/wv	+/+	wv/wv
P 3	1.96	1.25	n.t.	n.t.
P 5	7.8	n.t.	191	n.t.
P 7	6.8	3.9	185	220
P14	11.7	11.3	137	171
P22	59.2	64.0	213	173
P45	44.1	43.1	24.3	323

n.t. = not tested.
The results represent the average value of at least 5 determinations per age group.

In the weaver mutant mouse, where taurine levels were found to be reduced (Roffler-Tarlov and Turey, 1983) and granule cell neurons die before maturation, CSAD activity of the cerebellum was elevated, particularly at times of increased cell death [postnatal day 4 (P4)] (Figure 2) and was significantly higher than in normal mice (P21) at the time of synapse formation between Purkinje cell dendrites and granule cell axons, which in wv/wv is perturbed. This result may be the consequence of impaired synapse formation and the increase in glutamate and taurine in the extracellular space.

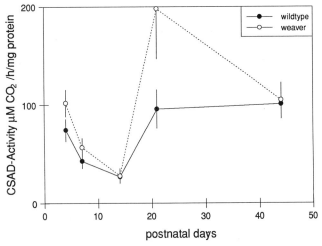

Figure 2. CSAD activity of the mouse cerebellum is developmentally regulated. Depending on the protein concentration per cerebellum, 1-4 mouse cerebella from one litter were pooled and analyzed per age group. The data represent the average of 3-7 sets of tissue per age group. The error-bars show the range of variation among parallel measurements.

In cultures of seven day old mouse cerebellar cells CSAD activity changed as a function of time (Figure 3). When compared to cerebellar tissue the activity was enhanced after dissociation (injury) of the tissue into single cells. The activity was significantly induced in the presence of 10^{-2} M β-alanine, a taurine-uptake blocker, whereas extracellular taurine (10^{-2} M) had no significant inductive effect, suggesting that intracellular taurine levels might be regulated according to cellular needs; if cells are depleted of taurine by β-alanine, synthesis is induced. Significant increases in enzyme activity were observed when excitotoxic concentrations of glutamate (10^{-3} M) were added. This suggests to us that CSAD activity is induced in order to increase the taurine concentration to levels necessary to balance extracellular glutamate and thus prevent excitotoxic cell death. Of course, more experiments are needed to substantiate this teleological interpretation.

In early postnatal kitten cerebellum and cerebrum CSAD activity was barely measurable whereas the activity in liver was easily measurable (Table 2), confirming previous results (Sturman, 1988).

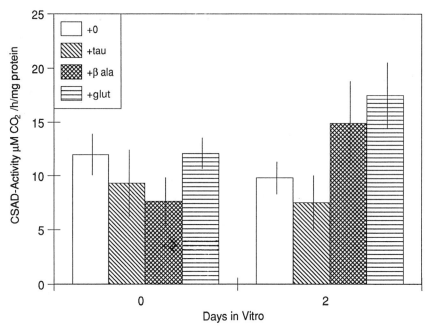

Figure 3. Induction of CSAD-activity in mouse cerebellar cultures. Cerebellar cells of 7-day-old mice were cultured for 2 days in the presence of taurine (10^{-2} M), β-alanine (10^{-2} M) or glutamate (10^{-3} M) compared with no addition.

Table 2. CSAD activity in kittens from mothers fed 0.05% taurine

Age	Cerebellum	Cerebrum	Liver
P 5	0.89	0	60.6
P 7	1.99	0	25.1

The results represent the average values of 2 determinations per age group of at least three animals.

In vitro, CSAD activity can be significantly induced in the cat (Figure 4). However, certain distinct differences exist between cat and mouse in the utilization and regulation of taurine:

1. Similar to mouse, early postnatal cerebellar tissue of kittens has significantly lower CSAD activity than dissociated cells while trypsinization alone had no effect suggesting that injury through trituration and possibly taurine depletion induces CSAD-activity.

2. CSAD activity increases with time in culture (3 days *in vitro*).

3. The temporal increase in CSAD activity of cultured cells is observed in the presence of 10^{-2} M taurine.

4. β-Alanine, which induces taurine-release and blocks taurine-uptake, induces maximal CSAD activity after one day in culture. This suggests that CSAD is transiently activated after taurine-depletion. Preliminary evidence shows similar effects in the presence of GES (not shown).

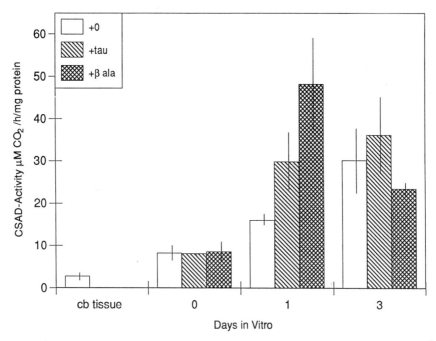

Figure 4. Induction of CSAD activity in early postnatal kitten cerebellar cells in culture. $2 - 4 \times 10^7$ cells per condition were analyzed for CSAD activity after 0, 1 and 3 days *in vitro*. Cerebellar cells were incubated throughout the culture period with 10^{-2} M taurine or ß-alanine and compared with cultures without addition.

These results clearly demonstrate that CSAD can be activated in cat cerebellar cells after injury (cell preparation). Also, both in kittens and mice, 10^{-3} M glutamate induces CSAD activity, supporting our hypothesis that the stoichiometric relationship between glutamate and taurine maintains the integrity of neuronal cells and tissue. Our model (Figure 1) also predicts a relationship between extracellular and intracellular concentrations of both amino acids. And here might lie the difference between mouse and cat. While in the mouse extracellular taurine does not induce or suppress CSAD activity, CSAD activity was induced by extracellular taurine in kitten. Considering the role of β-alanine and its effect on the induction of CSAD activity, we hypothesize that in cat the balanced relationship between extracellular and intracellular concentrations of taurine is controlled by synthesis rather than uptake.

In conclusion, we have shown that in mouse taurine synthesis is developmentally regulated. This synthesis can be stimulated by depletion of intracellular taurine concentration (β-alanine, GES), by dissociation of cerebellar tissue into single cells, through a genetic defect as in the weaver mutant mouse and through excitotoxic concentrations of glutamate. All of these conditions lead to the release of intracellular taurine and, therefore, to the alteration of the stoichiometric relationship between glutamate and taurine which, we postulate, is required to maintain the integrity of neuronal tissues. Therefore, physiological concentrations of taurine outside the cell help maintain the proper balance between taurine and glutamate thereby preventing cell death.

This hypothesis also applies to cats which exhibit rather low CSAD activity *in vivo*, but as shown here, exhibit significant activity *in vitro*. However, the regulation of intracellular taurine concentration appears to be different from that of the mouse since extracellular taurine appears to regulate (induce) CSAD activity in cerebellar cells. This indicates that kitten cerebellar cells regulate intracellular taurine concentration through biosynthesis. However, it might also be the response to injury since in the presence of extracellular taurine kitten cerebellar cells *in vitro* do not survive longer than 18 h. (Trenkner and Sturman, 1991).

REFERENCES

Almarghini, K., Remy, A., and Tappaz, M., 1991, Immunocytochemistry of the taurine biosynthesis enzyme, cysteine sulfinic acid decarboxylase, in the cerebellum: evidence for a glial localization, *Neuroscience* 43:111.

Chan-Palay, V., Lin, C.T., Palay, S., Yamamoto, M., and Wu, J. Y., 1982, Taurine in the mammalian cerebellum: demonstration by autoradiography with ^{3}H taurine and immunocytochemistry with antibodies against the taurine-synthesizing enzyme, cysteinesulfinic acid decarboxylase, *Proc. Natl. Acad. Sci. USA* 79:2695.

De La Rosa, J. and Stipanuk, M.H., 1985, Evidence for a rate-limiting role of cysteine sulfinate decarboxylase activity in taurine biosynthesis in vivo, *Comp. Biochem. Physiol.* 81B:565.

Griffith, O.W., 1983, Cysteinesulfinate metabolism, *J. Biol. Chem.* 258:1591.

Huxtable, R.Y., 1992, The physiological actions of taurine, *Physiological Reviews* 72:101.

Jacobsen, J.G., Thomas, L.L., and Smith, L.H., 1964, Properties and distribution of mammalian L-cysteine sulfinate carboxylases, *Biochim. Biophys. Acta* 85:103.

Pasantes-Morales, H. and Schousboe, A., 1988, Volume regulation in astrocytes: a role for taurine as osmoeffector, *J. Neurosci. Res.* 20:505.

Pasantes-Morales, H. and Schousboe, A., 1989, Release of taurine from astrocytes during potassium-evoked swelling, *Glia* 2:45.

Rakic, P. and Sidman, R.L., 1973, Weaver mutant mouse cerebellum: Defective neuronal migration secondary to abnormality of Bergmann glia, *Proc. Natl. Acad. Sci. USA* 70:240.

Remy, A., Henry, S., and Tappaz, M., 1990, Specific antiserum and monoclonal antibodies against taurine biosynthesis enzyme cysteine sulfinate decarboxylase: identity of brain and liver enzyme, *J. Neurochem.* 54:870.

Roffler-Tarlov, S. and Graybiel, A.M., 1984, Weaver mutation has differential effects on the dopamine-containing innervation of the limbic and non-limbic striatum, *Nature* 307:62.

Roffler-Tarlov, S. and Turey, M., 1982, The content of amino acids in developing cerebellar cortex and deep cerebellar nuclei of granule cell deficient mutant mice, *Brain Res.* 247:65.

Sotelo, C., 1975, Anatomical, physiological and biochemical studies of cerebellum from mutant mice: II Morphological study of cerebellar cortical neurons and circuits in the weaver mouse, *Brain Res.* 94:19.

Sturman, J.A., 1979, Taurine in developing rat brain: changes in blood brain barrier, *J. Neurochem.* 32:811.

Sturman, J.A., 1988, Taurine in development, *J. Nutr.* 118:1169.

Sturman, J.A. and Messing, J.M., 1992, High dietary taurine effects on feline tissue taurine concentrations and reproductive performance, *J. Nutr.* 122:82.

Sturman, J.A., Moretz, R.C., French, J.H., and Wisniewski, H.M., 1985, Taurine deficiency in the developing cat: persistence of the cerebellar external granule cell layer, *J. Neurosci. Res.* 13:405.

Trenkner, E., 1991, Cerebellar cells in culture, in: "Culturing Nerve Cells," G. Banker and K. Goslin, eds., MIT Press, p. 283.

Trenkner, E., 1990, The role of taurine and glutamate during early postnatal cerebellar development of normal and weaver mutant mice, in: "Excitatory Amino Acids and Neuronal Plasticity, Y. Ben-Ari, ed., Plenum Press, Adv. Exp. Med. Biol 268:239-244.

Trenkner, E., Matten, M.E., and Sidman, R.L., 1978, Ether-soluble serum components affect *in vitro* behavior of immature cerebellar cells in weaver mutant mice, *Neuroscience* 3:1093.

Trenkner, E. and Sidman, R.L., 1977, Histogenesis of mouse cerebellum in microwell cultures: cell reaggregation and migration, fiber- and synapse formation, *J. Cell Biol.* 75:915.

Trenkner, E. and Sturman, J.A., 1991, The role of taurine, β-alanine and glutamate in the survival and function of cerebellar cells in cultures of early postnatal cat, *Int. J. Develop. Neurosci.* 9:77.

Wu, J.Y., 1982, Purification and characterization of cysteic acid and cysteine sulfinic acid decarboxylase and L-glutamate decarboxylase from bovine brain, *Proc. Natl. Acad. Sci. USA* 79:4270.

AMINO ACID INTERACTION WITH TAURINE
METABOLISM IN CATS

Elke A. Trautwein and K.C. Hayes

Foster Biomedical Research Laboratory
Brandeis University
Waltham, MA 02254

INTRODUCTION

Taurine deficiency represents an important nutritional problem in cats because they cannot synthesize appreciable quantities of taurine, even though substantial taurine is required for biological functions, e.g. bile metabolism and vision[1,2]. Although some taurine is synthesized from methionine and cysteine, the main pathway of taurine biosynthesis appears to be limited by low activity of cysteine sulfinic acid decarboxylase. Thus, a dietary source of taurine is considered essential for maintaining normal health in cats, rendering taurine an essential amino acid for this species[3,4]. Feeding cats purified, taurine-free diets depletes plasma taurine to barely detectable levels[5-8]. Supplementation with sulfur amino acids failed to improve plasma taurine status substantially (i.e. clinically)[8,9]. By contrast, a large intake of dietary cystine (5%) restored plasma and platelet taurine concentrations to approximately 50% of normal, and taurine concentrations in white and red blood cells approximated those in taurine-supplemented cats[6]. However, this high level of dietary cystine caused debilitating neurological symptoms which resulted in death of some cats[6].

In earlier studies, retinal degeneration appeared in cats fed diets formulated with casein, whereas diets containing lactalbumin or egg albumin were protective[10]. This suggested that protein quality or amino acid availability may have an impact on taurine metabolism. With this concept in mind, we found that feeding a reduced level of protein (25% casein) in a taurine-free diet supplemented with arginine, threonine and cystine (relatively more abundant in lactalbumin and egg albumin than casein) mitigated the degree of plasma taurine depletion[11]. Others found that 0.2% cysteic acid added to a purified taurine-free diet normalized taurine levels in plasma and whole blood of cats[12]. The above observations supported the notion that certain aspects of protein quality and quantity, including availability of specific essential amino acids, could affect taurine metabolism.

Thus, in a previous study[13] we explored the possibility that adding limiting amino acids (arginine and threonine) in addition to methionine and cysteine to a purified casein-based diet would improve taurine status of taurine-depleted cats. Supplementing methionine (0.3%), cystine (0.5%), arginine (0.1%) and threonine (0.2%)

substantially increased plasma taurine concentrations from $10 \pm 13\,\mu\text{mol/L}$ at the end of depletion to $29 \pm 15\,\mu\text{mol/L}$ after 150 days of supplementation. Whole blood taurine concentrations rebounded from $67 \pm 54\,\mu\text{mol/L}$ to $146 \pm 85\,\mu\text{mol/L}$. These data suggested that the combination of arginine and threonine with sulfur amino acids either improved taurine synthesis or affected taurine utilization. In the present study, we examined the impact of individual supplements of these amino acids on plasma and whole blood taurine status during more severe taurine depletion.

MATERIAL AND METHODS

Eighteen adult cats, nine males (5 castrated) and nine females (all unspayed) weighing 3285 ± 890 g (mean $\pm$ SD) were included in the study. The cats represented a closed colony of domestic shorthairs, raised from birth on purified diets and housed in stainless steel cages in a temperature and light-controlled room. All experimental animal procedures were approved by the IACUC. Water was provided ad libitum and food intake was allocated daily according to the calculated energy requirement (60 kcal per kg body weight per day). Individual food intake was assessed daily. Cats were weighed frequently and body weight curves as well as food intake measurements were compiled as an index of food intake.

All cats received a taurine-free diet based on 35% casein (Table 1) to deplete their plasma and whole blood taurine pools for 40 days. Subsequently, in period 1 all cats received the taurine-free purified diet supplemented with arginine (0.1%) *and* threonine (0.2%) for 50 days. In period 2 cats were fed the same diet supplemented with methionine (0.3%) and cystine (0.5%) with *either* threonine (9 cats) *or* arginine (9 cats) for 60 days. In period 3 threonine or arginine were removed from the diet and all cats received a purified diet supplemented only with sulfur amino acids for another 50 days. Finally, in period 4 all cats received a diet supplemented with methionine, cystine, threonine and arginine for 120 days.

Throughout the experimental periods blood samples were taken in the morning before feeding but without deliberate fasting at the following time points: at the end of the taurine depletion period; at day 20 and 50 within period 1; at day 25 and 60 within period 2; at day 25 and 50 within period 3 and at days 25, 50, 80 and 120 within period 4. On the day of bleeding, cats were sedated with a 12-25 mg i.v injection of ketamine hydrochloride (Ketaset®, Aveco Co Inc., Fort Dodge, IA) and 2-3 mL of blood was drawn from the jugular vein into a disposable plastic syringe containing approximately $100\,\mu\text{l}$ of 10% EDTA as an anticoagulant. The blood was immediately transferred into a plastic tube and kept at room temperature prior to plasma separation. All blood collecting and handling adhered to these techniques which proved the most reliable in our recent analysis of procedures[14]. Recent findings in humans[14] and cats[4,13] suggest that a combination of both plasma and whole blood taurine are more predictive of the taurine status than plasma taurine alone. Thus, plasma and whole blood taurine were analyzed by high performance liquid chromatography (HPLC) as previously described[14,15].

Repeated-measures analysis and one-factor analysis of variance (ANOVA) were utilized to evaluate diet and gender effects on plasma and whole blood taurine concentrations throughout the experiment. Results were considered significant at the 95% confidence level.

RESULTS

Following the taurine deprivation period, plasma and whole blood taurine concentrations declined to $12 \pm 9\,\mu\text{mol/L}$ and $129 \pm 60\,\mu\text{mol/L}$, respectively (Table 2).

Table 1. Composition of purified diets

Ingredient	Taurine free	plus Threonine and Arginine	plus Methionine Cystine Threonine	plus Methionine Cystine Arginine	plus Methionine and Cystine	plus Methionine Cystine Threonine Arginine
			g/100 g diet (dry basis)[1]			
Casein	35.0	34.7	34.0	34.1	34.2	33.9
Beef tallow	25.0	25.0	25.0	25.0	25.0	25.0
Cornstarch	16.0	16.0	16.0	16.0	16.0	16.0
Sucrose	10.0	10.0	10.0	10.0	10.0	10.0
Cellulose	5.7	5.7	5.7	5.7	5.7	5.7
Mineral-Mix[2]	6.4	6.4	6.4	6.4	6.4	6.4
Vitamin-Mix[3]	0.6	0.6	0.6	0.6	0.6	0.6
Choline chloride	0.3	0.3	0.3	0.3	0.3	0.3
Agar	1.0	1.0	1.0	1.0	1.0	1.0
Methionine			0.3	0.3	0.3	0.3
Cystine			0.5	0.5	0.5	0.5
Threonine		0.2	0.2			0.2
Arginine		0.1		0.1		0.1

[1] Diets were fed as gels, prepared by premixing the 1% agar with 100 mL of simmering water to form a gel to which the remaining ingredients were added.

[2] Buffington-Rogers Salt Mix, F 8551, BioServ, Frenchtown, NJ

[3] Hayes Cat Vitamin Mix contains (g/kg): dextrin, 944; inositol, 10; α tocopheryl acetate (500 IU/g), 20; niacinamide, 8; calcium pantothenate, 5; retinyl acetate (500000 IU/g), 5; riboflavin, 1.6; thiamin-HCl, 0.8; pyridoxine-HCl, 0.8; folic aicd, 0.8;; menadione, 0.1; cholecaliferol (400000 IU/g) 0.625; biotin, 0.04; cyanocobalamin, 0.03.

17

Table 2. Plasma and whole blood taurine concentration after amino acid supplementation

Diet/amino acid supplementation	Time in days	Plasma taurine	Whole blood taurine
		μmol/L	
pre-depletion[1]		119 ± 31	397 ± 121
Taurine depletion (end of period)	40	12 ± 9[ab]	129 ± 60[a]
Period 1. plus threonine <u>and</u> arginine	20	3 ± 4[a]	108 ± 77
	50	2 ± 3[bcd]	74 ± 75[ab]
Period 2. plus methionine, cystine and threonine <u>or</u> arginine[2]	25	11 ± 14[c]	124 ± 69[bc]
	60	10 ± 12[d]	93 ± 51[d]
Period 3. plus methionine <u>and</u> cystine	25	6 ± 4	83 ± 56[c]
	50	3 ± 3[cfgh]	53 ± 34[defg]
Period 4. plus methionine, cystine, threonine <u>and</u> arginine	25	11 ± 11[e]	78 ± 45
	50	12 ± 10[f]	124 ± 61[e]
	80	10 ± 7[g]	109 ± 50[f]
	120	25 ± 23[h]	142 ± 63[g]

Values are mean ± SD. Means sharing a common superscript are significantly different ($p<0.05$) using repeated-measures ANOVA and Fisher's protected least significant difference test.

[1] Cats received a commercial dry cat food (Alpo Premium, Alpo Petfoods, Allentown, PA) prior to taurine depletion.
[2] Since no significant difference (using an unpaired student's t-test) was found in plasma and whole blood taurine concentrations between threonine <u>or</u> arginine in combination with methionine and cystine, values were combined.

Period 1. Supplementing the taurine-free diet with only threonine *and* arginine resulted in the continued decline in both plasma and whole blood taurine (Table 2). After 50 days on this regimen, plasma taurine had decreased to 2 ± 3 μmol/L and whole blood taurine was reduced to 74 ± 75 μmol/L.

Period 2. Adding either threonine <u>or</u> arginine in combination with methionine and cystine significantly increased plasma taurine (10 ± 12 μmol/L after 60 days) and whole blood taurine (93 ± 51 μmol/L after 60 days). Student's t-test (unpaired) was used to determine whether plasma and whole blood taurine concentrations differed in response to either threonine or arginine supplementation (period 2). No differential effect was detected. For example, threonine plus methionine and cystine produced a plasma taurine concentration of 8 ± 4 μmol/L after 60 days whereas adding arginine in combination with the sulfur amino acids resulted in a concentration of 11 ± 17 μmol/L. Whole blood taurine concentrations were 94 ± 34 μmol/L for threonine vs 92 ± 66 μmol/L for arginine. Accordingly, the threonine and arginine data were pooled and treated as one dietary manipulation to allow further evaluation of the data by repeated-measures and factorial ANOVA.

Period 3. When threonine and arginine were removed and only methionine and cystine remained as supplements in the diet, both plasma and whole blood taurine concentrations again declined significantly (Table 2). After 50 days plasma taurine had again decreased to 3 ± 2 μmol/L and the whole blood taurine concentration was depressed to $53 \pm 34 \ \mu\text{mol/L}$, the lowest concentration achieved throughout the entire study.

Period 4. Resupplementing with arginine and threonine in addition to the sulfur amino acids resulted in the steady progressive increase in both plasma and whole blood taurine concentrations by day 120 to levels significantly elevated above depletion and higher than values during any of the previous treatment periods. (Table 2). Analysis of variance (ANOVA) using repeated-measures revealed a highly significant difference in plasma and whole blood taurine concentrations due to dietary amino acid availability.

Fluctuations in the taurine concentrations were not related to changes in food intake or body weight. As depicted in Figure 1, body weight did not change significantly during the different feeding periods. Analysis of variance (ANOVA) did not reveal a significant difference in body weight between dietary treatments, even though body weights between male and female cats were significanty different with male cats being heavier than females (Figure 1).

When the plasma and whole blood taurine concentrations were evaluated on the basis of gender, differences in the rate of depletion for plasma and whole blood taurine concentrations were not significant, although female cats tended to deplete more slowly and maintained somewhat higher whole blood taurine concentrations (Figure 2).

DISCUSSION

The present study adds a new dimension to our understanding of the plasma and whole blood taurine status in cats. Our findings indicate that methionine and cystine supplementation alone failed to generate clinically improved plasma and whole blood taurine status in taurine-depleted cats, even though these sulfur amino

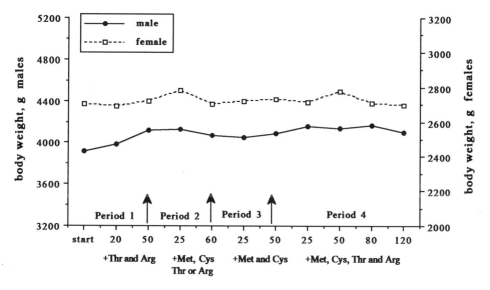

Figure 1. Body weight of male and female cats during the course of the study. Values are mean $\pm$ SD with n = 9 cats at each timepoint. Analysis of variance indicated no significant difference in body weights due to dietary treatment.

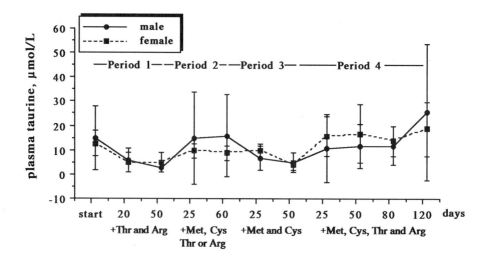

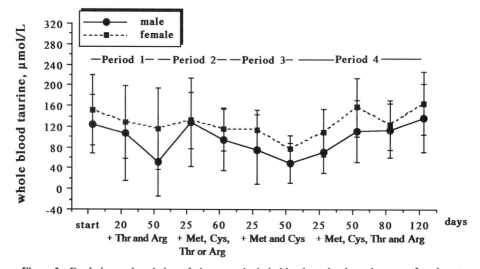

Figure 2. Depletion and repletion of plasma and whole blood taurine in male versus female cats.

acids represent direct precursors in taurine biosynthesis. In earlier studies[8,9], supplementation of a casein-based diet with methionine and/or cystine also failed to improve plasma taurine substantially (i.e. clinically). Supplementing our cats with threonine and arginine combined also had no effect on taurine status. In contrast, additional supplementation with either threonine *or* arginine in combination with sulfur amino acids exerted a modest positive effect on the plasma and whole blood taurine pools. The increase was even more pronounced when both threonine and arginine were fed along with the sulfur amino acids (period 4). However, the duration (60 days) of supplementation with either threonine *or* arginine (period 2) may not have been long enough to elicit the full effect of either alone.

The findings in this study corroborate our earlier experiments on arginine and threonine supplementation[11], where feeding a taurine-free diet with a reduced level of casein (25%), but supplemented with threonine, arginine and cystine, attenuated the severity of plasma taurine depletion. More recently we found that supplementing threonine and arginine as well as sulfur amino acids in a taurine-free purified diet noticeably enhanced plasma and whole blood taurine in taurine-depleted cats[13]. In fact, plasma taurine increased from 10 ± 13 μmol/L at the end of taurine depletion to 29 ± 15 μmol/L after 150 days of amino acid supplementation and whole blood taurine concentration was restored from 67 ± 54 μmol/L to 146 ± 85 μmol/L. In the present study plasma and whole blood taurine experienced a similar repletion, although more severe taurine depletion was present initially in this study (plasma and whole blood taurine concentrations were depressed to 3 ± 2 μmol/L and 53 ± 34 μmol/L, respectively).

Taken together these data imply that the combination of arginine and threonine with sulfur amino acids seemed to have a positive effect on the pool of taurine available for metabolic functions. Improving the protein quality presumably either facilitated taurine synthesis, modulated enterohepatic conservation of taurine, or enhanced kidney tubular reabsorption of taurine. The fact that improvement only occurred in the presence of excess sulfur amino acids suggests that the effect was on taurine synthesis, i.e. improved conversion of cysteine to taurine was facilitated by threonine and arginine. Modifying dietary protein quality may have influenced the activity of enzymes involved in taurine synthesis, e.g. cysteine sulfinic acid decarboxylase (CSAD). A recent finding in rats suggests that high protein intake can depress CSAD activity[16]. Although these data suggest that superior protein quality (i.e amino acid availability) has an impact on taurine metabolism, the specific mechanism underlying the positive effect of threonine or arginine remains unknown. It is also possible that enhancing the availability of other essential amino acids (e.g. leucine or lysine) in conjunction with the sulfur amino acids would affect taurine metabolism.

In contrast to our earlier study[13] that indicated a gender effect in the rate of taurine depletion, the present study failed to reveal differences in the decline or rebound in plasma and whole blood taurine between male and female cats (Figure 2). However, as in the previous study, female cats tended to maintain higher whole blood taurine concentrations compared to males. This is in agreement with our earlier findings that female cats maintained three times more whole blood taurine after 100 days of taurine deprivation and had twice the concentration of whole blood taurine after 150 days of supplementation with methionine, cystine, threonine and arginine[13].

The plasma taurine pool depleted more rapidly than the whole blood taurine pool and after supplementation with methionine, cystine, threonine and arginine the plasma taurine pool was restored more slowly than whole blood taurine. Therefore, changes in taurine status (depletion and repletion) are differently reflected in the plasma and whole blood taurine pool. The whole blood taurine pool is more stable than the plasma taurine pool and intracellular depletion (whole blood taurine represents approximately 80% intracellular taurine) lags behind the decline in plasma taurine. Consequently, the whole blood taurine concentration appears to be a more reliable index of taurine status, and the taurine status is best evaluated by measuring both concentrations.

ACKNOWLEDGMENTS

This work was supported in part by a grant from Alpo Petfoods, Allentown, PA. Dr. Trautwein was supported by a Fellowship from Deutsche Forschungsgemeinschaft, Bonn, Germany.

REFERENCES

1.   K.C. Hayes, Taurine nutrition, *Nutr. Res. Rev.* 1:99 (1988).
2.   K.C. Hayes, and E.A. Trautwein, Taurine deficiency syndrome in cats, Veterinary Clinics of North America: *Small Animal Practice* 19:403 (1989).

3.   K. Knopf, J.A. Sturman, M. Armstrong, and K.C. Hayes, Taurine: an essential nutrient for the cat, *J. Nutr.* 108: 773 (1978).

4.   J.R. Morris, Q.R. Rogers, and L.M. Pacioretty, Taurine: an essential nutrient for cats, *J. Small Animal Pract.* 31:502 (1990).

5.   S.Y. Schmidt, E.L. Berson, and K.C. Hayes, Retinal degeneration in cats fed casein, I.Taurine deficiency, *Invest. Ophthalmol. Visual Sci.* 15: 47 (1976).

6.   S.A. Laidlaw, J.A. Sturman, and J.D. Kopple, Effect of dietary taurine on plasma and blood cell taurine concentrations in cats, *J. Nutr.* 117: 1945 (1987).

7.   J.A. Sturman, D.K. Rassin, and K.C. Hayes, Taurine deficiency in the kitten: Exchange and turnover of [^{35}S] taurine in brain, retina, and other tissues, *J. Nutr.* 108: 1462 (1978).

8.   J.A. O'Donnell, Q.R. Rogers, and J.G. Morris, Effect of diet on plasma taurine in the cat, *J. Nutr.* 111: 1111 (1981).

9.   E.L. Berson, K.C.Hayes, A.R. Rabin, S.Y. Schmidt, and G. Watson, Retinal degeneration in cats fed casein. II.Supplementation with methionine, cystine, or taurine, *Invest. Ophthalmol. Visual Sci.* 15: 52 (1976).

10. K.C. Hayes, A.R. Rabin and E.L. Berson, An ultrastructural study of nutritionally induced and reversed retinal degenerations in cats, *Am. J. Pathol.* 78:504 (1975).

11. K.C. Hayes, A. Pronczuk, A.E. Addessa, and Z.F. Stephan, Taurine decreases platelet aggregation in cats and humans, *Am. J. Clin. Nutr.* 49: 1211 (1989).

12. S.E. Edgar, J.G. Morris, and Q.R. Rogers, Cysteic acid as a precursor of taurine in cats, *FASEB J.* 4: A798 abs.(1990).

13. E.A. Trautwein, and K.C. Hayes, Gender and dietary amino acid supplementation influence the plasma and whole blood taurine status of taurine-depleted cats, *J. Nutr.* 121: S173 (1991).

14. E.A. Trautwein and K.C. Hayes, Taurine concentrations in human plasma and whole blood: estimation of error from intra- and interindividual variation and sampling technique, *Am. J. Clin. Nutr.* 52: 758 (1990).

15. E.A. Trautwein, and K.C. Hayes, Evaluating taurine status: Determination of plasma and whole blood taurine concentration, *J. Nutr. Biochem.* 2: 571 (1991).

16. A.A. Jerkins, L.E. Bobroff, and R.D. Steele, Hepatic cysteine sulfinic acid decarboxylase activity in rats fed various levels of dietary casein, *J. Nutr.* 119:1593 (1989).

THE EFFECT OF DIETARY SUPPLEMENTATION WITH
CYSTEIC ACID ON THE PLASMA TAURINE
CONCENTRATION OF CATS MAINTAINED ON A
TAURINE-RESTRICTED DIET

K.E. Earle and P.M. Smith

Waltham Centre for Pet Nutrition
Freeby Lane
Waltham-on-the-Wolds, Melton Mowbray
Leicestershire LE14 4RT UK

INTRODUCTION

The number of interconnecting pathways involved in the synthesis of taurine and the effect of the addition of various precursor molecules are still relatively unknown in the cat. Earlier work at the Waltham Centre for Pet Nutrition (WCPN) showed that the addition of 17.0 g/kg fresh weight L-cystine to a taurine free semi-purified diet (basal diet contained 6.4 g/kg fresh weight L-cystine) resulted in higher plasma taurine levels than those on the basal taurine-free diet, although the differences were not statistically significant[1]. Studies with rats have shown that the addition of ^{14}C-cysteine to a liver perfusate resulted in the formation of ^{14}C-taurine[2]. When the work was repeated with cats, only three out of six animals showed an ability to synthesise ^{14}C-taurine. The measured rate of hepatic taurine synthesis was found to be twice as rapid in the rat as in the cat liver. For an explanation of these differences we must look more closely at the taurine anabolic pathways.

The enzyme cysteine dioxygenase was found to have lower activity in cat tissues compared to rat tissues[3] and be unaffected by taurine deficiency[4]. As a consequence there is reduced synthesis of cysteine sulphinic acid (CSA) from the cystine and cysteine in the diet of the cat (see Figure 1) as compared to the rat. There are a number of divergent pathways which occur at this point including the decarboxylation, oxidation or transamination of CSA. A number of authors have noted a lower activity of the enzyme, cysteine sulphinic acid decarboxylase (CSAD), in the liver of the cat. Hepatic CSAD activity was found to be 200 times greater in a 250 g male rat than seen in a six week old kitten[5], and both liver and brain CSAD activity were found to be 19 times higher in a six week old kitten than a juvenile cat (5 months) and 73 times higher than an adult cat[6]. This reduced activity of cysteine sulphinic decarboxylase in the cat had previously been reported as the primary rate-limiting step in

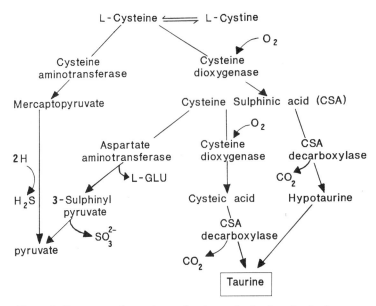

Figure 1. Interconnecting pathways for the synthesis of taurine in the cat.

taurine synthesis. It has also been reported that the primary route for the formation of taurine for most other animals is via this CSAD enzyme[7,8]. This single enzyme, in the liver of the mouse and the guinea pig[9] and the brain of the rat and human[10,11], is known to catalyse both the decarboxylation of CSA and cysteic acid. If the same were true in the liver of the cat then the addition of cysteic acid to the diet of a cat with taurine deficiency, would also result in an increase in plasma taurine concentration.

The presence of an alternative taurine synthetic pathway, either directly from CSA or via sulphinyl pyruvate, would seem feasible as in earlier experiments some cats did show a limited response to the addition of cystine and cysteine in their diet[1,2]. The enzyme glutamate decarboxylase can also utilise CSA as a substrate and may therefore reduce the amount of CSA available for decarboxylation to hypotaurine[7,12]. A further pathway for the formation of taurine includes the synthesis from inorganic sulphate via phosphoadenosine phosphosulphate, but this is not thought to be of importance in the cat[2].

This overview of the taurine synthetic pathway points towards the role of L-cysteic acid as a precursor to taurine in the cat, where the predominant pathway for the formation of taurine may be dependent on individual tissues. The formation of taurine may be possible in a cat fed a taurine restricted diet if the substrate concentration is adjusted to overcome the reduced activity of the cysteine dioxygenase and cysteine sulphinic acid decarboxylase enzymes.

The aim of this study was to measure the effect of the addition of 2.0 g/kg fresh weight L-cysteic acid on the circulating plasma taurine concentration of adult cats fed a taurine restricted, canned diet. The taurine restricted diet was formulated to deliver < 40 mg taurine/kg BW/d (1325.3 mg/kg DM) to the cats which, in canned cat food, is known to result in depletion of plasma taurine concentration[16,18]. A quantity of 2.0 g/kg fresh weight L-cysteic acid was added to the diet to represent a normal dietary intake of amino acid from a canned cat food; mean concentration of individual sulphur amino acids, (Table 1), was 2.9 g/kg fresh weight. An earlier study using a semi purified diet containing 0 g/kg taurine, showed that the addition of 2.0 g/kg fresh weight cysteic acid, to the diet, satisfied the taurine requirement of the cat[17].

METHODS

A total of twelve British, domestic, short haired, adult cats (six castrated males, six entire females) all born and reared at WCPN were used in this study. They had been vaccinated against feline infectious enteritis, feline calici virus, and feline viral rhinotracheitis and had been reared on commercial cat food since weaning. The group were fed to appetite on a taurine-restricted diet made from meat based materials which had been canned and cooked at 129°C for sixty minutes (see Table 1). A mean daily food intake per cat and individual weekly bodyweights were recorded for a period of eighty four days. Plasma taurine concentration was measured on fasting blood samples using an amino acid analyzer[18] every two weeks for the duration of the study.

Day 0 - 14 inclusive, basal diet only
Day 15 - 56 inclusive, basal diet + L-cysteic acid
Day 57 - 84 inclusive, basal diet only

The L-cysteic acid (BDH Chemicals, Warwicks, UK) was added to the diet in the cattery (post processing) and mixed evenly using a laboratory grade food mixer (Silverson, Chesham, Bucks, UK) to ensure even distribution. It was not possible to monitor individual food intake within the group. All data were expressed as mean +/- S.D. and comparisons were made by the Student's t test for paired data.

Table 1. Nutrient content of the basal, canned diet*

	g/kg fresh weight
Moisture	834.0
Protein	91.0
Acid - ether extract	51.0
Ash	21.0
Predicted ME	707.0 kcal (2970 kJ)

Essential fatty acids per kilogram (3 analyses): linolenic acid 0.20 g; linoleic acid 0.60 g; arachidonic acid 1.2 g.
Minerals per kilogram (3 analyses): calcium 2.8 g; phosphorus 3.0 g; sodium 1.9 g; potassium 1.8 g; magnesium 14.0 g; iron 47.0 g; copper 6.8 mg; manganese 6.8 mg; zinc 11.0 mg.
Vitamins per kilogram (3 analyses): Vitamin A 27,000IU; vitamin E 10.0 mg; thiamin 1.3 mg; riboflavin 4.5 mg; niacin 16.0 mg; pyridoxine 2.0 mg; pantothenic acid 7.5 mg; folic acid 0.90 mg; vitamin B12 130 μg; choline 1.1 g.
Amino acids per kilogram (3 analyses): Methionine 4.5 g; cystine 1.4 g; lysine 4.3 g; arginine 2.9 g; phenylalanine + tyrosine 5.8 g; threonine 3.5 g; leucine 6.3 g; isoleucine 2.8 g.
Taurine: 0.22 g/kg
* The formulation of this canned product was based on meat, poultry, offals, wheat flour and fortified with vitamins and minerals.

RESULTS

The mean bodyweights of the cats remained fairly stable over the 84 day period and at the end of the trial their mean weight was 3.8 ± 0.8 kg versus 4.2 ± 1.0 kg on day zero. Food was offered to the cats ad-libitum throughout the trial and there were no significant differences in food or energy intake from week to week during the study.

The taurine content of the basal, canned diet was 0.22 g/kg fresh weight; Table 2 records the mean daily taurine intake (mg/kg BW). Throughout the twelve weeks of the study the mean dietary taurine intake was less than 23.0 mg/kg BW. Fluctuations in intake were due to the need for cats to undergo an eighteen hour fast prior to blood sampling. It is worth noting that these figures are mean values and therefore the intake of certain individuals may be higher or lower than the reported values, especially in weeks 3, 5, 7, 9. 10, 11 and 12 when the standard deviation on the mean values were high.

Fasting plasma values were recorded every fortnight (Figure 2) for individual cats. At the beginning of the trial the mean plasma taurine concentration for all twelve cats was 141.8 ± 71.6 μmol/L, and there was no significant difference between females and males (149.5 ± 37.6 vs 134 ± 98.5 μmol/L, ns.). After two weeks of ingesting the taurine restricted diet the plasma taurine values for all twelve cats had decreased and the mean value was now 39.2 ± 26.1 μmol/L, this value being statistically significantly lower than at day 0 (p < 0.01); there was no difference between males and females (27.3 ± 15.5 vs. 51.0 ± 30.3 μmol/L, ns.).

On day 15 the basal diet was supplemented with 2.0 g/kg fresh weight L-cysteic acid and this supplementation regime was continued for six weeks. After two weeks on the supplemented diet the mean plasma taurine level had increased significantly to 122.8 ± 85.1 μmol/L (p < 0.01) and after a further four weeks the mean value was 153.1 ± 117.4 μmol/L. There was no significant difference in mean plasma taurine concentrations on days 28, 42 and 56. There was no significant sex difference between the cats in their response to dietary cysteic acid supplementation, however, the six female cats had consistently higher circulating plasma taurine levels than the male cats (day 28: 155.8 ± 94.2 vs. 89.7 ± 66.7 μmol/L,ns.; day 42: 149.2 ± 91.9 vs. 92.3 ± 72.9 μmol/L,ns.; day 56: 184.5 ± 145.5 vs. 122.3 ± 82.6 μmol/L,ns.). On day 57 the L-cysteic acid was withdrawn and the basal diet was fed solus. After two weeks of solus feeding the mean plasma taurine level had fallen significantly to 55.6

Table 2. Weekly means (± S.D.) for taurine (mg/kg BW/d) and cysteic acid intake (g/cat/d) throughout the trial period (n=12).

Weeks of Trial	Taurine Intake (mg/kg/BW/d)	L-Cysteic Acid Intake (g/kg/BW/d)
1	22.7 ± 0.06	0.00
2	21.8 ± 0.35	0.00
3	21.9 ± 0.30	0.197 ± 0.004
4	21.9 ± 0.33	0.198 ± 0.010
5	22.4 ± 0.33	0.193 ± 0.003
6	21.9 ± 0.30	0.195 ± 0.004
7	21.9 ± 0.31	0.182 ± 0.003
8	21.4 ± 0.26	0.108 ± 0.004
9	21.9 ± 0.31	0.00
10	21.8 ± 0.32	0.00
11	21.7 ± 0.32	0.00
12	21.8 ± 0.32	0.00

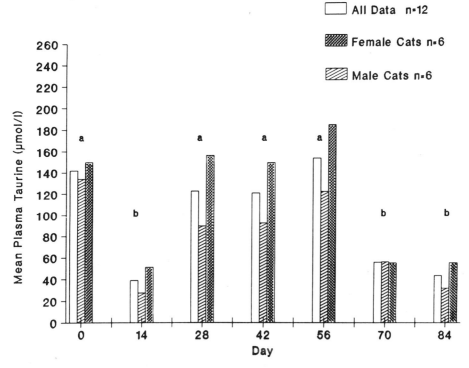

Figure 2. Fasting plasma taurine concentrations in male and female cats.

$\pm$ 34.7 μmol/L (p < 0.01) and at the end of trial had dropped further to a mean value of 43.3 $\pm$ 34.3 μmol/L. There was no statistically significant difference in mean plasma taurine concentrations on days 70 and 84.

Looking more closely at the data of individual cats they seem to fall into two groups: Responders (Figure 3) whose plasma taurine concentrations increased significantly with the addition of L-cysteic acid to their diet and non-responders (Figure 4) who showed little response.

Figure 3 shows the data from the "responder" cats; eight in total, of which four were male and four female. Their initial plasma taurine concentration of 159.8 $\pm$ 78.8 μmol/L fell substantially to only 52.0 $\pm$ 22.3 μmol/L (p < 0.01) after two weeks on the taurine restricted diet and there were no significant differences between males and females (35.3 $\pm$ 12.2 vs 68.8 $\pm$ 16.2 μmol/L,ns.). Following supplementation of the diet with cysteic acid on day 15 there was a significant increase in the mean plasma taurine level by day 28 (52.0 $\pm$ 22.3 vs 160.6 $\pm$ 79.6 μmol/L, p < 0.01) which may not have reached its peak when the cysteic acid was withdrawn at day 57. On days 28, 42 and 56 the female cats had higher mean plasma taurine concentrations than the male cats but these differences were not statistically significant (day 28: 205.8 $\pm$ 69.4 vs 115.5 $\pm$ 67.2 μmol/L,ns.; day 42: 197.0 $\pm$ 67.5 vs 130.8 $\pm$ 54.3 μmol/L,ns., day 56: 259.0 $\pm$ 114.4 vs 165.8 $\pm$ 58.9 μmol/L,ns.). Within two weeks (day 70) of the withdrawal of L-cysteic acid the mean plasma taurine had dropped significantly to 65.5 $\pm$ 37.1 μmol/L (p < 0.01) and this continued to fall until at the end of the trial the mean value was 59.0 $\pm$ 31.6 μmol/L. There was no statistically significant difference in mean plasma taurine concentration on days 70 and 84.

Figure 4 shows the data from the "non responder" cats; four in total, of which two were male and two female. Their initial mean plasma taurine concentration of 105.8 $\pm$ 41.3 μmol/L decreased significantly to 13.5 $\pm$ 4.2 μmol/L (p < 0.01) after the first

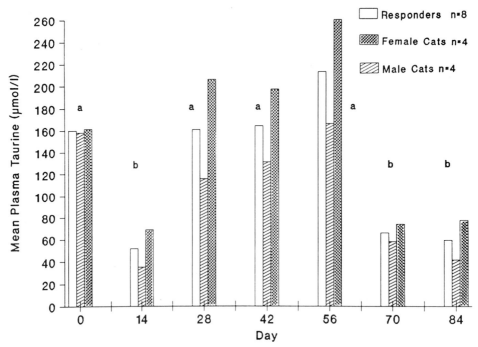

Figure 3. Plasma taurine concentrations of cats responding to dietary L-cysteic acid.

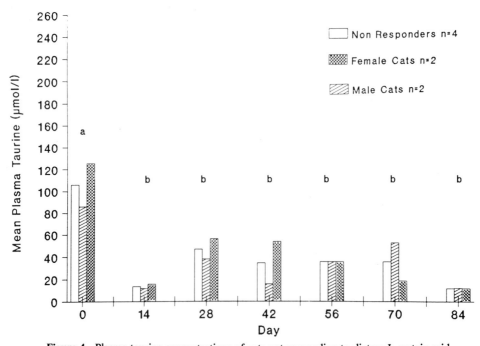

Figure 4. Plasma taurine concentrations of cats not responding to dietary L-cysteic acid.

two weeks on the taurine restricted diet, but stabilised around 35.5 ± 19.4 μmol/L after six weeks of supplementing their diet with L-cysteic acid. There were no statistically significant differences between female and male cats throughout this trial (day 0: 125.5 ± 51.6 vs 86.0 ± 29.7 μmol/L,ns.; day 14: 15.5 ± 4.9 vs 11.5 ± 3.5 μmol/L,ns.; day 28: 56.0 ± 5.7 vs 38.0 ± 25.5 μmol/L,ns.; day 42: 53.5 ± 33.2 vs 15.5 ± 2.1 μmol/L,ns.; day 56: 35.5 ± 4.9 vs 35.5 ± 33.2 μmol/L,ns.; day 70: 18.5 ± 7.8 vs 52.5 ± 4.9 μmol/L,ns.; day 84: 11.5 ± 2.1 vs 12.0 ± 1.4 μmol/L,ns.). Withdrawing the L-cysteic acid on day 57 had very little effect on plasma taurine levels for the following two weeks; their mean plasma level measured at day 70 was 35.5 ± 20.3 μmol/L as compared to 35.5 ± 19.4 μmol/L on day 56. Continuing the feeding of the taurine restricted diet solus for a further two weeks resulted in a mean level of 11.8 ± 1.5 μmol/L by day 84. These values were the lowest recorded for any cat during the trial, but they were not statistically significant from the mean value on days 14, 28, 42, 56 and 70.

The mean L-cysteic acid intake (g/kg BW/d) for weeks three to eight inclusive are shown in Table 2. The mean daily L-cysteic acid intakes ranged from 0.108 - 0.198 g/kg BW depending on whether or not the cats had been fasted. During the first two and last four weeks of the study when the plasma taurine concentrations were decreasing the total sulphur amino acid input was 2.1 g/cat/d. In the intervening six weeks the addition of L-cysteic acid to the diet increased the total sulphur amino acid input to 2.8 g/cat/d and resulted in an increase in the plasma taurine concentration of eight out of twelve cats. The addition of 2.0 g/kg diet, fresh weight L-cysteic acid to the basal diet increased the total sulphur amino acid input (including taurine) by 30%.

Figure 5 shows the pattern of changing plasma taurine concentrations for both responders (n=8) non-responders (n=4). At the beginning of the trial there was no

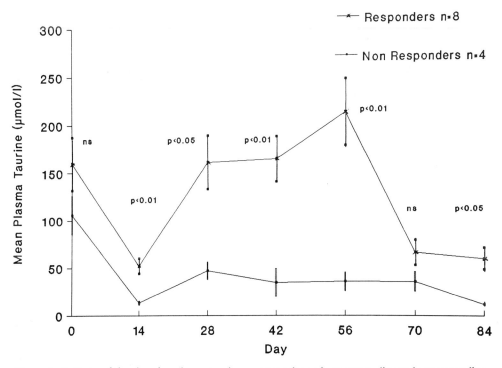

Figure 5. Patterns of the changing plasma taurine concentrations of cats responding and not responding to dietary L-cysteic acid.

significant difference between the fasting plasma taurine concentration of these two groups (159.8 ± 78.8 vs 105.8 ± 41.3 μmol/L,ns.). After two weeks on the basal diet the responders had maintained a significantly higher plasma taurine concentration than the non-responders (52.0 ± 22.3 vs 13.5 ± 4.2 μmol/L, p < 0.01). Following dietary supplementation with L-cysteic acid on days 15 through 56 the responder cats maintained significantly higher plasma taurine concentrations than the non-responder cats. Withdrawal of the L-cysteic acid on day 57 resulted in a drop of plasma levels, especially noticeable for the responder cats, until on day 70 there were no significant differences between the groups.

DISCUSSION

The mean calorie intakes of the cats were close to the recommended level of 70-90 kcal/kg BW[12] and consequently there were no significant changes in bodyweight. The addition of L- cysteic acid to the diet for six weeks did not affect the level of energy intake, therefore it was assumed not to be adversely affecting the acceptance of the test diet.

The mean daily taurine intake from the basal diet was consistently less than 23 mg/kg BW which has previously been identified as below the minimum daily requirement to maintain plasma taurine concentration above 40 μmol/L when feeding commercial canned cat food[15,16,17,19]. During the initial two weeks of the study all recorded plasma taurine values decreased, but only six out of the twelve actually fell below 40 μmol/L. As the aim of the study was to observe the effect of adding L-cysteic acid on the overall level of plasma taurine, it was not considered essential for all cats to show true taurine deficiency prior to supplementation. From day 15 to day 56 (inclusive) the total sulphur amino acid content of the diet was increased by the addition of L-cysteic acid and subsequently the plasma taurine concentration of "responder" animals increased.

The cats fell into two distinct groups, responders and non-responders, and there were no age or sex differences between the two groups. Our results were similar to those observed by an earlier group where only three out of the six cats used were able to form ^{14}C-taurine from ^{14}C-cysteine[2]. This concept of responders and non-responders were also consistent with earlier work at WCPN where a small number of animals had low levels of plasma taurine (29, 25, 39 μmol/L respectively) but showed no signs of retinal lesions[1] over an extended period.

The "responder" cats in this study had increased their synthesis of taurine as measured by an increase in plasma taurine concentration when given between 0.1 and 0.2 g/kg BW/d L-cysteic acid in their diet. One explanation for this effect could be the stimulation of a secondary pathway utilising L-cysteic acid or alternatively the addition of L-cysteic acid as an additional substrate to overcome the low activity of CSAD. For some unexplained reason the "non-responders" were unable to facilitate this response during the six week trial period, however, given an extended period of time their plasma concentrations may have increased. Conversely, it is not uncommon for certain animals to maintain a plasma concentration of 40-60 μmol/L for a long period of time when given a taurine restricted diet[1], with no deleterious effect.

CONCLUSION

The plasma taurine concentration of the cats in this study were shown to be responsive to their overall dietary taurine intake. When intake fell below 23 mg/kg BW plasma concentration declined. The addition of 2.0 g/kg fresh weight L-cysteic acid to a taurine restricted diet gave rise to an increase in the circulating level of taurine in three-quarters of the group; the other four cats showed very little response. In the non-responder group, although the plasma taurine level did not increase, they also did not continue to fall below 35 μmol/L until the L-cysteic acid was withdrawn. There was evidence that the circulating plasma taurine levels were higher for females than males although differences were not statistically significant[20].

These data may indicate the ability of the cat to utilise a biosynthetic pathway for taurine via cysteic acid, however, in vitro analysis of specific enzyme activities would be needed to confirm these results. Alternatively the addition of cysteic acid may have increased the overall substrate availability for the taurine synthetic pathway thereby improving the flow of nutrients for taurine formation.

SUMMARY

The biochemical impairment in the taurine anabolic pathway of the cat has not yet been fully elucidated; however, a number of key enzymes are known to have reduced activity in the cat compared to the rat. There are a series of possible routes resulting in the formation of taurine, one of which is the decarboxylation of cysteic acid. The aim of this study was to investigate the effect of cysteic acid, as a precursor, on the circulating concentration of taurine. A group of twelve adult cats was fed a basal, canned diet containing 0.22 g taurine/kg fresh weight for a period of eighty-four days. The diet was supplemented with 2.0 g/kg fresh weight L-cysteic acid on day fifteen through fifty six and plasma taurine concentration was measured every two weeks throughout the study. The results showed that when the dietary intake of taurine was inadequate to maintain the plasma concentration above 40 μmol/L, the addition of L-cysteic acid to the diet gave rise to an increase in plasma taurine concentration in some cats. Eight of the twelve cats showed a significant rise in plasma taurine after dietary supplementation for six weeks (52.0 $\pm$ 22.3 vs 212.4 $\pm$ 97.9 μmol/L, $p < 0.01$) and a subsequent decrease in plasma levels when the cysteic acid was withdrawn (65.6 $\pm$ 37.1 μmol/L). The other four cats showed no significant rise in plasma taurine after six weeks supplementation (13.5 $\pm$ 4.2 vs 35.5 $\pm$ 19.4 μmol/L, ns.). However, withdrawal of the cysteic acid resulted in a subsequent decrease in circulating levels of taurine (11.8 $\pm$ 1.5 μmol/L, ns.). These data indicate that the addition of cysteic acid to a taurine-restricted, canned diet will, in some cats, result in the biosynthesis of taurine. The plasma taurine concentration of the remaining cats, although not apparently increasing significantly, was maintained at a slightly higher constant level until the cysteic acid was withdrawn. These results suggest that cats are able to synthesise taurine via an alternative pathway utilising L-cysteic acid as a precursor, although the efficiency of this process differs considerably among individual animals.

REFERENCES

1. K.C. Barnett and I.H. Burger, Taurine deficiency retinopathy in the cat, *J.S.A.P.* 21:521-534 (1980).
2. M.G.M. Hardison, C.A. Wood, and J.H. Proffitt, Quantification of taurine synthesis in the rat and cat liver, *Proceedings of the Society for Experimental Biology and Medicine* 155:55-58 (1977).
3. J. De la Rosa and M.H. Stipanuk, Evidence for a rate limiting role of cysteinesulfinate decarboxylase activity in taurine biosynthesis in vivo, *Comp. Biochem. Physiol* 81B(3):565-571 (1985).
4. L.A. Rentschler, L.L. Hirschberger, and M.H. Stipanuk, Response of the kitten to dietary taurine depletion: Effects of renal reabsorption, bile acid conjugation and activities of enzymes involved in taurine synthesis, *Comp. Biochem. Physiol.* 84B(3):319-325 (1980).
5. J. De la Rosa, M.R. Drake, and M.H. Stipanuk, Metabolism of cysteinesulfinate in rat and cat hepatocytes, *Am. J. Clin. Nutr.* 117:549-558 (1987).
6. J.A. Worden and M.H. Stipanuk, A comparison by species, age and six of cysteinesulfinate activity and taurine concentration in liver and brain of animals, *Comp. Biochem. Physiol.* 82B:233-239 (1985).
7. J.G. Jacobsen and L.H. Smith, Biochemistry and physiology of taurine and taurine derivatives, *Physiol. Rev.* 48(2):423-511 (1968).
8. J. Awapara, The metabolism of taurine in the animal, *in*: "Taurine", R. Huxtable and A. Barbeau, eds., pp. 8-11, Raven Press, New York (1976) 8-11.

9.   F. Chatagner, Lefauconnier, and C. Portemer, On the formation of hypotaurine in various tissues of different species, *in*: "Taurine", R. Huxtable and A. Barbeau A. eds., pp. 67-71, Raven Press, New York (1976).

10. S.C. Datta, Re-evaluation of taurine levels and distribution of cysteic acid decarboxylase in developing human fetal brain regions, *J. Neurochem.* 50(4):999-1002 (1968).

11. S.S. Oja and P. Kontro, Taurine, *in*: "Handbook of Neurochemistry. Metabolism in the Nervous System", 2nd edition, A. Lajtha, ed., 3,501-533 (1982).

12. D.B. Hope, Pyridoxal phosphate as the coenzyme of the mammalian decarboxylase for L-cysteine sulphinic and L-cysteic acids, *J. Biochem.* 59:497-500 (1955).

13. "Nutrient Requirements of Cats", Revised Edition, National Academy Press, Washington D.C., USA (1986).

14. G.D. Aguirre, Retinal degeneration associated with the feeding of dog foods to cats, *J.A.V.M.A.* 172:791-796 (1978).

15. P.D. Pion, M.D. Kittleson, Q.R. Rogers, and J.G. Morris, Myocardial failure in cats associated with low plasma taurine: A reversible cardiomyopathy, *Science* 237:764-768 (1987).

16. P.D. Pion, M.D. Kittleson, Q.R. Rogers, and J.G. Morris, Cardiomyopathy in the cat and its relation to taurine deficiency, *in*: "Current Veterinary Therapy", R.W. Kirk and W.B. Saunders, eds., pp. 251-262, 10th Ed., W.B. Saunders Company, Harcourt Brace Jovanovich Inc, Philadelphia (1989).

17. S.E. Edgar, J.G. Morris, Q.R. Rogers, and M.A. Hickman, In vivo conversion of cysteic acid to taurine in the cat, *J. Nutr.* in press (1991).

18. K.E. Earle and P.M. Smith, The effect of dietary taurine content on the plasma taurine concentration of the cat, *Br. J. Nutr.* 66:227-235 (1991).

19. J.A. Cook, Q.R. Rogers, and J.G. Morris, Urinary and faecal excretion of taurine by cats fed commercial canned diets, *FASEB J.*, A1617 (1988).

20. E.A. Trautwein and K.C. Hayes, Sex and dietary amino acid supplementation influence the plasma and whole blood taurine status of taurine-depleted cats, *J. Nutr.* in press (1991).

THE METABOLIC BASIS FOR THE TAURINE
REQUIREMENT OF CATS

James H. Morris and Quinton R. Rogers

Department of Physiological Sciences
School of Veterinary Medicine
University of California
Davis, CA 95616

INTRODUCTION

Taurine is an essential dietary nutrient for cats because the rate of synthesis under most conditions is less than the rate of loss from the body. A deficiency of taurine results in pathological states affecting a number of organ systems, (Morris et al., 1990). Definition of the minimal dietary requirements of cats for taurine has been particularly challenging. Unlike many other nutrients, the loss of taurine from the body is dependent on the composition of the diet, and the method by which the diet is processed. The minimal concentrations of taurine in the diet to satisfy each physiological function of taurine at this time are unknown and likely to be somewhat different for each function. For example, two clinical conditions attributed to taurine deficiency in cats, feline central retinal degeneration (FCRD) and dilated cardiomyopathy do not always occur together in the same cat, each can occur without presence of the other (Pion et al., 1987). The purpose of this paper is to briefly review two important factors that determine the requirement for this nutrient, i.e. endogenous synthesis of taurine and loss of taurine from the body.

SYNTHESIS OF TAURINE

The sulfur amino acids cystine and methionine, are the normal dietary precursors for the synthesis of taurine by mammals. Dietary cystine is rapidly converted to cysteine in the body, but the concentration of cysteine in cells is normally low (10-100 μM). The metabolism of sulfur amino acids, and cysteine in particular, is complex. Some of the pathways of sulfur amino acid and sulfur metabolism in general were reviewed by Cooper (1983) and Huxtable (1986). Cats do not appear to be able to synthesize taurine from serine and sulfate (via 3'-phosphoadenosine-5'phosphosulfate or PAPS) as proposed by Martin et al., (1972) for rats. Cats have a high dietary requirement for sulfur amino acids (SAA) compared to other mammals and one might expect that their rate of synthesis of taurine would be high. A schematic outline of the pathways of mammalian cysteine metabolism and taurine synthesis is presented in Figure 1. An alternate pathway for taurine synthesis via

Taurine, Edited by J.B. Lombardini *et al.*
Plenum Press, New York, 1992

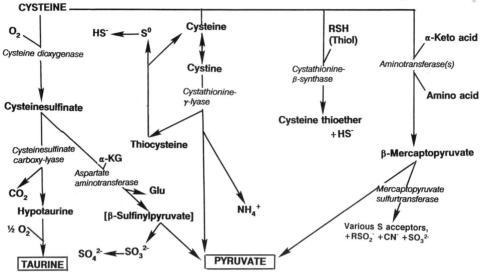

Figure 1. Metabolism of cysteine in animals. Note that despite the many pathways there is a limited number of end products of cysteine metabolism:- pyruvate, ammonia, sulfur (in various oxidative states), carbon dioxide, and taurine.

cysteamine ($H_2NCH_2CH_2SH$)has been proposed (Cavallini et al., 1978), but has not been demonstrated in cats, and appears to be minor or non-existent in mice (Weinstein et al., 1988) so will not be discussed.

Cysteine is metabolized by at least four pathways, of which only the cysteinesulfinate pathway (CS) involves the oxidation of the sulfhydryl group of cysteine and the production of taurine. The other pathways involve desulfhydration of cysteine and production of pyruvate, ammonia and sulfur in various states of oxidation. These pathways are important sources of sulfur in various oxidative states for biosynthetic reactions, as mammals are unable to reduce sulfate, in addition to the benefit of supplying pyruvate. The CS-pathway has been regarded as the major route of cysteine metabolism in rats (Yamaguchi et al., 1973). Some estimates have indicated that about 70-85% of the cysteine metabolism occurs via the CS pathway (Yamaguchi et al., 1973; Stipanuk and Rotter, 1984), but there is considerable evidence supporting a less important role for this pathway. Labelling studies where cysteine metabolism has been measured with and without the addition of CS have shown only mild depression in the yields of the labelled product, indicating the importance of alternative pathways. Wainer (1964) showed that the addition of unlabelled CS to isolated rat mitochondria had little effect on the metabolism of [^{35}S] cysteine to $^{35}SO_4$. Similarly Simpson and Freedland (1975) reported that the addition of CS to the perfusate of rat livers depressed the incorporation of the label from [U-^{14}C] cysteine into glucose by only 30 %. They concluded that about two-thirds of the gluconeogenesis from cysteine was by the cysteine desulfhydration pathway and one-third via the CS-dependent pathway. Drake et al., (1987) reported that in rat hepatocytes over 80% of the $^{14}CO_2$ arising from the metabolism of [1-^{14}C] cysteinesulfinate could be accounted for as taurine and hypotaurine, less than 10% of $^{14}CO_2$ arising from [1-^{14}C] cysteine could be accounted for as taurine and hypotaurine.

The first step in the CS pathway, involves oxidation of cysteine by cysteine dioxygenase (EC 1.13.11.20) to cysteinesulfinate, which is then decarboxylated by L-cysteinesulfinate carboxy-lyase (EC 4.1.1.29) (CSD) to hypotaurine. Spontaneous oxidation of hypotaurine produces taurine. The activity of cysteine dioxygenase (CD) is inducible (Kohashi et al., 1978; Stipanuk, 1979; Daniels and Stipanuk, 1982; Yamaguchi et al., 1985; Hosokawa et. al 1988). A 10 fold increase in the CD activity of the liver of rats occurred when the protein content of the diet was increased from 20 to 40%, or when 1% cysteine was added to a 2.5% casein diet (Kohashi et al., 1978). Similarly, Hosokawa et al.,(1988) reported that the addition of 0.36% methionine to 18% soy or casein based diets induced a 20 fold increase in CD activity. In contrast to rats, the hepatic activity of CD in cats is low. For cats fed a taurine-free diet, CD had about half the activity of CSD (Rentschaler et al., 1986). We also have found only low concentrations of CD in cat liver, and suggest that this pathway is not a main route of cysteine metabolism in cats and have concluded that the low activity of CD is the major limitation to the synthesis of taurine in cats.

The activity of CSD is often proposed as the rate-limiting enzyme of taurine synthesis (De La Rosa and Stipanuk, 1985). In cat liver the activity of this enzyme is low, being considerably less than that of rats (De La Rosa et al., 1987; Rentschaler et al., 1986). Park (1991), measured the activity of CSD in cats and found higher activity in kidney than liver, but the activities in all tissues were low. Hepatic CSD activity was significantly higher in cats that had received a diet containing no taurine than cats which had been fed a taurine-adequate diet. A high level of protein (600 g/kg diet) had no effect on hepatic activity of CSD compared to a low level (200 g/kg diet). However, the kidneys of cats receiving the taurine-free diet had higher activities of CSD than kidneys from cats receiving a taurine adequate diet, In contrast to liver, a high protein diet resulted in an increase in renal CSD activity over that found after feeding a low protein diet. The response of hepatic CSD to dietary protein in cats is in marked contrast to that reported by Jerkins et al., (1989) for rats. They found that CSD activity in rat liver was inversely related to the protein level in the diet. Activity in the liver of rats consuming a 600 g casein /kg diet was only one fourth that in the liver from rats given a diet containing 180 g casein/kg.

Cysteinesulfinate, besides being decarboxylated, may be transaminated leading to the formation of β-sulfinylpyruvate which spontaneously decomposes to pyruvate and sulfite. The enzyme catalyzing this reaction appears to be identical with aspartate aminotransferase. When mice were given labelled CS in vivo, about 85 % of the CS was decarboxylated and 15% transaminated, (Griffith, 1983). In cat liver, the activity of cysteinesulfinate aminotransferase was reported by Park (1991) to be in excess of 10^3 times greater than that of CSD per g tissue. Further work needs to be done to determine the importance of this pathway in vivo in cats and its contribution to reducing the availability of CS for taurine synthesis.

Cysteine is also metabolized by non-CS-pathways which generally lead to the production of pyruvate and are collectively referred to as "cysteine desulfhydrase" pathways (Weinstein et al., 1988). The enzymes of these pathways include cysteine aminotransferase (EC 2.6.1.3) in conjunction with β-mercaptopyruvate sul-furtransferase (EC 2.8.1.2), L-cysteine hydrogensulfide-lyase (cysteine desulfhydrase) or cystathionine γ-lyase (EC 4.4.1.1.) and cysteine synthase or cystathionine β-synthase (EC 4.2.1.22). In the transamination pathway, cysteine is transaminated to the α-keto acid, β-mercaptopyruvate, which by transsulfuration or desulfuration is further degraded to pyruvate and sulfur, (as H_2S or a more oxidized state). Cysteine aminotransferase (CAT) occurs in both the mitochondria and cytosol of rat liver (Ubuka, 1977b,1978) and catalyses the transamination of other amino acids besides L-cysteine, (L-aspartate, L-alanine sulfinic acid and L-cysteic acid, (Akahori et al., 1987)). As the K_m of purified CAT for aspartate is much lower (0.5-1.6 mM) than for cysteine (22 mM), it was suggested by Ubuka et al., (1978), that mitochondrial and cytosolic CAT are identical to aspartate aminotransferase. The enzyme mercap-topyruvate sulfurtransferase transfers the sulfur of β-mercaptopyruvate to sulfite to form thiosulfate, or to other sulfur acceptors such as thiol groups, to form persulfides, or to CN⁻ to form thiocyanates. The rate limiting step in the transamination pathway in rats appears to be the aminotransferase reaction, (Ubuka et al., 1977a). The highest activity of the desulfhydration pathway was found in the heart and liver of rats and cats (Stipanuk and King, 1982).

Cystathionine γ-lyase (cysteine desulfhydrase) catalyses the desulfuration of disulfides including cystine. Cystine in the presence of water is cleaved by cys-tathionine γ-lyase in a β-disulfide elimination reaction to produce thiocysteine, pyruvate, and ammonia. Thiocysteine may react with cystine or other thiols to form thiocystine and cysteine, or H_2S, disulfides, or sulfur(S^0) or SSO^{-23}. As inhibition of cystathionine γ-lyase with proparglyglycine has been shown by Stipanuk and Beck (1982) to result in a large decrease in production of H_2S in rat liver homogenates, it indicates that this is a major pathway of cysteine desulfhydration.

Cystathionine β-synthase catalyses the general substitution of thiol groups. Besides the familiar homocysteine-serine reaction, this enzyme catalyses the reaction of cysteine with thiol compounds to form H_2S and a thioether. Park et al.,(1991b) measured cysteine desulfhydration activity in tissues of cats fed one of four diets (two levels of protein, 200 or 600 g protein/kg diet, with two levels of taurine, 0 or 1.5 g/kg diet). Activities were measured by the production of $H_2{}^{35}S$ from [^{35}S] cysteine in the presence or absence of α-ketoglutarate. All tissues examined had activity, but liver and kidney had the highest activity/g tissue. Muscle because of its large mass would represent the highest total activity. A significant finding was that the direct pathway of desulfhydration (absence of α-ketoglutarate) accounted for 81 to 88% of the activity in cats. The high protein diet resulted in a greater total hepatic desulfuration activity because of the 70% larger liver in cats fed the high protein diets, whereas the taurine treatments had a highly significant positive effect on the activity/g liver.

Taurine Synthesis from Cysteic Acid

Although homogenates of rat brain are capable of producing [^{35}S]cysteic acid from [^{35}S]methionine (Peck and Awapara, 1967) and cysteic acid may be produced from the oxidation of CS, and act as a precursor for synthesis of taurine (Jacobsen and Smith 1963), we have been unable to detect cysteic acid in the brain or other tissues of cats. Rat liver homogenates have been shown to decarboxylate cysteic acid to taurine (Blaschko et al., 1953), but cat liver was reported not to contain the decarboxylase (Blaschko, 1942; Hope 1955). Human liver homogenates were also reported not to decarboxylate cysteic acid, but human brain, dog and rat brain and liver homogenates were active (Jacobsen and Smith 1963). Cysteic acid can be readily produced commercially by treatment of keratinous waste materials which are high in cystine(e.g. feathers) with alkali and peroxide. If cysteic acid could be decarboxylated by cats even in extrahepatic tissues, it could serve as a replacement for taurine in the diet. Edgar et al., (1991), reported that cats fed diets devoid of taurine with graded levels of cysteic acid had a progressive increase in the concentration of taurine in plasma and whole blood with each increment of cysteic acid added to the diet. These results along with increased urinary taurine excretion which paralleled dietary cysteic acid intake indicate that cysteic acid can be decarboxylated by cats and serve as a precursor of taurine. It is not clear whether cysteic acid is decarboxylated by CSD or by a separate enzyme. The bulk of the evidence suggests that there is only one enzyme with different activities towards both substrates (Jacobsen et al., 1964; Guion-Rain and Chatagner, 1972).

Limitation of Taurine Synthesis in Cats

It has been repeatedly suggested, e.g. Hayes (1988), that "the rate-limiting step in taurine biosynthesis is the decarboxylation of cysteine sulfinate by cysteinesulfinate decarboxylase". While this generalization may apply for some comparisons between species, it is apparently not true for cats, and the hepatic activity of CSD in rats is inversely correlated with the urinary excretion of taurine, which is an index of taurine synthesis (Jerkins et al., 1989). Taurine synthesis in cats is primarily limited by the low activity of cysteine dioxygenase, which results in a low flux through cysteinesulfinate. While the activity of CSD in cats is low, apparently it is adequate to decarboxylate cysteic acid, and the low activity does not appear to be the basis for cats being unable to synthesize adequate taurine. Moreover, the activity of this enzyme in the liver of cats is similar to that in humans (Jacobsen et al., 1964), and adult humans do not require a dietary source of taurine. This information leads us to suggest that the main functional difference between humans and cats does not involve synthesis, but the ability of the former to conserve taurine by using glycine conjugation of the bile acids when taurine is limiting.

Information is not available to estimate the fluxes in the various pathways of cysteine metabolism in cats. However, Stipanuk and King (1982) and Stipanuk and Beck (1982) measured the enzymes of desulfhydration in cats and rats and reported that cysteine transamination appeared to play a greater role in cats than rats. Stipanuk et al., (1991), have reported the following activities of some of the key enzymes of cysteine metabolism in rat kidney tubules [nmol product $\cdot$ min^{-1} $\cdot$ mg tubules (dry weight)$^{-1}$)] cysteine dioxygenase, 1.9; cysteinesulfinate decarboxylase, 0.7; aspartate aminotransferase, 495 and γ-cystathionase, 2.2. These authors concluded that in rat enterocytes and renal cortical tubules approximately half of the total cyst(e)ine catabolism occurred by cysteinesulfinate-independent desulfhydration and as much as half of the cysteinesulfinate-dependent metabolism involved transamination of the cysteinesulfinate to yield pyruvate and sulfite. It appears even in rats the flux along the CS pathway leading to taurine is minor compared to alternate routes of cysteine metabolism. This is consistent with the known end products of methionine and cysteine metabolism which are mainly pyruvate sulfate and ammonia and not taurine.

In Vivo Synthesis of Taurine in Cats

As cysteine is the precursor for taurine synthesis, it would be expected that the concentration of SAA in the diet would affect the amount of taurine synthesized and its concentration in the plasma of cats. Soon after Hayes and coworkers showed that taurine was essential in the diet of cats for the maintenance of the eye, we took two groups each of four kittens and gave them amino acid based diets (34 % crystalline amino acids) devoid of taurine. One diet contained 1.1 % L-methionine and 0.8 % cystine, and the other diet contained the same level of methionine but no cystine. While the SAA requirements at that time were unknown, in retrospect both diets contained levels of SAA well in excess of the requirements for maximal growth. Cats were fed these diets for 9 months and periodic plasma samples were taken for taurine analysis and the eyes were ophthalmoscopically examined for FCRD. Growth was normal, the mean ($\pm$ SE) body weight of the cats was 3.2 $\pm$ 0.5 kg, (one male cat was 5.2 kg) at the end of 9 months and none of the cats showed clinical signs of FCRD. Plasma concentrations of taurine in all cats were low, but not significantly different due to the level of cystine in the diet, (+ cystine group, 12 $\pm$ 5; – cystine group, 11 $\pm$ 3 μmol/L). It would appear that the plasma concentration required to produce FCRD is either lower than the level recorded in this experiment, or the time of deficiency has to be greater than 9 months.

In another experiment, groups of six kittens were fed purified diets containing one of two levels of SAA (9 or 16 g/kg diet dry matter (DM)) along with four levels of taurine (0, 250, 375 or 500 mg taurine/kg diet DM) for 17 months. The diets contained 350 g protein/kg DM derived from a mixture of 50:50 isolated soy protein and casein and were fed *ad libitum*. A dietary SAA concentration of 9 g/kg is 1.2 times the minimal requirement of kittens for maximal growth, (National Research Council 1986). The cats were periodically examined ophthalmoscopically for FCRD and at the conclusion of the experiment electroretinograms were done on all cats. The cats in the 0 and 250 mg taurine/kg groups were necropsied and the eyes were examined histologically for abnormalities, and the taurine concentration in the retina and chorionic retinal pigment epithelium were measured (Rogers, Morris and Buyukmihci, unpublished data).

The concentration of taurine in the plasma of the cats reflected both the levels of taurine and SAA in the diets. By 9 months the concentrations of taurine in the plasma of cats fed the high SAA diets were approximately double those of the cats fed the low SAA diet, irrespective of the level of taurine in the diet. These differences persisted until the experiment was terminated at 17 months (Table 1).

By 9 months two cats in the 0 tau-9 g SAA group showed clinical FCRD and by 17 months four of the remaining 5 cats on the study in this group had FCRD. None of the other cats including those receiving the 0 tau-16 g SAA diet showed clinical signs of FCRD. All remaining cats in the 0 tau-9 g SAA/kg group had reduced rod and cone electroretinogram (ERG) responses, as well as 3 of the 4 cats in the 0 tau-16 g SAA group. A slight reduction in the ERG response of 2 out of 6 cats in the 250 mg taurine/kg diet group was found, but all other cats gave normal responses. Histological changes were found only in one group of cats, those receiving the 0 tau-9 g SAA diet in which four cats had extensive abnormalities and one thinning of the

Table 1. Effect of level of dietary taurine and sulfur amino acids (SAA) on the concentration of taurine in plasma at 9 and 17 months, and in the retina and chorionic retinal pigment epithelium (CREP) at 17 months.

Dietary Tau mg/kg diet	Dietary SAA g/kg diet	Sampling months[a]	Plasma Tau μmol/L	Retinal Tau μmol/g	CREPE Tau μmol/g
0	9	9	2.7[b]	--[c]	--
0	9	17	2.3	28.3	2.5
0	16	9	6.1	--	--
0	16	17	5.0	53.9	10.3
250	9	9	6.9	--	--
250	9	17	7.4	45.9	15.8
250	16	9	16.1	--	--
250	16	17	11.1	46.7	15.7
375	9	9	12.9	--	--
375	9	17	10.2	--	--
375	16	9	26.6	--	--
375	16	17	31.7	--	--
500	9	9	18.4	--	--
500	9	17	13.8	--	--
500	16	9	30.3	--	--
500	16	17	23.8	--	--

[a]months from commencement of the diet.
[b]mean value from 4-6 cats.
[c]not determined.

cells only (Table 2). Analysis of the retina and chorionic retinal pigment epithelium from cats fed 0 taurine showed approximately twice the concentration of taurine in these tissues from cats receiving 16 g SAA/kg diet versus those cats receiving 9 g SAA/kg diet.

These two studies demonstrated that adequate taurine can be synthesized *de novo* by growing cats consuming purified diets well supplied with SAA for the prevention of clinical signs of FCRD, but not enough to sustain full ERG responses. These findings cannot be applied to cats consuming commercial type diets as will be discussed later.

Table 2. Effect of level of dietary sulfur amino acids and taurine on clinical FCRD, ERG and histopathology of cat eyes.

Taurine (mg/kg)	SAA (g/kg)	Clinical FCRD	Reduced ERG Rod	Reduced ERG Cone	Histopathology abnormalities
0	9	4/5	5/5	5/5	4[a]/5[+]; 1[b]/5
	16	0/4	3/4	3/4	0/4[c]
250	9	0/4	0/4	0/4	0/4
	16	0/6	2/6*	2/6*	0/6
375	9	0/6	0/6	0/6	0/1
	16	0/5	0/5	0/5	0/1
500	9	0/5	0/5	0/5	0/1
	16	0/5	0/5	0/5	0/1

* somewhat reduced,
+ cats necropsied per group,
[a] extensive abnormalities,
[b] thinning of cells only,
[c] i.e no lesions in 4 cats.

LOSS OF TAURINE FROM THE BODY

Taurine is not degraded by mammalian enzymes, the major pathway of metabolism is via conjugation to form bile acids. Minor metabolites of taurine include dipeptides such as γ-glutamyl-taurine, (Varga et al., 1984). When a normal nutrient balance study is done on cats consuming diets adequate in taurine, the sum of the quantity of taurine recovered in feces and urine is considerably less than the taurine ingested, even disregarding synthesis. The reason for this disparity between taurine ingested and taurine excreted must be due to microbial degradation in the gut. Anaerobic bacteria deconjugate the bile acids and degrade taurine (Fellman et al., 1978; Ishimoto et al., 1983) as well as modifying the structure of the steroid moiety (Hylemon, 1985) to produce secondary bile acids. The metabolism in the gut is therefore important in determining the taurine status of cats.

Gastrointestinal Loss of Taurine

Secretion of bile results in a constant loss of taurine from the body. Factors that either enhance bile secretion, or prevent recycling of the taurine by binding taurocholic acid, or increase deconjugation and degradation of taurine by changes in the flora of the gut, will increase loss of taurine from the body. The contribution of some of these factors as determinants of taurine status of cats have been measured. Hickman et al., (1990), showed that the extent of oxidation of a pulse labelled dose of [^{14}C] taurine was dependent on diet composition and the method of processing the diet. Other studies have shown that cats consuming commercial canned type diets before heat processing are able to maintain higher plasma concentrations of taurine than cats fed the same diets after heat processing (canned). Taurine recovered by water extraction of commercial canned diets before and after thermal processing is identical, so taurine is not bound or unavailable for absorption. The reason the heat processed diets do not maintain the same taurine concentrations in the plasma of cats as the diet before heat processing appears to be due to either greater bile production, exposing more taurine to microbial degradation, or a shift in the gut microflora to a population that has a greater capacity to deconjugate and degrade taurine.

A number of comparisons of casein based and isolated soy protein based purified diets has been made in our laboratory, and routinely cats fed casein diets have higher plasma concentrations of taurine than cats fed soy-based diets. Similar responses in blood cholesterol have been observed in other animals fed soy and casein based diets. That is, rats fed casein diets have higher plasma cholesterol levels than rats fed soy diets, (Choi et al., 1989). The cholesterol lowering fraction in soybean protein appears to be a high molecular weight residue left after digestion of soybean protein isolate with proteases, (Sugano et al., 1990). The hypocholestrolemic response of various protease digestion products was highly correlated with their ability to increase fecal steroid excretion.

We have found that cats fed heat processed purified diets have lower plasma taurine concentrations than cats fed non-heat processed diets. The effect of heat processing can be enhanced if a small portion of the starch component is replaced by glucose. These glucose-containing heat processed diets are dark brown in color, due to the formation of Maillard reaction products. One would also expect that these heat processed diets would have a lower digestibility, at least in the upper small intestine, than the non-heat processed diets.

The taurine concentration in the plasma of cats is also reduced when 20 g of the plasma-cholesterol-reducing-resin, Cholestyramine® is added per kg diet. Cholestyramine® binds taurocholic acid and increases the loss of taurine from the body. From the above observations it appears that dietary components or digestion products could deplete the body pool of taurine by at least two mechanisms:
1. Binding bile acids which are transported in the bound state from the absorption sites of the upper gut, to regions of the gut of lower absorptive capacity and/or more intense microbial activity (lower ileum and colon) where they are degraded.
2. Increasing secretion of bile by stimulation of gall bladder contraction through release of cholecystokinin, either due to a secretogogue effect of amino acids, or by CCK releasing peptide and/or monitor peptide as a result of upper intestinal accumulation of indigestible protein.

Urinary Loss of Taurine

Most of the free taurine in the diet is probably absorbed and homeostasis is achieved by regulation of renal clearance of taurine. Up and down regulation of cat kidneys was examined by Park et al., (1989), who used brush-border membrane vesicles from the renal proximal tubules of kittens and showed that maximal renal up and down regulation occurred in 6 and 2 weeks respectively after changes were made in the concentration of taurine in the diet. Renal adaptation of the kittens to changes in dietary taurine occur with modification of both V_{max} and K_m. Park, et al., (1991a), also showed that dietary acid and alkali loading did not alter the ability of the renal proximal tubule to transport taurine.

Taurine is present in urine in both the free and bound form and measurement of total taurine requires hydrolysis before analysis by ion exchange resin or HPLC methods. The proportion of bound taurine varies with the diet and the amount of taurine excreted. Taurine is known to be used as a conjugate to excrete various xenobiotics including phenylacetic acid (James et al., 1972) and some nutrients such as vitamin A. For a fixed basal diet taurine excretion has been reported to be closely correlated with dietary taurine intake (Glass et al., 1991) and the taurine:creatinine ratio has been suggested for use as an index of taurine status.

Fetal and Lactational Losses

Inadequate intake of taurine by the queen leads to reproductive failure resulting in abortions, fetal resorptions and birth of kittens with developmental abnormalities (Sturman et al., 1986, 1987; Sturman and Messing, 1991). In studies conducted at University of California, Davis, taurine deficient queens came into estrus, mated, and apparently had a normal pregnancy up to about day 27 of gestation. After this time there appeared to be a breakdown in normal association of the placenta and endometrium. Taurine-deficient queens that went to full term had plasma progesterone levels that were consistently lower than normal, but relaxin levels were similar to those in normal queens. Queens that aborted had low progesterone values by day 25 with a concomitant reduction in relaxin levels (Dieter et al., 1988)

Sturman and Messing (1991) reported a low concentration of taurine in the milk of queens fed taurine deficient diets and concluded that postnatal supply of taurine in the mother's milk had a greater impact than intrauterine environment on the taurine concentration of kittens at 8 weeks. Hickman (unpublished data) measured taurine concentration in the blood of queens fed a normal commercial expanded diet during pregnancy and lactation. Blood and plasma concentrations of taurine were stable during pregnancy, but plasma values decreased markedly during lactation. These observations indicate that the demand for taurine during lactation is much greater than that required for the products of conception, and lactation is the most demanding physiological state for taurine, as well as for other nutrients.

CONCLUSIONS

Cats have a dietary requirement for taurine due to inadequate synthesis and to the requirements for taurine to conjugate bile acids. Cats are unable to use the strategy of many other mammals of switching to the glycine conjugation and conserving taurine. Taurine synthesis by cats appears to be primarily limited by the high flux through the "desulfhydrase pathways" and a low flux through the cysteinesulfinate pathway, rather than to a low activity of the enzyme cysteinesulfinate decarboxylase. As cats cannot metabolically regulate the loss of taurine (through taurocholic acid) via the gastrointestinal tract, factors that interrupt the enterohepatic recycling of taurine cause a greater demand for taurine. Important factors include changes in the rate of secretion of bile, and the microbial deconjugation and degradation of taurine.

The determination of a "safe" minimal dietary requirement of taurine for cats has been particularly challenging, as loss of taurine from the body is dependent on diet composition. While minimal requirements of taurine for various physiological states of cats are not known, the metabolic basis for the requirement varying with diet is at least partially understood. Expanded dry diets containing 1200 mg taurine/kg dry matter and canned diets containing 2500 mg taurine/kg dry matter would appear to meet the requirements of the majority of cats. Alternatively, adequacy of the level of taurine in a diet can be assessed by feeding the diet and measuring equilibrium plasma and whole blood taurine concentrations. Maintenance of group mean concentration of taurine in plasma > 40 μmol/L and whole blood >250 μmol/L would also indicate adequacy. Unfortunately equilibrium taurine concentrations in plasma and blood take many months to be achieved, but the direction of changes in plasma and blood concentrations of taurine can generally be determined by two months. These values then give an indication of the taurine status of the diet. Plasma concentration of taurine show greater variation than whole blood concentrations when the diet is adequate in taurine, thus blood concentrations are more reliable indices of status, but are slower to respond to dietary changes than plasma concentrations.

REFERENCES

Akahori,S., Ejiri, K., Kanemori, H., Kudo, T., Sekiba,T., Ubuka, T. and Akagi, R., 1987, Transamination of L-cysteine sulfinate in the growing rat, *Acta Med. Okayama* 41:279-283.

Blaschko, H. 1942, L(-)-Cysteic acid decarboxylase, *Biochem. J.* 36:571-574.

Blaschko, H., Datta, S.P and Harris, H., 1953, Pyridoxin deficiency in the rat: Liver L-cysteic acid decarboxylase activity and urinary amino-acids, *Brit. J. Nutr.* 7:364-371.

Cavallini, D., Duprè, S., Frederici, G., Solinas, S., Ricii, G., Antonucci, A., Spoto, G. and Matarese, M., 1978, Isethionic acid as a taurine co-metabolite. *in*: "Taurine and Neurological Disorders," R. Huxtable and A. Barbeau, ed., pp.29-34, Raven Press, New York.

Choi, Y.-S., Goto, S., Ikeda, I. and Sugano, M., 1989, Interaction of dietary protein, cholesterol and age on lipid metabolism of the rat, *Br. J. Nutr.* 61:531-543.

Cooper, A.J.L., 1983, Biochemistry of sulfur-containing amino acids, *Ann. Rev. Biochem.* 52:187-222.

Daniels, K.M. and Stipanuk, M.H., 1982, The effect of dietary cysteine level on cysteine metabolism in rats, *J. Nutr.* 112:2130-2141.

De La Rosa, J. and Stipanuk, M.H., 1985, Evidence for a rate-limiting role of cysteinesulfinate decarboxylase activity in taurine biosynthesis in vivo, *Comp. Biochem. Physiol.* 81B:565-571.

De La Rosa, J., Drake, M.R. and Stipanuk, M.H., 1987, Metabolism of cysteine and cysteinesulfinate in rat and cat hepatocytes, *J. Nutr.* 117:549-558.

Dieter, J.A., Stewart, D.R., Haggerty, M.A., and Stabenfeldt, G.H., 1988, Pregnancy failure in cats associated with dietary taurine deficiency, Abstract #271, Annual Meeting of Society for the Study of Reproduction, Seattle, Wash. Aug 1-4,1988.

Drake, M.R., de La Rosa, J. and Stipanuk, M.H., 1987, Metabolism of cysteine in rat hepatocytes, *Biochem. J.* 244:279-286.

Edgar, S.E., Morris, J.G. Rogers, Q.R., and Hickman, M.A., 1991, In vivo conversion of cysteic acid to taurine in cats, *J. Nutr.* 121:S183-184.

Fellman, J.H., Roth, E.S. and Fujita, T.S., 1978, Taurine is not metabolized to isethionate in mammalian tissue, *in*: "Taurine and Neurological Disorders," A. Barbeau and R.J. Huxtable, ed., Raven Press, New York.

Glass, E.N., Odle, J., Baker, D.H. and Czarnecki-Maulden, G.L., 1991, Development of a renal adaptive response assay as a measure of taurine bioavailability in adult cats, *FASEB J.* 5:(4) A591, Abst 1284.

Griffith, O.W., 1983, Cysteinesulfinate metabolism, *J. Biol. Chem.* 258:1591-1598.

Guion-Rain, M-C. and Chatagner, F., 1972, Rat liver cysteine sulfinate decarboxylase: some observations about substrate specificity, *Biochim. Biophys. Acta*, 276:272-276.

Hayes, K.C., 1988, Taurine nutrition, *Nutr. Res. Rev.* 1:99-113.

Hickman, M.A., Rogers, Q.R. and Morris, J.G., 1990, Effect of processing on the fate of dietary [14C]taurine in cats, *J. Nutr.* 120:995-1000.

Hope, D.B., 1955, Pyridoxal phosphate as the coenzyme of the mammalian decarboxylase of L-cysteine sulphinic and L-cysteic acids, *Biochem. J.* 59:497-500.

Hosokawa, Y., Niizeki, S., Tojo, H., Sato, I. and Yamaguchi, K., 1988, Hepatic cysteine dioxygenase activity and sulfur amino acid metabolism in rats: possible indicators in the evaluation of protein quality, *J. Nutr.* 118:456-461.

Huxtable, R.J., 1986, Taurine and the oxidative metabolism of cysteine, *in*: "Biochemistry of Sulfur," Biochemistry of the Elements, vol 6, Plenum Press, New York.

Hylemon, P.B., 1985, Metabolism of bile acids in the intestinal microflora. *in*: "Sterols and Bile Acids," H. Danielsson and J. Sjövall, ed., Chap 12 New Comprehensive Biochemistry, vol 12, Elsevier, Amsterdam.

Ishimoto, M., Kondo, H., Enami, M., and Yazawa, M., 1983, Sulfite formation by bacterial enzymes from taurine and benzenesulfonate, *in*: "Sulfur Amino Acids Biochemical and Clinical Aspects," K. Kuriyama, R.J. Huxtable, and H. Iwata, ed., Arthur R. Liss, New York. pp 393-394.

Jacobsen, J.G. and Smith, L.H., 1963, Comparison of decarboxylation of cysteine sulphinic acid-1-14C and cysteic acid-1-14C by human, dog and rat liver and brain, *Nature* 200:575-577.

Jacobsen, J.G., Thomas, L.L., and Smith, L.H., 1964, Properties and distribution of mammalian L-cysteine sulfinate carboxy-lyase, *Biochim. Biophys. Acta* 85:103-116.

James, M.O., Smith, R.L., Williams, R.T. and Reidenberg, M., 1972, The conjugation of phenylacetic acid in man, sub-human primates and some non-primate species, *Proc. R. Soc. Lond.* B.182:25-35.

Jerkins, A.A., Bobroff, L.E., and Steele, R.D., 1989, Hepatic cysteine sulfinic acid decarboxylase activity in rats fed various levels of dietary casein, *J. Nutr.* 119:1593-1597.

Kohashi, N., Yamaguchi, Y., Hosokawa, Y., Kori, Y., Fujii, O., and Ueda, I, 1978, Dietary control of cysteine dioxygenase in rat liver, *J. Biochem.* 84:159-168.

Martin, W.G, Sass, N.L., Hill, L., Tarka, S., and Truex, R., 1972, The synthesis of taurine from sulfate. IV. An alternative pathway for taurine synthesis by the rat, *Proc. Soc. Exp. Biol. Med.* 141:632-633.

Morris, J.G., Rogers, Q.R., and Pacioretty, L.M., 1990, Taurine: an essential dietary nutrient for cats, *J. Sm. Anim. Pract.* 31:502-509.

National Research Council, 1986, Nutrient Requirements of Cats, National Academy Press, Washington D.C.

Park, T., Rogers, Q.R., Morris, J.G., and Chesney, R.W., 1989, Effect of dietary taurine on renal taurine transport by proximal brushborder membrane vesicles in the kitten, *J. Nutr.* 119:1452-1460.

Park, T., 1991, Regulation of taurine transport across the renal proximal tubule brush border membrane and cysteine metabolism in cats, Ph.D. thesis, University of California, Davis.

Park, T., Rogers, Q.R., Morris, J.G., and Morris, J.P.G., 1991a, Dietary acid and alkali loading do not alter taurine uptake by renal proximal tubule brush border membrane vesicles in kittens, *J.Nutr.* 121:215-222.

Park, T., Jerkins, A.A., Steele, R.D., Rogers, Q.R., and Morris, J.G., 1991b, Effect of dietary protein and taurine on enzyme activities involved in cysteine metabolism in cat tissue, *J. Nutr.* 121:S181-S182.

Peck, E.J. and Awapara, J., 1967, Formation of taurine and isethionic acid in rat brain, *Biochim. Biophys. Acta* 141:499-506.

Pion, P.D., Kittleson, M.D., Rogers, Q.R., and Morris, J.G., 1987, Myocardial failure in cats associated with low plasma taurine: a reversible cardiomyopathy, *Science* 237:764-768.

Rentschaler, L.A., Hirschberger, L.L., and Stipanuk, M.H., 1986, Response of the kitten to dietary taurine depletion: Effects on renal reabsorption, bile acids conjugation and activities of enzymes involved in taurine synthesis, *Comp. Biochem. Physiol.* 84B:319-325.

Simpson, R.C. and Freedland, R.A., 1975, Relative importance of the two major pathways for the conversion of cysteine to glucose in the perfused rat liver, *J. Nutr.* 105:1440-1446.

Stipanuk, M.H., 1979, Effect of excess dietary methionine on the catabolism of cysteine in rats, *J. Nutr.* 109:2126-2139.

Stipanuk, M.H. and King, K.M., 1982, Characteristics of the enzyme capacity for cysteine desulfhydration in cat tissue, *Comp. Biochem. Physiol.* 73B:595-601.

Stipanuk, M.H. and Beck, P.W., 1982, Characterization of the enzymatic capacity for cysteine desulphhydration in liver and kidney of the rat, *Biochem. J.* 206:267-277.

Stipanuk, M.H. and Rotter, M.A., 1984, Metabolism of cysteine, cysteinesulfinate and cysteinesulfonate in rats fed adequate and excess levels of sulfur-containing amino acids, *J. Nutr.* 114:1426-1437.

Stipanuk, M.H., De La Rosa, J., and Hirschberger, L.L., 1990, Catabolism of cyst(e)ine by rat cortical tubules, *J. Nutr.* 120:450-458.

Sturman, J.A., Gargano, A.D., Messing, J.M., and Imaki, H., 1986, Feline maternal taurine deficiency: effect on mother and offspring, *J. Nutr.* 166:655-667.

Sturman, J.A., Palackal, T., Imaki, H.,Moretz, R.C. French, J., and Wisniewski, H.M., 1987, Nutritional taurine deficiency and feline pregnancy outcome, *in*: "The Biology of Taurine. Methods and Mechanisms," R.J. Huxtable, F. Franconi, and A. Giotti, eds., *Adv. Exp. Med.* 217, pp 113-124.

Sturman, J.A. and Messing, J.M., 1991, Dietary taurine content and feline reproduction and outcome, *J. Nutr.* 121:1195-1203.

Sugano, M., Goto, S., Yamada, Y., Yoshida, K., Hashimoto, Y., Matsuo, T., and Kimoto, M., 1990, Cholesterol-lowering activity of various undigested fractions of soybean protein in rats, *J. Nutr.* 120:977-985.

Ubuka, T., Yuasa, S., Ishimoto, Y., and Shimomura, M., 1977a, Desulfuration of L-cysteine through transamination and transsulfuration in rat liver, *Physiol. Chem. Phys.* 9:241-246.

Ubuka, T., Umemura, S., Ishimoto, Y., and Shimomura, M., 1977b, Transamination of L-cysteine in rat liver mitochondria, *Physiol. Chem. Phys.* 9: 91-96.

Ubuka, T., Umemura, S., Yuasa, S., Kinuta, M., and Watanabe, K., 1978, Purification and characterization of mitochondrial cysteine aminotransferase from rat liver, *Physiol. Chem. Phys.* 10:483-500.

Varga, V., Török, K., Feuer, L., Gulyás, J., and Somogyi, J., 1985, γ-Glutamyltransferase in the brain and its role in the formation of γ-L-glutamyl-taurine, *in*: "Taurine: Biological Actions and Clinical Perspectives," S.S. Oja, L. Ahtee, P. Kontro, and M.K. Paasonen, eds., Progress in Clinical and Biological Research 119. Alan R. Liss, Inc., New York.

Wainer, A., 1964, The production of sulfate from cysteine without the formation of free cysteinesulfinate, *Biochem. Biophys. Res. Comm.* 16:141-144.

Weinstein, C.L., Haschemeyer, R.H., and Griffith, O.W., 1988, In vivo studies of cysteine metabolism. Use of D-cysteinesulfinate, a novel cysteinesulfinate decarboxylase inhibitor, to probe taurine and pyruvate synthesis, *J. Biol.* Chem. 263:16568-16579.

Yamaguchi, K., Sakakibara, S., Asamizu, J.Z. and Ueda,I. 1973, Induction and activation of cysteine oxidase of rat liver. II. The measurement of cysteine metabolism in vivo and the activation of in vivo activity of cysteine oxidase. Biochim. Biophys. Acta 297:48-59.

Yamaguchi, K., Hosokawa, Y., Kohashi., N., Kori, Y., Sakakibara, S., and Ueda, I., 1978, Rat liver cysteine dioxygenase (cysteine oxidase). Further purification, characterization and analysis of the activation and inactivation, *J. Biochem.* (Tokyo) 83: 479-491.

Yamaguchi, K., Hosokawa, Y., Niizeki, S., Tojo, H., and Sato, I., 1985, Nutritional significance of the cysteine dioxygenase on the biological evaluation of dietary protein in growing rats, *Prog. Clin. Biol. Res.* 179:23-32.

INTESTINAL TAURINE AND THE ENTEROHEPATIC CIRCULATION

OF TAUROCHOLIC ACID IN THE CAT

Mary Anne Hickman, James G. Morris, and Quinton R. Rogers

Department of Physiological Sciences
School of Veterinary Medicine
University of California
Davis, CA 95619

INTRODUCTION

Taurine's most well defined role is the conjugation of bile salts in the liver. After synthesis from cholesterol, bile salts are conjugated with taurine and/or glycine before secretion into the bile canaliculi[1]. When taurine is limiting, most mammals have the ability to conjugate bile salts with glycine, exceptions being the dog[2] and the cat[3,4], which conjugate bile salts almost exclusively with taurine. The percentage of total bile salts conjugated with taurine is determined by both the hepatic taurine concentration and the affinity of the bile salt conjugase for glycine and taurine[2,5-7]. Taurine is the preferred substrate in most species with 90 percent taurine conjugation occurring in the rat when hepatic taurine and glycine concentrations are equal[7]. Hepatic taurine depletion in rats, caused by the infusion of cholic acid[7] or by feeding guanidinoethanesulfonic acid[5], results in a substantial increase in the amount of bile salts conjugated with glycine. In species that cannot conjugate bile salts with glycine (the dog and cat), hepatic taurine depletion leads instead to an increase in the proportion of unconjugated bile salts, the majority being free cholic acid[2,3].

Conjugated bile salts are released from the gallbladder when contraction is stimulated by cholecystokinin (CCK), a duodenal-jejunal hormone secreted into the portal circulation in response to the presence of intestinal fat and amino acids[8]. The presence of bile salts in the duodenum-jejunum act to inhibit further CCK secretion[9]. Bile salts that are released into the small intestine will be mostly reabsorbed by active and passive transport processes in the jejunum and ileum. Reabsorbed bile salts are then returned to the liver via the portal vein to complete the enterohepatic circulation. Anaerobic bacteria in the terminal ileum and colon disrupt this cycle, by deconjugating taurine and glycine and by converting primary bile salts into secondary bile salts by various modifications of the steroid nucleus[10]. Studies in humans, using radioactively labelled taurocholic acid, estimate that 40% of taurine conjugated bile salts undergo deconjugation each day[11]. Taurine released by deconjugation may be reabsorbed from the intestine or undergo further bacterial degradation and loss from the body taurine pool. The efficiency of the enterohepatic circulation is affected by numerous factors, including the amount and type of dietary protein and fiber.

Rats and rabbits fed diets containing casein excrete less bile salts in their feces, absorb greater amounts of dietary cholesterol and have higher plasma cholesterol than animals fed soy protein diets[12,13]. The mechanism responsible for these effects appears to be multifaceted, but a major factor appears to be enhanced intestinal binding of bile salts and increased fecal bile salt loss in animals fed soy compared to casein diets[12-14]. A peptide formed by the in vitro digestion of soy protein with peptidases has been isolated and found to bind bile salts[14,15]. When this peptide is fed to rats it causes an even greater lowering of plasma cholesterol and greater fecal excretion of acidic and neutral steroids than diets with isonitrogenous amounts of soy protein[14]. Studies in our laboratory have found that cats fed diets containing casein have greater plasma and whole blood taurine concentrations than cats fed soy protein diets (unpublished data).

The amount and type of dietary fiber can also affect the enterohepatic circulation of bile salts. Differential bile salt binding to various dietary fibers has been demonstrated in vitro and could result in increased losses of fecal bile salts[16,17], similar to the effects of cholestyramine[18]. Water-soluble fibers such as pectin, guar-gum and konjac mannan are viscous polysaccharides that do not bind bile salts but act by entrapping bile salts in the intestine and increasing the total amount of bile salt present in the aqueous intestinal contents as compared to fiber-free diets or diets containing water-insoluble fibers[19]. Rats fed diets containing water-soluble fibers have increased hepatic bile salt secretion, increased fecal bile salt excretion, increased activity of cholesterol 7 α-hydroxylase and decreased hepatic taurine concentration[20-22]. The decreased hepatic taurine concentration is most likely the result of increased taurine conjugation of new bile salts, synthesized in response to the greater fecal losses. These studies suggest that dietary factors which cause increased losses of bile salts may also induce increased intestinal taurine losses. Intestinal taurine loss, including loss due to the inefficiency of the enterohepatic circulation of taurine conjugated bile salts, has been examined to a limited extent in the cat. The objectives of the following studies were to determine the effects of diet and taurine status on taurine metabolism in this important area.

Our studies became more focused on gastrointestinal taurine loss following the discovery in 1986 that adult cats fed canned commercial diets with taurine concentrations 3-4 fold higher than the 1986 National Research Council recommendation of 400 mg/kg for kittens, developed abnormally low plasma and tissue taurine concentrations and had associated clinical abnormalities including feline central retinal degeneration and dilated cardiomyopathy[23,24]. The National Research Council's taurine recommendation was based on studies of kittens fed purified diets and did not account for nutrient interactions or processing effects that occur in commercial diets. Attempts to determine taurine requirements with traditional balance studies, which measured fecal and urinary taurine excretion in cats fed purified and commercial diets, failed to account for 30-50 percent of ingested taurine[25]. These balance discrepancies could be the result of taurine degradation by intestinal bacteria[26,27] which would cause an underestimation of fecal taurine loss.

Further experiments revealed that the commercial diets most consistently resulting in decreased plasma and tissue taurine concentrations were heat-processed canned formulations[24]. It appeared that heat-processing neither destroyed nor resulted in the binding of taurine to dietary components, as the taurine was fully water extractable. Heat-processing also did not increase the urinary excretion of taurine, another possible mechanism of increased taurine loss. When cats fed commercial heat-processed diets became taurine-depleted, urinary excretion decreased to very low levels, indicating active conservation of taurine by the kidney. With this sequence of findings in both commercial and purified diets, intestinal taurine metabolism and loss became the focus of further studies.

TAURINE DEGRADATION BY INTESTINAL BACTERIA

The objective of the first study was to determine the fate, including bacterial catabolism to CO_2, of a oral pulse dose of [^{14}C]taurine in cats fed a commercial diet in two forms, heat-processed and preserved by freezing, and in two purified diets, containing 0 and 1325 mg taurine/kg (0 and 10,600 μmol/kg)[28]. The analyzed taurine concentration of the commercial diet was 1070 mg/kg dry matter (8560 μmol/kg). The frozen-preserved commercial diet had previously been shown to maintain plasma taurine concentration when fed to cats, whereas the heat-processed diet did not. The excretion of ^{14}C in CO_2, urine, and feces was determined. Significant quantities of $^{14}CO_2$ were produced (Table 1), with greater amounts excreted by cats fed the heat-processed commercial diet (9.4% of initial dose) than by those fed the frozen-preserved diet (0.09%), indicating extensive taurine degradation by the intestinal microflora. Purified diet groups were intermediate between the two commercial groups. Excretion of ^{14}C in urine peaked at 24 h and was highest for cats fed the frozen-preserved commercial diet and the 1325 mg taurine/kg diet. Excretion of ^{14}C in feces was highest for cats fed the two commercial diet formulations, with peak amounts at 48-72 h. These results indicated that substantial amounts of taurine can be degraded by intestinal microorganisms and that processing affected the digestive and/or absorptive process in a manner that increased this catabolism in the cats fed the heat-processed compared to the frozen diet. The degradation of taurine by intestinal bacteria could explain the inability of traditional balance studies to account for 30-50% of ingested taurine in cats fed commercial and purified diets.

Table 1. Percentage of ^{14}C from an oral pulse dose of [^{14}C]taurine recovered from cats given four different diets[1]

Diet	CO_2[2]	Feces[3]	Urine[3]	CO_2+ Feces
Commercial heat-processed	9.41 ± 0.95[a]	11.7 ± 1.1[ab]	6.6 ± 0.8[a]	21.1 ± 1.0[a]
Commercial frozen	0.09 ± 0.03[b]	12.6 ± 3.9[a]	31.1 ± 2.5[b]	12.7 ± 3.9[b]
Purified 1325 mg taurine/kg	1.55 ± 0.70[c]	3.4 ± 0.3[c]	25.9 ± 5.6[b]	4.9 ± 0.9[c]
Purified 0 mg taurine/kg	3.75 ± 0.28[d]	5.8 ± 1.0[bc]	10.4 ± 1.9[a]	9.6 ± 0.9[bc]

[1] All values are means ± SEM for four cats. Means in a column not sharing a common superscript letter are significantly different at $p < 0.05$ as determined by LSD tests.
[2] Cumulative total for 5 d.
[3] Cumulative total for 14 d.

TAURINE BALANCE DETERMINED BY ILEAL TAURINE FLUX

The objective of the second study was to measure intestinal taurine loss and taurine balance in cats fed commercial and purified diets by measuring taurine in ileal digesta instead of feces[29]. A technique was developed for permanent ileal cannulation of cats, to allow daily sampling of ileal digesta. By collecting samples from the terminal ileum, where bacterial numbers are significantly lower than in the colon (10^7 versus 10^{10}-10^{11}/g digesta)[30], it was anticipated that intestinal degradation of taurine by microorganisms would be substantially reduced. The response to the same four diets used in the first study was evaluated, a commercial diet processed by heat or preserved by freezing and purified diets containing 1225 or 0 mg/kg taurine.

Five day balance trials were performed on d 3-7 with measurement of food intake and taurine in urine and ileal digesta in two groups of cats. Group 1 cats were fed the 1225 mg taurine/kg purified diet and Group 2 cats the 0 mg taurine/kg diet for 6-8 mo prior to the start of ileal collections. Group 1 cats were then fed all four diets in a rotating order to determine taurine balance in replete cats. Group 2 cats were fed the 0 taurine diet to evaluate taurine balance in taurine-depleted cats fed no taurine.

Substantial quantities of both free and bound taurine were found in the terminal ileal digesta of cats fed all diets (Table 2). Significantly greater quantities of total taurine (free + bound) were found in ileal digesta from cats fed the heat-processed than frozen-preserved diet (205% versus 101% of the average daily taurine intake, respectively) with calculated taurine balances of -609 versus -212 μmol/d, respectively (Table 3). These observations were consistent with the results of the first study where greater amounts of $^{14}CO_2$ were recovered from cats given a pulse-labeled dose of [^{14}C]taurine when fed the heat-processed than frozen diet. Taurine present at the terminal ileum which was not reabsorbed in the colon would be available for degradation by anaerobic bacteria.

Flux of taurine at the ileum was significantly less when cats were fed the purified than commercial diets. The quantity of taurine found in ileal digesta was not significantly different in taurine-replete cats fed the 1225 or 0 mg/kg taurine diets. This suggests that the majority of the taurine appearing at the terminal ileum is of endogenous origin. Taurine-depleted cats had significantly lower amounts of taurine in ileal digesta with a taurine balance of -77 μmol/d. The source(s) of the large amounts of free and bound taurine appearing at the terminal ileum of cats fed all diets was not determined in this study but the bound taurine possibly represented taurine conjugated bile salts. Therefore, the next two studies focused on the turnover of taurine conjugated bile salts in the cat.

EFFECT OF DIETARY PROTEIN SOURCE ON TAUROCHOLIC ACID KINETICS

The first turnover study examined the effect of dietary protein source (soy versus casein), and taurine status, on kinetics of [24-^{14}C] and [taurine-2-^3H]taurocholic acid in taurine-replete and taurine-depleted cats[39]. Taurine-replete cats were fed 1500 mg taurine/kg purified diets containing either 435 g/kg casein (1500 Cas) or soy protein (1500 Soy) in a cross-over design. Taurine-depleted cats were fed the soy protein diet with no taurine (0 Soy). Specific activity of [^{14}C and ^3H]taurocholic acid in bile was determined for 6 d following a pulse dose of dual-labelled taurocholic acid. Common bile duct cannulae were surgically placed in the cats 1-2 wk prior to the start of the experiment, to allow periodic sampling of bile without interruption of normal bile flow.

Dietary protein source, soy versus casein, was found to affect kinetics of the enterohepatic circulation of both the taurine and cholic acid moieties of taurocholic acid in cats. Taurocholic acid pool size was significantly greater in cats fed the 1500 Soy diet than cats fed the 1500 Cas or 0 Soy diet (Table 4). Total entry rate, irreversible loss rate and recycling rate of [taurine-2-^3H]taurocholic acid tended to be higher and the irreversible loss rate of [24-^{14}C]taurocholic acid was greater (p < 0.06) in cats fed the 1500 Soy than 1500 Cas diet. This is consistent with the greater ability of soy protein to bind bile salts and increase fecal loss than casein as previously demonstrated[13-15].

The non-linearity of the ln specific activity (SA) versus time curves for [^3H-2-taurine]taurocholic acid found in taurine-replete cats indicated that at least two pools of taurine participated in the enterohepatic circulation of taurocholic acid. In addition to the pool of taurine in taurocholic acid, cats may have a substantial pool of taurine that is formed as intestinal taurocholic acid is deconjugated. Some of this taurine would

Table 2. Taurine in ileal digesta in cats fed four different diets[1]

Diet	n	Ileal Taurine			Ileal Taurine		
		Free	Bound	Total	Free	Bound	Total
		% dietary taurine intake			umol · d⁻¹ · g diet⁻¹		
Commercial (heat processed)							
Days 3-7	4	96 ± 11[a]	109 ± 19[a]	205 ± 29[a]	8.1 ± 0.9[a]	9.2 ± 1.6[a]	17.3 ± 2.4[a]
Days 24-28	6	99 ± 11[a]	51 ± 10[b]	149 ± 14[a]	10.5 ± 1.1[a]	5.3 ± 1.1[a]	15.8 ± 1.5[a]
Commercial (frozen)	4	43 ± 6[b]	58 ± 21[b]	101 ± 18[b]	4.5 ± 0.7[b]	6.1 ± 2.2[ab]	10.6 ± 1.9[b]
Purified (1225 Tau)							
Group 1[2]	6	32 ± 5[b]	40 ± 11[b]	72 ± 16[c]	3.5 ± 0.5[bc]	4.3 ± 1.2[b]	7.8 ± 1.6[c]
Purified (0 Tau)							
Group 1[2]	6				2.5 ± 0.5[c]	4.4 ± 0.7[b]	6.9 ± 1.0[c]
Group 2[3]	4				0.5 ± 0.1[d]	0.3 ± 0.1[c]	0.8 ± 0.2[d]

[1] All values are means ± SEM. Means in a column not sharing a common superscript letter are significantly different (p < 0.05) as determined by ANOVA and LSD test.
[2] Taurine-replete cats.
[3] Taurine-depleted cats.

49

Table 3. Taurine intakes and taurine losses on days three to seven in cats fed four different diets[1,2]

Diet	n	Dietary taurine intake	Ileal taurine flux	Urinary taurine loss	Taurine balance
			μmol/d		
Commercial (heat processed)	4	485 ± 54[a]	969 ± 108[a]	133 ± 24[a]	-609 ± 91*[a]
Commercial (frozen)	4	749 ± 129[bc]	772 ± 205[a]	189 ± 43[a]	-212 ± 141[b]
Purified (1225 Tau)	6	577 ± 52[ac]	462 ± 110[b]	131 ± 18[a]	- 15 ± 105[b]
Purified (0 Tau)					
Group 1[3]	6	0	440 ± 87[b]	134 ± 15[a]	-574 ± 98*[a]
Group 2[4]	4	0	66 ± 15[c]	11 ± 3[b]	- 77 ± 18*[b]

[1] Balance was calculated as dietary intake - (fecal loss calculated from ileal values + urine loss).
[2] All values are means ± SEM. Means in a column not sharing a common superscript letter are significantly different (p<0.05) as determined by ANOVA and LSD test. * Significantly different from zero as determined by a Student's t test.
[3] Taurine-replete cats.
[4] Taurine-depleted cats.

Table 4. Taurocholic acid kinetics in taurine-replete and taurine-depleted cats fed 1500 Cas, 1500 Soy or 0 Soy diets[1]

Parameter	Isotope	Diet		
		1500 Cas	1500 Soy	0 Soy
Pool size	^{3}H-Tau	188.1 ± 45.6*[a]	404.2 ± 58.9+[a]	219.8 ± 37.0[a]
μmol	^{14}C-Cholyl	173.6 ± 24.7*[a]	344.4 ± 38.7+[a]	215.8 ± 36.7[a]
Total entry rate, μmol/d	^{3}H-Tau	401.6 ± 38.3	675.8 ± 179.1	
Irreversible loss rate	^{3}H-Tau	318.0 ± 30.7[a]	444.5 ± 68.8+[a]	56.4 ± 6.2[a]
μmol/d	^{14}C-Cholyl	220.4 ± 45.9[a]	356.0 ± 43.3+[b]	120.0 ± 11.7[b]
Recycling rate μmol/d	^{3}H-Tau	83.6 ± 11.8	231.4 ± 112.2	

[1] All values are means ± SEM. * Significantly different from 1500 Soy group with p < 0.05 (one-way analysis of variance). + Significantly different from 0 Soy group with p < 0.05 (one-way analysis of variance). Means within each kinetic parameter determined using ^{14}C compared to ^{3}H, not sharing a common superscript letter are significantly different at p < 0.05 (paired t-test).

be reabsorbed from the intestine and mixed with the large pool of free taurine found in the blood and tissues. The remainder of the taurine would be available for degradation by intestinal microorganisms. The finding of a lower irreversible loss rate of [³H-2-taurine]taurocholic acid in cats fed casein than soy protein diets could explain the observations from our laboratory that cats fed casein have higher plasma and whole blood taurine concentrations than cats fed soy protein diets.

Dietary protein source also significantly affected the composition of the bile salt pool (Table 5). Cats fed the 1500 Cas diet had a lower fraction of taurocholic acid and higher fractions of taurodeoxycholic and taurochenodeoxycholic acids than cats fed the 1500 Soy diet. The difference may be related to differential binding of the various bile salts to casein and soy protein, different synthesis rates, and/or different rates of intestinal bacterial modification.

Taurine status had substantial effects on kinetics of the enterohepatic circulation, with decreased synthesis and loss of both the taurine and cholic acid moieties of taurocholic acid in taurine-depleted compared to taurine-replete cats (Table 4). Taurocholic acid pool size in depleted cats was almost one-half the size of that found in taurine-replete cats fed 1500 Soy (Table 5), although total pool sizes were similar due to substantial amounts of free cholic acid. No recycling of [³H]taurocholic acid was evident for the taurine-depleted cats (ln SA curves were linear), likely due to the very low concentrations of taurine found in plasma and tissues. The irreversible loss rate of taurine from the taurocholic acid pool therefore, can be used as estimate of total body taurine turnover, being approximately $56\,\mu$mol/d. This value would also be an estimate of taurine synthesis in the cat assuming that insignificant amounts of unlabelled taurine are available from depleted tissue pools and thus the only source of unlabeled taurine is endogenous synthesis.

EFFECT OF DIETARY HEAT-PROCESSING ON TAUROCHOLIC ACID KINETICS

A final study examined the effects of dietary heat-processing on kinetics of [24-¹⁴C] and [taurine-2-³H]taurocholic acid in taurine-replete cats fed a single formulation of a commercial diet in two forms, heat-processed and frozen, in a cross-over design. The frozen-preserved diet maintained significantly greater plasma and whole blood taurine concentrations when fed to cats than the heat-processed diet. Specific activity of [¹⁴C and ³H]taurocholic acid in bile was determined for 6 d following a pulse dose of dual-labelled taurocholic acid.

Dietary heat-processing had a significant effect on the kinetics of the taurine moiety of taurocholic acid. The irreversible loss rate of [³H-2-taurine]taurocholic acid was significantly higher in cats fed the heat-processed than frozen commercial diet

Table 5. Bile salt composition in taurine-replete and taurine-depleted cats fed 1500 Cas, 1500 Soy or 0 Soy Diets[1]

Diet	Taurocholate	Taurodeoxy-cholate	Taurocheno-deoxycholate	Glycocholate	Cholate
		Fraction of total bile salts			
1500 Cas	0.84 ± 0.013*	0.12 ± 0.010*	0.040 ± 0.003*	0.006 ± 0.006	N.D.[2]
1500 Soy	0.88 ± 0.016+	0.09 ± 0.016+	0.025 ± 0.004+	0.005 ± 0.004*	N.D.
0 Soy	0.63 ± 0.125	0.01 ± 0.010	0.017 ± 0.010	0.030 ± 0.004	0.31 ± 0.124

[1] All values are means ± SEM. * Significantly different from 1500 Soy group with $p < 0.05$ (two-way analysis of variance). + Significantly different from 0 Soy group with $p < 0.05$ (one-way analysis of variance).
[2] None detected

(Table 6). When compared to the previous kinetic study utilizing purified diets, the heat-processed commercial diet caused a substantially higher irreversible loss rate of [^{3}H-2-taurine]taurocholic acid, while the frozen diet had a similar irreversible loss rate to the 1500 Soy diet. Increased enterohepatic cycling of taurocholic acid, with subsequent bacterial degradation of the taurine released by deconjugation, could explain the greater conversion of ^{14}C-labelled taurine to $^{14}CO_2$ and the increased amounts of taurine found at the terminal ileum of cats fed the heat-processed compared to frozen-preserved commercial diets.

Table 6. Taurocholic acid kinetics in cats fed a commercial diet in two forms, heat-processed and frozen[1]

Parameter	Isotope	Diet	
		Heat-processed	Frozen
Pool size	^{3}H-Tau	838.6 ± 141.1[a]	923.2 ± 112.3[a]
μmol	^{14}C-Cholyl	722.6 ± 117.5[a]	808.7 ± 105.3[a]
Total entry rate, $\mu mol/d$	^{3}H-Tau	854.4 ± 106.1	517.8 ± 132.7
Irreversible loss rate	^{3}H-Tau	719.3 ± 99.1[a]	423.5 ± 63.3[a*]
$\mu mol/d$	^{14}C-Cholyl	470.4 ± 75.0[b]	373.5 ± 48.3[a]
Recycling rate $\mu mol/d$	^{3}H-Tau	135.1 ± 79.6	94.3 ± 94.3

[1] All values are means ± SEM. * Significantly different from heat-processed group with $p < 0.05$ (t-test). Means within each kinetic parameter determined using ^{14}C compared to ^{3}H, not sharing a common superscript letter are significantly different at $p < 0.05$ (paired t-test).

CONCLUSIONS

The results of these studies demonstrate that substantial amounts of taurine are lost by cats via the intestine, associated with bacterial taurine degradation and the enterohepatic circulation of taurine conjugated bile salts. Intestinal loss of taurine is affected by the type of diet (commercial vs purified), and the type of processing of the diet, (heat-processed vs frozen). Intestinal taurine loss is also affected by the source of dietary protein and the taurine status of the cat. The increased irreversible loss of the taurine moiety of taurocholic acid and the greater degradation of taurine by intestinal bacteria in cats fed heat-processed than frozen diets, could account for the inability of some heat-processed commercial diets to maintain adequate plasma and tissue taurine concentrations[23,24]. However, further research will be required to determine the mechanism(s) responsible for this increased intestinal taurine loss. Several possible mechanisms include: increased bacterial deconjugation of bile salts due to changes in location and/or numbers of intestinal microflora, increased secretion of bile salts due to changes in release of cholecystokinin (CCK), and/or increased binding or entrapment of bile salts by heat-processed dietary components. Dietary factors have previously been shown to affect the number and location of intestinal bacteria, including bacteria that deconjugate taurine-conjugated bile salts[32,33].

The extent of taurine deconjugation and further degradation depends on the location, number and species of intestinal bacteria. While it is generally assumed that the majority of bacterial deconjugation occurs in the terminal ileum and colon where bacterial numbers are greatest[10,34,38], normal rats fed purified diets had 35 and 50 percent unconjugated bile salts in jejunum and ileum, respectively[20]. Due to the large numbers of bacteria normally present in the cecum and colon, bile salts found in the feces of normal animals are almost entirely unconjugated[10,34,35]. In contrast, germ-free rats excrete only conjugated bile salts in their feces[36,37] and rats whose bile flow is diverted from the intestine into the tail vein, have no evidence of bile salt deconjugation as measured by [^{3}H]glycine degradation[38]. If germ-free rats are inoculated with a single species of bacteria that can deconjugate bile salts, total fecal bile salt excretion is doubled by 9 days and the percentage of conjugated bile salts in feces decreases 40-60 percent[36]. Similarly, a five-fold increase in the turnover of [^{14}C]cholic acid was demonstrated in normal compared to germ-free rats[37], evidence that bacterial bile salt transformation increases the turnover of bile salts and contributes to the inefficiency of the enterohepatic circulation.

Felids appear to be unique mammals with a dietary requirement for taurine due to both low endogenous taurine synthesis and the obligatory conjugation of bile salts with taurine. Taurine losses due to the inefficiency of the enterohepatic circulation are not constant but vary depending on the composition of the diet and the method of processing. The contribution of intestinal taurine loss to overall taurine balance has previously been unappreciated, due largely to an underestimation of taurine degradation by intestinal microorganisms. The formulation of feline diets that maintain adequate tissue taurine concentrations will require consideration of the intestinal metabolism of taurine, including absorption, bacterial degradation and factors affecting the enterohepatic circulation of taurine conjugated bile salts.

REFERENCES

1. W.H. Elliott, Metabolism of bile acids in liver and extrahepatic tissues, in: "Sterols and Bile Acids (New Comprehensive Biochemistry)", H. Danielsson and J. Sjövall, eds., Elsevier, Amsterdam, pp. 303-329 (1985).
2. E.R.L. O'Máille, T.G. Richards, and A.H. Short, Acute taurine depletion and maximal rates of hepatic conjugation and secretion of cholic acid in the dog, J. Physiol. 180:67-79 (1965).
3. L.A. Rentschler, L.L. Hirschberger, and M.H. Stipanuk, Response of the kitten to dietary taurine depletion: Effects on renal reabsorption, bile acid conjugation and activities of enzymes involved in taurine synthesis, Comp. Biochem. Physiol. 84B:319-325 (1986).
4. J.A. Sturman, D.K. Rassin, K.C. Hayes, and G. E. Guall, Taurine deficiency in the kitten: Exchange and turnover of [^{35}S] taurine in brain, retina, and other tissues, J. Nutr. 108:1462-1476 (1978).
5. J. De La Rosa and M.H. Stipanuk, The effect of taurine depletion with guanidinoethanesulfonate on bile acid metabolism in the rat, Life Sci. 36:1347-1351 (1985).
6. D.A. Vessey, The biochemical basis for the conjugation of bile acids with either glycine or taurine, Biochem. J. 174:621-626 (1978).
7. W.G.M. Hardison and J.H. Proffitt, Influence of hepatic taurine concentration on bile acid conjugation with taurine, Am. J. Physiol. 1:E75-E79 (1977).
8. J.R. Malagelada, V.L.W. Go, E.P. DiMagno, and W.H.J. Summerskill, Interactions between intraluminal bile acids and digestive products on pancreatic and gallbladder function, J. Clin. Invest. 52:2160-2165 (1973).
9. G. Gomez, J.R. Upp, F. Lluis, R.W. Alexander, G.J. Poston, G.H. Greeley, Jr. and J.C. Thompson, Regulation of the release of cholecystokinin by bile salts in dogs and humans, Gastroenterology 94:1036-1046 (1988).
10. T. Midtvedt and A. Norman, Bile acid transformations by microbial strains belonging to genera found in intestinal contents, Acta Path. Microbiol. Scandinav. 71:629-638 (1967).
11. G.W. Hepner, J.A. Sturman, A.F. Hofmann, and P.I. Thomas, Metabolism of steroid and amino acid moieties of conjugated bile acids in man. III. Choyltaurine (taurocholic acid), J. Clin. Invest. 52:443-440 (1973).
12. M.W. Huff. and K.K. Carroll, Effects of dietary protein on turnover, oxidation, and absorption of cholesterol, and on steroid excretion in rabbits, J. Lipid Res. 21:546-558 (1980).
13. Y.-S. Choi, S. Goto, I. Ikeda, and M. Sugano, Interaction of dietary protein, cholesterol and age on lipid metabolism of the rat, Br. J. Nutr. 61:531-543 (1989).
14. M. Sugano, S. Goto, Y. Yamada, K. Yoshida, Y. Hashimoto, T. Matsuo, and M. Kimoto, Cholesterol-lowering activity of various undigested fractions of soybean protein in rats, J. Nutr. 120:977-985 (1990).

15. S. Makino, H. Nakashima, K. Minami, R. Moriyama, and S. Takao, Bile acid-binding protein from soybean seed: Isolation, partial characterization and insulin-stimulating activity, *Agric. Biol. Chem.* 52:803-809 (1988).

16. D. Kritchevsky and J.A. Story, Binding of bile salts in vitro by nonnutritive fiber, *J. Nutr.* 104:458-462 (1974).

17. E.W. Pomare, K.W. Heaton, T.S. Low-Beer, and H.J. Espiner, The effect of wheat bran upon bile salt metabolism and upon the lipid composition of bile in gallstone patients, *Digestive Diseases* 21:521-526 (1976).

18. Y. Imai, S. Kawata, M. Inaoa, S. Miyoshi, Y. Minami, Y. Matsuzawa, K. Uchioa, and S. Tarui, Effect of cholestyramine on bile acid metabolism in conventional rats, *Lipids* 22:513-516 (1987).

19. K. Ebihara and B.O. Schneeman, Interaction of bile acids, phospholipids, cholesterol and triglyceride with dietary fibers in the small intestine of rats, *J. Nutr.* 119:1100-1106 (1989).

20. T. Ide and M. Horii, Predominant conjugation with glycine of biliary and lumen bile acids in rats fed on pectin, *Br. J. Nutr.* 61:545-557 (1988).

21. T. Ide, K. Takashi, M. Horri, T. Yamamoto, and K. Kawashima, Contrasting effects of water-soluble and water-insoluble dietary fibers on bile acid conjugation and taurine metabolism in the rat, *Lipids* 25:335-340 (1990).

22. T. Ide, M. Horri, K. Kawashima, and T. Yamamoto, Bile acid conjugation and hepatic taurine concentration in rats fed on pectin, *Br. J. Nutr.* 62:539-550 (1989).

23. P.D. Pion, M.D. Kittleson, Q.R. Rogers, and J.G. Morris, Myocardial failure in cats associated with low plasma taurine: A reversible cardiomyopathy, *Science* 237:764-768 (1987).

24. J.G. Morris, Q.R. Rogers, and L.M. Pacioretty, Taurine: An essential nutrient for cats, *J. Small Anim. Pract.* 31:502-509 (1990).

25. J.A. Cooke, Q.R. Rogers, and J.G. Morris, Urinary and fecal excretion of taurine by cats fed commercial canned diets, *FASEB J.* 3:A1617 (1989).

26. K. Shiekh, Taurine deficiency and retinal defects associated with small intestine bacterial overgrowth, *Gastroenterology* 80:1363 (1981).

27. K. Ikeda, H. Yamada, and S. Tanaka, The bacterial degradation of taurine, *J. Biochem.* 54:312-316 (1963).

28. M.A. Hickman, Q.R. Rogers, and J.G. Morris, Effect of processing on fate of dietary [^{14}C]taurine in cats, *J. Nutr.* 120:995-1000 (1990).

29. M.A. Hickman, Q.R. Rogers, and J.G. Morris, Taurine balance is different in cats fed purified and commercial diets, *J. Nutr.*, in press (1992).

30. S.L. Gorbach, Progress in gastroenterology, intestinal microflora, *Gastroenterology* 60:1110-1129 (1971).

31. M.A. Hickman, M.L. Bruss, J.G. Morris, and Q.R. Rogers, Kinetics of the enterohepatic circulation of taurocholic acid are affected by dietary protein source, soybean versus casein, and taurine status in cats, *J. Nutr.*, in press (1992).

32. S.D. Feighner and M.P. Dashkevicz, Effect of dietary carbohydrates on bacterial cholytaurine hydrolase in poultry intestinal homogenates, *Appl. Environ. Micro.* 54:337-342 (1988).

33. M. Winitz, R.F. Adams, D.A. Seedman, P.N. Davis, L.G. Jayko, and J.A. Hamilton, Studies in metabolic nutrition employing chemically defined diets, *Am. J. Clin. Nutr.* 23:546-559 (1970).

34. M.J. Hill and B.S. Drasar, Degradation of bile salts by human intestinal bacteria, *Gut* 9:22-27 (1968).

35. S. Hayakawa, Microbiological transformation of bile acids, *Adv. Lipid Res.* 11:143-192 (1973).

36. T. Chikai, H. Nakao, and K. Uchida, Deconjugation of bile acids by human intestinal bacteria implanted in germ-free rats, *Lipids* 22:669-671 (1987).

37. B.E. Gustafsson, S. Bergstsröm, S. Lindstedt, and A. Norman, Turnover and nature of fecal bile acids in germfree and infected rats fed cholic acid-24-^{14}C, *Proc. Soc. Exp. Biol. Med.* 94:467-471 (1957).

38. S. Borgström, L. Krabisch, M. Lindstrom, and J. Lillienau, Deconjugation of bile salts: Does it occur outside the contents of the intestinal tract in the rat?, *Scand. J. Clin. Lab. Invest.* 47:543-549 (1987).

URINARY EXCRETION OF TAURINE AS A FUNCTION
OF TAURINE INTAKE: POTENTIAL FOR ESTIMATING
TAURINE BIOAVAILABILITY IN THE ADULT CAT

Jack Odle[1], Eric N. Glass[1,2], Gail L. Czarnecki-Maulden[3]
and David H. Baker[1]

[1]Department of Animal Sciences and Division of Nutritional
    Sciences, University of Illinois, Urbana, IL 61801
[2]Present Address: Cornell University, School of Veterinary
    Medicine, Ithaca, NY 14853
[3]Friskies Research, Nestec, Inc., St. Louis, MO 64503

ESSENTIALITY OF TAURINE TO THE FELID

Taurine is an essential nutrient for felids (Schmidt et al., 1976). When cats are fed diets containing inadequate taurine, plasma and tissue levels of taurine become depleted. If the dietary deficiency is prolonged, the depleted status can result in overt clinical ailments including central retinal degeneration (Schmidt et al., 1976; Knopf et al., 1978; Anderson et al., 1979), dilated cardiomyopathy (Pion et al., 1987) as well as reproductive failure (Sturman et al., 1986). Taurine essentiality for felids is likely a consequence of impaired denovo synthesis coupled with a high physiological demand. Specifically, cats have a relatively low activity of cysteinesulfinic acid decarboxylase which is the putative rate limiting enzyme in the conversion of methionine/cystine to taurine (Wordon and Stipanuk, 1985). The need for dietary taurine is further exacerbated by an obligatory requirement for the synthesis of taurine-conjugated bile acids (Figure 1). Unlike other species, cats conjugate bile acids almost exclusively with taurine and do not substitute glycine for taurine when taurine availability becomes limited (Rabin et al., 1976).

The biochemical mechanism by which taurine deficiency culminates in the noted clinical manifestations is not precisely known. Since cats rely almost exclusively on taurine for bile acid conjugation, it could be argued that the ailments are due to a secondary deficiency of lipids (ie, retinol, essential fatty acids, etc) due to impairment of their absorption. However, this hypothesis has not been substantiated to date (Lehmann, et al., 1990), and direct etiological involvement of taurine is suspected, possibly involving regulation of ion-balance across the cell membrane. Taurine may also help maintain membrane integrity through antioxidant effects (Wright, et al., 1986).

Taurine, Edited by J.B. Lombardini *et al.*
Plenum Press, New York, 1992

55

TAURINE REQUIREMENT

Although a minimal dietary taurine requirement of 400 mg/kg diet has been established (National Research Council, 1986), commercial diets formulated on this basis may result in suboptimal taurine status. Indeed, taurine deficiency has been observed in cats fed commercial diets (Pion et al., 1987), and interestingly, dilated cardiomyopathy was observed in cats fed allegedly sufficient taurine according to the established requirement. These observations suggest that the established requirement is too low and/or that other factors are affecting the bioavailability of dietary taurine.

Controversy over dietary requirements for taurine is attributable, in part, to the response parameter measured as well the source of the dietary taurine. Plasma taurine concentration is the most commonly monitored response criterion since it can be assessed non-invasively; however, different tissue pools of taurine are known to turn over at different rates (Sturman, et al., 1978). Consequently, depending on the length of time the diet is fed, plasma levels may not represent concentrations present in important tissues. Nonetheless, clinical deficiency has been consistently observed when plasma taurine concentrations fall below 20 ηmole per ml (Pion et al., 1987), and current recommendations suggest that plasma concentrations be at or above 60 ηmole per ml.

The amount of dietary taurine required to maintain adequate taurine status seems to depend on the source of dietary taurine. This connotes that the bioavailability of dietary taurine is not a constant. Indeed, Douglas et al. (1990) showed that cats fed a canned diet require over twice as much taurine as those fed a dry diet to maintain adequate plasma taurine concentration. Thus, a very significant effect of dietary ingredients and/or matrix has been imputed and is the basis for the current recommendation that dry commercial diets be formulated to contain 1000-1200 mg/kg diet while canned diets should contain 2200-2500 mg/kg dry matter. The reason for this large difference is unknown, but is under active investigation (see Baker and Czarnecki-Maulden, 1991 for current review).

Published analytical values for the taurine content of dietary ingredients commonly used in the formulation of pet foods is limited. This is probably due in part to the variation observed among suppliers and even among lots from a given supplier. Based upon our analyses (Glass and Czarnecki-Maulden, 1990), fish products rank among the richest sources of taurine. For example, condensed fish solubles (Menhaden) typically contain greater than 15,000 mg/kg dry matter. In contrast, fish meal typically contains less than half this amount. Hydrolyzed protein concentrates from poultry contain on the order of 7,000 mg/kg dry matter, whereas poultry meal may contain only half of this amount. Feather meals contain less than 600 mg/kg. Among meat products, beef and liver digests contain roughly 1400-2600 mg/kg dry matter, while meat & bone meal averages 1200 mg/kg and blood meal contains less than 500 mg/kg dry matter. Dry diets typically contain poultry meal, fish meal and/or digests of fish, liver or poultry as significant sources of taurine, whereas canned diets are heat processed and often include visceral organs such as lungs, spleen, liver, fish & poultry by-products as well as deboned beef and whole-poultry carcasses.

The observed difference between canned and dry diets does not appear to result from production of taurine-Maillard products formed during the thermal processing of canned diets. However, Hickman et al. (1990) showed that thermal processing significantly increases taurine oxidation by the intestinal mircoflora. Douglas et al (1990) showed greater urinary excretion of ^{15}N from labeled taurine in canned diets (80 % of dose) than from dry diets (50 % of dose), suggesting that perhaps renal taurine reabsorption is also decreased when canned diets are fed.

Collectively, these data beg the question: Does taurine bioavailability vary among feed ingredients, and is it influenced by thermal processing? In order to answer this, we have been working to establish a non-invasive bioavailability assay based upon renal taurine excretion by adult cats. The conceptual basis for this approach as well as some initial experiments are described below.

DEVELOPMENT OF A RENAL-BIOAVAILABILITY ASSAY

CONCEPTUALIZATION

The conceptual framework for consideration of urinary taurine excretion as an index for relative taurine bioavailability assessment using adult cats is depicted in Figures 1 and 2. Figure 1 illustrates the enterohepatic recirculation of taurine conjugated in bile acids (TC). Upon stimulation by cholecystokinin, the gall bladder contracts and delivers TC to the proximal small intestine which, together with dietary taurine, flow to the distal small intestine where they are absorbed from the ileum and jejunum respectively. Upon entering via portal blood, TC is efficiently taken up by the liver to be recycled to the gall bladder, whereas free taurine is distributed to systemic tissues. Like other metabolites in the plasma, taurine is filtered by kidney glomeruli and is then actively reabsorbed from the proximal tubule in a concentration dependent manner. Figure 2 illustrates, in general terms, the relationship between urinary excretion of a metabolite and the tubular load of that metabolite. Tubular load is defined as the rate at which the metabolite filters through the glomerular membrane (mg/min). As the plasma concentration of the metabolite (eg., taurine) increases, the tubular load increases, and in accordance with Figure 2, a point (threshold) is reached where the tubular load exceeds the rate of reabsorption, and the metabolite appears in the urine. As the tubular load (plasma concentration) increases further, urinary excretion increases markedly and proportionally.

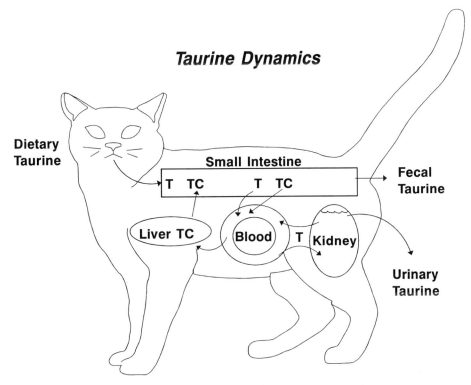

Figure 1. Schematic diagram illustrating conceptual basis for the use of urinary taurine excretion as an index for taurine bioavailability. See text for further description. T = taurine, TC = taurine conjugated in bile acids.

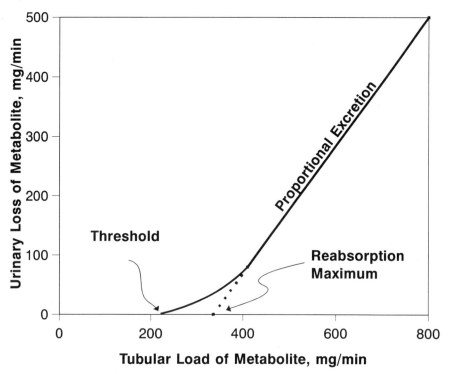

Figure 2. Renal physiological dogma showing relationship between renal tubular load of a plasma metabolite and its excretion in the urine.

RENAL ADAPTATION TO TAURINE INTAKE

As described above, the kidney plays a primary role in regulating the taurine pools in the body, with urine being a major excretory route when taurine intake is high. Beyond the natural drop in urinary taurine excretion that occurs as taurine status is diminished (as per Figure 2) an additional renal adaptation has been characterized in both rodents and cats wherein taurine reabsorption from the proximal tubule is further attenuated when taurine status is low. Initial studies (Rozen et al., 1979; Friedman et al., 1981) using kidney brush-border-membrane vesicles from mice and rats showed an increase in Vmax for taurine resorption rate in animals fed diets containing low taurine compared with animals fed high taurine. They found no detectable difference in Km (taurine affinity of transporter) however. These results suggested an increase in the number and/or capacity of resorption transport sites without an increase in affinity. Also, congruent with the relationship in Figure 2, urinary taurine excretion was dramatically increased in rats fed diets high in taurine (184 μmole/mg creatinine) compared with control rats (6 μmole/mg creatinine) and was remarkably decreased in animals fed low taurine (0.6 μmole/mg creatinine) (Friedman et al., 1981).

More recently, Park et al. (1989) reported effects of dietary taurine on renal taurine transport in 18 to 22-week-old kittens. The kittens were fed diets containing 0, 1500, or 10,000 mg taurine/kg diet. Applying similar kinetic methodology to the study of kidney brush-border-membrane vesicles, they reported a significant increase in Vmax and a decrease in Km for taurine resorption in preparations derived from animals fed taurine-free diets. Thus, both an increase in the number and/or activity of transport sites as well as a greater affinity for taurine by the tubules was inferred. Furthermore, after only a 1-week adaptation period to the diets, there was a 15-fold higher urinary excretion of taurine by kittens fed the intermediate taurine diet compared to those fed the diet devoid of taurine, and a further 10-fold increase in animals fed the high taurine diet over those fed the intermediate level. After 5 weeks, the differences increased to 23- and 14-fold respectively. Although only a few levels of taurine were fed in this study, it indicates that cats appear to follow the principle depicted in Figure 2.

For application in determination of taurine bioavailability, our goal was to exploit this general physiological response of the kidney by supplementing a basal diet with sufficient crystalline taurine so as to establish a tubular taurine load in the lower end of proportional region (Figure 2). Incremental urinary taurine excretion elicited by further additions of taurine from natural feedstuffs could then be compared to additions of crystalline taurine as an index of relative bioavailability. Effective implementation of this bioassay required knowledge of the time course of adaptation of urinary taurine excretion to perturbations in taurine intake as well as the level of taurine intake beyond which urinary excretion markedly increases. Thus, two experiments were performed (Glass et al., 1992) to provide this information.

The first experiment was designed to determine the length of time required for urinary taurine excretion to reach a steady state following alteration of taurine intake. Six adult domestic cats were fed a casein-dextrose diet supplemented with 1000 mg crystalline taurine/kg. After 10 d, a urine sample was taken by cystocentesis and three cats were abruptly switched to a diet supplemented with 2000 mg taurine/kg while the other three animals were switched to a taurine-free diet. Serial cystocentesis urine samples were collected for 8 d. The diets were then abruptly reversed according to a crossover design such that cats previously receiving the 2000 mg/kg diet were fed the taurine-free diet and those fed the taurine-free diet were fed 2000 mg/kg. And again, urine was sampled serially for an additional 8 d. Statistical analysis of the time course indicated that urinary taurine excretion stabilized in 2 d when animals were switched from a high-taurine diet to a low-taurine diet. In contrast, 7 d were required to reach steady state when taurine intake was increased. Consequently, in subsequent work, we have allowed a minimum of 7 d for adaptation.

The second experiment was designed to determine more precisely the level of taurine intake required to significantly increase urinary taurine excretion. Eighteen adult cats were blocked by weight and were fed a casein-dextrose diet supplemented with one of six levels of crystalline taurine ranging from 0 to 2000 mg/kg diet. After a 7-d adaptation period, total urine excretion was collected for 5 d. Immediately following the collection, an additional urine sample was taken by cystocentesis for comparison to total collection data, and blood was also sampled for determination of taurine concentration. Consistent with the general dogma previously discussed (Figure 2), urinary taurine excretion increased biphasically as taurine intake increased. Fitting a two-slope broken line model (Robbins, et al., 1979) to the data revealed a significant breakpoint at a taurine intake of 96 μmole/(kg BW$\bullet$d) with a slope after the breakpoint being 15-fold greater than that observed before the breakpoint. This was true for urinary excretion data from both total collection and cyctocentesis sampling, since these data were highly correlated (Figure 3.) Urine collection by cystocentesis requires much less labor than total collection and would therefore be the method of choice under most circumstances.

In addition to analysis of free taurine, subsamples of urine were also hydrolyzed in 6 N HCl for determination of total and "bound" taurine content (ie., bound = total - free). As taurine intake increased, urinary bound taurine excretion increased only slightly (from approximately 20 to 27.3 μmole/(kg BW$\bullet$d). This comprised essentially 100 % of the taurine excreted by animals fed the taurine-free diet. The presence of bound taurine in the urine has been reported by others (Hickman, et al., 1990), but, its chemical nature has not been elucidated to date.

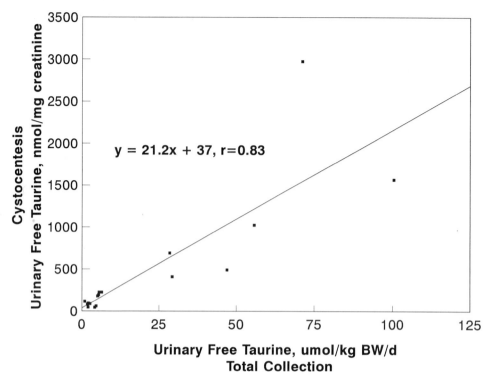

Figure 3. Correlation between urinary taurine excretion determined by cystocentesis (ηmole/mg creatinine) and taurine excretion determined by total quantitative collection (ηmole/kg BW/d).

Plasma taurine concentration increased linearly from 30 to 150 ηmole/ml as taurine intake increased, while the concentration in whole blood increased asymptotically to 450 ηmole/ml. Interestingly, the taurine intake corresponding to 95 % of the asymptotic value corresponded precisely to the inflection point observed for urinary excretion.

DISCUSSION

Collectively, these data suggest that a 7-d adaptation period is sufficient for urinary taurine excretion to respond to an alteration in taurine intake and that urinary taurine excretion increases markedly as taurine intake exceeds 96 μmole/(kg BW•d). Future research directed toward assessment of taurine bioavailability in natural feedstuffs could proceed by supplementing a basal diet with sufficient crystalline taurine to provide intakes greater than 96 μmole/(kg BW•d). Additional increments of taurine could be supplemented from natural ingredients and the increment in urinary taurine excretion could be compared to that obtained from additional crystalline taurine supplementation. Subsequently, slope-ratio or standard curve analysis (Anderson, et al., 1978; Funk et al., 1990) could be performed to compute a relative bioavailability index. This procedure was followed in a recent adult-cat study from our laboratory (unpublished data), and a commercially-available poultry meal though containing 2,812 mg *analyzed* taurine per kg, was found to provide essentially no *bioavailable* taurine.

Bioavailability data for common dietary ingredients would be of great value to the pet food industry, since a substantial portion of the taurine contained in commercial diets is of natural origin. Furthermore, the effects of thermal processing could be evaluated so that accurate quantitative contributions of the taurine contained in fresh and processed feedstuffs could be used in the formulation of commercial diets.

SUMMARY

Urinary taurine excretion increases markedly when excess taurine is consumed. Experiments were designed to characterize this response in an attempt to develop an assay system for taurine bioavailability in common cat foods using an adult cat model. Initial studies investigated the time course of changes in urinary taurine excretion in response to alterations in taurine intake. The rate of urinary taurine excretion decreased rapidly when cats were switched from a casein diet supplemented with 0.2% crystalline taurine to a diet containing no supplemental taurine, reaching steady-state in 2 d. In contrast, urinary taurine excretion by cats switched from low to high taurine did not plateau until 6 to 7 d. Subsequently, cats ($n = 18$) were fed a casein diet containing graded levels of crystalline taurine (0, 0.025, 0.05, 0.10, 0.15 or 0.20%). After a 7-d adjustment period, urinary taurine excretion was quantified over a 5-d collection period and also by cystocentesis, and blood taurine levels were measured on d 6. Plasma taurine increased linearly ($r = 0.88$) as taurine intake increased, while whole-blood taurine increased asymptotically, reaching 95% of maximum concentration at a taurine intake of 93 μmole/(kg body weight• d). The rate of urinary taurine excretion increased only slightly as taurine intakes increased to 96 μmole/(kg body weight• d), but increased markedly (15-fold) thereafter. The same pattern was observed whether urinary taurine excretion was expressed as μmole/(kg body weight• d) from total urine collection or as μmole/g creatinine from cystocentesis. At highest taurine intake, urinary taurine excretion accounted for 78% of the taurine consumed. A small amount of urinary taurine was present in "bound" form (ie, released by acid hydrolysis) which constituted 100 % of urinary taurine excretion at low taurine intakes but only 26 % at high taurine intakes. Urinary taurine excretion (at taurine intakes above 96 μmole/(kg BW• d) did not increase when a poultry meal product was supplemented into a commercial cat diet, while taurine excretion increased substantially when a similar level of taurine was provided as crystalline taurine.

ACKNOWLEDGEMENTS

This research was supported in part by funds from Friskies Research, Nestec, Inc., St. Joseph, MO and Biokyowa, Inc., St. Louis, MO.

REFERENCES

Anderson, P. A., Baker, D. H., Corbin, J. E. and Helper, L. C., 1979, Biochemical lesions associated with taurine deficiency in the cat, *J. Anim. Sci.* 49:1227-1234.

Anderson, P. A., Baker, D. H., and Mistry, S. P., 1978, Bioassay determination of the biotin content of corn, barley, sorghum and wheat, *J. Anim. Sci.* 47:654-659.

Baker, D. H., and Czarnecki-Maulden, G. L., 1991, Comparative nutrition of cats and dogs, *Annu. Rev. Nutr.* 11:239-263.

Douglass, G. M., Fern, E. B., and Brown, R. C., 1990, Feline plasma and whole blood taurine levels as influenced by commercial dry and canned diets. *Waltham International Symposium on the Nutrition of Small Companion Animals, Davis, CA, September 4-8,* p. 64.

Friedman, A. L., Albright, P.W., and Chesney, R. W., 1981, Dietary adaptation of taurine transport by rat epithelium, *Life Sci.* 29:2415-2419.

Funk, M. A., Hortin, A. E., and Baker, D. H., 1990, Utilization of D-methionine by growing rats, *Nutr. Res.* 10:1029-1034.

Glass, E. N. and Czarnecki-Maulden, G. L., 1990, Taurine concentrations in different food products, *FASEB J.* 4:A799 (abst).

Glass, E. N., Odle, J., and Baker, D. H., 1992, The effects of taurine intake on urinary taurine excretion in the domestic, *J. Nutr.* (in press).

Hickman, M. A., Rogers, Q. R., and Morris, J. G., 1990, Effect of processing on fate of dietary [^{14}C] taurine in cats, *J. Nutr.* 120:995-1000.

Knopf, K., Sturman, J. A., Armstrong, M. and Hayes, K. C., 1978, Taurine: an essential nutrient for the cat, *J. Nutr.* 108:773-778.

Lehmann, A., Knutsson, L., and Bosaeus, I., 1990, Elevation of retinol levels and suppression of alanine aminotransferase activity in the liver of taurine-deficient kittens, *J. Nutr.* 120:1163-1170.

National Research Council, "Nutrient Requirements of Cats," Natl. Acad. Sci., Washington, DC. (1986).
Park, T., Rogers, Q. R., Morris, J. G., and Chesney, R. W., 1989, Effect of dietary taurine on renal taurine transport by proximal tubule brush-border-membrane vesicles in the kitten, *J. Nutr.* 119:1452-1460.

Park, T., Rogers, Q.R., Morris, J.G. and Chesney, R.W., 1989, Effect of dietary taurine on renal taurine transport by proximal tubule brush-border-membrane vesicles in the kitten, *J. Nutr.* 119: 1452-1460.

Pion, P. D., Kittleson, D. D., and Rogers, Q. R., Morris J. G., 1987, Myocardial failure in cats associated with low plasma taurine: a reversible cardiomyopathy, *Science* 237:764-768.

Rabin, B., Nicolosi, R. J., and Hayes, K. C., 1976, Dietary influence on bile acid conjugation in the cat, *J. Nutr.* 106:1241-1246.

Robbins, K. R., Norton, H. W., and Baker, D. H., 1979, Estimation of nutrient requirements from growth data, *J. Nutr.* 109:1717-1714.

Rozen, R., Tenenhouse, H. S., and Scriver, C. R., 1979, Taurine transport in renal brush-border-membrane vesicles, *Biochem. J.* 180:245-248.

Schmidt, S. Y., Berson, E. L., and Hayes, K. C., 1976, Retinal degeneration in cats fed casein. I. Taurine deficiency, *Invest. Opthalmol.* 15:47-52.

Sturman, J. A., Rassin, D. K., Hayes, K. C., and Gaull, G. E., 1978, Taurine deficiency in the kitten: exchange and turnover of [^{35}S]taurine in brain, retina, and other tissues, *J. Nutr.* 108:1462-1476.

Sturman, J. A., Gargano, A. O., Massing, J. M., and Imaki, H., 1986, Feline maternal taurine deficiency: effect on mother and offspring, *J. Nutr.* 116:655-667.

Wright, C. E., Tallan, H. H., and Lin, Y. Y., 1986, Taurine: biological update, *Ann. Rev. Biochem.* 55:427-453.

Wordon, J. A. and Stipanuk, M. H, 1985, A comparison by species, age and sex of cysteinesulfinate decarboxylase activity and taurine concentration in liver and brain of animals, *Comp. Biochem. Physiol. B* 82:233-239.

DILATED CARDIOMYOPATHY ASSOCIATED WITH TAURINE
DEFICIENCY IN THE DOMESTIC CAT: RELATIONSHIP TO
DIET AND MYOCARDIAL TAURINE CONTENT

Paul D. Pion[1], Mark D. Kittleson[1], Mary L. Skiles[1],
Quinton R. Rogers[2], James G. Morris[2]

Department of Medicine[1]
Department of Physiological Sciences[2]
School of Veterinary Medicine
University of California
Davis, CA 95616

Dilated (or congestive) cardiomyopathy (DCM) was not widely recognized in cats until the early 1970's[1]. The clinical findings, response to treatment, and the gross and microscopic pathologic findings of this disease have been described[2,3,4,5].

An association between DCM in pet cats, low plasma taurine concentration, and diet was first reported in 1987[6]. A recent report suggests a similar association between taurine deficiency and myocardial failure in foxes[7], and another report independently confirmed the findings of the original report in cats[8]. The implications of these studies for human patients with DCM still needs further investigation[9].

Taurine is an essential amino acid for the cat. The majority of pet cats studied with DCM and taurine deficiency were eating commercial cat foods that contained taurine in sufficient amounts to meet or exceed the published requirement for dietary taurine for cats[10]. Many commercial cat food manufacturers responded to this finding by increasing the taurine content in the foods they produce. It has since become clear that no single value can be assigned to the dietary taurine requirement for cats. The taurine requirement for cats is now known to depend upon the type of food (canned vs dry) and many other factors which are currently a focus of study[11,12].

In this paper we summarize clinical, laboratory, and epidemiologic data which support our assertion that diet-induced taurine deficiency was (and may still remain) a major factor in the etiopathogenesis of many cases of myocardial failure in the domestic cat. A major purpose of this report is to illustrate the dramatic effect that the use of taurine in the management of feline DCM has had upon the expected prognosis for feline patients with DCM and more importantly, the effect that the addition of more taurine to commercial cat foods has had upon the health of pet cats in the United States.

Taurine, Edited by J.B. Lombardini *et al.*
Plenum Press, New York, 1992

PROSPECTIVE AND RETROSPECTIVE CLINICAL STUDIES:

Prospective Population Selection

Between November of 1986 and April of 1988 37 cats with moderate to severe DCM were identified at the University of California, Davis Veterinary Medical Teaching Hospital. Echocardiographic criteria for inclusion in this study were a left ventricular short axis end-systolic dimension (ESD) > 14 mm and a shortening fraction (SF) $\leq$ 28% in the absence of evidence for underlying congenital or acquired cardiac disease.

Retrospective Population Selection

Reviewing the medical records of cats examined at UC Davis between January 1980 and October 1986, 33 cats were identified that had an echocardiographically confirmed diagnosis of DCM.

Diet Histories and Survival Data

Diet histories were obtained by repeatedly interviewing owners of cats in the prospective study population and by review of clinical records and telephone interviews with owners of cats in the retrospective study population. Diet histories were available for all 37 cats in the prospective study population and 25 of the 33 cats studied retrospectively.

Survival data were tabulated from review of the clinical records and telephone follow up with owners of cats in both the retrospective and prospective populations.

RESULTS OF CLINICAL STUDIES

Age, sex, and breed parameters were similar in the retrospective and prospective populations. Clinical signs in the 37 prospective study cats are summarized in Table 1.

With two exceptions, the list of clinical signs in these cats were typical of what might be observed in a patient with DCM, independent of species. The unusual clinical findings were that 27% of cats in the study population had feline central retinal degeneration and 27% (not necessarily the same cats) had a prior history of feline urologic syndrome (FUS). Both of these clinical signs/syndromes have been reported to occur in less than 2% of the pet cat population[13].

Diet Histories

Of particular interest is that 35 of 62 (56%) cats (10/25 or 40% in the retrospective population and 25/37 or 68% in the prospective population) with known diet histories were primarily being fed the same brand of food (Brand A). In a survey of 150 cats evaluated at the same clinic, only 25% of cats were being primarily fed Brand A. By comparison, of the other 4 brands of food identified as the primary diet fed to cats with DCM, the next largest proportion of cats represented was 4/62 or 6% of the DCM population for a food (Brand B) that was fed to 21% of the survey population (without DCM).

It is significant to note that the diet which was most commonly fed to affected cats (Brand A) is a diet which is often prescribed as long-term dietary therapy for the treatment of feline urological syndrome (see above).

Table 1. Clinical signs in 37 prospectively studied cats.

CLINICAL SIGN	% OF CATS
Gallop	86
Dyspnea	76
Pleural Effusion	70
Murmur	32
Retinal Degeneration (FCRD)[1]	27
FUS (historical)[2]	27
Systemic Thromboembolism	22
Pulmonary Edema	14

[1]Feline Central Retinal Degeneration; [2]Feline Urologic Syndrome

Plasma Taurine Concentrations

Plasma taurine concentrations from the 37 prospectively studied cats over the treatment period are depicted in Figure 1.

The median initial (period 0) plasma taurine concentration for the 37 prospectively studied cats with DCM was 9.8 (range, 1 to 114) nmol/ml. Thromboembolism[15] was the only clinical finding in these cats that was found to be significantly related to plasma taurine concentration.

Plasma taurine concentration was significantly (Wilcoxin Rank Sum Test, $P < 0.05$) higher in eight cats with aortic thromboembolism (median = 24, range, 8-114 nmol/ml) than in 29 cats without thromboembolism (median = 6, range, 1-42 nmol/ml). Plasma taurine concentration in clinical cats without thromboembolism and in laboratory cats fed the same diet were similar (data not shown) adding credence to owner provided diet histories.

Six cats (four with and two without thromboembolism) had plasma taurine concentrations > 25 nmol/ml. In each case, the history suggested a plausible explanation for why the cats' plasma taurine concentration might have recently been

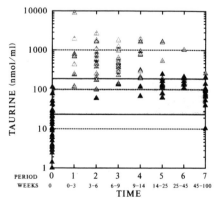

Figure 1. Plasma taurine concentrations (drawn on a log scale for clarity) in 37 cats with DCM before (period 0), during, and after taurine supplementation. The range of time (in weeks) corresponding to each treatment period is given in the inset at the bottom of the figure. Solid triangles represent samples collected while cats were not receiving taurine supplementation. The open triangles represent samples collected while cats were receiving taurine supplementation. The shaded region represents the 95% confidence interval determined from 194 client-owned cats surveyed in the Fall of 1989[14].

elevated.[a]

Other than one cat that survived 29 months after diagnosis of DCM and eventually entered the prospective study, plasma taurine concentrations were not available in cats in the retrospective population. Plasma taurine concentration in this cat was low (10.7 nmol/ml).

THERAPY, CLINICAL RESPONSE AND SURVIVAL

Clinical management of cats in the retrospective population was in agreement with what has been reported for medical management of DCM in cats and other species at the time. Primary oral drug therapy in most of these cats was limited to digoxin and furosemide. Digoxin was prescribed for 26 of 31 cats with known outcomes. Furosemide was prescribed for 28 cats.

Management of cats in the prospective population was designed to minimize the number of drugs prescribed and the duration of drug exposure. Cats with radiographic evidence of pleural effusion or pulmonary edema were treated with furosemide (31 cats) and captopril (18 cats). Digoxin was not prescribed for most cats and taurine, 500 or 1000 mg/day was prescribed for all cats. All medications other than taurine were withdrawn after radiographic and clinical signs of congestive heart failure had subsided. Of seven cats being given digoxin when initially examined, one died during initial hospitalization, and digoxin was discontinued in the other six cats at 0, 2, 4, 6, 9 and 14 weeks after initiating taurine supplementation. Furosemide was withdrawn 7 ± 4 (range, 3 to 16) weeks after initiating taurine and captopril was withdrawn 7 ± 3 (range, 2 to 13) weeks after initiating taurine.

Systolic murmurs resolved in 8 of 12 cats with murmurs within 27 weeks of initiation of taurine supplementation. Resolution of the gallop sound was documented in all surviving cats. All owners of cats which survived at least 2 weeks reported that their cats' attitude and appetite had dramatically improved during the first weeks of treatment. In general, taurine supplementation was discontinued after a demonstrably improved, near normal echocardiogram (end-systolic dimension < 12 mm, shortening fraction > 33%) was recorded.

Mean echocardiographic parameters in the population of surviving cats returned to within the range of clinical normality within 25 weeks of beginning taurine supplementation (Figure 2).

A marked divergence between the survival curves of the retrospective and prospective study populations was observed after the first weeks of observation (Figure 3, generalized Savage test[16], $P < 0.001$).

---

[a] The two cats without thromboemboli were identified while screening asymptomatic cats in the household of affected cats. Both cats had an audible gallop, one had a soft systolic murmur, and both had a history of feline urologic syndrome. In both cats, there was a delay of approximately one month between diagnosis of the first cat (in which we were able to document a low plasma taurine concentration) in the household and evaluation of the second cat, during which time both owners changed the diets of all cats in their respective households to ones we recommended as not inducing taurine deficiency.

Of the four cats with thromboemboli and plasma taurine concentrations > 25 nmol/ml, three were examined in the acute phase of their embolic episode. They were depressed with heart rates < 160 beats/min, and two were hyperkalemic (serum potassium concentrations >5.5 mEq/l) which is thought to be a result of release of intracellular potassium from ischemic skeletal muscle as the affected limb begins to regain perfusion. Serum potassium concentration was not evaluated in the third cat. The fourth cat was referred after being treated with cage rest and heparin for one week, during which time the owners had force fed the cat baby food and a commercial dietary supplement.

66

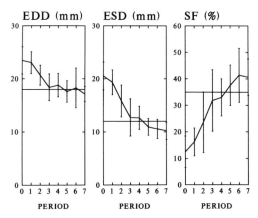

Figure 2. End-diastolic and end-systolic dimensions and shortening fraction in cats with DCM before (period 0) and after taurine supplementation. The time periods are the same as in Figure 1. Error bars represent 1 standard deviation about the mean. Horizontal lines represent the clinical limits of normality for these measures in cats at the University of California - Davis.

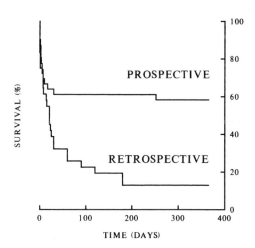

Figure 3. Percent of cats remaining alive vs time (in days) after diagnosis of DCM in the retrospective (not taurine treated) and prospective (taurine treated) populations. The outcome in two cats in the retrospective and one cat in the prospective populations are unknown (i.e., n=36 for the prospective and n=31 for the retrospective populations).

EPIDEMIOLOGY

To assess whether the recent increase in the taurine content of many commercial cat foods in the United States had a significant effect on the taurine status and the occurrence of dilated cardiomyopathy in cats, three studies were performed.

A diet history and a blood sample for determination of plasma taurine concentration were collected from 194 cats from two veterinary hospitals in 1989, 2 years after additional taurine was added to most commercial cat food formulations.

Results of this survey were compared to those of a similar survey performed in 1987.[b] The rank sum of plasma taurine concentration determined in survey cats in 1989 was not significantly different (Kruskal-Wallis test, P>0.05) from the value determined in 1987 cats. However, when only those cats whose diet was composed of cat foods that were found to be associated with DCM and low plasma taurine concentration in 1987 (see above) were evaluated a different result was obtained. Of three commercial diets fed to five or more cats, two were found to be associated with a significantly (t-test, P<0.05) higher mean plasma taurine concentration in 1989 cats than in 1987 cats fed the same commercial food.

If diet-induced taurine deficiency is the major cause of DCM in pet cats in the United States then the increase in taurine content of commercial diets should have been temporally associated with a reduction in the occurrence of DCM in pet cats in the United States. We retrospectively evaluated the records of all cats presenting to two private radiologic referral veterinary practices for echocardiography between January of 1985 and September of 1989. The percentage of cats presenting for echocardiography that were found to have DCM[c] declined from 28% (61 of 221 echocardiographic diagnoses made) in 1986 to 6% (12 of 207) in 1989. The occurrence of idiopathic left ventricular hypertrophy (hypertrophic cardiomyopathy) was unchanged during the same time period (Figure 4).

Concurrent with the above studies, we followed a single household of 27 cats fed predominantly the commercial cat food which was most commonly fed to cats in the clinical DCM studies discussed above (Brand A). In April of 1987 cats in this household were changed from the "old" (taurine depleting) formulation of this food to the "new" (not taurine depleting) formulation of this food which had approximately 70% more (0.25% vs 0.15%) taurine than the "old" formulation. Plasma taurine concentration increased significantly in these cats (paired t-test, p<0.0005) between November 1986 and July 1987 (Figure 5). In this household there had been eight deaths due to dilated cardiomyopathy between 1975 and 1986. Two cats had dilated cardiomyopathy in November of 1986 which, after taurine supplementation, resolved by July of 1987. No new cases were diagnosed in 1987, 1988 or 1989.

TAURINE DEFICIENCY MYOCARDIAL FAILURE IN LABORATORY CATS

Previous reports have documented a reduction in myocardial mechanical function as determined by echocardiography in cats fed several forms of taurine depleting foods (TAU+, 0.14%; TAU-, 0.0%; CANNED, 0.15%)[17,18]. The CANNED food was similar in composition to commercial canned diets that were fed to many of the prospective and retrospectively studied client-owned cats with DCM reported above. Despite the fact that the CANNED diet contained approximately the same concentration of taurine as the TAU+ diet and 3 times the concentration in Sturman's control diet (0.05%)[18], it was a taurine depleting diet.

An unexplained phenomenon is the nonuniform response of cats to dietary taurine deficiency; in our previous report[17] approximately 30% of taurine depleted

---

[b] The period of sample collection began prior to the date when the first pet food manufacturers began increasing the taurine content in certain cat foods and continued for a short period after these new foods were produced. It is therefore possible that some cats surveyed were being fed foods produced after manufacturers began increasing the taurine content in these foods.

[c] The diagnoses of HCM and DCM were based upon subjective and objective echocardiographic parameters. It might be argued that retrospective analysis of echocardiographic reports alone is not as reliable of as source of information as direct examination of the patient or review of all case material. It should therefore be noted that the person assigning diagnoses to the information provided is a board certified veterinary cardiologist who was blinded to the identification of the cat and the year the examination took place.

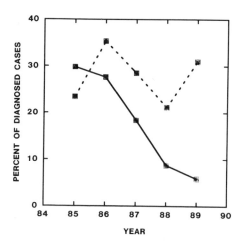

Figure 4. Percent of cats examined at two veterinary hospitals between 1985 and 1989 with an echocardiographic diagnosis of dilated cardiomyopathy (■ – ■) and hypertrophic cardiomyopathy (■ --- ■). Total number of cases per year were: 1985 (n=94); 1986 (n=221); 1987 (n=267); 1988 (n=265); 1989 (n=207).

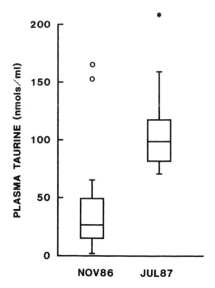

Figure 5. Plasma taurine concentrations in November 1986 and July 1987 in a household of cats fed almost exclusively one commercial cat food. November 1986: n=27. July 1987: n=27. The box encloses the middle 50% of the data. The center line of the box is the median. The whiskers denote the range unless there are outliers, in which case, the whiskers extend 1.5 times the range of the box. The "*" and "o" indicates data points beyond the whiskers. In July of 1987, when the latter of the paired samples was obtained, none of the cats were being supplemented with additional taurine other than that provided in the reformulated commercial food.

cats demonstrated echocardiographic signs of myocardial failure. In an attempt to understand this heterogeneity, tissue from these cats were subjected to light and electron microscopic analysis (n=3 or 4 per diet group), and tissue taurine determination (n=7 per diet group).

69

Light and electron microscopy provided no explanation for the heterogeneity of response to taurine depletion. All observed lesions (endocardial fibrosis, mild myocyte hypertrophy, myocardial fibrosis, lipid deposition) that were observed were considered nonspecific.

Results of tissue taurine analyses are summarized in Table 2. Mean TAU concentration in myocardium, skeletal muscle and liver at the time of death from cats fed the purified diet containing 0% taurine (TAU-) were not significantly different from that measured in tissues from cats fed the commercial-type diet containing 0.15% taurine (CANNED). Taurine concentration in tissues from both of these taurine depleted groups was significantly ($P < 0.01$, analysis of variance) lower than in the control (TAU+) group. Four of 11 TAU- cats and 5 out of 9 CANNED diet cats developed myocardial failure. We hypothesized that a difference in the magnitude of myocardial taurine depletion within the TAU- and CANNED diet groups might explain why some taurine depleted cats develop myocardial failure while others do not. However, the data did not support this hypothesis. Tissue TAU concentration in cats with MF fed the TAU- and CANNED diets were not significantly different from cats in the same group without MF. In fact, the trend in this data was toward higher, not lower myocardial taurine concentration in taurine depleted cats with myocardial failure.

Table 2. Tissue taurine concentration (mean $\pm$ 1 sd) after up to 850 days on the specified diet in umol/g wet weight in cats with and without myocardial failure (MF) fed control (TAU+) and taurine depleting (TAU-, CANNED) diets as described in the text and in reference 17. The superscripted numbers represent the number of data points per cell. Taurine analysis were performed using a Beckman amino acid analyzer; homogenates were diluted in 9 volumes of distilled water followed by precipitation of proteins with 3% SSA.

DIET	MF	LV	RV	SKEL MUS	LIVER
TAU+	N^9	12 ± 4^7	8.7 ± 2^7	2.5 ± 1.1^7	16 ± 3^6
TAU-	Y^4	1.6 ± 1.0^3	1.1 ± 0.5^3	0.4 ± 0.1^3	1.4 ± 0.1^2
	N^7	0.5 ± 0.2^4	0.5 ± 0.3^4	0.3 ± 0.2^4	2.8 ± 1.2^4
CANNED	Y^5	4.2 ± 1.4^5	2.3 ± 2.4^5	1.0 ± 0.4^5	1.0 ± 0.1^2
	N^4	1.9 ± 1.4^3	1.4 ± 1.3^3	0.5 ± 0.5^3	1.0 ± 0.1^3

Y=yes; N=no; SKEL MUS=skeletal muscle. MF was defined as LV end-systolic diameter > 13 mm with SF < 28%.

DISCUSSION

The clinical, laboratory, and epidemiologic data presented in this paper expand upon previous reports of dilated cardiomyopathy associated with taurine deficiency in the domestic cat[6,8,17]. The number of cats in the clinical population reported in this paper is larger than in prior reports[6,17]. In order to ensure a more uniform and accurate disease categorization more mildly affected (12<ESD<14, 28<SF<35) patients included in previous reports were excluded from this analysis.

The striking association between DCM and the brand of commercial diet fed to clinical cats became apparent very early in the course of the prospective clinical studies. The added finding of similar diet histories in cats in the retrospective study, an estimate of the percentage of cats eating implicated diets within the clinical populations evaluated, finding similar plasma taurine concentrations in laboratory cats fed similar diets, and finding that there were reasonable explanations for why not all cats in the clinical population had low plasma taurine concentrations fill important gaps in our prior studies[6,17.]

Two revealing clinical findings in retrospect were a high incidence of feline central retinal degeneration and a history of feline urologic syndrome. That many of the cats with DCM and taurine deficiency had central retinal degeneration is not unexpected[19]. However, it still remains to be explained why not all cats with DCM and taurine deficiency had overt retinal lesions. Although not directly related to the effects of taurine on myocardial or retinal function, the finding of a large percentage of cats with feline urologic syndrome is very interesting. It now appears reasonable to conclude that feeding specific diets, many of which were formulated for the treatment of FUS and now known to be taurine-depleting lead to taurine deficiency, FCRD, and diet-induced DCM in some cats.

The improved survival (Figure 3) in cats with DCM treated with taurine (prospective study) compared to a similar population not treated with taurine (retrospective study) is the first demonstration of a lasting cure for a large number of cases of DCM in any species. There was no indication that the clinical and echocardiographic improvement were temporary or dependent upon continued administration of high doses of taurine. As can be seen in Figure 1, the plasma taurine concentration in most cats, after discontinuing taurine supplementation, decreased to and remained within normal limits.

In cats, dogs, and humans, administration of positive inotropic agents, diuretics, and vasodilators have been the mainstay of palliative therapy for DCM. None has effectively reversed the underlying myocardial defect which many have claimed to be irreversible[20,21]. Many of the cats in the prospective study received captopril whereas few of those in the retrospective population did. This and the greatly reduced use of digoxin in the prospective population might be considered confounding variables. However it is extremely unlikely that digoxin administration was deleterious to feline patients with DCM[22] or that captopril could have been responsible for the dramatic and continual improvement in myocardial function and clinical signs observed in this study.

The epidemiologic studies presented were designed to determine the effect of adding more taurine to many commercial cat foods on the "taurine status" of pet cats. Although it has been qualitatively obvious to the veterinary profession that the occurrence of feline DCM in the United States has decreased to a small fraction of what it was before 1987 these studies represent the first attempt at quantifying the magnitude of the decrease.

No significant change in the distribution of plasma taurine concentration in the clinical populations evaluated was detected between 1987 and 1989 when all cats evaluated were included in the analysis. However, analysis of a subset of the data, including only cats eating specific diets known to be taurine depleting prior to 1987, did detect a statistically significant increase in the plasma taurine concentration in cats eating these foods. A similar increase in plasma taurine content was documented in a single household of cats fed a diet which was documented to be taurine depleting in 1986 and early 1987. A statistically significant decrease in the occurrence of DCM was temporally associated with manufacturers altering the taurine content of commercial cat foods that were known to be taurine depleting.

As is true for reports of naturally occurring DCM in cats and other species, the majority of cats did not demonstrate significant light or electron microscopic myocardial lesions[20,23]. The nonspecific histologic lesions observed in these cats did not provide any additional information regarding the myocardium's response to taurine depletion or the etiopathogenesis of myocardial failure in taurine depleted cats.

From the myocardial taurine content data we conclude that cats fed the two taurine depleting diets (a purified diet containing 0% taurine or a commercial-type diet containing 0.15% taurine) had significantly lower myocardial taurine concentrations at the time of death than cats fed the control purified diet containing 0.14% taurine. Although the number of available analyses are too few to make definitive conclusions, the data strongly suggest that among cats fed taurine depleting diets, the degree of myocardial taurine depletion evident at the time of death does not explain why certain cats develop myocardial failure and others do not. It is possible that the higher myocardial taurine concentration observed in taurine deficient cats with myocardial failure as compared to taurine deficient cats without myocardial failure represent a compensatory response in the myocardium of cats with myocardial failure similar to what has been documented for humans with heart disease[24], and experimental animals exposed to a pressure overload[25]. That cats fed the CANNED food had a significantly higher myocardial taurine concentration than cats fed the TAU- diet and yet had a similar incidence of myocardial failure further suggests that factors other than taurine may be involved.

Feline DCM has been proposed as a useful model of DCM in humans[3]. Recently Moise[7] reported an association between taurine deficiency and myocardial failure in foxes and Kittleson et al.[26] reported data suggesting that a small subpopulation of canine patients with DCM are taurine deficient. It was suggested that administration of taurine may prove beneficial in some foxes and some dogs with DCM. Certainly, in light of the ever growing list of myocardial cellular and whole organ functions affected in vitro by taurine it is not surprising that we are now finding that myocardial dysfunction is evident in taurine deficient individuals of certain species. The cat, being an animal which has a need for exogenous taurine to maintain adequate stores, is an ideal species for studying these processes. That we are beginning to recognize similar conditions in the dog, a species which in general does not require taurine in the diet to maintain adequate stores, suggests that more work needs to be done on humans with DCM to determine if there is in fact a subpopulation whose DCM results from taurine deficiency and/or might respond to taurine supplementation.

In summary, the data presented here and in prior reports clearly associates DCM in cats with low plasma taurine concentration. We and others[27] have documented that myocardial concentration is reduced in these cats. However the mechanistic relationship between myocardial taurine concentration, other co-factors, and the onset of myocardial failure in cats remains unresolved.

ACKNOWLEDGEMENTS

Supported by the National Institutes of Health (HL02107), The Robert H. Winn Foundations, and a Grant-In-Aid from the American Heart Association (California Affiliate).

We thank Dr. Margaret Billingham (Stanford University) for providing her time and expertise with regard to the histopathological evaluation of myocardial tissues.

REFERENCES

1. S.K. Liu, Congestive heart failure in the cat, *J. Am. Vet. Med. Assoc.* 156:1319-1330 (1970).
2. S.K. Liu, Pathology of feline heart disease, *Vet. Clin. N. Am.* 7:323-339 (1977).
3. S.K. Liu and L.P. Tilley, Animal models of myocardial diseases, *Yale J. Biol. Med.* 53:191-211 (1980).
4. J.F. Van Vleet, V.J. Ferrans, and W.E. Weirich, Pathologic alterations in hypertrophic and congestive cardiomyopathy of cats, *Am. J. Vet. Res.* 41:2037-2048 (1980).
5. P.F. Lord, A. Wood, L.P. Tilley, and S.K. Liu, Radiographic and hemodynamic evaluation of cardiomyopathy of cats, *J. Am. Vet. Med. Assoc.* 164:154-165 (1974).
6. P.D. Pion, M.D. Kittleson, Q.R. Rogers, and J.G. Morris, Myocardial failure in cats associated with low plasma taurine: a reversible cardiomyopathy, *Science* 237:764-768 (1987).
7. N.S. Moise, L.M. Pacioretty, F.A. Kallfelz, M.H. Stipanuk, J.M. King, and R.F. Gilmour, Dietary taurine deficiency and dilated cardiomyopathy in the fox, *Am. Heart J.* 121:541-547 (1991).
8. D.D. Sisson, K.H. Knight, C. Helinski, P.R. Fox, B.R. Bond, N.K. Harpster, N.S. Moise, P.M.

Kaplan, J.D. Bonagura, G. Czarnecki, and D.J. Schaefer, Plasma taurine concentrations and M-mode echocardiographic measures in healthy cats and in cats with dilated cardiomyopathy, *J. Vet. Internal Med.* 5:232-238 (1991).

9.   A. Tenaglia and R. Cody, Evidence for a taurine-deficiency cardiomyopathy, *Am. J. Cardiol.* 62:136-139 (1988).

10. Q.R. Rogers, D.H. Baker, K.C. Hayes, P.T. Kendall, and J.G. Morris, "Nutrient Requirements of Cats," National Academy Press, Washington, pp. 14-15 (1986).

11. Article by Hickman, Morris and Rogers in this volume.

12. Article by Morris and Rogers in this volume.

13. C.A. Osborne, J.M. Kruger, G.R. Johnston, and D.J. Polzin, Feline lower urinary tract disorders, *in*: "Textbook of Veterinary Internal Medicine", S.J. Ettinger, ed., Saunders, Philadelphia, PA p. 2058 (1989).

14. M.L. Skiles, P.D. Pion, D.W. Hird, M.D. Kittleson, B.S. Stein, J. Lewis, and M.E. Peterson, Epidemiologic evaluation of taurine deficiency and dilated cardiomyopathy in cats, *J. Vet. Internal Med.* 4:117 (1990).

15. P.D. Pion, Feline aortic thromboemboli and the potential utility of thrombolytic therapy with tissue plasminogen activator, *Vet. Clinics No. Am.* 18:79-86 (1988).

16. J.E. Lawless, "Statistical Models and Methods for Lifetime," John Wiley & Sons, New York, NY (1982).

17. P.D. Pion, M.D. Kittleson, Q.R. Rogers, and J.G. Morris, Taurine deficiency myocardial failure in the domestic cat, *in*: "Functional Neurochemistry of Taurine," H. Pasante-Morales, ed., alan R. Liss, Inc., New York, NY, pp. 423-430 (19xx).

18. P.D. Pion, J.D. Sturman, Q.R. Rogers, M.D. Kittleson, and K.C. Hayes, Feeding diets that lower plasma taurine concentrations causes reduced myocardial mechanical function in cats, *FASEB* 2:A1616 (1988).

19. K.C. Hayes, R.E. Carey, and S.Y. Schmidt, Retinal degeneration associated with taurine deficiency in the cat, *Science* 188:949-951 (1975).

20. J. Wynne and E. Braunwald, The cardiomyopathies and myocarditides, *in*: "Heart Disease," E. Braunwald, ed., W.B. Saunders, Philadelphia, PA, pp. 1414-1417 (1988).

21. P.R. Fox, Feline myocardial disease and canine myocardial disease, *in*: "Canine and Feline Cardiology," Churchill Livingston, New York, pp. 435-493 (1988).

22. C.E. Atkins, P.S. Snyder, B.W. Keene, J.E. Rush, and S. Eicker, Efficacy of digoxin for treatment of cats with dilated cardiomyopathy, *J. Am. Vet. Med. Assoc.* 196:1463-1469 (1990).

23. A.G. Rose and W. Beck, Dilated (congestive) cardiomyopathy: a syndrome of severe cardiac dysfunction with remarkably few morphological features of myocardial damage, *Histopathology* 9:367-379 (1985).

24. R. Huxtable and R. Bressler, Elevation of taurine in human congestive heart failure, *Life Sci.* 14:1353-1359 (1974).

25. W.H. Newman, C.J. Frangakis, D.S. Grosso, and R. Bressler, A relation between myocardial taurine content and pulmonary wedge pressure in dogs with heart failure, *Physiol. Chem. & Physics.* 9:259-263 (1977).

26. M.D. Kittleson, P.D. Pion, L.A. DeLellis, and A.H. Tobias, Dilated cardiomyopathy in American Cocker Spaniels: taurine deficiency and preliminary results of response to supplementation, *J. Vet. Internal Med.* 5:123 (1991).

27. P.R. Fox and J.A. Sturman, Evaluation of myocardial taurine in spontaneous feline myocardial diseases and healthy cats fed synthetic taurine modified diets, *J. Vet. Internal Med.* 4:117 (1990).

REDUCTION OF INTRINSIC CONTRACTILE FUNCTION OF THE LEFT VENTRICLE BY TAURINE DEFICIENCY IN CATS

Mark J. Novotny[1] and Patricia M. Hogan[2]

[1]Department of Anatomy and Physiology
[2]Department of Companion Animals
Atlantic Veterinary College, University of
Prince Edward Island, Charlottetown, Prince
Edward Island, CANADA C1A 4P3

Dilated cardiomyopathy (DCM) is characterized by ventricular systolic dysfunction, cardiac chamber dilation, elevated filling pressures, and myocardial hypertrophy with increased heart weight[1,2]. Historically, cats had a high incidence of dilated cardiomyopathy and often presented clinically with end-stage heart failure[3]. The initial report by Pion et al.[4] of an association between low plasma taurine concentrations and DCM was based largely on clinical cases. Additional evidence of this association has recently been reported by Sisson et al.[5] from a multicenter study of clinical cases of DCM in cats. In the present study a more direct cause-and-effect relationship between taurine deficiency and myocardial dysfunction was sought through use of a purified taurine-free diet to induce taurine deficiency *in vivo*. Isolated, perfused left ventricular preparations were subsequently used to evaluate cardiac function *in vitro*. Results from these studies have been previously reported elsewhere[6].

MATERIALS AND METHODS

Random-source, mature mixed-breed female cats were used in these experiments. Prior to the onset of the study, cats were subjected to a health-conditioning program that consisted of immunization against feline panleukopenia, rhinotracheitis and calici viruses, treatment for internal and external parasites, and negative testing for feline leukemia virus infection. Normal cardiac function was determined through thoracic auscultation, electrocardiography (lead II), thoracic radiography, and M-mode and 2-D echocardiography. The duration of the housing of cats in the animal care facilities prior to the onset of the experiment ranged from 2 to 8 months, during which cats were fed a nutritionally balanced commericial diet (the dry formulation of Science Diet Feline Maintenance, Hills Pet Products, Topeka, KS). The nutritional status of all cats prior to arrival in our animal care facilities was unknown.

At the onset of the study, cats were randomly divided into two groups of six each.

At this point the diet of all animals was changed to a commercially-prepared, purified, taurine-deficient feline diet (Bioserv, Frenchtown, NJ). The diet was certified to contain less than 0.5 mg/kg taurine. Cats in the first group (n=6) received 1000 mg of crystalline taurine (Sigma, St. Louis, MO) orally once daily throughout the duration of the study. Cats in the second group (n=6) received no taurine replacement. Each cat was evaluated monthly through physical examination, thoracic auscultation and radiography, and echocardiography to seek clinical signs of myocardial dysfunction. Ten cats were maintained on this protocol for a period of 8 months. The remaining two cats, one from each group, were maintained on the program for 6 months (see RESULTS). At monthly intervals, blood samples were collected in heparin and plasma separated and stored at -20°C for measurement of plasma taurine concentrations.

At the completion of either the 6 or 8 month *in vivo* feeding trial, each cat was anesthetized with an intraperitoneal injection of sodium pentobarbital, 30 mg/kg, and heparinized with sodium heparin. The heart was exposed through a ventral midline thoracotomy and excised. The aortic stump was cannulated for perfusion of the coronary circulation using the Langendorff technique[7]. Krebs-Henseleit bicarbonate-buffered solution was used for all heart perfusions. Antegrade perfusion of the coronary arteries was achieved by retrograde perfusion of the aortic stump at a constant perfusion pressure of 60 mm Hg. Coronary flow was measured by collecting timed samples of the coronary effluent. Coronary perfusion pressure was monitored with a pressure transducer (Gould P23, Gould Inc., Oxnard, CA).

Isovolumic left ventricular pressure development (LVP) was measured with a latex balloon inserted through an incision in the left atrium. The balloon cannula was attached to a pressure transducer. Left ventricular $\pm dP/dt_{max}$ values in mmHg/s were obtained using an electronic differentiator (Gould Inc., Oxnard, CA). Left ventricular systolic and diastolic pressure, $\pm dP/dt$, and coronary perfusion pressure were continuously recorded on a 3 channel physiograph recorder (Gould Model 8188). Hearts were paced from the right atrial appendage at a rate of 150 beats/min with an electronic stimulator (Grass S88F, Grass Instruments Co., Quincy, MA) and miniature electrodes impaled in the epicardial surface.

Following onset of coronary flow, preparations were allowed a 30 min *in vitro* stabilization period with an end-diastolic pressure held constant at 8 mm Hg through adjustment of the LV balloon volume. At the end of the 30 min stabilization period, the following measurements were made: left ventricular developed pressure, $\pm dP/dt_{max}$, time-to-peak left ventricular developed pressure, time-to-90% left ventricular relaxation, time-to-positive dP/dt_{max}, time-to-negative dP/dt_{max} and coronary flow rate. Using a 0.5 ml Hamilton glass syringe, the intraventricular balloon was then inflated with water over a range of volumes, producing end-diastolic pressures from 0 to 28 mmHg. The balloon volume was recorded at each increment. Measurements of inotropic indices, left ventricular developed pressure and $\pm dP/dt_{max}$, were made 1 min after a volume change, and these values were plotted against end-diastolic pressure to create left ventricular function curves. Also, changes in left ventricular volume were plotted against end-diastolic pressure to yield left ventricular diastolic compliance curves. Following assessment of basal myocardial function, 5 hearts from each group were allowed to equilibrate for an additional 20 min while being perfused with Krebs-Henseleit solution containing 10 mM taurine. At the end of the 20 min period, coronary flow rates and left ventricular function measurements were repeated.

RESULTS

Within three months from the onset of the study, the plasma taurine concentrations in cats not receiving taurine replacement decreased to negligible levels and remained so throughout the duration of the study[6]. Although cats receiving 1000 mg of taurine daily experienced fluctuations in plasma taurine concentrations from pre-study baseline values, these plasma concentrations remained at or above established normal plasma concentration for cats[4].

One cat from the taurine-deficient group developed cardiomegaly and pulmonary edema, detected by thoracic radiography, by six months. This cat, along with one cat from the taurine-supplemented group, was subjected to *in vitro* assessment of myocardial function at this time. All other cats were maintained on the dietary program for the 8 month interval, as outlined in MATERIALS and METHODS.

On average, electrically paced coronary-perfused hearts isolated from taurine-deficient cats did not achieve isovolumic left ventricular developed pressure (LVP) and maximal velocity of left ventricular pressure rise ($+dP/dt_{max}$) and fall ($-dP/dt_{max}$) of the magnitude recorded in hearts from cats receiving taurine replacement (Table 1). The ranges of left ventricular developed pressures for supplemented cats and deficient cats were 92-124 mmHg and 25-112 mmHg, respectively. All but one heart from the taurine-deficient group generated left ventricular developed pressures and $\pm dP/dt_{max}$ values less than corresponding values from the supplemented group. The heart from the deficient group that maintained normal systolic contractile function (as assessed by LVP = 112 mmHg, $+dP/dt_{max}$ = 1200 mmHg/s, and $-dP/dt_{max}$ = 960 mmHg/s) did, however, display evidence of left ventricular chamber compliance failure as noted below. Mean values for time-to-peak left ventricular developed pressure, time-to-90% left ventricular relaxation, and time-to-$\pm dP/dt_{max}$ were not significantly different between the two groups.

Left ventricular function curves were generated by adjusting the volume of the ventricular balloon so that diastolic pressures varied between 0 and 28 mmHg. The results are presented in Figure 1, where inotropic indices are plotted against end-diastolic pressure. Left ventricular contractile performance improved in both groups of hearts as preload was increased. However, function curves derived from the taurine-deficient group were shifted downward and to the right, indicating inotropic failure. Even when diastolic volume was increased to maximally effective preload, inotropic indices from the taurine-deficient group remained significantly less than corresponding values from the supplemented group.

Table 1. Basal functional performance of perfused hearts isolated from taurine-supplemented and taurine-deficient cats

	Supplemented	Deficient
LVP (mm Hg)	107 ± 6	66 ± 15*
$+dP/dt_{max}$ (mm Hg/s)	1103 ± 38	718 ± 172*
$-dP/dt_{max}$ (mm Hg/s)	930 ± 46	587 ± 129*
$+/-dP/dt_{max}$ ratio	1.19 ± 0.03	1.20 ± 0.04
TPP (msec)	130 ± 5	125 ± 3
RT-90 (msec)	149 ± 8	137 ± 4
Time-to-max $+dP/dt$ (msec)	68 ± 3	70 ± 3
Time-to-max $-dP/dt$ (msec)	104 ± 8	94 ± 4
CFR (ml/min)	25 ± 2	25 ± 3
CVR (mm Hg/ml/min)	2.5 ± 0.2	2.6 ± 0.4

End-diastolic pressure = 8 mm Hg; each value is the mean ± SE of data from six hearts. LVP, developed left ventricular systolic pressure; $\pm dP/dt_{max}$, maximal rate of LV pressure rise and fall; TPP, time-to-peak LVP; RT-90, time-to-90% relaxation; CFR, coronary flow rate; CVR, coronary vascular resistance. * Significantly different at $p < 0.05$; Student's unpaired t-test. From Novotny et al., 1991[6].

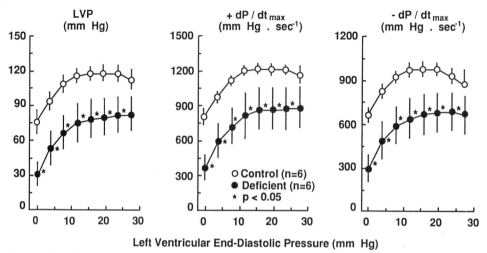

Figure 1. Isovolumic LV function curves generated by perfused hearts from taurine-deficient and taurine-supplemented (control) cats. Each symbol represents the mean ± SE. LVP, developed LV systolic pressure; +dP/dt$_{max}$, maximal rate of LV pressure increase; -dP/dt$_{max}$, maximal rate of LV pressure fall. From Novotny et al., 1991[6].

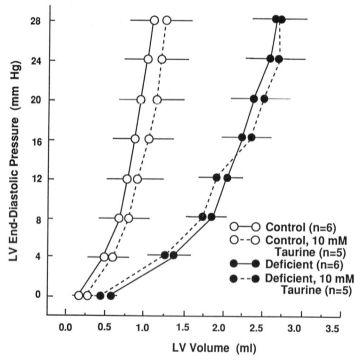

Figure 2. Left ventricular end-diastolic pressure-volume relationships of perfused hearts from taurine-deficient and taurine-supplemented (control) cats before (n = 6) and after (n = 5) perfusion with 10 mM taurine. Each symbol represents the mean with representative SE. From Novotny et al., 1991[6].

Hearts from taurine-deficient cats showed a significant increase in left ventricular diastolic chamber compliance or distensibility, as demonstrated by a shift to the right and flattening of the diastolic pressure-volume curve (Figure 2). The compliance curve generated from the one taurine-deficient cat with normal inotropic indices was comparable to compliance curves from other taurine-deficient cats. For example, left ventricular volume at a diastolic pressure of 0 mmHg was 0.7 ml (group mean = 0.6 $\pm$ 0.1 ml). At a diastolic pressure of 12 mmHg, the volume was 2.2 ml (group mean = 2.1 $\pm$ 0.2 ml), and at a diastolic pressure of 28 mmHg, the volume was 2.7 ml (group mean = 2.8 $\pm$ 0.3 ml).

As shown in Table 2, the 5 hearts from the taurine-deficient group experienced 18-22% increases in inotropic indices after 20 min of perfusion with modified Krebs-Henseleit + 10 mM taurine. This represented a significant enhancement of left ventricular contractility. Hearts from the taurine-supplemented group experienced a slight (3-4%) increase in inotropic indices that was not statistically significant.

Table 2. Functional performance of isovolumic hearts before and after taurine perfusion

TAURINE-DEFICIENT:	Baseline	Taurine (10 mM)
LVP (mm Hg)	74 $\pm$ 14	88 $\pm$ 14*
+dP/dt$_{max}$ (mm Hg/s)	810 $\pm$ 156	960 $\pm$ 161*
-dP/dt$_{max}$ (mm Hg/s)	660 $\pm$ 113	804 $\pm$ 123*
CFR (ml/min)	27 $\pm$ 3	29 $\pm$ 3

TAURINE-SUPPLEMENTED:	Baseline	Taurine (10 mM)
LVP (mm Hg)	108 $\pm$ 7	112 $\pm$ 11
+dP/dt$_{max}$ (mm Hg/s)	1120 $\pm$ 38	1160 $\pm$ 78
-dP/dt$_{max}$ (mm Hg/s)	932 $\pm$ 51	964 $\pm$ 79
CFR (ml/min)	25 $\pm$ 2	28 $\pm$ 3

End-diastolic pressure = 8 mm Hg; each value is the mean $\pm$ SE of data from five hearts.
* Significantly different at $p < 0.05$; Student's paired t-test. From Novotny et al., 1991[6].

DISCUSSION

The present study using isovolumic left ventricular preparations from taurine-deficient and taurine-supplemented cats demonstrated that intrinsic myocardial dysfunction develops in hearts of taurine-deficient cats. Inotropic failure and increased left ventricular chamber distensibility (chamber compliance) are findings compatible with dilated cardiomyopathy[1,2]. These findings support a cause-and-effect relationship between dietary taurine deficiency and dilated cardiomyopathy in cats and are in agreement with those of Pion et al.[4].

Advantages of the current experimental approach are that it represents a prospective study, utilizing a taurine-supplemented control group of cats and that it provides unique assessment of cardiac mechanical behavior independent of vascular, humoral, and neurogenic modulation of cardiac performance observed with intact subjects. Five of 6 taurine-deficient cats demonstrated diminished left ventricular contractility as evident from decreased values of inotropic indices (systolic developed pressure and $\pm$dP/dt_{max}). Another advantage of using isolated perfused hearts to assess ventricular function is the ability to generate ventricular function curves, testing the inotropic response of the heart to different levels of preload. The downward shift of the function curves in the taurine-deficient group provides further evidence for myocardial failure. Even at maximally effective preload, inotropic indices from the taurine-deficient hearts failed to approach corresponding values from the taurine-replacement group.

A further advantage of the experimental model is provision for determination of left ventricular diastolic chamber compliance. A shift in diastolic pressure-volume relationship can occur during pathological states associated with ventricular wall hypertrophy, ventricular chamber dilation, or altered composition of the ventricular wall[8]. Cardiac chamber dilation, myofibrillar hypertrophy, and myofibrillar fibrosis are common features of dilated cardiomyopathy and may account for the shift of the pressure-volume curve in hearts from taurine-deficient cats[1,2].

The observation of normal inotropic indices accompanied by a shift of the compliance curve to the right in one taurine-deficient cat suggests that during the progressive development of taurine-deficient dilated cardiomyopathy a shift in diastolic compliance may precede systolic contractile failure. Further study is needed, however, to verify the progression of pathophysiologic changes. Recently we have observed, through echocardiographic evaluation of cardiac structure and function *in situ*, that left ventricular chamber dilation generally precedes reduction of contractile indices (fractional shortening, mean velocity of circumferential fiber shortening, ejection fraction).

That all six taurine-deficient cats developed some evidence of dilated cardiomyopathy might suggest that dietary taurine deficiency consistently results in myocardial dysfunction. Again, in a recent study, we have observed that several cats maintained on the taurine-deficient diet for up to 18 months failed to develop echocardiographic evidence of dilated cardiomyopathy. Pion et al. recently reported similar observations[9]. Isolated perfused heart preparations may be more sensitive in detecting decreased intrinsic contractility and increased chamber compliance *in vitro* as compared with echocardiographic evaluation of cardiac structure and function *in vivo* and this may be a plausible explanation for the existence of normal echocardiographic findings in our recent studies. One additional difference that existed between our original research and our recent study was that the nutritional status of the six cats from the original study was unknown prior to the time of arrival in our animal care facilities. In contrast, cats in our recent study were known to have been maintained on diets supporting adequate plasma taurine concentrations prior to the onset of these studies. Perhaps the nutritional status of cats prior to the arrival at our research facility was a contributing factor in our ability to readily induce DCM in our original work.

It was interesting to note that diminished myocardial function resulting from chronic deficiency of taurine was partially reversed by acute replenishment, while perfusion of control hearts with taurine had essentially no effect on myocardial contractility. This finding suggests that one or more of the pathophysiological processes operative during taurine deficiency cardiomyopathy is at least partially surmountable by taurine replacment.

In summary, in our study utilizing the isolated heart model, we have demonstrated that dietary taurine deficiency induces decreases in left ventricular contractility and increases in left ventricular chamber distensibility. These findings support the hypothesis that dietary taurine deficiency in the cat is one cause of dilated cardiomyopathy. This feline model of dilated cardiomyopathy could be useful in the study of pathophysiologic processes during dilated cardiomyopathy, serving as an animal model of the human disease state. Furthermore, this model may be useful in the study of therapeutic approaches to the management of dilated cardiomyopathy.

ACKNOWLEDGEMENTS

This study was support by grants from the Natural Sciences and Engineering Research Council of Canada and the Heart and Stroke Foundation of Canada.

REFERENCES

1. A. Becker, Pathology of cardiomyopathies, in: "Cardiomyopathies: Clinical Presentation, Differential Diagnosis, and Management", J.A. Shaver, ed., F.A. Davis Co., Philadelphia, pp. 9-31 (1988).
2. J.T. Fallon, Myocarditis and dilated cardiomyopathy: different stages of the same disease, in: "Contemporary Issues in Cardiovascular Pathology", B.F. Waller, ed., F.A. Davis, Co., Philadelphia, pp. 155-162 (1988).
3. L.P. Tilley, Feline cardiology, Vet. Clin. North Am. Small Anim. Pract., 6:415-432 (1976).
4. P.D. Pion, M.D. Kittleson, Q.R. Rogers, and J.G. Morris, Myocardial failure in cats associated with low plasma taurine: a reversible cardiomyopathy, Science, 237:764-767 (1987).
5. D.D. Sisson, D.H. Knight, C. Helinski, et al., Plasma taurine concentrations and M-mode echocardiographic measures in healthy cats and in cats with dilated cardiomyopathy, J. Vet. Intern. Med., 5:232-238 (1991).
6. M.J. Novotny, P.M. Hogan, D.M. Paley, and H.R. Adams, Systolic and diastolic dysfunction of the left ventricle induced by dietary taurine deficiency in cats, Am. J. Physiol., 261 (Heart Circ. Physiol. 30):H121-H127 (1991).
7. H.J. Doring and H. Dehnert, Technical set up for constant-perfusion of the heart, in: "The Isolated Perfused Heart According to Langendorff", Biomesstechnik-Verlag, March, FRG, pp. 17-23 (1987).
8. B.S. Lewis and M.S. Gotsman, Current concepts of left ventricular relaxation and compliance, Am. Heart J., 99:101-112 (1980).
9. P.D. Pion, Taurine deficiency myocardial failure in cats: experimental data, in: "Scientific Proceedings, AAHA 57th Annual Meeting," American Animal Hospital Association, Denver, p. 673 (1990).

"ACTIVATION" OF ALVEOLAR LEUKOCYTES ISOLATED FROM CATS FED TAURINE-FREE DIETS

Georgia B. Schuller-Levis and John A. Sturman

New York State Institute for Basic Research in
Developmental Disabilities
Staten Island, NY 10314

INTRODUCTION

Taurine (2-aminoethane sulfonic acid) is a ubiquitous amino acid present in most mammalian tissues and cells with particularly high concentrations in tissues exposed to elevated levels of pro-oxidants[1]. The beneficial effects of this amino acid have been attributed to its detoxifying, antioxidant, and membrane stabilizing properties[2,3]. Exposure of man and experimental animals to oxidant gases such as hyperoxia, NO_2 or ozone results in pulmonary injury[4,7]. The morphological changes that occur in pulmonary tissue include destruction of capillary endothelial cells, edema, hypertrophy and hyperplasia of the bronchiolar epithelium, bronchiolization of the alveolar duct epithelium and an influx of macrophages and polymorphonuclear leukocytes (PMNs) into the alveolar air spaces[4-8]. Metabolic changes that accompany oxidant injury to the lung include membrane lipid peroxidation[9,10] and depletion of intracellular antioxidants such as glutathione[11], ascorbic acid[12], and vitamin E[13]. Recent studies have demonstrated that dietary supplementation of antioxidant vitamins such as vitamin C[14] and E[15] protects against oxidant induced lung injury. Similarly, dietary taurine supplementation has been reported to protect against bronchiolar damage and fibrosis induced by NO_2[16], bleomycin[2] and paraquat[17]. Banks et al.[18,19] found that taurine protected rat alveolar macrophages *in vitro* from ozone induced lipid peroxidation and membrane leakage. In response to oxidant injury they found an increase in cytoplasmic taurine in alveolar macrophages. The mechanism by which taurine exerts its protective effect is unknown. The present studies were aimed at determining if alveolar lavage leukocytes from cats fed taurine-deficient diets display any changes in the production of reactive oxygen intermediates (ROI).

MATERIALS AND METHODS

Animals and Diets

Domestic female cats were used for these studies. All cat diets were completely defined and sufficient in all nutrients with the exception of taurine (added at 0.05%) when it was being tested and was prepared in pellet form by Bioserve (Frenchtown, N.J.). Water was provided *ad libitum*. All cats used in this study had been fed their respective diets for at least one year before this study.

Collection of Lymphoid Tissue

Cats were anaesthetized with 1 ml ketamine (100 mg/ml) i.m. followed by 3 ml nembutal (50 mg/ml) i.p. The thoracic cavity was opened and blood was drawn by cardiac puncture into an EDTA coated syringe. Alveolar cells were collected following a modification of the method of Holt[20]. Briefly, a cannula attached to a 60 ml syringe was inserted into the trachea and 50 ml PBS was injected into lungs. Another 50 ml volume of PBS was injected, the lungs were massaged to loosen cells and 50 ml of lavage fluid was withdrawn. This procedure was repeated until 500 ml lavage fluid was collected. Lung lavage cells were centrifuged at 250 xg, washed once with PBS, and counted in a hemacytometer. The yield was typically greater than 1 x 10^8 cells/cat.

Flow Cytometric Analysis of Oxidation Burst in Phagocytic Cells

Measurement of the capacity of feline phagocytic cells to respond to PMA with an oxidative burst was also done by a flow cytometric technique involving the oxidation of a non-fluorescent precursor to a fluorescent dye[21]. DCFH is a non-fluorescent chemical that, in the presence of H_2O_2, is oxidized to dichlorofluorescin (DCF). Phagocytic alveolar lavage cells (5 x 10^5 cells) were incubated with DCFH (10 μM) and PMA (10^{-6} g/ml) in a final volume of 1 ml for 20 min at 37°C in a shaking water bath. Following incubation, flow cytometry was performed on cells to determine the percentage of cells that had oxidized DCFH to DCF. In order to demonstrate that oxidation of DCFH was due to H_2O_2, control tubes were incubated with catalase (100 U/ml) at 37°C for 10 min prior to incubation with DCFH and PMA.

Superoxide Anion Assay

Superoxide production in response to stimulation by PMA was measured in alveolar lavage cells from control and taurine-deficient cats by reduction of cytochrome C as described by McCord and Fridovich[22]. The positive controls include cells with Cytochrome C (0.4 ng/ml, Sigma, St. Louis, MO) for quantitation of baseline O_2^- production. The negative control contains cells, cytochrome C, PMA (5-170 nM) and 34 μg/ml superoxide dismutase (SOD, 3570 U/mg; Sigma). Tubes were incubated in a shaking 37°C water bath for 15 min., centrifuged at 900 x g for 5 min, supernatants collected and read on an Ultrospec II spectrophotometer (LKB, Gaithersburg, MD) at 550 nm. Superoxide anion production was calculated by the following formula:

$$\frac{ABS_{550} (Sample) - ABS_{550} (SOD\ control)}{E_{cyto\ C}} \times 1000 = nM\ O_2^-\ produced$$

where: $E_{cyto\ C}$ = 21.1 AU/m

Amino Acid Analysis

We used a sensitive reverse-phase HPLC assay for taurine. The method used precolumn derivatization with phenyl isothiocyanate and the following buffers: solvent A, 10 mM NaH_2PO_4, 2 mM EGTA, and 5% acetonitrile, pH 6.0; solvent B, 60% acetonitrile. The pumping system was a Spectra Physics 8800 ternary HPLC pump, the column was a 4.6 mm X 25 cm BakerBond WP C-18 maintained at 34°C, amino acid derivative was detected at 254 nm with an LDC spectroMonitor D, and quantified using Nelson Analytical 2600 chromatography software with an IBM PC-AT. A refrigerated Waters 712 WISP allows the automatic analysis of up to 96 samples.

Identification of Cell Surface Markers by Fluorescence

VMRD Inc., (Pullman, Washington), a company that specializes in reagents for basic immunological and veterinary research, provided commercially available monoclonal antibodies to a variety of white cell surface antigens on feline leukocytes.

Dilutions of the monoclonal antibodies were added to two million alveolar cells for 45 minutes at 4°C. Alveolar cells were washed twice and stained with the FITC-conjugated goat anti-mouse Ig antibodies (Jackson ImmunoResearch, Westgrave, PA). Cells were fixed overnight in 3% paraformaldehyde then washed twice and analyzed by flow cytometry using an EPICS PROFILE.

Production of anti-taurine antiserum

Antiserum to taurine was produced by injection of taurine-glutaraldehyde-bovine serum albumin (T-GA-BSA) into rabbits by a modification of previously described methods[23]. The immunogen T-GA-BSA was produced by glutaraldehyde conjugation of taurine (Sigma Chemical Co., St. Louis, MO) to BSA (Sigma) at taurine: BSA ratio of 1:3 (w/w). Male white rabbits (Hazelton Research Products Inc., Denver, CO) were initially injected subcutaneously at numerous different sites or intramuscularly with 500 μg of T-GA-BSA in complete Freund's adjuvant (CFA). Animals were subsequently injected monthly with immunogen in incomplete Freund's adjuvant for 8 months. Rabbits were bled prior to initial immunization and, following the third injection, at regular intervals (7 days and 21 days post injection) after each injection. Antisera were tested for titer to T-GA-BSA by ELISA.

RESULTS

Alveolar cells were lavaged as described in the methods section, typically yielding > 1 x 10^8 cells. Differential counts and determination of taurine concentration were performed (Table 1). Analysis of taurine concentrations indicated that alveolar cells from taurine-deficient cats contained 25-60% less taurine than control cats. In addition to alveolar cells, a lower concentration of taurine was measured in the lungs of cats fed taurine-fed diets (2.1 $\pm$ 1.6 μmol/g wet weight) compared to cats fed taurine-supplemented diets (8.3 $\pm$ 2.6 μmol/g wet weight) (n=16). There appeared

Table 1. Differential count and taurine concentration of feline alveolar cells.

Diet	Macrophages	PMN	Other[a]	Taurine[b]
control	70	25	5	590
control	66	34	0	501
deficient	79	4	17	378
deficient	77	19	4	226
deficient	85	5	10	247

[a] including basophil/mast cells, lymphocytes and pneumocytes.
[b] taurine concentration in nM/10^6 cells.

to be a higher percentage of PMNs in alveolar cells from control cats than from taurine-deficient cats. This is similar to previous findings in blood that taurine deficiency reduced the proportion of PMNs[24].

We analyzed the effects of taurine deficiency on the functional activity of alveolar lavage cells measuring H_2O_2, superoxide anion and chemiluminescence (CL). Using DCFH and flow cytometry, we found that feline lung lavage cells produce significant levels of H_2O_2. Lung lavage cells from taurine-deficient cats showed an increased percentage of responsiveness to PMA (22.3 ± 3.7, n=3) as compared to taurine-supplemented cats (14.0 ± 2.8, n=3). We measured superoxide anion in PMA induced alveolar lavage cells. PMA (5 nM) induced superoxide anion production was 0.33 nM O_2^- in cells from a control cat and 8.82 nM O_2^- in cells from a taurine-deficient cat (Table 2). Using a CL assay we found that both normal and taurine-deficient cats produced ROI in response to both PMA and opsonized zymosan. However, cells from taurine-deficient cats reached the peak response more rapidly than cells from taurine-supplemented cats, although the actual peak cpms were similar in both groups (Table 3). There was a rapid decline in counts in the taurine-deficient group, whereas the cats fed diets supplemented with taurine reached the peak response more slowly and maintained steady cpms throughout the assay. Thus, the kinetics of the response were very different for each group. All three assays indicate an increase in ROI in lavage cells from taurine-deficient cats compared to cats fed diets supplemented with taurine.

In further studies we used specific monoclonal antibodies in combination with flow cytometry to analyze expression of granulocyte/macrophage markers and class II major histocompatibility antigens on lung macrophages. We found no significant difference in the percentage of alveolar exudate cells expressing the granulocyte/macrophage marker in 6 cats tested (3 cats fed taurine-free diets and 3 cats fed taurine-supplemented diets). We found that 42% of the cells from cats fed a taurine-supplemented diet expressed class II antigens. In contrast, 72% of the cells from the taurine-deficient cats expressed this antigen (Figure 1).

Table 2. PMA-induced superoxide anion production by lung lavage cells.

Diet	5 nM[a]	50 nM	100 nM	170 nM
control	0.33[b]	2.08	ND[c]	1.75
deficient	8.82	7.01	7.44	7.44

[a] PMA concentration.
[b] nMO_2^-.
[c] Not done.

Table 3. PMA and opsonized zymosan induced chemiluminescence in lung lavage cells.

Diet	Zymosan		PMA	
	time[a]	cpm[b]	time	cpm
control	48	115,000	7	60,000
deficient	48	200,000	2	75,000

[a] time (minutes) to reach peak response.
[b] peak counts per minute.

We have raised a specific antibody to taurine which was used immunocytochemically to identify cellular distribution of taurine in leukocytes. Note the intense staining of the alveolar lavage macrophages from the cat fed the taurine-supplemented diet (Figure 2). This is in contrast to the weak staining of the alveolar lavage cells from the cat fed the diet deficient in taurine.

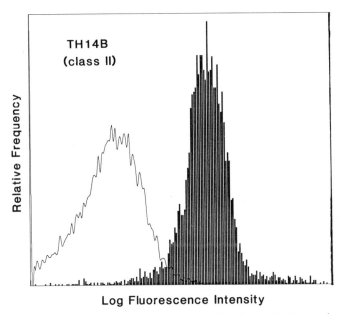

Figure 1. Flow cytometric analysis of the expression of class II surface molecules on alveolar lavage cells isolated from cats fed taurine-free diets (shaded) and taurine-supplemented diets (unshaded). Non-specific fluorescence was subtracted from each sample.

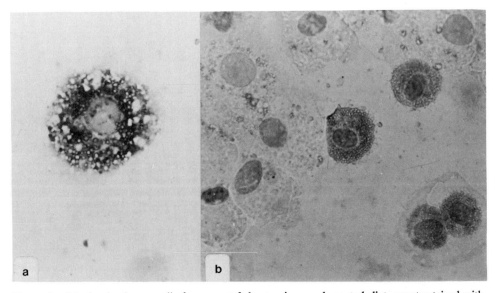

Figure 2. (a) alveolar lavage cells from a cat fed a taurine-supplemented diet, counterstained with hematoxylin. (b) alveolar lavage cells from a cat fed a taurine-free diet, counterstained with hematoxylin.

DISCUSSION

Our data show a decrease in intracellular taurine concentration in the alveolar lavage cells as well as the lungs of cats fed taurine-free diets compared to taurine-supplemented cats. This decrease in taurine is also apparent using immunocyto-chemistry to localize taurine in alveolar lavage cells from the taurine-deficient cats.

Our data demonstrate an increase in class II molecules on alveolar leukocytes isolated from cats fed taurine-free diets compared to cells isolated from cats fed taurine-supplemented diets. The class II molecules are important in macrophage mediated activation of T cells and are expressed in high levels on activated macrophages[25-27]. It is of particular interest that the increase in class II major histocompatibility antigen has been demonstrated not only in alveolar cells, but also in the spleen and peripheral blood leukocytes in taurine-deficient cats (data not shown). These data suggest that taurine-depletion may be associated with activation of macrophages and/or other leukocytes.

In addition, these studies show there is increased production of reactive oxygen intermediates in alveolar lavage cells isolated from cats fed taurine-deficient diets. This may be due to insufficient taurine to scavenge these potent metabolites.

Taurine is found in higher concentrations in tissues that have potential for oxidant exposure[1]. Human leukocytes and neutrophils contain high intracellular concentrations of taurine[28,29]. Taurine appears to act as a trap for hypochlorous acid (HOCL) produced by the myeloperoxidase-hydrogen peroxide-chloride system of macrophages and PMNs, forming the long-lived oxidant chlorotaurine, which is much less reactive and less toxic than HOCL[30]. Therefore, the taurine in leukocytes may protect cells from chlorinated oxidants[3] and may explain the increase in reactive oxygen intermediates in macrophages and PMNs isolated from cats fed diets deficient in taurine. In addition, taurine may stabilize the leukocyte membrane and a deficiency of taurine may make the leukocyte more vulnerable to damage. Moreover, it is likely that taurine acts by more than one action physiologically. Our data indicate high levels of taurine in alveolar lavage cells from cats fed diets supplemented with taurine. We hypothesize that taurine functions to prevent terminal activation and release of cytotoxic mediators by lung macrophages and epithelial cells. Thus, a deficiency of taurine will indeed cause an activation of leukocytes, as evidenced by our data which show an increase in ROI, as well as an increase in class II antigen.

SUMMARY

Taurine is a ubiquitous amino sulfonic acid in mammals, present in high concentrations in tissues, including those exposed to elevated levels of oxidants. Experiments were designed to examine the consequences of taurine deficiency on production of ROI in leukocytes isolated from the lungs and blood of cats fed taurine-deficient diets. Cats were maintained on taurine-free or taurine-supplemented diets for at least 12 months at which time taurine deficiency was evident. To analyze alveolar cells, lungs were lavaged to recover lung macrophages and PMNs. Lung lavage fluid from cats contained macrophages and PMNs, although taurine deficiency was associated with a decrease in the percentage of PMNs in the lungs. This is similar to our findings in blood that taurine deficiency reduced the proportion of PMNs. Taurine measurements revealed 2.1 ± 1.6 μmol/g wet wt of taurine in the lungs from cats fed a taurine-deficient diet versus 8.3 ± 2.6 in lungs from cats fed a diet supplemented with taurine (n=16). The effects of taurine deficiency on the functional activity of lung macrophages and PMNs were analyzed including the production of ROI. Alveolar leukocytes from cats fed taurine-deficient diets produced more superoxide anion in response to phorbol myristate acetate than cats fed taurine supplemented diets. Similar results were obtained using a chemiluminescence assay. Using the highly specific H_2O_2 indicator dye, dichlorofluorescin, and flow cytometry we found that alveolar leukocytes made more H_2O_2 than cells from cats fed taurine-supplemented diets. Forty-two percent of the cells from cats fed a taurine-supplemented diet expressed class II antigens. In contrast, 72% of cells from the taurine-deficient cats expressed this antigen. We hypothesize that taurine functions to prevent terminal activation and release of cytotoxic mediators by lung macrophages. Thus, a deficiency of taurine will indeed cause an activation of leukocytes, as evidenced by our data which show an increase in ROI, as well as an increase in class II antigen.

ACKNOWLEDGEMENTS

We thank Mrs. Joan Maffei for secretarial assistance and Mr. Cliff Meeker for technical help.

REFERENCES

1. C.P.L. Lewis, G.M. Cohen, and L.L. Smith, The identification and characterization of an uptake system for taurine into rat lung slices, *Biochem. Pharmac.* 39:431 (1990).
2. Q. Wang, S.N. Giri. D.M. Hyde, and J.M. Nakashima, Effects of taurine on bleomycin-induced lung fibroblasts in hamsters, *PSEBM* 190:330 (1989).
3. C.E. Wright, T.T. Lin, Y.Y. Lin, J. Sturman, and G. Gaull, *in*: "Taurine: Biological Actions and Clinical Perspectives," Alan R. Liss Inc., New York, NY (1985).
4. C.K. Chow, M.Z. Hussain, C.E. Cross, D.L. Dungworth, and M.G. Mustafa, Effect of low levels of ozone on rat lungs. I. Biochemical responses during recovery and reexposure, *Exp. Mole. Pathol.* 25:182 (1976).
5. J.D. Crapo, B.E. Barry, H.A. Foscue, and J. Shelburne, Structural and biochemical changes in rat lungs occurring during exposures to lethal and adaptive doses of oxygen, *Amer. Rev. Rspir. Dis.* 122:123 (1980).
6. M.G. Mustafa and D.F. Tierney, Biochemical and metabolic changes in the lung with oxygen, ozone, and nitrogen dioxide toxicity, *Amer. Rev. Respir. Dis.* 118:1061 (1978).
7. C.G. Plopper, C.K. Chow, D.L. Dungworth, M. Brummer, and T.J. Nemeth, Effect of low level of ozone on rat lungs. II. Morphological responses during recovery and re-exposure, *Exp. Mol. Pathol.* 29:400 (1978).
8. P.J. Rombout, J.A. Dormans, M. Marra, and G.J. Van Esch, Influence of exposure regimen on nitrogen dioxide-induced morphological changes in the rat lung, *Environ. Res.* 41:466 (1986).
9. C.K. Chow and A.L. Tapell, Effect of low levels of ozone on rat lungs. I. Biochemical responses during recovery and re-exposure, *Exp. Mol. Pathol.* 25:182 (1976).
10. M. Sagai, T. Ichinose, H. Oda, and K. Kubota, Studies on biochemical effects of nitrogen dioxide. II. Changes of the protective systems in rat lungs and of lipid peroxidation by acute exposure, *J. Toxicol. Environ. Health* 9:153 (1982).
11. A.J. Delucia, M.G. Mustafa, M.Z. Hussain, and C.E. Cross, Ozone interaction with rodent lung. III. Oxidation of reduced glutathione and formation of mixed disulfides between protein and nonprotein sulfhdryls, *J. Clin. Invest.* 55:794 (1975).
12. C.C. Kratzing and R.J. Willis, Decreased levels of ascorbic acid in lung following exposure to ozone, *Chem. Biol. Interact.* 30:53 (1980).
13. B.D. Goldstein, R.M. Buckley, R. Cardenas, and O.J. Balchum, Ozone and vitamin E, *Science* 169:605 (1970).
14. M.J. Evans, L.J. Cabral-Anderson, V.P. Decker, and G. Freeman, The effects of dietary antioxidants on NO2-induced injury to type 1 alveolar cells, *Chest* 80:55 (1981).
15. N.M. Elsayed and M.G. Mustafa, Dietary antioxidants and the biochemical response to oxidant inhalation. I. Influence of dietary vitamin E on the biochemical effects of nitrogen dioxide exposure in rat lung, *Toxicol. & Appl. Pharmacol.* 66:319 (1982).
16. R.E. Gordon, A.A. Shaked, and D.F. Solano, Taurine protects hamster bronchioles from acute NO_2-induced alterations, *Am. J. Pathol.* 125:585 (1986).
17. K. Izumi, R. Nagata, T. Motoya, J. Yamashita, et al., Preventive effect of taurine against acute paraquat intoxication in beagles, *Japan J. Pharmacol.* 50:229 (1989).
18. M.A. Banks, D.W. Porter, W.G. Martin, and V. Castranova, Effects of in vitro ozone exposure on peroxidative damage, membrane leakage, and taurine content of rat alveolar macrophages, *Tox. and Appl. Pharmacol.* 105:55 (1990).

19. M.A. Banks, D.W. Porter, W.G. Martin, and Castranova, Ozone-induced lipid peroxidation and membrane leakage in isolated rat alveolar macrophages: protective effects of taurine, *J. Nutr. Biochem.* 2:308 (1991).

20. P.G. Holt, Alveolar macrophages I. A simple technique for the preparation of high numbers of viable alveolar macrophages from small laboratory animals, *J. Immunol. Methods*, 27:189 (1979).

21. L. Kobzik, J.J. Godleski, and J.D. Brain, Oxidative metabolism in the alveolar macrophage: Analysis by flow cytometry, *J. Leuk. Biol.* 47:295 (1990).

22. J.M. McCord and I. Frederich, Superoxide dismutase: an enzymic function of erythrocuprein (hemocuprein), *J. Biol. Chem.* 244:6049 (1969).

23. N. Lake and C. Verdon-Smith, Immunocytochemical localization of taurine in the mammalian retina, *Curr. Eye Res.* 8:163 (1989).

24. G.B. Schuller-Levis, P.D. Mehta, R. Rudelli, and J. Sturman, Immunologic consequences of taurine deficiency in cats, *J. Leuk. Biol.* 47:321 (1990).

25. D.I. Beller, J.M. Kiely, and E.R. Unanue, Regulation of macrophage populations. I. Preferential induction of Ia-rich peritoneal exudates by immunologic stimuli, *J. Immunol.* 124:1426 (1980).

26. R.M. Steinman, N. Nogueira, M.D. Witmer, J.D. Tydings, and I.S. Mellman, Lymphokine enhances the expression and synthesis of Ia antigens on cultured mouse peritoneal macrophage, *J. Exp. Med.* 152:1248 (1980).

27. G.B. Mackaness, The influence of immunologically committed lymphoid cells on macrophage activity in vivo, *J. Exp. Med.* 129:973 (1969).

28. P. Soupart, "Amino Acid Pools," J.T. Holden, ed., pp. 220-262, Elsivier/North-Holland, Amsterdam (1962).

29. D.B. Learn, V.A. Fried, and E.L. Thomas, Taurine and hypotaurine content of human leukocytes, *J. Leukocyte Biology* 48:174 (1990).

30. E.L. Thomas, M.B. Grisham, and M.M. Jefferson, Myeloperoxidase-dependent effect of amines on functions of isolated neutrophils, *J. Clin. Invest.* 72:441 (1983).

HIGH DIETARY TAURINE AND FELINE REPRODUCTION

John A. Sturman and Jeffrey M. Messing

Department of Developmental Biochemistry
Institute for Basic Research in Developmental Disabilities
Staten Island, NY 10314

INTRODUCTION

Taurine is now well established as an essential nutrient for cats and may also be a conditionally essential nutrient for some other species, especially during development (Sturman, 1988). There is now a large body of literature dealing with the effects of taurine deficiency (Hayes, 1988; Sturman, 1988; Hayes and Trautwein, 1989; Sturman, 1990). Little attention has been paid to potential effects of taurine supplementation, however. In humans and cynamolgus monkeys which conjugate bile acids with taurine and glycine, the proportion of bile acids conjugated with taurine increases as the amount of taurine in the diet is increased (Sjovall, 1960; Haslewood, 1967; Hofmann and Small, 1967; Schersten, 1971; Sturman et al., 1975; Hayes et al., 1980). Such changes influence absorption from the gut, and beneficial effects of dietary taurine supplementation have been reported in chronic and acute hepatitis (Matsuyama et al., 1983; Nakashima et al., 1983), drug-induced liver disease (Attili et al., 1984), cirrhosis (Kroll and Lund, 1966), myotonia (Durelli et al., 1983), cystic fibrosis (Darling et al., 1985; Belli et al., 1987; Colombo et al., 1988; Thompson, 1988) and epilepsy (Barbeau and Donaldson, 1973; Barbeau and Donaldson, 1974; Bergamini et al., 1974; Fukuyama and Ochiai, 1982) although not in retinitis pigmentosa (Reccia et al., 1980). In addition, taurine has been added to commercial infant formulas and pediatric parenteral solutions in recent years because of mounting evidence of subtle abnormalities in visual function resulting from its absence (Sturman, 1986). A recent study reported adverse effects in the guinea pig comprising of fatty changes in the liver accompanied by changes in the lipid content after 14 days of oral administration of taurine (Cantafora et al., 1986). Sudden death syndrome in broiler chickens is reduced by supplementing their diet with taurine (Campbell and Classen, 1989). Other recent reports implicated taurine deficiency in feline dilated cardiomyopathy, and demonstrated its reversal by nutritional taurine therapy if treated in time (Pion et al., 1987; Pion et al., 1988; Pion et al., 1990; Novotny et al., 1991; Fox and Sturman, 1992). This successful treatment led to the fortification of commercial cat foods, which already contained taurine, with additional taurine. Although this has resulted in the virtual disappearance of this condition, no systematic studies have been reported on the long term effects of a high taurine diet. The results of such a study are reported here.

Taurine, Edited by J.B. Lombardini *et al.*
Plenum Press, New York, 1992

Female cats were fed completely defined purified diets containing 0.05%, 0.2%, or 1% taurine for at least 6 months prior to breeding as described in detail elsewhere (Sturman and Messing, 1992). Breeding performance was evaluated and taurine concentrations in tissues and fluids of adults and offspring measured.

The high taurine diet had no effect on appetite, food consumption, weight gain, or estrus cycle of the adult females. The reproductive performance, if anything, was slightly better in the females fed the high taurine diet; the proportion of pregnancies reaching term, and the number of kittens surviving to weaning per term pregnancy was slightly greater for the cats fed 1% taurine than those fed 0.05% or 0.2% taurine although none of these trends was statistically significant (Table 1). The growth rates of the kittens from females fed the different amounts of taurine were not significantly different although the greatest was achieved by the kittens from females fed the 0.05% taurine diet (Figure 1). This observation is supported by examination of the birth weights and 8-week-old weights of all kittens in this study (Table 2). The kittens at birth weigh more from females fed the greatest amount of taurine, whereas the reverse is true at 8 weeks of age. The brain weights of kittens from mothers fed 1% taurine were significantly greater than those of the other diet groups, both at birth and at 8 weeks of age. The concentration of taurine in the milk of the lactating females was greater in those fed the highest amounts of dietary taurine and generally increased during lactation (Figure 2).

Table 1. Outcome of pregnancies from females fed a purified diet supplemented with various amounts of taurine.

Diet (% taurine)	0.05	0.2	1.0
Pregnancies	73	24	38
To term	64	20	37
Kittens stillborn[1]	12	9	4
Kittens live[1]	218	65	125
Survivors[2]	154	44	99
% Pregnancies to term	88	83	97
# Kittens/ term pregnancy[3]	3.6	3.7	3.5
# Survivors/ term pregnancy	2.41	2.20	2.68

[1] From term pregnancies.
[2] Alive at weaning at 8 weeks after birth.
[3] Includes live and stillborn kittens.

Tissue taurine concentrations in adult cats fed the high taurine diet over an extended period of time (average 2.5 years) were greater in soft tissues and some muscles than controls, but not in retina or brain. Despite spending the entire gestation period in a taurine-enriched environment, newborn kittens from mothers fed 1% taurine had few tissues with significantly higher taurine concentrations. By weaning at 8 weeks after birth, such kittens had many tissues with greater taurine concentrations, including most brain regions. By 12 and 20 weeks after birth, most tissues had significantly greater taurine concentrations. Some representative values for tissues at different ages are provided in Table 3.

Taken together, these results indicate that the fully mature cat brain is largely resistant to significant increases in taurine concentration by consuming a high taurine

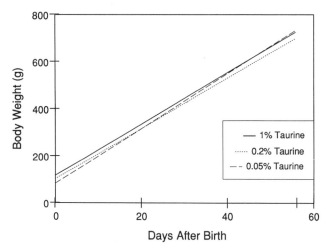

Figure 1. Growth curves of kittens from females fed 0.05%, 0.2%, or 1.0% taurine. The curves are derived from the twice-weekly weights of all kittens included in this study using a standard computer program for linear regression. Correlation coefficients are 0.87, 0.85 and 0.86, respectively.

Table 2. Body and brain weight of newborn and 8-week-old kittens from females fed a purified diet supplemented with various amounts of taurine.

Diet (% taurine)	0.05	0.2	1.0
Newborn			
Body	105.6 ± 30.3	111.6 ± 23.8	113.4 ± 20.8 [1]
Brain	4.86 ± 1.14(23)	4.44 ± 1.10(11)	5.41 ± 0.90[2](9)
8-Week-old			
Body	749 ± 142	722 ± 141	699 ± 147 [3]
Brain	21.7 ± 1.8(28)	22.1 ± 0.9(13)	23.0 ± 1.0[4](10)

Each value represents the mean (in g) ± SD of the body weights of all kittens used in this study and of the number of brain samples in parentheses.
Significance was determined using Student's t test.
[1] Significantly greater than 0.05% (P < 0.01).
[2] Significantly greater than 0.2% (P < 0.05).
[3] Significantly smaller than 0.05% (P < 0.05).
[4] Significantly greater than 0.05% (P < 0.05).

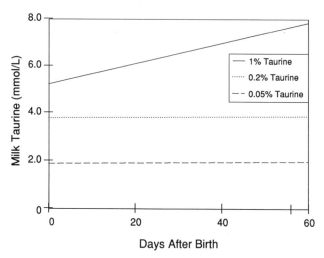

Figure 2. Concentration of taurine in milk of lactating females fed 0.05%, 0.2%, or 1% taurine. The curves are derived from the twice-weekly milk samples from all females included in this study using a standard computer program for linear regression. Correlation coefficients are 0.03, 0.03 and 0.05, respectively.

Table 3. Tissue taurine concentrations in kittens of different ages.

Age and diet	Liver	Lung	Biceps
	μmol/g wet weight		
Newborn 0.05%	9.37 ± 3.95*	8.54 ± 2.31	9.58 ± 2.51
1%	12.9 ± 3.0	9.35 ± 2.19	9.41 ± 4.52
8 Weeks 0.05%	13.1 ± 4.4	9.73 ± 3.58	10.6 ± 4.8
1%	13.1 ± 3.0	10.5 ± 1.6	13.7 ± 3.7
12 Weeks 0.05%	12.5 ± 4.1*	11.1 ± 2.2*	10.1 ± 3.7*
1%	20.1 ± 3.4	14.9 ± 2.3	19.3 ± 3.2
20 Weeks 0.05%	9.16 ± 2.92*	14.9 ± 10.3	9.38 ± 1.35*
1%	18.2 ± 0.8	15.2 ± 2.0	14.4 ± 2.8
Adult 0.05%	8.50 ± 3.33*	8.28 ± 2.6*	6.35 ± 1.62*
1%	17.2 ± 6.3	11.8 ± 2.1	11.4 ± 3.1

* Significantly different, $P < 0.05$.

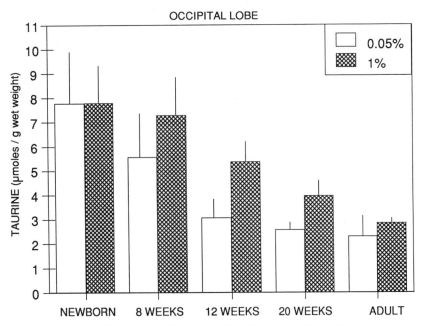

Figure 3. Taurine concentration in occipital lobe as a function of diet and age.

diet over a long period of time, as might be expected. Less expected was the observation that the fetal cat brain was also resistant to increases in taurine concentration, despite the immaturity of the blood brain barrier during gestation. Even more surprising in light of this observation was the apparent decrease in resistance to increases in brain taurine concentrations in young juveniles at 8 wk, 12 wk, and 20 wk after birth, when brain development has been largely completed and the blood brain barrier is fully mature. The high dietary intake provided to the kittens during lactation had a greater impact than the in utero environment of the mothers consuming the high taurine diet. Perhaps the explanation of these results is that the higher taurine concentrations in the blood prevent the normal decrease in brain taurine concentrations during development, illustrated for occipital lobe and cerebellum (Figures 3 and 4). Olfactory bulb, which has a much greater taurine concentration than other regions is not affected (Figure 5), nor is the retina which has an extremely high taurine content. It would be of interest to know when the adult property of resistance to increases in brain taurine concentrations is reached, and whether other compounds besides taurine can increase in juvenile kitten brain.

A number of reports in the literature have linked dietary taurine metabolism to dietary protein content. Mature rhesus monkeys do not appear to be dependent on dietary taurine to maintain their body taurine pools (although rhesus monkey infants do, Sturman et al., 1988) unless their diet is deficient in protein (Neuringer et al., 1979; Neuringer et al., 1985). Supplementary dietary taurine given to lactating mice fed a protein-deficient diet increased the neonatal survival, but had no effect on lactating mice fed a protein-sufficient diet (van Gelder and Parent, 1981). Further data obtained from this same animal model showed that a limited period of undernutrition had a permanent effect on the levels of certain amino acids, including taurine, in the adult cerebellum, and that these changes were modified by taurine supplementation (van Gelder and Parent, 1982). Weanling rats fed a low-protein diet have reduced taurine concentrations in plasma and retina and abnormal retinal function (depressed a and b waves in the electroretinogram) (Bankson and Russell, 1988). Dietary taurine supplementation normalized the taurine concentrations but

95

resulted in further impairment of visual function. Injection of taurine, but not of sodium chloride or valine, into fertilized chicken eggs resulted in increased taurine concentrations in heart and brain, and hatchlings with severe ataxia, reduced muscle strength and impaired motor coordination (van Gelder and Belanger, 1988). There are some significant differences in the lipid composition of liver from adults fed

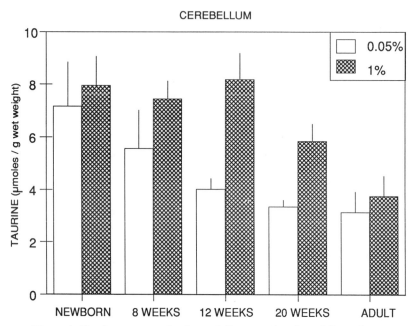

Figure 4. Taurine concentration in cerebellum as a function of diet and age.

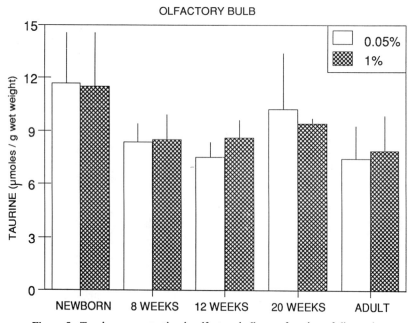

Figure 5. Taurine concentration in olfactory bulb as a function of diet and age.

Table 4. Lipid composition in liver of adult cats fed 0.05% or 1% taurine.

Diet (% taurine)	0.05	1
	mol/g wet weight	
Cholesteryl esters	3.44 ± 1.51	2.69 ± 1.75
Triglycerides	1.78 ± 1.34	6.45 ± 1.46[1]
Free fatty acids	19.57 ± 8.29	6.18 ± 2.18[1]
Cholesterol	5.25 ± 1.62	7.11 ± 1.53[2]
Phosphatidylethanolamine	4.89 ± 1.1	3.90 ± 0.86[3]
Phosphatidylcholine	7.85 ± 2.48	6.36 ± 2.53
Sphingomyelin	1.85 ± 0.42	1.76 ± 0.38
Total lipid	44.6 ± 7.9	34.5 ± 4.8 [3]

[1] Significantly different (P < 0.001).
[2] Significantly different (P < 0.01).
[3] Significantly different (P < 0.05).

0.05% and 1% taurine (Table 4) and in the fatty acid distribution within the lipid classes (Cantafora et al., 1991). The consequences of these differences, if any, are not obvious, and at this stage our studies provide no evidence of ill effects produced by prolonged feeding of high taurine diets to adult cats or on their offspring.

ACKNOWLEDGEMENTS

This research was supported by the New York State Office of Mental Retardation and Developmental Disabilities, NIH grant HD-16634, and a grant from the Sanford Foundation. We are grateful to Sharon Mathier for art work, to Ann Parese for secretarial assistance, and to members of the IBR Animal Colony Facility for practical help throughout this study.

REFERENCES

Attili, A.F., Angelico, M., Alvaro, D., Marin, M., De Santes, A., and Capocaccia, L., 1984, Reproduction in serum transaminases during taurine administration in patients with chronic active hepatitis, *Gastroenterology* Abs. 86:1017.

Bankson, D.D., and Russell, R.M., 1988, Protein energy malnutrition and taurine supplementation: Effects on vitamin A nutritional status and electroretinogram of young rats, *J. Nutr.* 118:23.

Barbeau, A., and Donaldson, J., 1973, Taurine in epilepsy, *Lancet* ii, 387.

Barbeau, A., and Donaldson, J., 1974, Zinc, taurine, and epilepsy, *Arch. Neurol.* 30:52.

Belli, D.C., Levy, E., Darling, P., Leroy, C., Lepage, G., Giguere, R., and Roy, C.F., 1987, Taurine improves the absorption of a fat meal in patients with cystic fibrosis, *Pediatrics* 80:517.

Bergamini, L., Mutani, R., Delsedime, M., and Durelli, L., 1974, First clinical experience on the antiepileptic action of taurine, *Europ. Neurol.* 11:261.

Campbell, G.L., and Classen, H.L., 1989, Effect of dietary taurine supplementation on sudden death syndrome in broiler chickens, *Can. J. Anim. Sci.* 69:509.

Cantafora, A., Mantovani, A., Masella, R., Mechelli, M., and Alvaro, D., 1986, Effect of taurine administration on liver lipids in guinea pig, *Experientia* 42:407.

Cantafora, A., I. Blotta, S.S. Rossi, A.F. Hofmann, and J.A. Sturman, 1991, Dietary taurine content changes liver lipids in cats, *J. Nutr.* 121:1522.

Colombo, C., Arlati, S., Curcio, L., Maiavacca, R., Garatti, M., Ronchi, M., Corbetta, C., and Giunta, A, 1988, Effect of taurine supplementation on fat and bile acid absorption in patients with cystic fibrosis, *Scand. J. Gastroenterol.* 23:151.

Darling, P.B., Lepage, G., Leroy, C., Masson, P., and Roy, C.C., 1985, Effect of taurine supplements on fat absorption in cystic fibrosis, *Pediat. Res.* 19:578.

Durelli, L., Mutani, R., and Fassio, F., 1983, The treatment of myotonia evaluation of chronic oral taurine therapy, *Neurology* 33:599.

Fox, P.R., and Sturman, J.A., 1992, Myocardial taurine concentrations in cats with spontaneous feline myocardial disease and healthy cats fed synthetic taurine-modified diets, *Amer. J. Vet. Res.* in press.

Fukuyama, Y., and Ochiai, Y., 1982, Therapeutic trial by taurine for intractable childhood epilepsies, *Brain Dev.* 4:63.

Haslewood, G.A.D., 1967, Bile Salts, Halsted Press, New York.

Hayes, K.C., 1988, Taurine nutrition, *Nutr. Res. Rev.* 1:99.

Hayes, K.C., Stephan, Z.V., and Sturman, J.A., 1980, Growth depression in taurine-depleted monkeys, *J. Nutr.* 110:2058.

Hayes, K.C., and Trautwein, E.A., 1989, Taurine deficiency syndrome in cats, *Clin. Nutr.* 19:403.

Hofmann, A.F., and Small, D.M., 1967, Detergent properties of bile salts: correlation with physiological function, *Ann. Rev. Med.* 18:333.

Kroll, J., and Lund, E., 1966, *Dan. Med. Bull.* 13:173.

Matsuyama, Y., Morita, T., Higuchi, M., and Tsujii, T., 1983, The effect of taurine administration on patients with acute hepatitis, *Prog. Clin. Biol. Res.* 125:461.

Nakashima, T., Sano, A., Nakagawa, Y., Okuno, T., Takimo, T., and Kuriyama, K., 1983, Effect of taurine on the course of drug-induced chronic liver disease, *Prog. Clin. Biol. Res.* 125:472.

Neuringer, M., Denney, D., and Sturman, J., 1979, Reduced plasma taurine concentration and cone electroretinogram amplitude in monkeys fed a protein-deficient semipurified diet, *J. Nutr.* Abs. 109(6):xxvi.

Neuringer, M., Sturman, J.A., Wen, G.Y., and Wisniewski, H.M., 1985, Dietary taurine is necessary for normal retinal development in monkeys, *in:* "Taurine: Biological Actions and Clinical Perspectives," S.S. Oja, L. Ahtee, P. Kontro, and M.K. Paasonen, eds., Alan R. Liss, New York.

Novotny, M.J., Hogan, P.M., Paley, D.M., and Adams, H.R., 1991, Systolic and diastolic dysfunction of the left ventricle induced by dietary taurine deficiency in cats, *Amer. Physiol. Soc.* :H121.

Pion, P.D., Kittleson, M.D., Rogers, Q.R., and Morris, J.G., 1987, Myocardial failure in cats associated with low plasma taurine: Reversible cardiomyopathy, *Science* 237:764.

Pion, P.D., Kittleson, M.D., Rogers, Q.R., and Morris, J.G., 1990, Taurine deficiency myocardial failure in the domestic cat, *in* "Taurine: Functional Neurochemistry, Physiology, and Cardiology," (H. Pasantes-Morales, D.L. Martin, W. Shain, and R. Martin del Rio, eds., Wiley Liss, New York.

Pion, P.D., Sturman, J.A., Rogers, Q.R., Kittleson, M.D., Hayes, K.C., and Morris, J.G., 1988, Feeding diets that lower plasma taurine (TAU) concentrations causes reduced myocardial mechanical function in cats, *FASEB* Abs. 2:A1617.

Reccia, R., Pignalosa, B., Grasso, A., and Campanella, G., 1980, Taurine treatment in retinitis pigmentosa, *Acta Neurol.* 35:132.

Schersten, T., 1971, Bile acid conjugation, *in:* "Metabolic Conjugation and Metabolic Hydrolysis," W.H. Fishman, ed., Academic Press, New York.

Sjovall, J., 1960, Bile acids in man under normal and pathological conditions, Bile acids and steroids 73, *Clin. Chim. Acta* 5:33.

Sturman, J.A., 1986, Taurine in infant physiology and nutrition, *Excerpta Medica,* Special Issue, Pediatrics & Nutrition Review, 1.

Sturman, J.A., 1988, Taurine in development, *J. Nutr.* 118:1169.

Sturman, J.A., 1990, Taurine deficiency, *in:* "Taurine: Functional Neurochemistry, Physiology, and Cardiology," H. Pasantes-Morales, D.L. Martin, W. Shain and R. Martin del Rio, eds., Wiley-Liss, New York.

Sturman, J.A., Hepner, G.W., Hofmann, A.F., and Thomas, P.J., 1975, Metabolism of [^{35}S]taurine in man, *J. Nutr.* 105:1206.

Sturman, J.A., Messing, J.M., Rossi, S.S., Hofmann, A.F., and Neuringer, M.D., 1988, Tissue taurine content and conjugated bile acid composition of rhesus monkey infants fed a human infant soy-protein formula with or without taurine supplementation for 3 months, *Neurochem. Res.* 13:311.

Sturman, J.A., and Messing, J.M., 1992, High dietary taurine effects on feline tissue taurine concentrations and reproductive performance, *J. Nutr.* 122:82.

Thompson, G.N., 1988, Assessment of taurine deficiency in cystic fibrosis, *Clin. Chim. Acta* 171:233.

van Gelder, N.M., and Belanger, F., 1988, Embryonic exposure to high taurine: A possible nutritional contribution to Friedreich's ataxia, *J. Neurosci. Res.* 20:383.

van Gelder, N.M., and Parent, M., 1981, Effect of protein and taurine content of maternal diet on the physical development of neonates, *Neurochem. Res.* 6:539.

van Gelder, N.M., and Parent, M., 1982, Protein and taurine of maternal diets during the mouse neonatal period: Permanent effects on cerebellar-brainstem amino acid levels in mature offspring, *Neurochem. Res.* 7:987.

TAURINE DISTRIBUTION IN THE CAT MUSCLE: AN IMMUNOHISTOCHEMICAL STUDY

Octavio Quesada, Peimin Lu, and John A. Sturman

Department of Developmental Biochemistry
Institute for Basic Research in Developmental Disabilities
Staten Island, NY   10314

INTRODUCTION

Taurine is present in large amounts in every kind of muscle tissue (Jacobsen and Smith, 1968). In the heart, taurine accounts for up to 50% of the total free amino acid pool, and in skeletal and smooth muscles it reaches mM concentrations (Kocsis et al., 1976; Huxtable et al., 1979; Sturman and Messing, 1991). Many different effects of taurine on the general performance and on electrophysiological parameters of the heart and skeletal muscle have been described. The former includes an antiarrhythmic action, positive inotropy at low external calcium concentrations, negative inotropy at normal to high calcium concentrations, and antagonism of the so-called "calcium paradox", (for review, see Huxtable and Sebring, 1983). In skeletal muscle, it has been reported that taurine affects electrophysiological and biochemical parameters including hyperpolarization of muscle fibers and a decrease of the action potential duration (Gruener et al., 1976), an increase of the chloride membrane conductance (Conte-Camerino et al., 1989), and an attenuation of the cholinergic response (Lehmann and Hamberger, 1983). All these studies have been performed under *in vitro* conditions, adding exogenous taurine to the system.

Although such studies give important clues to the possible natural action of taurine on these muscles it is often impossible to distinguish between its physiological role and a pharmacological effect. In order to shed some light on the role of taurine into the scheme of muscle physiology, we studied the distribution of the endogenous taurine pool in different muscle tissues of the cat using immunocytochemical techniques.

METHODS

Adult cats from the IBR colony were used in this study. Some animals were sedated with sodium pentobarbital (Nembutal) and intracardiacally perfused with 2.5% glutaraldehyde/1% formaldehyde in 0.1 M phosphate buffer (pH 7.4) for 15 min. After perfusion, samples from heart, intestine and different skeletal muscles

Taurine, Edited by J.B. Lombardini *et al.*
Plenum Press, New York, 1992

were taken and post-fixed 2-3 hrs in the same fixative mixture, then washed thoroughly with buffer and stored at 4°C. Other animals were exsanguinated following an overdose of sodium pentobarbital, and small tissue samples were immediately dissected and immersed in the same fixative solution for 3-4 hr, then washed and stored in phosphate buffer at 4°C.

Antibodies raised against taurine, glutamate or GABA conjugated to bovine serum albumin with glutaraldehyde (BSA-G-aa), were raised in rabbits by the method described by Campistron et al. (1986). Very briefly, 500 μg of immunogen emulsified with Freund's complete adjuvant were subcutaneously injected into rabbits every 4 weeks. Blood was collected 7 and 21 days after each immunization. The sera were characterized for titer and cross-reactivity by the ELISA method, and those with high, stable titre of reactivity were selected. Purification was afforded by preadsorbing the sera with all the other immunogens, removing residual cross-reactivity.

Paraffin-embedded sample blocks were prepared by conventional methods and 6 μm thick serial sections were prepared for immunocytochemical studies. Specimens were deparaffined and rehydrated and successively exposed to the following solutions: a) 3% normal goat serum in 4% BSA in PBS, for 30 min; b) 3% H_2O_2 in absolute methanol for 20 min; c) antiserum 1:300 in PBS, for 16 hr; d) biotin conjugated antirabbit IgG, 1:100 for 1 hr; e) avidin conjugated horseradish peroxidase, 1:100 for 1 hr; f) 3,3'-diamino benzidine, 0.05% in tris buffer with H_2O_2 0.005% for 2-4 min.

Denervated and stimulated muscles

In two deeply anesthetized cats, the left sciatic nerve was severed at the thigh level 20 min prior to intracardiac fixation and in two other deeply anesthetized cats the left sciatic nerve was electrically stimulated for 15 min prior to and during intracardiac fixation by passing train pulses (50 Hz, 4V), applied every 2 sec by means of two silver coated electrodes. Individual pulse duration was 1 μsec. After perfusion, samples of muscles from both hind limbs were taken, postfixed and embedded in the same block. In this way, both experimental and control tissue samples were processed identically throughout the immunohistochemical procedures.

Other species

In order to evaluate whether or not the TLI (taurine-like immunoreactivity) distribution found in cat skeletal muscle is species specific or a general phenomenon, we also examined the TLI distribution in skeletal muscle of rats and mice. Adult male Wistar rats and adult male VM mice were anesthetized and prepared for fixation by intracardiac perfusion. Just before the perfusion started, the left leg was clamped just below the hip and removed to provide samples of perfusion-fixed and immersion-fixed muscle. Both biceps femoris muscles were dissected and immediately fixed by immersion 4 hr (left) or postfixed 2 hr (right).

RESULTS AND DISCUSSION

Taurine-like immunoreactivity (TLI) was evenly distributed in the cardiac pectinate muscle and the smooth muscle of the small intestinal wall (Figure 1). The immunoreaction occurs in every fiber, being stronger in the cardiac tissue. Although the analysis was not brought beyond the light microscopy resolution, no evidence was found of an intracellular compartmentation of the taurine pool in any of the three muscle types examined (cardiac, smooth and skeletal).

The TLI distribution in skeletal muscle sharply contrasts with that observed in cardiac and smooth muscle. There is strong reactivity in some fibers while others nearby contain little or no taurine (Figure 2). Cross sections of every striated muscle

Figure 1. Cat muscle sections stained with the taurine antiserum from the cardiac pectinate muscle, 120x (a), and the smooth muscle of the small intestinal wall, 60x, (b).

examined (biceps, triceps, biceps femoris, fascia lata, gastrocnemius, plantaris, soleus and extensor digitorum longus) stained with the taurine antiserum, have a checkerboard appearance (Figures 2a,c), while in longitudinal sections, immunoreactivity (or its absence) can be followed all along each fiber (Figure 2b).

The uneven TLI distribution is not a consequence of the fixation process as it is observed both in perfusion-fixed tissue (Figures 2a,b,d; Figures 3 and 4) and in immersion-fixed tissue (Figure 2c). Taurine-negative fibers are not damaged cells as they become fully stained with hematoxylin (Figure 2d).

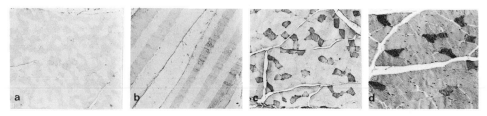

Figure 2. Cat skeletal muscle sections stained with the taurine antiserum fixed by intracardiac perfusion (a,b and d) or by immersion (c). Plantaris muscle: cross section, 30x, (a); longitudinal section, 60x, (b). Biceps femoris muscle: cross section, 30x, (c); cross section counterstained with hematoxylin, 30x, (d).

The TLI observed is truly marking the taurine endogenous pool as it does not appear when the immune serum is preincubated along with the immunogen (compare Figures 3a and b). Substitution of preimmune serum for immune serum does not produce any reaction (not shown), nor does exposure of tissue to an immune serum raised against BSA-G-GABA (Figures 3c,d).

These results are confusing in comparison to electrophysiological studies (Gruener et al., 1976; Conte-Camerino et al., 1987; 1989) which have reported that every single skeletal fiber is responsive to externally applied taurine, basically modifying chloride permeability, with no individual exceptions reported.

Reconciliation of these data comes from the TLI results obtained when the state of muscle contraction was experimentally modified. Figure 4 shows a cross section of the plantaris in a steady-state tone (a) and after a 15 min period of nerve stimulation (b) (see Methods) or after severing of the sciatic nerve (c). The ratio of taurine-positive and -negative fibers change in opposite directions as the muscle is contracted by stimulation or is relaxed by denervation. Similar results were observed for the gastrocnemius, although the basal ratio of positive/negative fibers is different.

These results suggest that the uneven TLI distribution is not a static condition

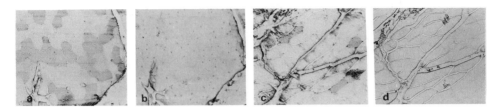

Figure 3. Serial sections from the cat gastrocnemius (a and b), and the biceps femoris (c and d). a) Immunostained for taurine, b) Idem as a) but preincubated with the taurine immunogen. Sections immunostained for taurine (c) and for GABA (d). All images 60x.

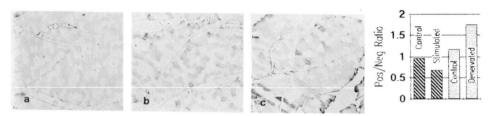

Figure 4. Taurine immunostaining of sections from the cat plantaris in a steady state (a), after nerve stimulation (b), or after denervation (c). See Methods for details. Histogram represents the taurine-positive/-negative fiber ratio obtained in these cats; each bar was calculated by counting 500 to 1000 fibers. All images 30x.

but rather a consequence of the state of contraction of each single fiber. Skeletal muscles develop force by recruiting contracted fibers, i.e. not all fibers contract at the same time in a given anatomically defined unit. The decrease of taurine-positive fibers after stimulation suggest that skeletal fibers release most of their taurine when they contract. As shown by Iwata et al. (1986), nerve stimulation of rat calf muscles is followed by increased taurine transport activity, regardless of the muscle phenotype. Our results are congruent and complementary to those as they suggest a massive mobilization of taurine to the extracellular space during contraction, which for a tissue with almost negligible *de novo* synthesis capacity (Yamaguchi et al., 1973), repletion of its endogenous pool through uptake should be critical. Furthermore, the massive release of taurine from contracted fibers will substantially increase the extracellular taurine which is then available for uptake.

Similarly, muscle relaxation under the extreme condition of denervation would allow every single fiber to accumulate taurine, as suggested by the increased number of TLI-positive cells observed. Iwata et al. (1986), report an increase in taurine transport and in the total taurine content after 4 days of denervation. Our results probably represent the early stages of the same process, i.e. the replenishment of all the empty (contracted) fibers.

Glutamate-like immunoreactivity (GLI) distribution in skeletal muscle shows some differences from that of TLI. GLI is present in every single fiber, however the intensity (reflection of glutamate concentration) varies. Since immunohistochemical techniques are semiquantitative, at best, we distinguished only two clearly different immunoreactive intensities, the lower present in every fiber. However, there are fibers showing stronger immunoreaction distributed regularly all over every cross section examined, (Figure 5b). Moreover, those strong glutamate positive fibers match almost completely with those taurine-positive cells, as observed in serial sections (compare Figures 5a and b). Cross reactions between the immunogens and antisera is precluded because: a) GLI persisted when a high proportion (1:100) of taurine immunogen was preincubated with glutamate antiserum (Figure 5c,d); b) the glutamate antiserum used was preabsorbed with BSA-G-taurine during purification, and checked by ELISA. c) preincubation of glutamate immunogen with the antiserum precluded the appearance of any reaction (not shown).

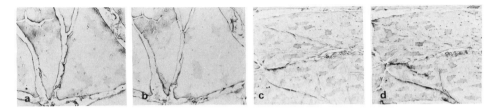

Figure 5. Serial sections of the cat biceps femoris immunostained for taurine (a) and glutamate (b). Other sections from the same muscle immunostained for glutamate (c and d) but (d) preincubated with the taurine immunogen. All images 30x.

Parallel movements of taurine and glutamate have been described in other tissues (van Gelder, 1985; Schousboe et al., 1991). In skeletal muscle, the description of increased taurine levels by Peterson et al. (1963) reports substantial increments in glutamic acid too. Although we have not yet analyzed the glutamate pool behavior in the stimulated and denervated muscles, the precise matching of taurine and glutamate positive fibers suggests parallel movements.

Finally, preliminary results are presented on TLI in rat and mouse skeletal muscle. Figure 6 shows the uneven distribution of TLI in the rat biceps femoris fixed by intracardiac perfusion (a and b), and a section from the same muscle fixed by immersion (c); and a section from the mouse biceps femoris fixed by immersion and stained with the glutamate antiserum (d).

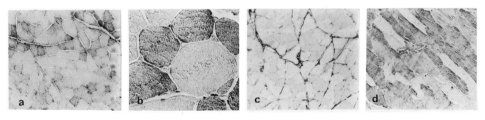

Figure 6. Sections of the biceps femoris from rat (a,b and c) and mouse (d). Sections from perfusion-fixed tissue, immunostained for taurine: (a,b,d). Section from immersion fixed tissue, immunostained for glutamate (c). a,c,d 60x; b 300x.

These results, even though preliminary, would indicate that the uneven distribution of the taurine pool in skeletal muscle is not restricted to the cat and is a general feature of this muscle.

ACKNOWLEDGEMENTS

Octavio Quesada is supported by a scholarship from the DGAPA from the National University of Mexico.

REFERENCES

Campistron, G., Geffard, M., and Bujis, R.M., 1986, Immunological approach to the detection of taurine and immunocytochemical results, *J. Neurochem.* 46:862.

Conte-Camerino, D., DeLucca, A., Mambrini, M., Ferrannini, E., Franconi, F., Giotti, A., and Bryant, S.H., 1989, The effects of taurine on pharmacologically induced myotonia, *Muscle and Nerve* 12:898.

Conte-Camerino, D., Franconi, F., Mambrini, M., Mitolo-Chieppa, D., Bennardini, F., Failli, P., Bryant, S.H., and Giotti, A., 1987, Effect of taurine on chloride conductance and excitability of rat skeletal muscle fibers, *in:* "The Biology of Taurine: Methods and Mechanisms," R.J. Huxtable, F. Franconi, and A. Giotti, eds., Plenum Press, New York.

Gruener, R., Bryant, H., Markowitz, D., Huxtable, R.J., and Bressler, R., 1976, Ionic actions of taurine on nerve and muscle membranes: electrophysiological studies, "Taurine" R.J. Huxtable and A. Barbeau, eds., Raven Press, New York.

Huxtable, R.J., Laird, N.E., and Lippincott, S.E., 1979, The transport of taurine in the heart and the rapid depletion of tissue taurine content by guanidinoethylsulfonate, *J. Pharmacol. Exp. Ther.* 211:465.

Huxtable, R.J., and Sebring, L.A., 1983, Cardiovascular actions of taurine, "Sulfur aminoacids: Biochemical and Clinical Aspects", K. Kuriyama, R.J. Huxtable, and H. Iwata, eds., Alan R. Liss, New York.

Iwata, H., Obara, T., Kim, B., and Baba, A., 1986, Regulation of taurine transport in rat skeletal muscle, *J. Neurochem.* 47:158.

Jacobsen, J.G., and Smith, L.L.H., 1968, Biochemistry and physiology of taurine and taurine derivatives, *Physiol. Rev.* 48:424.

Kocsis, J.J., Kostos, V.J., and Baskin, S.I., 1976, Taurine levels in the heart tissues of various species, "Taurine", R.J. Huxtable and A. Barbeau, eds., Raven Press, New York.

Lehmann, A., and Hamberger, A., 1983, Inhibition of chlinergic response by taurine in frog isolated skeletal muscle, *J. Pharm. Pharmacol.* 36:59.

Peterson, D.W., Lilyblade, A.L., and Lyon, I., 1963, Serine-ethanolamine-phosphate, taurine and free amino acids of muscle in hereditary dystrophy of the chicken, *Proc. Soc. Biol. Med.* 113:798.

Schousboe, A., Moran, J., and Pasantes-Morales, H., 1990, Potassium-stimulated release of taurine from cultured cerebellar granule neurons is associated with cell swelling, *J. Neurosci. Res.* 27:71.

Sturman, J.A., and Messing, J.M., 1991, Dietary taurine content and feline reproduction and outcome, *J. Nutr.* 121:1195.

van Gelder, N.M., and Barbeau, A., 1985, The osmoregulatory function of taurine and glutamic acid, "Taurine: Biological Actions and Clinical Perspectives", S.S. Oja, L. Ahtee, P. Kontro, and M.K. Paasonen, eds., Alan R. Liss, New York.

Yamaguchi, K., Sakakibara, S., Asamizu, J. and Ueda, I., 1973, Induction and activation of cysteine oxidase of rat liver II. Measurement of cysteine metabolism in vivo and the activation of in vivo activity of cysteine oxidase, *Biochim. Biophys. Acta* 297:48.

REVIEW: MYOCARDIAL PHYSIOLOGICAL EFFECTS OF TAURINE AND THEIR SIGNIFICANCE

Stephen W. Schaffer[1] and Junichi Azuma[2]

[1]University of South Alabama
School of Medicine
Department of Pharmacology
Mobile, AL 36688

[2]Department of Medicine III
School of Medicine
Osaka University
Osaka, Japan

INTRODUCTION

Recent interest in the role of taurine in the heart stems from the observation that cats fed a taurine deficient diet develop a cardiomyopathy[1]. This condition is reversible, as evidenced by the restoration of normal cardiac function upon dietary supplementation with taurine[2]. These studies have accelerated the search for a physiological function for taurine. Although several theories of taurine action have been proposed, none adequately accounts for all of the multiple properties and actions of taurine. Therefore, this review will address both the strengths and weaknesses of the present proposals, including the possibility that taurine may exhibit multiple physiological roles in the heart.

EFFECTS OF TAURINE ON MYOCARDIAL FUNCTION

Antiarrhythmic Activity of Taurine

Read and Welty[3,4] initially proposed that taurine altered ion movement in the heart. This conclusion was based on the observation that taurine prevented epinephrine-mediated efflux of potassium from intact dog heart while promoting its uptake into heart slices incubated in the presence of toxic doses of digoxin or epinephrine. Because taurine also abolished epinephrine- and digoxin-induced ventricular premature contractions, the authors suggested that the effects of taurine on potassium loss were linked to its antiarrhythmic activity. In a related study, Chazov et al.[5] reported that taurine suppressed arrhythmias in dogs administered toxic doses of strophanthin-K. Yet, in isolated guinea pig heart exposed to K^+-free buffer,

Taurine, Edited by J.B. Lombardini *et al.*
Plenum Press, New York, 1992

taurine had mixed results, aggravating some of the EKG changes associated with altered medium K^+ while delaying the development of irreversible fibrillation. These studies raised questions on the ability of taurine to modulate K^+ movement. While recent studies support the notion that taurine has antiarrhythmic activity, these effects usually have been linked to calcium movement rather than potassium loss[6-8]. Takahashi et al.[7] found that treatment of isolated myocytes with taurine protected the cells against arrhythmias induced by either elevating or lowering medium calcium content. Also attributed to calcium-dependent events are the electrophysiological changes evoked by drug-induced taurine depletion[8]. These data suggest that while it is generally accepted that taurine exhibits antiarrhythmic activity, the mechanism underlying this action still remains to be established.

Modulation of Myocardial Contraction by Taurine

Dietrich and Diacono[9] first recognized that taurine modulates myocardial contraction. They suggested that taurine exhibits ouabain-like activity, reducing myocardial contraction in the isolated rat heart while increasing it in guinea pig heart. This view was challenged by Schaffer et al.[10], who observed a positive inotropic effect of 10 mM taurine in the isolated working rat heart. The latter group also showed that the positive inotropic effect was highly dependent on the contractile status of the heart[11]. Hearts made hypodynamic by either replacement of the buffer with reduced levels of calcium or by inclusion of the calcium antagonist verapamil in the buffer exhibited enhanced responsiveness to taurine (Figure 1).

Most subsequent studies confirmed the dependency of the taurine response on extracellular calcium content and the presence of calcium antagonists[13-15], although one study observed a positive inotropic effect of taurine without any dependence on medium calcium or the presence of verapamil[12]. Because the latter study employed rabbit heart, as opposed to rat, guinea pig or frog heart, the possibility that the effects of taurine may be species dependent was raised. Another important issue discussed at the time was the physiological significance of taurine's positive inotropic effect. While Schaffer and coworkers[10,11,16] felt that the alterations in contraction associated with elevations in perfusate taurine content represented pharmacological actions of taurine, Franconi and coworkers[14,17] argued that since intracellular taurine levels were increased by this procedure, these effects were physiological. In support of her view, she found that there was a fairly good correlation between intracellular taurine levels and myocardial contraction. As seen in Figure 2, a direct linear relationship between intracellular taurine content and contractile force was observed in guinea pig ventricular strips perfused with medium containing low Ca^{2+} and varying taurine concentration. This relationship was further explored using various models of tissue taurine depletion. In one such study, Mozaffari et al.[18] reported that depletion of taurine with the taurine transport inhibitors, guanidinoethanesulfonate and ß-alanine, led to changes in myocardial metabolism without alterations in myocardial work output. Recognizing that cardiac work is a poor indicator of myocardial contractility, the status of mechanical function was reevaluated by measuring an index of contractility ($+dP/dt$) and relaxation ($-dP/dt$). These studies revealed that drug-induced taurine depletion led to a slight impairment in the ability of the heart to relax while having no effect on myocardial contractility[19]. A more complete study using papillary muscles of rats treated with guanidinoethanesulfonate to achieve a 75% reduction in tissue taurine levels was reported by Lake et al.[20]. In agreement with the earlier study by Schaffer et al.[19], they reported that isotonically contracting papillary muscles of taurine-depleted rats exhibited a reduction in the time to attain 50% shortening decline, an index of cardiac relaxation. However, the largest effects were noted in isometrically contracting muscle. As seen in Table 1, taurine depletion led to a decline in total tension development and the rate of tension development, but an increase in the duration of contraction. These data suggested that intracellular taurine levels modulate both systolic and diastolic function. Although the conclusions of the drug-induced taurine depletion studies are generally consistent with earlier reports, concern has been raised regarding the use of these drugs because of their adverse side effects, such as the depletion of cellular creatine phosphate content. This concern has stimulated the introduction of new procedures to assess the role of intracellular taurine content.

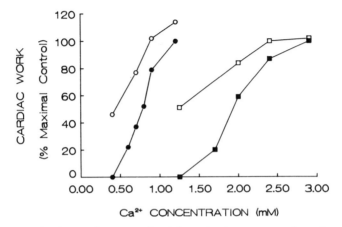

Figure 1. Calcium titration studies on perfused heart in the presence of taurine and verapamil. Perfusions were begun at 0.4 mM Ca^{2+} and perfusate Ca^{2+} concentration was increased stepwise. Cardiac work was monitored in control hearts (●) or hearts perfused with 10 mM taurine (O), 3 X 10^{-8} M verapamil (■), or both 10 mM taurine and 3 X 10^{-8} M verapamil (□). Values shown are means of 3 - 7 hearts. Reproduced from Chovan et al.[11] with permission of Molecular Pharmacology

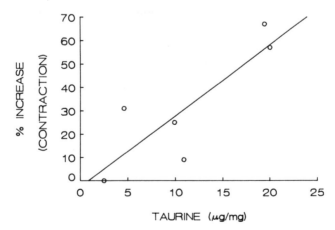

Figure 2. Relationship between intracellular taurine levels of guinea pig ventricle and contractile force. Ventricular strips were perfused with medium containing either 0.45 or 0.9 mM Ca^{2+}. Data obtained at higher medium Ca^{2+} were excluded because taurine is less effective at these Ca^{2+} concentrations. Shown is the linear correlation between intracellular taurine concentration and contractile force. The correlation coefficient for this relationship is r = 0.84. The data were recalculated by us from Franconi and coworkers[14,17].

Table 1. Effect of drug-induced taurine depletion on isometric contractions of isolated papillary muscles

Parameter	Control	Taurine-depleted
Total tension (g/mm^2)		
Basal rate	9.4 ± 0.5	7.1 ± 0.5*
Paired pulses at 200 msec	9.8 ± 0.4	7.3 ± 0.5*
After 3 min rest	9.8 ± 0.5	7.5 ± 0.5*
Rate of tension development (g/mm2/sec)		
Basal rate	100 ± 6	76 ± 7*
Paired pulses at 200 msec	99 ± 5	77 ± 7*
After 3 min rest	104 ± 5	82 ± 7
Duration of contraction (msec)		
Basal rate	462 ± 10	498 ± 15
After 3 min rest	471 ± 11	509 ± 21

Data represents means ± S.E.M. of 12 control and 13 taurine-depleted rats. Asterisks denote significant difference between the control and taurine-depleted groups (p < 0.05). Reproduced from Lake et al.[20] with permission of Can J Physiol Pharmacol.

    One of these new techniques is the chemically skinned fiber preparation[22,23]. Because the sarcolemma of this preparation is made completely permeable to various intracellular constituents, one can easily regulate both intracellular taurine and calcium content. Three important observations emerged from the studies using this preparation. First, taurine modulates calcium transport by the sarcoplasmic reticulum. Second, taurine increases the sensitivity of the myofibrils to calcium. Third, the alteration of caffeine-induced contractures by taurine is highly dependent on both medium taurine and calcium concentration. While the data of Steele et al.[22] and Franconi et al.[17] both revealed that taurine mediates a positive inotropic effect when medium Ca^{2+} content is low and induces a negative effect in the Ca^{2+}-overloaded heart, one major difference existed between the two studies. Rather than observing a linear relationship between intracellular taurine concentration and contractile function, Steele et al.[22] found after correcting for the osmotic effects of taurine a hyperbolic relationship. This pattern is consistent with nutritional studies which have shown that moderate decreases in intracellular taurine do not appreciably alter contractile function; only after taurine content has been decreased substantially, will contraction decline and a cardiomyopathy develop.

    Pion et al.[1] first recognized the clinical significance of taurine depletion on heart function. Their landmark 1987 study showing that a group of taurine-depleted cats exhibited improved myocardial function following dietary taurine supplementation led to the augmentation of commercial cat foods with taurine. Since the fortification of commercial cat foods with taurine, the incidence of dilated cardiomyopathy linked to taurine deficiency has dramatically declined[2]. These observations have stimulated the use of taurine in the treatment of congestive heart failure. Azuma et al.[21,38] recently showed that orally administered taurine benefited patients suffering from congestive heart failure. After 4 weeks of taurine therapy, the patients showed a significant improvement in dyspnea, palpitation, crackles, edema, cardiothoratic ratio on chest X-ray film and the measure of disease status defined by the New York Heart Association. These effects appeared to be related to a stimulation in myocardial contractile performance (Table 2).

Table 2. Effect of taurine on echocardiographic parameters

Parameter	Taurine Dose (g/day)	Number of Patients	Baseline	After
LVEDV (ml)	3	23	76 ± 16	183 ± 15
	6	19	149 ± 15	147 ± 16
Stroke volume (ml)	3	23	72 ± 7	78 ± 5
	6	19	62 ± 6	72 ± 6*
Cardiac output (l/min)	3	23	5.2 ± 0.4	5.9 ± 0.4*
	6	19	4.6 ± 0.5	5.6 ± 0.4*
Ejection fraction	3	23	51.3 ± 3.1	53.7 ± 2.7
	6	19	54.8 ± 4.6	61.4 ± 3.8*
Heart rate (min^{-1})	3	23	74.9 ± 3.9	76.9 ± 3.2
	6	19	75.9 ± 3.2	80.6 ± 5.0
LVET (msec)	3	23	280 ± 10	272 ± 11
	6	19	277 ± 12	279 ± 10
mVcf (sec^{-1})	3	23	0.828 ± 0.077	0.890 ± 0.068
	6	19	0.893 ± 0.084	1.037 ± 0.079*

Data are expressed as means ± S.E.M. and are reproduced from Azuma et al.[21] with permission of Kluwer Press.
LVEDV = Left ventricular end-diastolic volume; LVET = Left ventricular ejection time; mVcf = Mean velocity of circumferential shortening.
* Significant differences vs. baseline by paired t-test.

Cardioprotective Activity of Taurine

It is generally accepted that excessive accumulation of Ca^{2+} by the heart initiates a train of events resulting in cell damage. The most important Ca^{2+}-mediated changes include activation of cellular ATPases resulting in enhanced utilization of ATP; uptake of Ca^{2+} by the mitochondria leading to impaired oxidative ATP production; activation of Ca^{2+}-dependent phospholipases and proteases; and initiation of a Ca^{2+}-dependent state of contracture[24,25]. Therefore, interventions which influence the degree of intracellular Ca^{2+} overload, also alter the degree of Ca^{2+}-linked myocardial injury. In accordance with this theory, it has been shown that pharmacological doses of taurine provide protection against myocardial injury in several models of Ca^{2+} overload-induced heart failure, including the calcium paradox[26,27] isoprenaline-induced myocardial damage[28], doxorubicin-mediated cardiotoxicity[29], Na^+-promoted Ca^{2+} overload[30], myocardial hypoxia[31,32], congestive heart failure[33,34] and the cardiomyopathic hamster[35,36]. Moreover, drug-induced taurine depletion exacerbates the degree of myocardial dysfunction in the ischemic[19] and doxorubicin-treated[37] myocardium. These studies have raised the possibility that taurine might benefit patients suffering from ischemic heart disease.

Interaction between Taurine and Cardiovascular Drugs

Many of the interactions of taurine with various cardiovascular drugs are linked to changes in Ca^{2+} transport. Fujimoto and Iwata[39] found that taurine-mediated potentiation in the inotropic response to digitoxin was associated with an elevation in myocardial Ca^{2+} content. In the same study, they found that taurine enhanced the positive inotropic effect of low doses of noradrenaline. On the other hand, Welty et al.[40] have reported that taurine blocks the accumulation of tissue Ca^{2+} upon treatment with elevated levels of isoproterenol. In support of the latter study, Malchikova and Elizarova[42] found that taurine decreased tissue cAMP levels in the stressed myocardium. These conflicting observations may be related to competing effects of taurine on the activities of adenylate cyclase and phosphodiesterase[41,42]. Schaffer et al.[41] have reported that the effects of taurine on adenylate cyclase activity are complex. Under basal conditions, taurine has no effect on cAMP production. However, it inhibits the response to GTP (G protein coupling) while attenuating both the Ca^{2+}-dependent decline and the isoproterenol-mediated stimulation in adenylate cyclase activity.

It has been suggested that some of the effects of taurine on cardiovascular agents involve the modulation of membrane receptors. Lampson et al.[43] found that taurine alters glucose and glycogen metabolism of the heart in an insulin-dependent manner. Although taurine appears to interact with the insulin receptor[44], it remains to be established whether this interaction is linked to the metabolic effects of taurine.

An effect of taurine on the number of α-adrenoceptor binding sites has also been reported[45]. Because taurine reduces the sensitivity of guinea pig ventricular strips to the contractile response of the α-agonist, phenylephrine, it has been suggested that the contractile and receptor effects are causually related.

MOLECULAR BASIS UNDERLYING ACTIONS OF TAURINE

Several mechanisms have been proposed to account for the actions of taurine. One of the concerns has been to distinguish physiological and pharmacological actions of taurine. In the heart there presently is evidence supporting the involvement of taurine in osmoregulation, lipid metabolism, calcium transport and membrane function. Other effects, such as free radical scavenging, conjugation with various substances and inclusion in certain peptides are unlikely to be important in the heart.

Osmoregulation

Vislie and Fugelli[46] first recognized the importance of taurine as an osmoregulator in the heart. Based on the observation that ventricular taurine content of flounder heart was linearly related to plasma osmolality, they concluded that taurine is an osmotically active substance which prevents intracellular volume changes caused by alterations in plasma osmolality. This notion was strengthened by the observation of Schaffer et al.[47], who found that taurine transport in heart is Na^+-dependent and appears to involve a Na^+/taurine symport mechanism. However, the promotion of taurine transport by Na^+ does not appear to be the sole mechanism responsible for accumulation of taurine in hearts under osmotic control. Atlas et al.[48] found that the rate of taurine transport could be stimulated to the same degree by Na^+ and sucrose.

The initial study implicating taurine as an osmotic agent in mammalian heart was published by Thurston et al.[49], who found that myocardial taurine content increased 30% in young mice with chronic hypernatremia. Significantly, no other amino acid was affected by hypernatremic dehydration. The physiological significance of the osmotic effect of taurine is suggested by several pathological states. Streptozotocin-induced insulin-dependent diabetes in rats is associated with both an elevation in myocardial taurine content and a rise in plasma osmolarity[48]. Further, cellular water in the ischemic myocardium begins to accumulate about the same time that taurine is lost from the myocyte[50]. Because excessive cellular water accumulation has been implicated in myocardial ischemia injury, it is possible that some of the cardioprotective activity of taurine may involve osmoregulation.

Lipid Metabolism

Taurine is reported to stimulate glucose utilization of the normal isolated rat heart without significantly altering ß-oxidation of endogenous fatty acids[41]. However, in the taurine-depleted myocardium lipid metabolism appears to be notably altered[51]. Significant increases in myocardial levels of free CoA, free carnitine and long chain fatty acylcarnitine, as well as decreases in long-chain fatty acyl-CoA content, have been observed in drug-induced taurine depleted hearts. It is known that most long-chain fatty acylcarnitines are found in the cytoplasm where they are available for the synthesis of triacylglycerol. However, excessive levels of acylcarnitine can disrupt normal membrane function, including Ca^{2+} transport[52].

Modulation of phospholipid metabolism also appears to be an important physiological role of taurine. Two reactions are worthy of consideration because they can lead to altered Ca^{2+} transport. First, there is some evidence that taurine might affect phosphoinositide turnover. Lombardini and coworkers[53,54] have observed that taurine inhibits the phosphorylation of several cardiac and synaptosomal proteins by an endogenous protein kinase, which appears to belong to the protein kinase C group. One mechanism by which this might occur is through the modulation of phosphoinositide turnover. Recently, both Smith and Li[55] and Kuriyama et al.[56] have reported that taurine can influence this pathway, thereby affecting the amount of diacylglycerol available for the activation of protein kinase C. Another product of this reaction, IP_3, promotes the release of Ca^{2+} from intracellular stores.

The other phospholipid pathway affected by taurine is the methylation of phosphatidylethanolamine[57] (Figure 3). Unlike many other taurine-linked reactions, the biochemical basis underlying this reaction is more easily explained. The chemical structure of taurine resembles the charged headgroup of phosphatidylethanolamine. Therefore, it would be logical to assume that taurine could bind at the catalytic site designated for phosphatidylethanolamine, resulting in competitive inhibition of the methylation reaction. Table 3 reveals that taurine does inhibit the activity of sarcolemmal phospholipid methyltransferase, however, due to the complexity of the enzyme attempts to demonstrate competitive inhibition have been unfruitful. Phospholipid methyltransferase contains three catalytic sites, referred to as site I, site

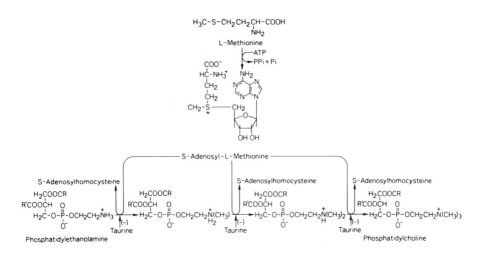

Figure 3. Reactions catalyzed by phospholipid methyltransferase.

Table 3. Effect of taurine on sarcolemmal phospholipid methyltransferase activity

Catalytic site	With taurine (pmol/mg/30 min)	With sucrose (pmol/mg/30 min)
Site I	0.65 ± 0.05*	0.93 ± 0.03
Site II	4.82 ± 0.50*	7.42 ± 0.37
Site III	99 ± 10*	135 ± 4

Activity of sarcolemmal phospholipid methyltransferase was assayed at catalytic sites I, II and III in medium containing either 10 mM sucrose or 10 mM taurine. Values shown are means ± S.E.M. of 3-4 different assays. Asterisks donate significant difference between activity in the presence of taurine and sucrose. Reproduced from Hamaguchi et al.[57] with permission of J Cardiovasc Pharmacol.

II and site III. The reaction favored at site I is the formation of phosphatidyl-N-monomethylethanolamine from phosphatidyl-ethanolamine. The primary products generated at the other two sites are phosphatidyl-N,N-dimethylethanolamine (site II) and phosphatidylcholine (site III). The methyltransferase reaction plays a very important role in the function of the myocardium. Gupta et al.[58] have found that the conversion of phosphatidylethanolamine to phosphatidylcholine dramatically affects the activity of certain Ca^{2+} transporters. Among the most dramatically affected transporters is the Na^+-Ca^{2+} exchanger, whose activity falls 75% in hearts exposed to buffer containing the methylating agent, methionine (see Figure 3). Taurine, by inhibiting the methyltransferase reaction, prevents the loss in Na^+-Ca^{2+} exchanger activity (Table 4). Moreover, the decrease in myocardial function associated with the loss in Na^+-Ca^{2+} exchanger activity also is restored by taurine (Table 4).

Table 4. Effect of taurine on Na^+-Ca^{2+} exchange and myocardial contractile function

Condition	Na^+-Ca^{2+} Exchange Activity (nmol/mg/sec)	Cardiac Work (kg-m/g dry wt/min)
Control	2.16 ± 0.4	0.43 ± 0.01
Methionine	0.50 ± 0.1*	0.37 ± 0.01*
Taurine	2.08 ± 0.03	0.43 ± 0.02
Taurine and Methionine	1.50 ± 0.1*	0.42 ± 0.02

Hearts were perfused with standard Krebs-Henseleit buffer for 20 min (Control), 15 min with standard buffer followed by 5 min with standard buffer supplemented with 300 μM L-methionine (Methionine), 20 min with standard buffer supplemented with 10 mM taurine (Taurine) or 15 min with standard buffer containing 10 mM taurine followed by 5 min with standard buffer containing 300 μM L-methionine and 10 mM taurine (Taurine and Methionine). Following the 20 min perfusion, the hearts were used in the isolation of sarcolemma in order to measure Na^+-Ca^{2+} exchanger activity. Cardiac work was measured after 20 min exposure to the indicated buffer. Reproduced from Hamaguchi et al.[57] with permission from J Cardiovasc Pharmacol.

The demonstration of an interaction between methionine and taurine is important, not only because of the significance of the phospholipid methyltransferase reaction in the heart, but also because it represents the first system in which the biochemical effects of taurine in the heart can be directly linked to a physiological event, such as myocardial contraction.

Calcium Transport

Many of the studies implicating taurine in the movement of calcium have utilized pharmacological doses of taurine. Typical of those studies is the work of Dolara et al.[59], who reported in 1973 that inclusion of taurine in the perfusion buffer delayed the washout of calcium from guinea pig hearts perfused with calcium-free buffer. Kinetic analysis of the calcium washout data led to the hypothesis that taurine increases the affinity of certain cellular components for calcium[60]. While subsequent studies indicated that taurine had little to no effect on calcium binding or accumulation by the sarcoplasmic reticulum, Chovan et al.[61] discovered a specific stimulation of calcium binding to the sarcolemma in the presence of taurine. Interestingly, taurine also reversed the inhibitory effects of cardiodepressants, such as verapamil and lanthanum, on calcium binding to the same sarcolemmal sites. The notion that these binding sites were important was reinforced by evidence that taurine treatment which protected the heart against damage caused by the calcium paradox also prevented excessive binding of calcium to the sarcolemmal sites[26]. The assumption was made that the size of this calcium pool would determine the amount of calcium entering the cell. Fura 2 studies have provided support for the idea that taurine increases $[Ca^{2+}]_i$ of the hypodynamic heart, but prevents excessive accumulation of intracellular calcium during the calcium paradox[62]. Nevertheless, interest in the size of the sarcolemmal Ca^{2+} pool has waned, as methods have developed to directly monitor Ca^{2+} transport.

The cardiomyopathic hamster model of heart failure has proved valuable in establishing a link between taurine action and calcium homeostasis. The Syrian hamster is genetically predisposed to undergo several discrete phases during their lifetime, with the final stage resulting in the development of a hypertrophic cardiomyopathy[63]. Between the age of 80-150 days, the myocardium enters a phase characterized by fibrosis and calcium deposition. Measurements of cellular calcium content and distribution using electron probe microanalysis have revealed that the cardiomyopathy is characterized by focal regions of necrosis and calcification[64]. During the latter phases of the disease, defects also develop in myocardial metabolism, such that perfusion of the heart with glucose containing buffer triggers an increase in diastolic $[Ca^{2+}]_i$ and a decrease in the amplitude of both $[Ca^{2+}]_i$ transients and myocardial contraction[65]. Defects in calcium handling by the sarcoplasmic reticulum and the sarcolemmal Na^+-Ca^{2+} exchanger and Ca^{2+} pump have been described[66,67]. Moreover, chronic verapamil therapy prevents the development of the contractile and biochemical lesions associated with the latter phases of the cardiomyopathy, indicating that calcium overload underlies many of the defects occurring in the cardiomyopathic hamster[68,69]. Although chronic taurine studies have not been performed with these animals, short-term treatment with pharmacological doses of taurine have led to a reduction in myocardial calcium content (Table 5). Also, reductions in cellular taurine content using the taurine transport inhibitor, guanidinoethane sulfonate, have been reported to cause a small, but significant decrease in cellular calcium levels (Table 5). The basis for the taurine-mediated improvement in calcium homeostasis of the cardiomyopathic hamster is not well understood. Welty and Welty[71] have proposed that taurine modulates the handling of calcium by the sarcolemma and mitochondria. Since verapamil prevents calcium overload in this model while taurine is capable of interacting with verapamil to alter calcium homeostasis, it is likely that taurine affects calcium influx across the sarcolemma.

Two sarcolemmal transporters contribute to calcium influx during excitation-contraction coupling, the calcium channel and the Na^+-Ca^{2+} exchanger. Yet, attempts to establish a direct physiological link between taurine and the activity of these transporters have largely been unsuccessful[72-74]. This has led to some confusion since

Table 5. Effect of taurine treatment on myocardial calcium content of Syrian hamsters

Strain	Heart Ca^{2+} (mEq/kg dry wt)
Cardiomyopathic Hamsters	
30 days old	16.0 ± 2.0
60 days old (- Taurine)	206.0 ± 60.4
60 days old (+ Taurine)	31.7 ± 5.9*
Normal Hamster	
30 days old	14.2 ± 4.1
60 days old (- Taurine)	18.6 ± 2.7
60 days old (+ Taurine)	18.8 ± 0.9
60 days old (+ GES)	14.0 ± 2.3*

Normal F$_1$B and cardiomyopathic BIO 14.6 hamsters were placed on one of the following *ad libitum* drinking regimes for 4 weeks: (i) tap water control (-Taurine), (ii) 0.1 M taurine (+Taurine), (iii) 1% guanidinoethanesulfonic acid (+GES). The indicated age reveals the time of calcium measurement. Values shown represent means ± S.E.M. of 5-9 animals. Asterisks indicate significant difference from the animals maintained on tap water. Reproduced from J.D. Welty[36,40] with permission of J. Mol. Cell. Cardiol.

taurine prevents calcium overload in heart failure models associated with excessive entry of calcium through the Na$^+$-Ca^{2+} exchanger[30]. The only rational explanation for this apparent discrepancy is that the effects of taurine on transsarcolemmal calcium flux presumably involve indirect actions on the transporters. One such effect was reported by Sperelakis et al.[74], who found that taurine stimulates the fast component of the Na$^+$ current, thereby elevating [Na$^+$]$_i$ and promoting Ca^{2+} influx through the Na$^+$-Ca^{2+} exchanger. Another indirect means of modulating Na$^+$-Ca^{2+} exchanger activity is through the regulation of the lipid microenvironment surrounding the transporter (see discussion above). A third possibility is the regulation of the transporter by protein phosphorylation. Although there is no evidence that taurine affects the phosphorylation of the Ca^{2+} channel or Na$^+$-Ca^{2+} exchanger *per se*, the fact that taurine affects the activity of certain protein kinases indicates that this idea is worthy of some consideration[53,54].

Effects of taurine do not appear to be restricted to the sarcolemma. As shown in Figure 4, taurine stimulates Ca^{2+}-induced, Ca^{2+}-efflux from the sarcoplasmic reticulum.

These results are in agreement with the study by Steele et al.[22], who found that taurine could activate Ca^{2+} release from the sarcoplasmic reticulum of skinned muscle fibers. The physiological consequences of this activity appears to depend upon the status of the sarcoplasmic reticulum. If Ca^{2+}-induced, Ca^{2+} release is promoted, but in the process the sarcoplasmic reticular Ca^{2+} stores are depleted, taurine will cause a reduction in contraction. This scenario appears to account for the negative effect of taurine in skinned muscle fibers subjected to a high sarcoplasmic reticular Ca^{2+} load. On the other hand, stimulation of Ca^{2+}-induced, Ca^{2+} release should increase contraction if the sarcoplasmic reticular Ca^{2+} stores are maintained. Steele et al.[22] have suggested that taurine will enhance sarcoplasmic reticular Ca^{2+} uptake when the sarcoplasmic reticular Ca^{2+} load is low. Thus, taurine could either increase or decrease myocardial contraction, depending upon intracellular calcium levels. In this respect, it is important to emphasize that taurine has no effect on calcium-independent release of sarcoplasmic reticular calcium, an effect implicated in the negative inotropic effect of doxorubicin.

An apparent weakness of the hypothesis proposed by Steele et al.[22] is the lack of *in vitro* evidence that taurine alters sarcoplasmic reticular Ca^{2+} uptake. Most

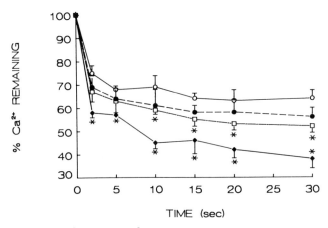

Figure 4. Effect of taurine on Ca^{2+}-induced, Ca^{2+} release from isolated junctional sarcoplasmic reticulum. The rate of Ca^{2+} release from $^{45}Ca^{2+}$-loaded sarcoplasmic reticular vesicles was assessed from the amount of $^{45}Ca^{2+}$ remaining within the vesicles. The release reaction was initiated by placing the vesicles in medium containing $10\,\mu M\,Ca^{2+}$ and the following concentrations of taurine: 0 mM (O); 10 mM (●); 30 mM (□) and 50 mM (♦). Asterisks denote significant difference between taurine and control ($p < 0.05$).

studies using isolated membrane preparations have concluded that taurine has little or no direct effect on sarcoplasmic reticular Ca^{2+} uptake[72,75,76]. Thus, effects of taurine on sarcoplasmic reticular Ca^{2+} uptake must be indirect, involving processes such as alterations in protein phosphorylation, the lipid-protein microenvironment and the calcium load on the sarcoplasmic reticulum.

Protein-Taurine and Phospholipid-Taurine Interactions

There seems little doubt that taurine is capable of interacting with both proteins and phospholipids. At least two taurine binding proteins have been identified in rat heart sarcolemma[77]. These proteins can be distinguished on the basis of binding characteristics and chromatographic properties[11,78]. One of the binding proteins was found by reconstitution studies to be identical to the taurine transport protein[47]. The other protein was proposed to mediate some of the actions of taurine[11]. Since these initial observations, several proteins have been identified which could modulate taurine function, including calmodulin[41,79,80], the insulin receptor[44,81] and the α-adrenergic receptor[45]. Yet, in each case definitive studies tying these proteins to the actions of taurine have been lacking, By contrast, there is little doubt that taurine's interaction with the muscle proteins contributes to its positive inotropic effect[22,23].

Considerable attention has been devoted to the interaction between taurine and phospholipids[82-86]. Sebring and Huxtable[82] found that taurine binding to phospholiposomes involves a low affinity, cooperative interaction primarily with neutral phospholipids. Because the characteristics of this interaction in phospholiposomes is similar to the pattern described by Kulakowski et al.[77] for the binding of taurine to isolated sarcolemma, Huxtable and Sebring[83] concluded that the association of taurine with the cell membrane must involve a taurine-phospholipid interaction rather than a protein-taurine interaction. This was considered significant because the low affinity taurine binding protein of rat heart sarcolemma had been implicated in the calcium modulating actions of taurine[11]. In support of this notion, taurine was found to regulate calcium binding to both phospholiposomes and isolated rat heart sarcolemma[61,84]. Based on these observations, Huxtable and Sebring[85] proposed that taurine serves as a membrane expander, forming ion-ion interactions with the zwitterionic head group portion of the phospholipids. The resulting charge redistribution causes a membrane conformational change, opening up new Ca^{2+} binding sites on the membrane and increasing the affinity of available sites for Ca^{2+}. These changes also were envisioned to impact other membrane functions, such as transport, receptor activity, membrane phosphorylation and activity of membrane-bound enzymes[86]. However, this idea has been challenged by some investigators. Contrary to the hypothesis of Huxtable and Sebring[85], taurine does not cause gross structural changes within the membrane[61]. Second, it is unclear if the taurine-mediated changes in sarcolemmal calcium binding translate into alterations in calcium transport. Third, the taurine-phospholipid bond is very weak and would be susceptible to displacement by other ions found *in vivo*. Finally, attempts to attribute taurine's activity solely to the phospholipid-taurine interaction ignores data suggesting that taurine can affect the activity of specific proteins.

A hypothesis which recognizes the importance of both phospholipids and proteins in the actions of taurine recently has emerged. With the identification of phospholipid methyltransferase as a site of taurine action, a model can be formulated taking into account the structural similarities between taurine and the neutral phospholipids. Figure 5 illustrates that similar ionic bonds can form between taurine (Figure 5B) or the head group of phosphatidylethanolamine (Figure 5A) and the charged groups on the enzyme. In the case of phospholipid methyltransferase, the binding of taurine to this site decreases the concentration of both free enzyme and the enzyme-phosphatidylethanolamine complex, leading to competitive inhibition. However, this model has wider implications. Most membrane proteins, enzymes and receptors are capable of binding the phospholipids. Many of these proteins are very sensitive to changes in the phospholipid bilayer adjacent to the protein. The cartoon in Figure 5C depicts a membrane protein which normally binds phosphatidyl-ethanolamine in the region of the headgroup, as well as in its nonpolar region. Displacement of phosphatidylethanolamine from its binding site by taurine could lead

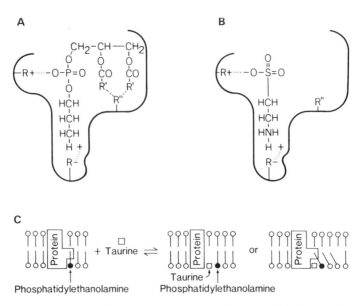

Figure 5. Model depicting the binding of taurine to the protein-phospholipid site on the surface of the membrane. Panel A. Binding site for phosphatidylethanolamine; Panel B. Binding of taurine to the phosphatidylethanolamine site; Panel C. Distortion of the membrane-bound protein or phospholipid bilayer by the displacement of phosphatidylethanolamine from its binding site by taurine.

to either distortion of the protein (because of an alteration in the binding of the nonpolar regions of the phospholipid with the protein) or a distortion of the phospholipid bilayer. Both effects would alter membrane function. Reconstitution studies using proteins which are very sensitive to the microenvironment of the neutral phospholipids may serve as a means of testing the validity of this new hypothesis.

REFERENCES

1. P.D. Pion, M.D. Kittleson, Q.R. Rogers and J.G. Morris, Myocardial failure in cats associated with low plasma taurine: A reversible cardiomyopathy, *Science* 237:764-768 (1987).
2. P.D. Pion, M.D. Kittleson, N.L. Skiles, Q.R. Rogers, and J.G. Morris, Taurine deficiency myocardial failure: incidence and relation to tissue taurine concentration, this book (1992).
3. W.O. Read and J.D. Welty, Effect of taurine on epinephrine and digoxin induced irregularities of the dog heart, *J. Pharmacol. Exp. Therap.* 139:283-289 (1963).
4. W.O. Read and J.D. Welty, Taurine as a regulator of cell potassium in the heart, *in*: "Electrolytes and Cardiovascular Diseases," E. Bajusz, ed, S. Karger, Basel/New York, 70-85 (1965).
5. E.I. Chazov, L.S. Malchikova, N.V. Lipina, G.B. Asafov and V.N. Smirnov, Taurine and electrical activity of the heart, *Circ. Res.* 34/35 SIII:III-11 - III-21 (1974).
6. G.X. Wang, J. Duan, S. Zhou, P. Li and Y. Kang, Antiarrhythmic action of taurine, this book (1992).
7. K. Takahashi, J. Azuma, N. Awata, A. Sawamura, S. Kishomoto, T. Yamagami, T. Kishi, H. Harada and S.W. Schaffer, Protective effect of taurine on the irregular beating pattern of cultured myocardial cells induced by high and low extracellular calcium ion, *J. Mol. Cell Cardiol.* 20:397-403 (1988).
8. N. Lake, M. de Roode and S. Nattel, Effects of taurine depletion on rat cardiac electrophysiology: In vivo and in vitro studies, *Life Sci.* 40:997-1005 (1987)..
9. J. Dietrich and J. Diacono, Comparison between ouabain and taurine effects on isolated rat and guinea-pig hearts in low calcium medium, *Life Sci.* 10:499-507 (1971).
10. S.W. Schaffer, J. Chovan and R.F. Werkman, Dissociation of cyclic AMP changes and myocardial contractility in taurine-perfused rat heart, *Biochem. Biophys. Res. Commun.* 81:248-253 (1978).
11. J.P. Chovan, E.C. Kulakowski, S. Sheakowski and S.W. Schaffer, Calcium regulation by the low-affinity taurine binding sites of cardiac sarcolemma, *Mol. Pharmacol.* 17:295-300 (1980).
12. M. Endoh, K. Ohkubo, H. Kushida and T. Hiramoto, Modulation of myocardial contractility by taurine: Absence of its interactions with the effects of low $[Ca^{2+}]_o$, verapamil, Bay K 8644, and α-and ß-adrenoceptor agonists in the rabbit papillary muscle, *in*: "Taurine and the Heart," H. Iwata, J.B. Lombardini and T. Segawa, eds, Kluwer Academic Publishers, Boston/Dordrecht/London, 51-73 (1989).
13. J.C. Khatter, P.L. Soni, R.J. Hoeschen, L.E. Alto and N.S. Dhalla, Subcellular effects of taurine on guinea pig heart, *in*: The Effects of Taurine on Excitable Tissues," S.W. Schaffer, S.I. Baskin and J.J. Kocsis, eds, Spectrum Press, New York, 281-293 (1981).
14. R. Bandinelli, F. Franconi, A. Giotti, F. Martini, G. Moneti, I. Stendardi and L. Zilletti, The positive inotropic effect of taurine and calcium and the levels of taurine in ventricular strips, *Br. J. Pharmacol.* 67:115-116P (1981).
15. A. Sawamura, J. Azuma, H. Harada, H. Hasegawa, K. Ogura, N. Sperelakis and S. Kishimoto, Protection by oral pretreatment with taurine against the negative inotropic effects of low-calcium medium on isolated perfused chick heart, *Cardiovasc. Res.* 27:620-626 (1983).
16. S.W. Schaffer, J. Kramer and J.P. Chovan, Regulation of calcium homeostasis in the heart by taurine, *Fed. Proc.* 39:2691-2694 (1980).
17. F. Franconi, F. Martini, I. Stendardi, R. Matucci, L. Zilletti and A. Giotti, Effect of taurine on calcium levels and contractility in guinea-pig ventricular strips, *Biochem. Pharmacol.* 31:3181-3185 (1982).
18. M.S. Mozaffari, B.H. Tan, M.A. Lucia and S.W. Schaffer, Effect of drug-induced taurine depletion on cardiac contractility and metabolism, *Biochem. Pharmacol.* 35:985-989 (1986).
19. S.W. Schaffer, S. Allo and M. Mozaffari, Potentiation of myocardial ischemic injury by drug-induced taurine depletion, *in*: "The Biology of Taurine," R.J. Huxtable, F. Franconi and A. Giotti, eds, Plenum Press, New York/London, 151-158 (1987).
20. N. Lake, J.B. Splawinski, C. Juneau and J.L. Rouleau, Effects of taurine depletion on intrinsic contractility of rat ventricular papillary muscles, *Can. J. Physiol. Pharmacol.* 68:800-806 (1990).
21. J. Azuma, H. Katsume, T. Kagoshima, K. Furukawa, N. Awata, T. Ishiyama, T. Yamagami, H. Ishikawa, H. Iwata, S. Kishimoto and Y. Yamamura, Clinical evaluation of taurine in congestive heart failure-A double blind comparative study using CoQ_{10} as a control drug, *in*: "Taurine and the Heart," H. Iwata, J.B. Lombardini and T. Segawa, eds, Kluwer Academic Press, Boston/Dordrecht/London, 75-97 (1989).

22. D.S. Steele, G.L. Smith and D.J. Miller, The effects of taurine on Ca^{2+} uptake by the sarcoplasmic reticulum and Ca^{2+} sensitivity of chemically skinned rat heart, *J. Physiol.* 422:499-511 (1990).

23. S. Galler, C. Hutzler and T. Haller, Effects of taurine on Ca^{2+}-dependent force development of skinned muscle fibre preparations, *J. Exp. Biol.* 152:255-264 (1990).

24. R.A. Chapman and J. Tunstall, The calcium paradox of the heart, *Prog. Biophys. Molec. Biol.* 50:67-96 (1987).

25. M. Borgers, The role of calcium in the toxicity of the myocardium, *Histochem. J.* 13:839-848 (1981).

26. J.H. Kramer, J.P. Chovan and S.W. Schaffer, The effect of taurine on calcium paradox and ischemic heart failure, *Am. J. Physiol.* 240:H238-H246 (1981).

27. K. Yamauchi-Takihara, J. Azuma, S. Kishimoto, S. Onishi and N. Sperelakis, Taurine prevention of calcium paradox-related damage in cardiac muscle, *Biochem. Pharmacol.* 37:2651-2658 (1988).

28. H. Ohta, J. Azuma, N. Awata, T. Hamaguchi, Y. Tanaka, A. Sawamura, S. Kishimoto and N. Sperelakis, Mechanism of the protective action of taurine against isoprenaline induced myocardial damage, *Cardiovasc. Res.* 22:407-413 (1988).

29. T. Hamaguchi, J. Azuma, N. Awata, H. Ohta, K. Takihara, H. Harada, S. Kishimoto and N. Sperelakis, Reduction of doxorubicin-induced cardiotoxicity in mice by taurine, *Res. Commun. Chem. Pathol. Pharmacol.* 59:21-30 (1988).

30. Y. Ihara, K. Takahashi, H. Harada, A. Sawamura, S.W. Schaffer and J. Azuma, Taurine attenuates contracture induced by low Na^+, high Ca^{2+} medium in perfused chick heart, this book (1992).

31. F. Franconi, I. Stendardi, P. Failli, R. Matucci, C. Baccaro, L. Montorsi, R. Bandinelli and A. Giotti, The protective effects of taurine on hypoxia (performed in the absence of glucose) and on reoxygenation (in the presence of glucose) in guinea-pig heart, *Biochem. Pharmacol.* 34:2611-2615 (1985).

32. A. Sawamura, N. Sperelakis and J. Azuma, Protective effect of taurine against decline of cardiac slow action potentials during hypoxia, *Eur. J. Pharmacol.* 120:235-239 (1986).

33. N. Awata, J. Azuma, T. Hamaguchi, Y. Tanaka, H. Ohta, K. Takihara, H. Harada, A. Sawamura and S. Kishimoto, Acute haemodynamic effect of taurine on hearts in vivo with normal and depressed myocardial function, *Cardiovasc. Res.* 21:241-247 (1987).

34. K. Takihara, J. Azuma, N. Awata, H. Ohta, T. Hamaguchi, A. Sawamura, Y. Tanaka, S. Kishimoto and N. Sperelakis, Beneficial effect of taurine in rabbits with chronic congestive heart failure, *Am. Heart. J.* 112:1278-1284 (1986).

35. J. Azari, P. Brumbaugh and R. Huxtable, Prophylaxis by taurine in the hearts of cardiomyopathic hamsters, *J. Mol. Cell. Cardiol.* 12:1353-1366 (1980).

36. M.J. McBroom and J.D. Welty, Effects of taurine on heart calcium in the cardiomyopathic hamster, *J. Mol. Cell. Cardiol.* 9:853-858 (1977).

37. H. Harada, B.J. Cusack, R.D. Olson, W. Stroo, J. Azuma, T. Hamaguchi and S.W. Schaffer, Taurine deficiency and doxorubicin: Interaction with the cardiac sarcolemmal calcium pump, *Biochem. Pharmacol.* 39:745-751 (1990).

38. J. Azuma, A. Sawamura, N. Awata, H. Hasegawa, K. Ogura, H. Harada, H. Ohta, K. Yamauchi, S. Kishimoto, T. Yamagami, E. Ueda and T. Ishiyama, Double-blind randomized crossover trial of taurine in congestive heart failure, *Curr. Therap. Res.* 34:543-557 (1983).

39. S. Fujimoto, H. Iwata and Y. Yoneda, Effect of taurine on responses to noradrenaline, acetylcholine and ouabain in isolated auricles from digitalized guinea pigs, *Japan J. Pharmacol.* 26:105-110 (1976).

40. M.C. Welty, J.D. Welty and M.J. McBroom, Effect of isoproterenol and taurine on heart calcium in normal and cardiomyopathic hamsters, *J. Mol. Cell Cardiol.* 14:353-357 (1982).

41. S.W. Schaffer, J.H. Kramer, W.G. Lampson, E. Kulakowski and Y. Sakane, Effect of taurine on myocardial metabolism: Role of calmodulin, in: "Sulfur Amino Acids: Biochemical and Clinical Aspects," K. Kuriyama, R.J. Huxtable and H. Iwata, eds, Alan R. Liss, Inc., New York, 39-50 (1983).

42. L.S. Malchikova and E.P. Elizarova, Taurine and cAMP content in the heart, *Medicina* (Moscow) 1:85-89 (1981).

43. W.G. Lampson, J.H. Kramer and S.W. Schaffer, Potentiation of the actions of insulin by taurine, *Can. J. Physiol. Pharmacol.* 61:457-463 (1983).

44. E.C. Kulakowski, J. Maturo and S.W. Schaffer, The low affinity taurine-binding protein may be related to the insulin receptor, in: "Taurine Biological Actions and Clinical Perspectives," S.S. Oja, L. Ahtee, P. Kontro and M.K. Paasonen, eds, Alan R. Liss, Inc., New York, pp. 127-136 (1985).

45. F. Franconi, F. Bennardini, R. Matucci, I. Stendardi, P. Failli, S. Manzini and A. Giotti, Functional and binding evidence of taurine inhibition of α-adrenoceptor effects on guinea-pig ventricle, *J. Mol. Cell Cardiol.* 18:461-468 (1986).

46. T. Vislie and K. Fugelli, Cell volume regulation in flounder (Platichthkys flesus) heart muscle accompanying an alteration in plasma osmolality, *Comp. Biochem. Physiol.* 52A:415-418 (1975).

47. S.W. Schaffer, E.C. Kulakowski and J.H. Kramer, Taurine transport by reconstituted membrane vesicles, in: "Taurine in Nutrition and Neurology," R.J. Huxtable and H. Pasantes-Morales, eds, Plenum Press, New York/London, pp. 143-160 (1982).

48. M. Atlas, J.J. Bahl, W. Roeske and R. Bressler, *In vitro* osmoregulation of taurine in fetal mouse hearts, *J. Mol. Cell Cardiol.* 16:311-320 (1984).

49. J.H. Thurston, R.E. Hauhart and E.F. Naccarato, Taurine: Possible role in osmotic regulation of mammalian heart, *Science* 214:1373-1374 (1981).

50. M.F. Crass III and J.B. Lombardini, Loss of cardiac muscle taurine after acute left ventricular ischemia, *Life Sci.* 21:951-958 (1977).

51. H. Harada, S. Allo, N. Viyuoh, J. Azuma, K. Takahashi and S.W. Schaffer, Regulation of calcium transport in drug-induced taurine-depleted hearts, *Biochim. Biophys. Acta* 944:273-278 (1988).

52. S.W. Schaffer, S. Allo, H. Harada, W. Stroo, J. Azuma and T. Hamaguchi, Mechanism underlying the membrane stabilizing activity of taurine, *in*: "Taurine and the Heart," H. Iwata, J.B. Lombardini and T. Segawa, eds, Kluwer Academic Publishers, Boston/Dordrecht/London, pp. 43-50 (1989).

53. J.B. Lombardini and S.M. Liebowitz, Taurine modifies calcium ion uptake and protein phosphorylation in rat heart, *in*: "Taurine and the Heart," H. Iwata, J.B. Lombardini and T. Segawa, eds, Kluwer Academic Publishers, Boston/Dordrecht/London, pp. 117-137 (1989).

54. Y-P Li and J.B. Lombardini, Taurine inhibits protein kinase C-catalyzed phosphorylation of specific proteins in a rat cortical P_2 fraction, *J. Neurochem.* 56:1747-1753 (1991).

55. S.S. Smith and J. Li, GABA receptor stimulation by baclofen and taurine enhances excitatory amino acid induced phosphatidylinositol turnover in neonatal rat cerebellum, *Neurosci. Lett.* 132:59-64 (1991).

56. K. Kuriyama, T. Hashimoto, M. Kimoti, Y. Nakamura and S-I Yamamoto, Taurine and receptor mechanisms in the heart: Possible correlates with the occurrence of ischemic myocardial damages, *in*: "Taurine and the Heart," H. Iwata, J.B. Lombardini and T. Segawa, eds, Kluwer Academic Publishers, Boston/Dordrecht/London, pp. 139-158 (1989).

57. T. Hamaguchi, J. Azuma and S. Schaffer, Interaction of taurine with methionine: Inhibition of myocardial phospholipid methyltransferase, *J. Cardiovasc. Pharmacol.* 18:224-230 (1991).

58. M.P. Gupta, V. Panagia and N.S. Dhalla, Phospholipid N-methylation-dependent alterations of cardiac contractile function by L-methionine, *J. Pharmacol. Exp. Therap.* 245:664-672 (1988).

59. P. Dolara, A. Agresti, A. Giotti and G. Pasquini, Effect of taurine on calcium kinetics of guinea-pig heart, *Eur. J. Pharmacol.* 24:352-358 (1973).

60. P. Dolara, A. Agresti, A. Giotti and E. Sorace, The effect of taurine on calcium exchange of sarcoplasmic reticulum of guinea pig heart studied by means of dialysis kinetics, *Can. J. Physiol. Pharmacol.* 54:529-533 (1976).

61. J.P. Chovan, E.C. Kulakowski, B.W. Benson and S.W. Schaffer, Taurine enhancement of calcium binding to rat heart sarcolemma, *Biochim. Biophys. Acta* 551:129-136 (1979).

62. K. Takahashi, H. Harada, S.W. Schaffer and J. Azuma, Effect of taurine on intracellular calcium dynamics of cultured myocardial cells during the calcium paradox, this book.

63. G. Jasmin and L. Proshek, Hereditary polymyopathy and cardiomyopathy in the Syrian hamster: I. Progression of heart and skeletal muscle lesions in the UM-X7.1 line, *Muscle Nerve* 5:20-25 (1982).

64. M. Bond, A-R Jaraki, C.H. Disch and B.P. Healy, Subcellular calcium content in cardiomyopathic hamster hearts in vivo: An electron probe study, *Circ. Res.* 64:1001-1012 (1989).

65. J. Wikman-Coffelt, T. Stefenelli, S.T. Wu, W.W. Parmley and G. Jasmin, $[Ca^{2+}]$ transients in the cardiomyopathic hamster heart, *Circ. Res.* 68:45-51 (1991).

66. J.T. Whitmer, P. Kumar and R.J. Solaro, Calcium transport properties of cardiac sarcoplasmic reticulum from cardiomyopathic Syrian hamsters (BIO 53.58 and 14.6): Evidence for a quantiative defect in dilated myopathic hearts not evident in hypertrophic hearts, *Circ. Res.* 62:81-85 (1988).

67. N. Makino, G. Jasmin, R.E. Beamish and N.S. Dhalla, Sarcolemmal Na^+-Ca^{2+} exchange during the development of genetically determined cardiomyopathy, *Biochem. Biophys. Res. Commun.* 133:491-497 (1985).

68. S.M. Factor, S. Cho, J. Scheuer, E.H. Sonnenblick and A. Malhotra, Prevention of hereditary cardiomyopathy in the Syrian hamster with chronic verapamil therapy, *J. Am. Coll. Cardiol.* 12:1599-1604 (1988).

69. J.M. Capasso, E.H. Sonnenblick and P. Anversa, Chronic calcium channel blockade prevents the progression of myocardial contractile and electrical dysfunction in the cardiomyopathic Syrian hamster, *Circ. Res.* 67:1381-1393 (1990).

70. M.J. McBroom and J.D. Welty, Effects of taurine and verapamil on heart calcium in hamsters and rats, *Comp. Biochem. Physiol.* 82C:279-281 (1985).

71. J.D. Welty and M.C. Welty, Effects of taurine on subcellular calcium dynamics in the normal and cardiomyopathic hamster heart, *in*: "The Effects of Taurine on Excitable Tissues," S.W. Schaffer, S.I. Baskin and J.J. Kocsis, eds, Spectrum Publications, New York, pp. 295-312 (1981).

72. S.W. Schaffer, P. Punna, J. Duan, H. Harada, T. Hamaguchi and J. Azuma, Mechanism underlying physiological modulation of myocardial contraction by taurine, this book (1992).

73. A. Sawamura, N. Sperelakis, J. Azuma and S. Kishimoto, Effects of taurine on the electrical and mechanical activities of embryonic chick heart, *Can. J. Physiol. Pharmacol.* 64:649-655 (1986).

74. N. Sperelakis, T. Yamamoto, G. Bkaily, H. Sada, A. Sawamura and J. Azuma, Taurine effects on action potentials and ionic currents in chick myocardial cells, *in*: "Taurine and the Heart," H. Iwata, J.B. Lombardini and T. Segawa, eds, Kluwer Academic Publications, Boston/Dordrecht/London, pp. 1-19 (1989).

75. M.L. Entman, E.P. Bornet and R. Bressler, The effect of taurine on sarcoplasmic reticulum, *Life Sci.* 21:543-550 (1977).

76. M.A. Remtulla, S. Katz and D.A. Applegarth, Effect of taurine on ATP-dependent calcium transport in guinea-pig cardiac muscle, *Life Sci.* 23:383-390 (1978).

77. E. Kulakowski, J. Maturo and S.W. Schaffer, Identification of taurine binding receptors from rat heart sarcolemma, *Biochem. Biophys. Res. Commun.* 80:936-941 (1978).

78. E.C. Kulakowski, J. Maturo and S.W. Schaffer, Solubilization and characterization of cardiac sarcolemmal taurine binding proteins, *Arch. Biochem. Biophys.* 210:204-209 (1981).

79. T. Segawa, Y. Nomura and I. Shimazaki, Possible involvement of calmodulin in modulatory role of taurine in rat cerebral ß-adrenergic neurons, *in*: "Taurine: Biological Actions and Clinical Perspectives," S.S. Oja, L. Ahtee, P. Kontro and M.K. Paasonen, eds, Alan R. Liss, Inc, New York, pp. 321-330 (1985).

80. S.W. Schaffer, S. Allo, H. Harada and J. Azuma, Regulation of calcium homeostasis by taurine: Role of calmodulin, *in*: "Taurine: Functional Neurochemistry, Physiology and Cardiology," H. Pasantes-Morales, D.L. Martin, W. Shain and R.M. Del Rio, eds, Wiley-Liss, New York, pp. 217-225 (1990).

81. J. Maturo and E.C. Kulakowski, Insulin-like activity of taurine, *in*: "The Biology of Taurine: Methods and Mechanisms," R.J. Huxtable, F. Franconi and A. Giotti, eds, Plenum Press, New York, pp. 217-226 (1987).

82. L.A. Sebring and R.J. Huxtable, Low affinity binding of taurine to phospholiposomes and cardiac sarcolemma, *Biochim. Biophys. Acta* 884:559-566 (1986).

83. R.J. Huxtable and L.A. Sebring, Taurine and the heart: The phospholipid connection, *in*: "Taurine and the Heart," H. Iwata, J.B. Lombardini and T. Segawa, eds, Kluwer Academic Publishers, Boston/Dordrecht/London, pp. 31-42 (1989).

84. L.A. Sebring and R.J. Huxtable, Taurine modulation of calcium binding to cardiac sarcolemma, *J. Pharmacol. Exp. Therap.* 232:445-451 (1985).

85. R.J. Huxtable and L.A. Sebring, Towards a unifying theory for the action of taurine, *Trends Pharmacol. Sci.* 7:481-485 (1986).

86. R.J. Huxtable, From heart to hypothesis: A mechanism for the calcium modulatory actions of taurine, *in*: "The Biology of Taurine: Methods and Mechanisms," R.J. Huxtable, F. Franconi and A. Giotti, eds, Plenum Press, New York, pp. 371-387 (1987).

SARCOLEMMAL ACTIONS OF TAURINE LINKED TO ALTERED PHOSPHOLIPID N-METHYLATION

Tomoyuki Hamaguchi[1,2], Junichi Azuma[1] and Stephen Schaffer[2]

[1]Department of Medicine III
Osaka University Medical School
Osaka, Japan

[2]Department of Pharmacology
University of South Alabama School of Medicine
Mobile, AL USA

INTRODUCTION

Taurine (2-aminoethanesulfonic acid) is found in very high concentrations in mammalian heart[1]. There has been numerous studies indicating that taurine administration attenuates the degree of myocardial injury commonly associated with various models of heart failure[2-5]. It has been proposed that the mechanism underlying the cardioprotective activity of taurine relates to its prevention of myocardial calcium overload. The means by which taurine modulates calcium, however, has not been elucidated. One theory which has been proposed as a unifying hypothesis of taurine action, suggests that the effects of taurine are mediated through its interaction with membrane phospholipids. Structurally and chemically many similarities exist between taurine and the head groups of the neutral phospholipids, such as phosphatidylethanolamine and phosphatidylcholine[7]. In addition, a fairly linear relationship has been found between the ratio of phosphatidylethanolamine/ phosphatidylcholine in the synaptosomal P_2B fraction of developing rat brain and taurine content[8].

Phosphatidylethanolamine and phosphatidylcholine are major components of membrane phospholipids. Three successive N-terminal methylations convert membrane phosphatidylethanolamine to phosphatidylcholine in a variety of tissues, including the heart[9]. These reactions are reported to play an important role in the regulation of a number of calcium transporters, such as the sarcolemmal Na^+-Ca^{2+} exchanger[10]. Moreover, the activity of the enzyme involved in the formation of phosphatidylcholine is affected in a number of myocardial disease states[11-14].

In the present paper, we wish to propose that taurine modulates phosphatidylethanolamine N-methylation of cardiac sarcolemmal membrane. This newly described action of taurine could be a mechanism by which taurine stabilizes the sarcolemma and regulates tissue calcium homeostasis.

Taurine, Edited by J.B. Lombardini *et al*.
Plenum Press, New York, 1992

METHODS

Hearts from male Wistar rats (250-300 g) were used in the experiments. Enriched sarcolemma was prepared from hearts according to the method of Pitts[15]. As previously reported, the standard marker enzymes, Na^+-K^+ ATPase, Ca^{2+}-stimulated ATPase and cytochrome c oxidase, as well as oxalate-facilitated Ca^{2+} uptake, were examined to verify that the preparations lacked significant mitochondrial and sarcoplasmic reticular contamination.

Phospholipid methyltransferase activity in sarcolemmal membranes was assayed according to the method described by Panagia et al.[9] Three catalytic sites have been identified on the enzyme, each site exhibiting a different substrate specificity. Site I favors the accumulation of phosphatidyl-N-monomethylethanolamine and was assayed by incubating 0.5 mg sarcolemmal protein in 50 mM Tris-glycylglycine buffer (pH 8.0) containing 1 mM $MgCl_2$ and 0.1 μM S-adenosyl-L-[^{3}H-methyl] methionine for 30 min at 37°C. The formation of phosphatidyl-N,N-dimethylethanolamine is favored in site II, which was assayed by a 30 min incubation of 0.5 mg sarcolemmal protein with 50 mM phosphate buffer (pH 7.0, 37°C) containing 10 μM S-adenosyl-L-[^{3}H-methyl] methionine but no magnesium. The final reaction favoring phosphatidylcholine formation occurs in 50 mM glycine buffer (pH 10.0, 37°C) containing 150 μM S-adenosyl-L-[^{3}H-methyl] methionine and 0.5 mg sarcolemmal protein and has been designated catalytic site III. In this study, reactions at all three sites were carried out in the presence of either 10 mM taurine or 10 mM sucrose, where sucrose was used to mimic the osmolar condition of taurine. Thirty minutes following the initiation of each N-methylation step, the reaction was terminated by addition of 3 ml chloroform/methanol/2N HCl (6:3:1 by volume). The lipids were extracted according to the procedure of Ganguly et al.[11] and radioactive phospholipid content determined.

Subsequently, the effect of taurine on the methylation reaction of sarcolemmal phospholipids was evaluated in hearts perfused on a standard working heart apparatus[17]. The perfusion medium consisted of Krebs-Henseleit buffer, pH 7.4, supplemented with 5 mM glucose, 2.5 U/l insulin and when desired either 300 μM methionine, 10 mM taurine or 300 μM methionine plus 10 mM taurine. In the cell, methionine serves as a substrate for the formation of the methyl donor, S-adenosyl-L-methionine, which cannot be used in the perfusion studies because it does not enter the cell readily[18]. All hearts were paced at 300 beats/min and perfused for a 15 min stabilization period prior to initiation of the experiment. Aortic pressure was monitored with a Statham P23Gb pressure transducer placed above the aorta. Cardiac work was calculated from coronary flow, aortic output and aortic pressure as described by Neely et al.[17]

In one group of studies, the hearts were initially perfused with normal Krebs-Henseleit buffer with or without 10 mM taurine for 15 min followed by perfusion with buffer supplemented with 300 μM L-[^{3}H-methyl]-methionine. When examining the rate of phospholipid methylation, perfusion with radioactive methionine was terminated after 5 min and N-methylated phospholipids were extracted from ventricular homogenates according to the method of Ganguly et al.[11] The extent of N-methylation was determined from the amount of tritium incorporated into the phospholipids. Similar studies were performed to examine the effect of taurine on methionine uptake by the heart. In those studies, hearts were perfused with radioactive methionine for 30 min. After washing the buffer out of the chamber of the heart, the ventricles were homogenized, sonicated and radioactivity counted.

Four experimental conditions were used to monitor changes in calcium transport activity. Control hearts were perfused for 25 min with standard Krebs-Henseleit buffer containing or lacking 10 mM taurine. A methionine group consisted of hearts perfused for 15 min with standard buffer followed by 10 min with standard buffer containing 300 μM L-methionine, while a fourth group of hearts was perfused for 15 min with both standard buffer containing 10 mM taurine followed by 10 min with standard buffer containing 10 mM taurine and 300 μM L-methionine. Hearts were removed from the cannula following the 20 min perfusion and used in the isolation of sarcolemma.

Sarcolemmal ATP-dependent Ca^{2+} uptake was determined by preincubating membrane (5 μg) at 37°C for 10 min in 0.5 ml of 20 mM Tris-MOPS buffer (pH 7.0) containing 160 mM KCl and the required amount of $^{45}Ca^{2+}$ and EGTA to produce 2 μM free Ca^{2+} concentration[19]. Accumulation of Ca^{2+} was initiated by adding 5 mM Tris-ATP (pH 7.0). After 1 min the reaction was terminated by adding 3.0 ml ice cold 20 mM MOPS buffer (pH 7.4) containing 160 mM KCl and 1.0 mM $LaCl_3$ and immediately filtering through Millipore filters. The filters were washed three times with 3.0 ml of the $LaCl_3$-containing arrest buffer, dried and then counted. Nonspecific binding was measured in the absence of ATP and subtracted from total accumulation to yield uptake. The data were expressed as nmol Ca^{2+} uptake/mg protein/min. The isolated membranes were also used in the determination of Na^+-Ca^{2+} exchanger activity assayed according to the method of Reeves and Sutko[20]. Briefly, sarcolemmal vesicles were loaded with 160 mM Na^+ by incubation with buffer A (160 mM NaCl, 20 mM MOPS, pH 7.4) for 60 min at 37°C. The exchange reaction was initiated by addition of 10 μg of membrane protein to 500 μl of 20 mM MOPS buffer (pH 7.4) containing 10 μM $^{45}Ca^{2+}$ and 160 mM KCl. After 1 sec the reaction was arrested by addition of 3 ml ice cold 20 mM MOPS buffer (pH 7.4) containing 160 mM KCl and 1 mM $LaCl_3$, followed by immediate filtration through Millipore filters. The filters were washed three times with the arresting buffer and then dried before counting for radioactivity. Nonspecific binding was assessed by replacing the reaction medium with buffer A, which was similar to the vesicle loaded buffer containing 160 mM KCl. In other words, nonspecific binding measures all membrane-associated $^{45}Ca^{2+}$ found in the absence of a sodium gradient across the membrane. The Na^+-Ca^{2+} exchange data were corrected for nonspecific Ca^{2+} binding and expressed as nmol Ca^{2+} accumulated/mg protein/sec. Sarcolemmal Na^+-K^+ ATPase activity was assayed according to the method of Makino et al[19].

Statistical analyses were performed by either Student's t test or analysis of variance followed by Bonferroni's method. A probability less than 0.05 was considered statistically significant.

RESULTS

Table 1 summarizes the activity of phospholipid methyltransferase for catalytic sites I, II and III of isolated sarcolemma incubated in medium containing or lacking 10 mM taurine. Inclusion of taurine in the reaction medium significantly decreased enzyme activity at all three catalytic sites. The greatest effect of taurine was to reduce the site II reaction by 35%, with the other two sites inhibited by about 30%.

Table 2 indicates the effect of taurine on uptake and incorporation of ^3H-methionine into myocardial phospholipids of the isolated heart. The rate of ^3H-methionine incorporation into the lipid fraction of the heart in the absence of taurine was 11.6 $\pm$ 0.2 nmol/g wet wt/hr. When the perfusion buffer contained taurine in addition to methionine, the rate of incorporation was reduced to 6.5 $\pm$ 0.8 nmol/g wet wt/hr. This effect was not caused by a taurine-mediated decrease in methionine uptake; total uptake of methionine by the isolated heart was 928 $\pm$ 126 nmol/g wet wt/hr in the absence of taurine and 913 $\pm$ 72 nmol/g wet wt/hr in hearts perfused with buffer containing taurine.

In agreement with Gupta et al.[18], we found that hearts perfused with buffer containing 300 μM L-methionine exhibited a decrease in myocardial function (Figure 1). Within 10 minutes following exposure to the methionine-containing buffer, cardiac work declined from 0.43 $\pm$ 0.01 to 0.37 $\pm$ 0.01 kg-m/g dry wt/min. Addition of 10 mM taurine to the methionine-containing buffer prevented the decrease in cardiac

Table 1. Effect of taurine on sarcolemmal phospholipid methyltransferase activity

Condition	Site I	Site II	Site III
		(pmol/mg protein/30 min)	
With sucrose	0.93 ± 0.03	7.42 ± 0.37	135 ± 4
With taurine	0.65 ± 0.05*	4.82 ± 0.50*	99 ± 10*

Activity of sarcolemmal phospholipid methyltransferase was assayed as described in METHODS. Catalytic site I activity refers to the reaction at 0.1 μM S-adenosylmethionine, which favors the conversion of phosphatidyl-ethanolamine to phosphatidyl-N-monomethylethanolamine. The formation of phosphatidyl-N-dimethylethanolamine occurs preferentially at catalytic site II in the presence of a moderate S-adenosylmethionine concentration of 10 μM. Phosphatidylcholine synthesis proceeds best with high S-adenosyl-methionine (150 μM) at catalytic site III. Values shown are means ± S.E.M. of 3-4 different assays. * denotes significant difference between activity in the presence of sucrose vs. taurine. Reproduced from Hamaguchi et al.[23] with permission of J. Cardiovasc. Pharmacol.

Table 2. Effect of taurine on methionine incorporation into cardiac tissue and phospholipids

Condition	Tissue	Phospholipids
	(nmol/g wet wt/hr)	
Methionine	928 ± 126	11.6 ± 0.2
Taurine plus Methionine	913 ± 72	6.5 ± 0.9*

Hearts were initially perfused with standard buffer for 15 min followed by perfusion with buffer supplemented with or without 10 mM taurine and containing 300 μM ^{3}H-methionine (Taurine plus Methionine or Methionine). Following the 20 min perfusion, the hearts were homogenized and then extracted with chloroform-methanol according to METHODS. The extent of N-methylation was determined from the amount of tritium incorporated into the extracted lipids. To examine the effect of taurine on methionine uptake by the heart, hearts were perfused with ^{3}H-methionine for 30 min. After washing the buffer out of the ventricles, hearts were homogenized, the homogenates sonicated and radioactivity of the extracts counted. Values shown represent means ± S.E.M. of 4-5 hearts. * denotes significant difference between Methionine and Taurine plus Methionine groups (p < 0.05). Reproduced from Hamaguchi et al.[23] with permission of J. Cardiovasc. Pharmacol.

function. Cardiac work was maintained at a level of 0.42 ± 0.02 kg-m/g dry wt/min in hearts perfused for 10 min with buffer containing both 10 mM taurine and 300 μM methionine while perfusion of hearts with 10 mM taurine alone had no effect on contractile function; cardiac work in the latter group was 0.43 ± 0.02 kg-m/g dry wt/min after a 10 min perfusion (Figure 1).

Because the effect of methionine on myocardial function is believed to involve changes in the activity of several key Ca^{2+} transporters[18], sarcolemmal membrane was isolated from hearts perfused in the presence or absence of 10 mM taurine and 300 μM methionine. Although perfusion of hearts with 300 μM methionine had no

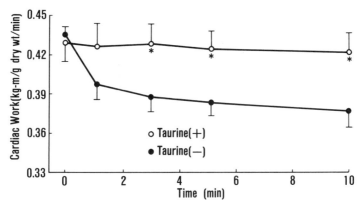

Figure 1. Effect of taurine on methionine-mediated negative inotropic effect. One group of hearts was perfused for 15 min with standard Krebs-Henseleit buffer followed by 10 min with buffer containing 300 μM methionine (O). The other group of hearts was perfused for 15 min with standard Krebs-Henseleit buffer supplemented with 10 mM taurine, followed by 10 min with buffer containing both 300 μM methionine and 10 mM taurine (O). Cardiac work was calculated from aortic pressure and cardiac output. Values represent means $\pm$ S.E.M. of 5-6 hearts. * denotes significant difference between taurine-treated and untreated groups. Reproduced from Hamaguchi et al.[23] with permission from J. Cardiovasc. Pharmacol.

apparent effect on sarcolemmal ATP-dependent Ca^{2+} transport or Na^+-K^+ ATPase activity, sarcolemmal Na^+-Ca^{2+} exchanger activity was 4-fold lower in membrane prepared from methionine-exposed hearts. Interestingly, taurine alone had no effect on the activity of the exchanger, but inclusion of 10 mM taurine in perfusion medium containing 300 μM methionine dramatically attenuated the methionine-linked loss in Na^+-Ca^{2+} exchanger activity (Table 3).

Table 3. Effect of taurine and methionine on sarcolemmal calcium transport

Group	Na^+-K^+ ATPase (μmol P_i/mg/hr)	ATP-dep. Ca^{2+} uptake (nmol/mg/min)	Na^+-Ca^{2+} Exchange (nmol/mg/sec)
Control	49.6 $\pm$ 4.3	22.1 $\pm$ 2.9	2.16 $\pm$ 0.4*
Methionine	50.8 $\pm$ 3.3	20.4 $\pm$ 2.2	0.50 $\pm$ 0.10
Taurine	50.3 $\pm$ 4.4	21.0 $\pm$ 1.8	2.08 $\pm$ 0.03*
Taurine + Methionine	48.9 $\pm$ 2.4	21.5 $\pm$ 2.6	1.52 $\pm$ 0.14*

Hearts were perfused with either standard Krebs-Henseleit buffer for 20 min (Control), 15 min with standard buffer followed by 5 min with standard buffer supplemented with 300 μM L-methionine (Methionine), 20 min with standard buffer supplemented with 10 mM taurine (Taurine) or 15 min with standard buffer containing 10 mM taurine followed by 5 min with standard buffer containing 300 μM L-methionine and 10 mM taurine (Taurine + Methionine). Following 20 min of perfusion, the hearts were used in the isolation of sarcolemma. Membrane Ca^{2+} transport activities were determined as described in METHODS. Values shown represent means $\pm$ S.E.M. of 4 hearts. * denotes significant difference from methionine group ($p < 0.05$). Reproduced from Hamaguchi et al.[23] with permission of J. Cardiovasc. Pharmacol.

DISCUSSION

Phospholipid methyltransferase catalyzes the incorporation of three methyl groups from S-adenosylmethionine into phosphatidylethanolamine, resulting in the formation of phosphatidylcholine. In this study, the rate of incorporation of ^{3}H-methyl groups from radioactive S-adenosylmethionine into sarcolemmal phospholipids in the absence of taurine was comparable to those reported previously[9]. However, in the presence of 10 mM taurine, N-methylation at all three sites of the methyltransferase enzyme was significantly inhibited. Because the intracellular taurine concentration of the heart is generally 10 mM or greater, taurine appears to be a physiological regulator of phospholipid N-methylation, thereby modulating the rate of phosphatidylcholine formation from phosphatidylethanolamine.

In addition to altering sarcolemmal phospholipid methyltransferase activity *in vitro* taurine also exerts inhibitory action on sarcolemmal phospholipid methyltransferase activity in perfused hearts. Gupta et al.[18] have shown that exposure of isolated rat hearts to methionine-containing buffer induces an initial negative inotropic effect lasting 10 min. In this study, 300 μM methionine caused a 15% decrease in cardiac work after 10 min which was blocked by taurine. Significantly, the ability of taurine to prevent the methionine-mediated negative inotropic effect correlated with its ability to attenuate the decrease in Na^+-Ca^{2+} exchanger activity observed in methionine-exposed hearts. In agreement with Gupta et al.[18], methionine-treated hearts exhibited only 25% of normal sarcolemmal Na^+-Ca^{2+} exchanger activity. By comparison, hearts perfused with buffer supplemented with both 10 mM taurine and 300 μM methionine contained approximately 75% of normal Na^+-Ca^{2+} exchanger activity.

The effects of membrane methylation on Na^+-Ca^{2+} exchanger activity have been attributed to both methylation of phospholipids and proteins. While Vermuri and Philipson[22] maintain that the methionine-mediated decrease in Na^+-Ca^{2+} exchanger activity is caused by the methylation of specific proteins in the membrane, Gupta et al.[18] have argued in favor of enhanced phospholipid methylation as the cause of the methionine effect. The present data are consistent with the involvement of phospholipid N-methylation in the regulation of Na^+-Ca^{2+} exchanger activity. Hearts perfused with buffer supplemented only with 300 μM ^{3}H-methionine led to a 4-fold drop in Na^+-Ca^{2+} exchanger activity associated with the incorporation of ^{3}H into the phospholipids at a rate of 11.6 $\pm$ 0.2 nmol/g wet wt/hr. By comparison, perfusion with buffer supplemented with both 300 μM ^{3}H-methionine and 10 mM taurine led to only a 23% decrease in sarcolemmal Na^+-Ca^{2+} exchanger activity and a rate of ^{3}H incorporation into myocardial phospholipids of 6.5 $\pm$ 0.9 nmol/g wet wt/hr. Thus, taurine appears to indirectly alter Na^+-Ca^{2+} exchanger activity by modulating the amount of phosphatidylethanolamine and phosphatidylcholine found in the sarcolemmal membrane.

The mechanisms underlying methionine-induced changes in heart function are complex. Figure 2 summarizes the participation of methionine and taurine in sarcolemmal phospholipid N-methylation of isolated heart. Methionine is taken up by the heart, where it is converted to S-adenosylmethionine. During the course of phosphatidylethanolamine methylation, the methyl group of S-adenosylmethionine is transferred from S-adenosylmethionine to the phospholipid, yielding one of the methylated phosphatidylethanolamine derivatives and S-adenosyl-L-homocysteine. Because phosphatidylethanolamine is required for maximal Na^+-Ca^{2+} exchanger activity, the methylation reaction leads to reduced exchanger activity. This in turn causes contractile function to fall as intracellular Ca^{2+} levels are affected. Taurine appears to antagonize methionine action on both Na^+-Ca^{2+} exchanger activity and contraction through inhibition of sarcolemmal phospholipid methyltransferase activity.

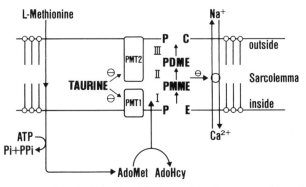

Figure 2. Schematic representation of effect of methionine and taurine on sarcolemmal phospholipid N-methylation. PMT = phospholipid methyltransferase; PE = phosphatidylethanolamine; PPME = phosphatidyl-N-monomethylethanolamine; PDME = phosphatidyl-N,N-dimethylethanolamine; PC = phosphatidylcholine; AdoMet = S-adenosyl-L-methionine; AdoHcy = S-adenosyl-L-homocysteine

Regulation of phospholipid N-methylation by taurine may explain the basis underlying the relationship between brain taurine levels and tissue phosphatidyl-choline/phosphatidyl-ethanolamine ratio. It also provides an explanation for the most widely accepted actions of taurine; namely, alterations of calcium transport and membrane stabilization.

ACKNOWLEDGMENTS

This work was supported in part by the Uehara Memorial Foundation.

REFERENCES

1. J.G. Jacobsen and L.H. Smith, Biochemistry and physiology of taurine and taurine derivatives, *Physiol. Rev.* 48:424-511 (1968).
2. M.J. McBroom and J.D. Welty, Effect of taurine on heart calcium in the cardiomyopathic hamster, *J. Mol. Cell Cardiol.* 9:853-858 (1977).
3. H. Ohta, J. Azuma, S. Ohnishi, N. Awata, K. Takihara and S. Kishimoto, Protective effect of taurine against isoprenaline-induced myocardial damage, *Basic Res. Cardiol.* 81:473-481 (1986).
4. T. Hamaguchi, J. Azuma, N. Awata, H. Ohta, K. Takihara, H. Harada, S. Kishimoto and N. Sperelakis, Reduction of doxorubicin-induced cardiotoxicity in mice by taurine, *Res. Commun. Chem. Pathol. Pharmacol.* 59:21-30 (1988).
5. K. Takihara, J. Azuma, N. Awata, H. Ohta, A. Sawamura, S. Kishimoto and N. Sperelakis, Taurine's possible protective role in age-dependent response to calcium paradox, *Life Sci.* 37:1705-1710 (1985).
6. J.H. Kramer, J.P. Chovan and S.W. Schaffer, Effect of taurine on calcium paradox and ischemic heart failure, *Am. J. Physiol.* 240:H238-H246 (1981).
7. R.J. Huxtable and L.A. Sebring, Cardiovascular actions of taurine, in: "Sulfur Amino Acids, K. Kuriyama, R.J. Huxtable and H. Iwata, eds, Alan R. Liss, New York, pp. 5-37 (1983).
8. R.J. Huxtable, S. Crosswell and D. Parker, Phospholipid composition and taurine content of synaptosomes in developing rat brain, *Neurochem. Int.* 15:233-238 (1989).
9. V. Panagia, P. Ganguly and N.S. Dhalla, Characterization of heart sarcolemmal phospholipid methylation, *Biochim. Biophys. Acta* 792:245-253 (1984).
10. V. Panagia, N. Makino, P.K. Ganguly and N.S. Dhalla, Inhibition of Na^+-Ca^{2+} exchange in heart sarcolemmal vesicles by phosphatidylethanolamine N-methylation, *Eur. J. Biochem.* 166:597-603 (1987).
11. P.K. Ganguly, K.M. Rice, V. Panagia and N. Dhalla, Sarcolemmal phosphatidylethanolamine in diabetic cardiomyopathy, *Circ. Res.* 55:504-512 (1984).
12. K. Okumura, V. Panagia, G. Jasmin and N.S. Dhalla, Sarcolemmal phospholipid N-methylation in

genetically determined hamster cardiomyopathy, *Biochem. Biophys. Res. Commun.* 27:31-37 (1987).

13.   K. Okumura, V. Panagia, R.E. Beamish and N.S. Dhalla, Biphasic changes in the sarcolemmal phosphatidylethanolamine N-methylation activity in catecholamine-induced cardiomyopathy, *J. Mol. Cell Cardiol.* 19:356-366 (1987).

14.   V. Panagia, K. Okumura, K.R. Stah and N.S. Dhalla, Modification of sarcolemmal phosphatidyl-ethanolamine N-methylation during heart hypertrophy, *Am. J. Physiol.* 253:H8-H15 (1987).

15.   B.J.R Pitts, Stoichiometry of sodium-calcium exchange in cardiac sarcolemmal vesicles, J. Biol. Chem. 254:6232-6235 (1979).

16.   H. Harada, S. Allo, N. Viyuoh, J. Azuma, K. Takahashi and S.W. Schaffer, Regulation of calcium transport in drug-induced taurine-depleted heart, *Biochim. Biophys. Acta* 944:273-278 (1988).

17.   J.R. Neely, H. Liebermeister, J. Battersby and H.E. Morgan, Effect of pressure development on oxygen consumption by isolated rat heart, *Am. J. Physiol.* 212:804-14 (1967).

18.   M.P. Gupta, V. Panagia and N.S. Dhalla, Phospholipid N-methylation-dependent alterations of cardiac contractile function by L-methionine, *J. Pharmacol. Exp. Therap.* 245:664-672 (1988).

19.   N. Makino, K.S. Dhalla, V. Elimban and N.S. Dhalla, Sarcolemmal Ca^{2+} transport in streptozotocin-induced diabetic cardiomyopathy in rats, *Am. J. Physiol.* 253:E202-E207 (1987).

20.   J.P. Reeves and J.L Sutko, Sodium-calcium ion exchange in cardiac membrane vesicles, *Proc. Natl. Acad. Sci. USA* 76:590-594 (1979).

21.   M.P. Gupta, N. Makino, K. Khatter and N.S. Dhalla, Stimulation of Na^+-Ca^{2+} exchange in heart sarcolemma by insulin, *Life Sci.* 39:1077-1083 (1986).

22.   R. Vemuri and K.D. Philipson, Protein methylation inhibits Na^+-Ca^{2+} exchange activity in cardiac sarcolemmal vesicles, *Biochim. Biophys. Acta* 939:503-508 (1988).

23.   T. Hamaguchi, J. Azuma and S. Schaffer, Interaction of taurine with methionine: Inhibition of myocardial phospholipid methyltransferase, *J. Cardiovasc. Pharmacol.* 18:224-230 (1991).

TAURINE EFFECTS ON IONIC CURRENTS IN
MYOCARDIAL CELLS

Nicholas Sperelakis[1], Hiroyasu Satoh, and
Ghassan Bkaily[2]

[1]Department of Physiology and Biophysics
University of Cincinnati College of Medicine
Cincinnati, OH 45267

[2]Department of Biophysics
University of Sherbrooke
Sherbrooke, Quebec, Canada

INTRODUCTION

Taurine has been reported to produce (a) a positive inotropic effect in heart muscle (Baskin & Finney, 1979; Dietrich & Diacono, 1971; Sawamura et al., 1983), (b) beneficial effects in treatment of congestive heart failure (Azuma et al., 1984), and (c) protective effects against calcium overload (Takihara et al., 1985, 1986) and cardiomyopathy (McBroom & Welty, 1977; Azari et al., 1980).

This section is a brief review of data previously published from our laboratory using conventional intracellular microelectrode recording techniques and recording of contractions. All data were on chick heart cells (old embryonic, and cultured cells prepared from young and old embryonic hearts) and on guinea pig papillary muscles.

Taurine exerts a positive inotropic effect in perfused embryonic (20-day-old) chick heart, without greatly affecting the slow Ca^{2+}-dependent action potentials (APs) accompanying the contractions, and it was suggested that the positive inotropic effect of taurine was not mediated through an increase in the slow inward Ca^{2+} current (Sawamura et al., 1986b). Taurine activated a fast transient component and blocked the slow sustained component of the TTX-insensitive (slow) Na^+ current in 3-day-old embryonic chick heart cells. Taurine stimulated a fast transient component of the inward Ca^{2+} current and inhibited a slow (sustained) component in both 3-day and 10-day-old embryonic cells (Sperelakis et al., 1989). In guinea pig myocytes, the effects of taurine on the Ca^{2+} slow current depended on the external Ca^{2+} concentration (Sawamura et al., 1990).

Taurine did not induce Ca^{2+}-dependent slow APs (Sawamura et al., 1986b), and did not elevate cyclic AMP level (Sawamura et al., 1983). Taurine (10 mM) also did not significantly increase the maximum rate of rise (max dV/dt), amplitude (APA), or duration (at 50% repolarization, APD_{50}) of the slow APs induced by isoproterenol

Taurine, Edited by J.B. Lombardini *et al.*
Plenum Press, New York, 1992

Table 1. Summary of the effects of taurine on the slow APs and concentration in isolated embryonic chick hearts perfused with 25 mM K^+ Tyrode's solution.

Induction of slow APs	± Taurine 10 mM	max dV/dt (V/s)	APA (mV)	APD_{50} (ms)	Developed tension
Isoproterenol (10^{-3} M)	Control	4.1 ± 0.6	58 ± 2	121 ± 12	100
	Taurine	4.0 ± 0.7	55 ± 3	106 ± 14	106 ± 18*
Histamine (10^{-4} M)	Control	3.3 ± 0.3	59 ± 2	144 ± 7	100
	Taurine	3.3 ± 0.4	55 ± 3	139 ± 10	129 ± 9*
TEA (10 mM)	Control	2.9 ± 0.3	64 ± 3	197 ± 22	100
	Taurine	2.8 ± 0.7	59 ± 3	238 ± 20*	146 ± 17*

Hearts isolated from 18-21 day embryos. The resting potentials in 25 mM $[K]_o$ were about -40 mV. Steady state was attained 7-13 min after addition of taurine. Values given are the mean ± S.E. * Statistically significant difference from control using paired t-test (p < 0.05). Data taken from Sawamura et al., 1986a.

or histamine (Table 1) (Sawamura et al., 1986b). In contrast, taurine exerted a significant positive inotropic effect (Table 1). On slow APs induced by tetraethylammonium (TEA, 10 mM) (to depress the outward K^+ current), taurine also had no effect on max dV/dt and APA; however, the APD_{50} was slightly, but significantly, increased (Table 1).

Taurine not only did not induce or stimulate slow APs, it actually exerted a transient depressant action on ongoing slow APs (Sawamura et al., 1986b). The transient depression of the slow APs occurred within 0.5 min and lasted for about 2-3 min; by 7 min, the response had spontaneously recovered back to the control levels. A summary of these data is given in Table 2.

Table 2. Transient depression of the slow APs by taurine in perfused embryonic (18-21 day) chick hearts and in cultured chick heart cells.

Slow AP Parameter	Control	Time After Taurine (10 mM)	
		0.5 - 2 min	7 - 13 min
A. Perfused hearts			
max dV/dt (%)	100%	63 ± 14*	94 ± 3
APA (%)	100%	66 ± 11*	94 ± 6
APD_{50} (%)	100%	66 ± 12*	100 ± 7
B. Cultured cells			
max dV/dt (V/s)	6.2 ± 1.0	4.7 ± 0.9*	6.2 ± 1.0
APA (mV)	54.2 ± 2.6	50.0 ± 3.3	52.4 ± 2.4
APD_{50} (ms)	70.6 ± 5.2	65.6 ± 5.0	77.0 ± 5.1

In A, taurine data expressed as percentage of the control values. Slow APs were induced by 10 mM TEA and elevated $[Ca]_o$ (4.5 mM). Data taken from Sawmura et al., 1986a.

130

There was an initial transient negative inotropic effect of taurine (over by 3 min), followed by a positive inotropic effect (still not maximal at 10 min). The contractions were slightly depressed in parallel with taurine depression of the slow APs. By 7 min, a clear positive inotropic effect was evident (Table 1). Therefore, these data suggested that the positive inotropic effect of taurine was not due to stimulation of the inward Ca^{2+} slow current ($I_{Ca(s)}$). There were no major differences in the positive inotropy of taurine related to the agent used to induce the slow APs and contractions (e.g., histamine or isoproterenol).

The positive inotropic effect of taurine (10 mM) also occurred in low $[Ca]_o$ (0.4 mM) solutions (Table 3). Table 3 also summarizes data on the normal fast APs, and shows that taurine did not significantly affect max dV/dt, APA, or APD_{50} of the fast APs and did not affect the resting potential.

Although taurine did not stimulate the slow APs under normal conditions, under hypoxic conditions taurine did exert a stimulant effect on guinea pig papillary muscle (Sawamura et al., 1986b). Hypoxia substantially depressed the slow APs within 5 min and blocked them by 20 min. Addition of 10 mM taurine partially restored the slow APs within 5 min. Taurine, when added in advance, was also able to protect against the depressant effect of hypoxia on the slow APs (Sawamura et al., 1986a). Since the functioning of the Ca^{2+} slow channels is dependent on metabolism (Sperelakis & Schneider, 1976) and is regulated by cyclic nucleotides and phosphorylation (Shigenobu & Sperelakis, 1972; Vogel & Sperelakis, 1981; Li & Sperelakis, 1983; Bkaily & Sperelakis, 1984; Sperelakis, 1984; Wahler & Sperelakis, 1985), this protective/restorative effect of taurine could be mediated indirectly by stimulating metabolism or phosphorylation or by a direct effect on one or more types of ion channels.

METHODS

Cell Isolation

Single heart cells were freshly isolated from 3-day, 10-day, and 17-day-old embryonic chick hearts (ventricles) by standard techniques, and used for whole-cell

Table 3. Steady-state effects of taurine on electrical and mechanical activities in isolated embryonic chick hearts perfused with low Ca^{2+} (0.4 mM) Tyrode's solution.

Activities	Control	Taurine (10 mM)
Electrical		
max dV/dt (V/s)	73.4 ± 3.8	70.8 ± 3.5
APA (mV)	97 ± 2	94 ± 2
APD_{50} (ms)	204 ± 10	199 ± 13
E_m (mV)	-80 ± 2	-77 ± 2
Mechanical		
Developed tension (%)	100	127 ± 12*

Hearts isolated from 18-21 day embryos. Steady-state was attained 7 - 13 min after addition of taurine. Values given are the mean ± S.E. *Statistically significant difference from control (P < 0.05). Data taken from Sawamura et al., 1986a.

voltage clamp (Satoh and Sperelakis, 1991b). The single ventricular cells were round. Single cells were dispersed from guinea pig ventricles and from rabbit aorta using collagenase (0.2%) and pronase (0.028%) in a HEPES-buffered saline solution.

Whole-cell voltage clamp recording

Whole-cell voltage clamp recordings were made using an Axopatch patch-clamp amplifier and standard techniques (Hamill et al., 1981). Patch pipettes were fabricated using a two-stage puller, and they had resistances of 1-3 MΩ. Experiments were carried out at room temperature (22° C). The amplitude of I_{Na} was determined as the difference between the peak current and the steady-state current level (at 80 ms). The amplitudes of I_{Ca} and I_K were determined as the difference between the peak current and zero current level. The data were stored and analyzed on an IBM-AT microcomputer using the PCLAMP analysis program. The current traces were filtered with a cut-off frequency of 2 KHz for plotting. All values given are means $\pm$ SEM.

Experimental Solutions

The cells were plated on glass coverslips and superfused with a modified Tyrode solution. The composition of this solution was (in mM): NaCl 137, KCl 5.4, CaCl$_2$ 0.1, MgCl$_2$ 1.0, NaH$_2$PO$_4$ 0.3, glucose 5.0, HEPES 5.0 (pH 7.4). To measure I_{Na}, the I_{Ca} and transient outward current (I_{to}) were blocked using CoCl$_2$ (3 mM) or MgCl$_2$ (2 mM) and 4-aminopyridine (4-AP) (3 mM) added to the bath solution. To examine I_{Ca}, TEA (10 mM) and 4-AP (3 mM) were added to the bath solution to block I_K. For I_K, Mn^{2+} (2 mM) was added to block I_{Ca}. Taurine (2-aminoethanesulfonic acid) was dissolved in the bath solution. The internal (pipette) solution for I_{Na} and I_{Ca} contained (in mM): CsOH 110, CsCl 20, MgCl$_2$ 2, EGTA 10, MgATP 5, creatine phosphate 5, aspartic acid 100, and HEPES 5 (pH 7.2). For I_K, the pipette solution contained (in mM): KOH 110, KCl 20, MgCl$_2$ 2, EGTA 10, MgATP 5, creatine phosphate 5, aspartic acid 150, and HEPES 5 (pH 7.2). The [Ca]$_i$ was adjusted to pCa 10 or pCa 7 according to the equations of Fabiato and Fabiato (1979) and Tsien and Rink (1980).

RESULTS

Ion Currents in Young and Old Embryonic Chick Heart Cells

The effects of taurine on the different ionic currents (slow Na$^+$, fast Na$^+$, slow Ca^{2+}, fast Ca^{2+}, and K$^+$) were studied in cultured single ventricular heart cells using whole-cell voltage clamp. The 3-day cells had resting potentials of -50 to -60 mV, and the slow-rising APs and the inward currents were sensitive to [Na]$_o$ and insensitive to TTX and Mn^{2+}. The slow Na$^+$ current ($I_{Na(s)}$) of 3-day cells was maintained for over 15 min without rundown. The 10-day and 17-day cells had resting potentials of -75 to -85 mV and fast-rising APs. The inward current responsible for this AP was blocked by TTX (Kojima & Sperelakis, 1984). I_{Ca} was maintained for up to 10 min without rundown, and was blocked by 2 mM Mn^{2+}.

Na^{2+} Current (TTX-insensitive)

The 3-day cells exhibited two types of I_{Na}: fast ($I_{Na(f)}$) and slow ($I_{Na(s)}$) (Table 4). The first type was activated from a holding potential (HP) of -80 mV using a relatively small depolarizing voltage step (VS), and showed fast transient activation and inactivation. The second type was activated from an HP of -80 or -50 mV, required larger voltage steps, and showed slower activation and inactivation. Some

Table 4. Summary of effect of taurine on single isolated embryonic chick (3-day, 10-day) heart cells

Current	Taurine (5 mM)
I_{Na} (TTX- and Mn^{2+}-insensitive	
Fast (transient) component	Stimulated
Slow (sustained) component	Unaffected (?)
I_{Na} (TTX-sensitive)	Inhibited (most cells)[+]
	Stimulated (some cells)
$I_{Ca(s)}$ (L-type)	Inhibited (normal $[Ca]_o$ & $[Ca]_i$)
	Stimulated (low $[Ca]_o$ & $[Ca]_i$)
$I_{Ca(f)}$ (T-type)	Stimulated

[+] Inhibited only (all cells) at 10-20 mM taurine.

cells showed both components, whereas other cells showed only one of the two types. We focused primarily on cells that exhibited only one type of I_{Na}.

In cells which showed only fast Na^+ current under control conditions, 5 mM taurine progressively and markedly increased the amplitude of the I_{Na}; steady-state was reached at about 5 min. (Table 4).

In control condition (bath containing 10^{-5} M TTX), the TTX-insensitive fast I_{Na} had a fast activation and inactivation (within 5 ms). Peak current was maximum at about -25 mV. Addition of 5 mM taurine rapidly (within 5 min) increased the amplitude of this current. The current/voltage relationship (I/V curve) showed that the voltage dependency of the channels was shifted to the left; that is, activation occurred at more negative potentials. The reversal potential (E_{rev}) for this current was at about +52 mV (near the usual Na^+ equilibrium potential, E_{Na}), and E_{rev} was not affected by taurine.

In cells that exhibited only a slow sustained component of I_{Na}, taurine (5 mM) rapidly (within 6 min) activated a fast transient component. Taurine increased the I/V curve amplitude over a wide voltage range without changing E_{rev}.

Ca^{2+} Currents

To determine whether taurine affected the Ca^{2+} currents, experiments were carried out in Na^+-free solution (Na^+ replaced by TEA and 4-AP). Figure 1A shows a typical experiment on a 3-day embryonic chick heart cell. The cell was stepped from an HP of -80 mV to +13 mV. A slow I_{Ca} was detected in about 20% of the cells tested. Addition of 5 mM taurine completely blocked the slow I_{Ca} within 2 min. Later (at about 3 min) there appeared a fast transient component of I_{Ca}, which reached a steady-state level at about 8 min (Figure 1A).

Figure 1B shows a typical experiment in a ventricular cell from a 10-day-old embryo. This cell showed only a slow I_{Ca} under control conditions (zero $[Na]_o$). Immediately after addition of 5 mM taurine, there was a decrease in $I_{Ca(s)}$ and the appearance of a fast $I_{Ca(f)}$ component (at 3 min). By 11 min, complete block of $I_{Ca(s)}$ occurred and a steady-state level of $I_{Ca(f)}$ was reached (Figure 1B). The data suggest that taurine blocks $I_{Ca(s)}$ and activates a fast I_{Ca} component in both 3-day- and 10-day-old embryonic chick heart cells. These findings are summarized in Table 4 and in Figure 12.

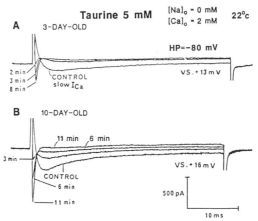

Figure 1. Taurine activation of a fast transient component in single 3-day (A) and 10-day-old (B) embryonic chick heart cells. A: In a cell from a 3-day embryonic chick superfused with a Na^+-free solution (containing 2.0 mM Ca^{2+} and 130 mM TEA), depolarizing voltage steps (VS) gave rise to an inward Ca^{2+} current that showed only a slow component ($I_{Ca(s)}$). Superfusion with 5 mM taurine activated a fast transient component over a period of 8 min., progressively decreased the amplitude of the slow component. B: Similar experiment as in part A with a cell from a 10-day embryonic chick. Addition of 5 mM taurine progressively activated a fast transient component over a period of 11 min. Temperature was 22° C. Taken from Sperelakis et al., 1989.

In 3-day-old embryonic cells, slow Ca^{2+} (L-type) current ($I_{Ca(s)}$) and I_K were produced at depolarizing steps from the HP of -40 mV (Figure 2A). Taurine (20 mM) inhibited $I_{Ca(s)}$ and I_K at the pipette pCa 7. In 4 of 15 cells, taurine induced a marked T-type Ca^{2+} current ($I_{Ca(f)}$) (Figure 2B-C). I-V relationships for $I_{Ca(s)}$ and $I_{Ca(f)}$ are given in Figure 2D.

In guinea pig myocytes, the Ca^{2+} current was slightly depressed by 20 mM taurine at a $[Ca]_o$ of 3.6 mM (Figure 3, left) (Sawamura et al., 1990). However, in 0.9 mM $[Ca]_o$, taurine exerted a very slight stimulatory effect (Figure 3, right). These effects of taurine were reversed upon washout. In addition, an outward K^+ current (I_{K1}) was increased by taurine.

To examine whether taurine exerts $[Ca]_i$-dependent actions on I_{Ca}, two $[Ca]_i$ levels were used: pCa 10 and pCa 7. At pCa 10, 20 mM taurine enhanced I_{Ca} by 60% at +10 mV (Figure 4A-B), but at pCa 7, the peak I_{Ca} was inhibited by 57% in 10 mM taurine, and was inhibited by 73% in 20 mM taurine (Figure 4C-D).

Taurine actually elevated $[Ca]_i$ from approximately 130 to 190 nM at 1 mM, and to 260 nM at 2 mM (Figure 5). The $[Ca]_i$ level in single heart cell was measured by using fura-2. The experiments were performed using a PTI microflurometer (Photon Technology International).

Fast Na^+ Current (TTX-sensitive)

Effects of taurine on the fast Na^+ current (I_{Na}) in 17-day-old embryonic chick ventricular myocytes were examined (Satoh and Sperelakis, 1991a). The myocytes were spherical (10-15 μm diameter), and the membrane capacitance was 9.8 ± 1.3 pF (n = 14). The experiments were performed at room temperature (22° C). The HP was -90 mV. The characteristics of I_{Na} were similar to those in other tissues,

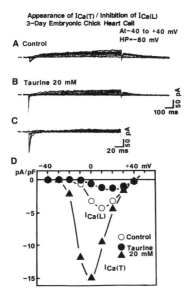

Figure 2. Appearance of T-type I_{Ca} and inhibition of L-type I_{Ca} in 3-day embryonic chick heart cells. A: Control. Test pulses (1 sec) were applied to -40 to +40 mV from the holding potential of -80 mV. B: Taurine 20 mM. C: Current traces at a fast time scale. D: Current/voltage curves for $I_{Ca(f)}$ and $I_{Ca(s)}$. Symbols used are control (open circles) and 20 mM taurine (filled circles for $I_{Ca(s)}$ and triangles for $I_{Ca(f)}$). (Satoh and Sperelakis, unpublished).

including sensitivity to TTX (10 μM). After the patch membrane was broken, peak I_{Na} initially increased, and then decreased and became stable within 3-5 min. The experiments on taurine were started after I_{Na} had stabilized. Taurine added to the bath, inhibited I_{Na} and shifted E_{rev} in the hyperpolarizing direction in a concentration-dependent manner (Figure 6). At 10 mM, taurine inhibited I_{Na} by 38.2 $\pm$ 4.3% (n = 10), and shifted E_{rev} by 10.2 $\pm$ 3.1 mV (n = 10). These effects of taurine were not reversed by 30 min washout. The I_{Na} inhibition induced by taurine is consistent with a report by Dumaine and Schanne (1991).

At low concentrations, taurine actually enhanced I_{Na} at 1 mM (in 3 of 8 cells) and 5 mM (in 4 of 10 cells); E_{rev} was still shifted in the hyperpolarizing direction by 5.7 $\pm$ 1.6 mV (n = 4) (Figure 7). The time course of inactivation (fitted as a single exponential) was not affected: 1.1 $\pm$ 0.5 ms (n = 13) in control, 1.2 $\pm$ 0.4 ms (n = 10) at 10 mM taurine. Taurine (10 mM) also did not modify the potentials of half-inactivation and half-activation: control of -86.4 $\pm$ 5.3 mV (n = 9) and -48.5 $\pm$ 4.4 mV (n = 17); in taurine, -89.3 $\pm$ 4.1 mV (n = 9) and -49.7 $\pm$ 4.7 mV (n = 10), respectively.

Intracellular application of taurine (10 or 20 mM) enhanced I_{Na} by about 7 - 15% (Figure 8) (Satoh and Sperelakis, 1991a). A subsequent depression was not observed. This experiment was done at pCa_i 10.

K⁺ Current

Outward K⁺ currents play an important role in regulation of repolarization in excitable membranes. Blockade of K⁺ channels prolongs the AP duration, and

Inhibition / Enhancement of I_{Ca}
Guinea Pig Ventricular Cells

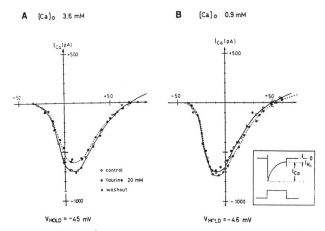

Figure 3. I/V relationships for the Ca^{2+} slow current recorded from single ventricular cells isolated from guinea pig heart bathed in 3.6 mM $[Ca]_o$ (left graph) or in 0.8 mM $[Ca]_o$ (right graph). Plotted are the control data in absence of taurine (o), in presence of 20 mM taurine for 7 min (•), and after washout for 10 min. (▲). The HPs were -45 mV (left graph) and -46 mV (right graph). Taken from Sawamura et al., 1990.

Enhancement (pCa 10)/ Inhibition (pCa 7) of I_{Ca}
3–Day Embryonic Chick Heart Cells

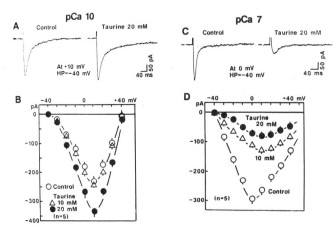

Figure 4. Dependency of $[Ca]_i$ on taurine modulation of I_{Ca}. A: Current traces at pCa 10 in control and 20 mM taurine. Test pulses were applied to +10 mV from a HP of -40 mV. B: Current/voltage curves for I_{Ca}. Symbols are: control (open circles), 10 mM taurine (triangles), and 20 mM taurine (filled circles). C: Current traces at pCa 7 in control and 20 mM taurine. Test pulses were applied to 0 mV from a HP of -40 mV. D: Current/voltage curves for I_{Ca}. Symbols are: control (open circles), 10 mM taurine (triangles), and 20 mM taurine (filled circles). (Satoh and Sperelakis, unpublished.)

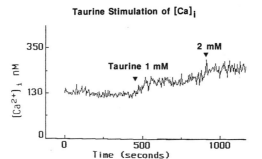

Taurine Stimulation of [Ca]ᵢ

Figure 5. Elevation of intracellular free [Ca] induced by external taurine. $[Ca]_i$ level of a single heart cell was measured using the fura-2 microfluorimetry technique. Addition of 1 mM taurine increased the level of $[Ca]_i$ within a few seconds. Increasing the concentration of taurine up to 2 mM further increased $[Ca]_i$. The experiments were carried out using a PTI (Photon Technology International) microfluorimeter (Bkaily et al., unpublished results).

depolarizes the membrane. In Figure 9, current density of the outward K⁺ current (I_K) at +90 mV in 10-day-old embryonic chick ventricular cells was activated from 15.0 ± 0.8 pA/pF to 24.7 ± 0.9 pA/pF by increasing Ca²⁺ from pCa 10 to pCa 7 in the pipette. The HP was -40 mV. Taurine (20 mM) increased I_K at pCa 10, whereas it decreased I_K at pCa 7 (Figure 9; Table 5).

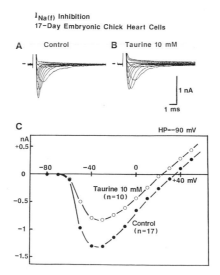

Figure 6. Inhibition of I_{Na} in 17-day-old embryonic chick ventricular myocytes. A-B: Family of current recordings taken from one myocyte before (control, A) and after addition of 10 mM taurine (B). The HP was −90 mV; the test pulses (at 10 mV increments) were applied from -60 mV to +50 mV. C: Current/voltage relationships for I_{Na} in the absence (filled circles) and presence (open circles) of 10 mM taurine. $[Ca]_i$ was pCa 10. To avoid the interference of other currents, $CoCl_2$ (3 mM) or $MgCl_2$ (2 mM) and 4-aminopyridine (4-AP) (3 mM) were added to the bath solution, and Cs⁺ was substituted for K⁺ in the pipette solution. The experiments were performed at room temperature (22° C). Points plotted are the mean $\pm$ SEM (n = 17 for control and 10 for taurine); the SEM bars are less than the thickness of the symbols. Note the shift to the left of E_{rev} produced by taurine and the decrease in the peak I_{Na} (at -30 mV). (Satoh and Sperelakis, 1991a)

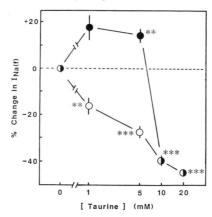

$I_{Na(f)}$ Inhibition
17-Day Embryonic Chick Heart Cells

Figure 7. Dose-response curve for changes in I_{Na} produced by exposure to different concentrations of taurine. In some cells, taurine (1 and 5 mM) enhanced I_{Na}, whereas in other cells, the same taurine concentrations produced inhibition of I_{Na}. Therefore, the data were separated into two groups for these concentrations. Both sets of data are plotted: cells in which I_{Na} was enhanced (filled circles) and cells in which I_{Na} was inhibited (unfilled circles). In 10 mM and 20 mM taurine, only inhibition was observed. The half-filled circles are comprised of both sets of data. The control value averaged -1.38 ± 0.88 nA (n = 17). The symbols give the mean values, and the vertical bars represent $\pm$ SEM. **: $p < 0.01$; ***: $p < 0.001$, with respect to control value. (Satoh and Sperelakis, 1991a)

In 10-day embryonic chick heart cells (HP of -50 mV), the time-dependent change in taurine action on I_K was examined (Figure 10A). Taurine (1 mM) gradually decreased I_K and reached steady-state 10 min after application (Figure 10A). I/V relationships for I_K in the absence and presence of taurine are given in Figure 10B.

Taurine (1 mM) also inhibited I_K in rabbit aortic vascular smooth muscle cells (HP of -50 mV) as in cardiac muscle cells (Figure 11). At about 20 min after taurine application, steady-state inhibition of I_K was reached. Addition of Ba^{2+} (10 mM) blocked I_K completely.

DISCUSSION

The present results show that in single 3-, 10-, and 17-day-old embryonic chick heart cells in culture, taurine (5 mM) activates two fast transient inward currents: (1) in Na^+-free solution, thus, a Ca^{2+} current, and (2) in Ca^{2+}-free solution, thus, a Na^+ current. Therefore, it appears that taurine activates a fast Ca^{2+} channel and a

$I_{Na(f)}$ Stimulation

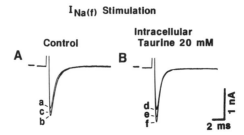

Figure 8. Effects on I_{Na} of intracellular application of taurine via the patch pipette. A-B: Change in peak I_{Na} produced by taurine. A: Current I_{Na} traces in control 0 min. (a), 1.5 min. (b), and 4.5 min. (c) after break of patch membrane. B: Current traces in 20 mM taurine applied internally after 0 min (d), 1 min. (e), and 5 min. (f). (Satoh and Sperelakis, 1991a)

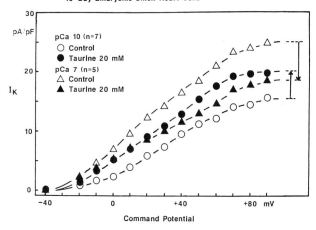

I_K Inhibition / Stimulation
10-Day Embryonic Chick Heart Cells

Figure 9. Dependency of I_K on $[Ca]_i$ in 10-day-old embryonic chick heart cells and effect of taurine. Test pulses (1 sec) were applied to -20 to +70 mV from a HP of -40 mV. In pCa 10 pipette solutions, taurine (20 mM) increased I_K (filled circles). In pCa 7, taurine (20 mM) decreased I_K (filled triangles). The actual values at several potentials are given in Table 5. (Satoh and Sperelakis, unpublished.)

Table 5. Effects of taurine (20 mM) on the outward K⁺ current in 10-day-old embryonic chick ventricular myocytes.

	+ 30 mV	+ 60 mV	+ 90 mV
pCa 10			
Control	6.0 ± 0.2 pA/pF	12.5 ± 0.5 pA/pF	15.0 ± 0.8 pA/pF
Taurine	10.8 ± 0.3	17.4 ± 0.4	19.4 ± 1.0
pCa 7			
Control	14.2 ± 0.1	20.2 ± 0.3	24.7 ± 0.9
Taurine	9.8 ± 0.2	14.1 ± 0.5	18.4 ± 0.8

Values are mean ± SEM. Test pulses (2 s) were applied from holding potential of -40 mV.

TTX-insensitive fast Na⁺ channel (Table 4). Taurine induced a fast Na⁺ current in 3-day-old embryo cells that did not exhibit this type of current in the absence of taurine. In addition, taurine accelerated the decay of the slow (sustained) Na⁺ current. In cells of 3-day-old and 10-day-old chick embryos that only showed a slow (sustained) component of Ca²⁺ current ($I_{Ca(s)}$), taurine blocked this current and induced a fast (transient) component of I_{Ca} ($I_{Ca(f)}$) (Table 4).

The rapid (e.g., 1 min) transient decrease of slow AP amplitude and max dV/dt by 1 mM taurine reported by Sawamura et al. (1986b) could be explained by the decrease of the slow $I_{Ca(s)}$. The spontaneous recovery of the AP amplitude and max dV/dt after a few minutes in continued presence of taurine could be explained by the activation of the fast (transient) Ca²⁺ current ($I_{Ca(f)}$). The blockade of $I_{Ca(s)}$ and the appearance of $I_{Ca(f)}$ might explain the lack of effect of taurine at steady state on the Ca²⁺-dependent slow APs in embryonic heart (Sawamura et al., 1986b).

Taurine Effect on I_K
10-Day Embryonic Heart Cell

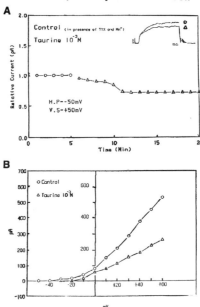

Figure 10. **A:** Time course of effects of taurine on the delayed outward K^+ current in a 10-day-old embryonic chick heart cell. (o): relative peak outward K^+ current recorded using Tyrode solution containing tetrodotoxin (TTX, 10^{-5} M) to block the fast Na^+ current and Mn^{2+} (2 mM) to block the T- and L-type Ca^{2+} currents. Addition of 1 mM taurine progressively decreased the peak K^+ current (within 5 min). **Inset current traces:** Traces of the K^+ currents recorded before and 6 min after addition of 1 mM taurine. Note that both the peak outward current and the tail current were decreased by taurine. **B:** I/V relationship of the delayed outward K^+ current recorded (HP of -50 mV) in presence of TTX and Mn^{2+} (o). The amplitude of the I/V curve was decreased by 1 mM taurine at all voltage steps used (Δ). (Bkaily et al., unpublished results)

Effect of Taurine on I_K
Rabbit Aortic Cell

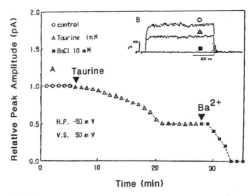

Figure 11. Time course of effect of taurine on the delayed outward K^+ current in a single vascular smooth muscle cell from rabbit aorta. o: Relative peak amplitude of the delayed outward K^+ current (I_K) recorded in presence of TTX (10^{-5} M) and Mn^{2+} (2 mM). Δ : Relative peak I_K after the addition of 1 mM taurine. There was a 50% decrease of the peak I_K amplitude. ■ : Addition of the K^+ blocker barium (10 mM) completely blocked the remaining I_K within 6 min. **B:** Current traces of the K^+ current recorded in presence of TTX and Mn^{2+}. Addition of taurine decreased the I_K and the tail current amplitudes by 50% of the control value. The remaining outward current was completely blocked by Ba^{2+}.

140

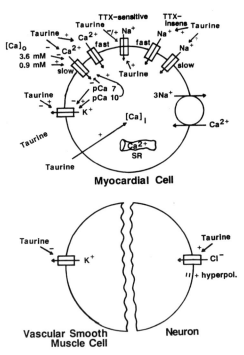

Figure 12. Upper Diagram: Diagrammatic summary of some possible actions of taurine on the sarcolemmal ion channels of chick ventricular myocardial cells, based on the results from the experiments described in this article. As depicted, taurine may stimulate or activate one type of TTX-insensitive (Mn^{2+}-insensitive) Na channel (responsible for the fast transient component of the I_{Na}). TTX-sensitive Na^+ channels were inhibited by external taurine, but stimulated slightly by internal taurine. Effects of taurine on I_{Ca} and I_K were dependent on $[Ca]_i$ and $[Ca]_o$. Taurine actually elevated $[Ca]_i$. **Lower diagrams:** Summary of some actions of taurine on ion channels of vascular smooth muscle cells (left) and neurons (right portion).

The increase or activation of a fast (TTX-insensitive) transient Na^+ current by taurine would increase $[Na]_i$ near the sarcolemma, which in turn may reverse the Na^+-Ca^{2+} exchange and thus allow Ca^{2+} to flow inside the cell by this pathway (Figure 12). This may suggest that the positive inotropic effect of taurine is not mediated through an increase in the inward slow Ca^{2+} current (Sawamura et al., 1986b and Sperelakis et al., 1989), but could be due to Ca^{2+} influx through the Na^+-Ca^{2+} exchanger, resulting from an increase in Na^+ influx via activation of the fast Na^+ current (Table 4; Figure 12).

On the other hand, taurine exerts a dual effect on the TTX-sensitive fast I_{Na}, with inhibition or stimulation at lower concentrations (1, 5 mM) and inhibition only at higher concentrations (10, 20 mM) (Figure 7). However, a dual action of taurine is not unique. For example, some local anesthetics have a dual action on max dV/dt ($I_{Na(f)}$): at low concentrations, max dV/dt is increased, and at high concentrations it is decreased (Bigger et al., 1968, 1970). The dual action of taurine may reflect two separate sites of action: (a) an action to directly stimulate the fast Na^+ channels, and (b) an indirect action to inhibit the channels. Another possibility is that taurine acts only on the fast Na^+ channels, stimulating them from the inner surface and inhibiting them from the outer surface. Since the shift of the reversal potential for I_{Na} in the hyperpolarizing direction (Figure 6) occurred when I_{Na} was increased or decreased (and therefore opposite changes in $[Na]_i$), the mechanism for the shift may be due to an additional effect of taurine.

Effect of taurine on peak I_{Ca} was dependent on both $[Ca]_o$ and $[Ca]_i$. At low $[Ca]_i$ or $[Ca]_o$, taurine stimulated I_{Ca}, whereas taurine inhibited I_{Ca} at high $[Ca]_i$ or $[Ca]_o$. When taurine was found to increase I_{Ca}, it resulted in an elevation of $[Ca]_i$, and vice versa. It is unknown whether taurine acts directly on Ca^{2+} channels. One regulator of I_{Ca} channels is the $[Ca]_i$ level, inhibition occurring at high $[Ca]_i$ levels.

The effect of taurine on total outward K^+ current was also dependent on $[Ca]_i$. Taurine stimulated I_K at low $[Ca]_i$ and inhibited I_K at high $[Ca]_i$. These results can explain the taurine action on the APs: APD prolongation at high $[Ca]_o$ and APD shortening at low $[Ca]_i$.

Taurine actually elevated $[Ca]_i$, presumably resulting from the increases in fast I_{Ca}, slow I_{Na}, and fast I_{Na} (in some cells), release of Ca^{2+} from SR, and stimulation of Na-Ca exchange. Thus, taurine may exert a positive inotropic effect by such mechanisms.

SUMMARY

The effects of taurine on the slow and fast Na^+ currents and slow and fast Ca^{2+} currents in cultured single ventricular cells from young (3-day-old) and old (10, 17 day) chick embryos were studied using the whole-cell voltage clamp technique. In single 3-day cells that showed only a TTX-insensitive fast (transient) Na^+ current ($I_{Na(f)}$), taurine (5 mM) rapidly increased the amplitude of this current. In single cells that showed only a typical slow (sustained) Na^+ current ($I_{Na(s)}$), taurine (5 mM) induced a fast transient component. A slow Ca^{2+} current ($I_{Ca(s)}$) was also present in the 3-day-old embryonic chick cells, and taurine inhibited this current and activated a fast transient component ($I_{Ca(f)}$). Taurine had similar actions in 10-day-old embryonic heart cells. Thus, in embryonic chick heart cells, taurine stimulates the TTX-insensitive fast transient Na^+ current and blocks the slow component. Taurine also activates a fast (transient) component of the Ca^{2+} current ($I_{Ca(f)}$). The activation of the TTX-insensitive $I_{Na(f)}$ may increase Ca^{2+} influx via Na^+-Ca^{2+} exchange. This may explain, in part, the positive inotropic effect of taurine in heart muscle, with relatively little effect on the Ca^{2+}-dependent slow APs.

Taurine (10 or 20 mM) added to the outside markedly inhibited TTX-sensitive I_{Na}, but slightly stimulated I_{Na} when added internally. $I_{Ca(s)}$ and I_K both were stimulated by external taurine at pCa 10 but inhibited at pCa 7. Taurine also inhibited I_K in aortic VSM cells. In contrast, taurine induced or stimulted $I_{Ca(f)}$.

Elevation of $[Ca]_i$ was induced by taurine. The elevation may result from the enhancement of $I_{Ca(f)}$ and possibly of Na^+-Ca^{2+} exchange, resulting in a positive inotropic effect.

It has been shown that in neurons, taurine increases the Cl^- current, resulting in hyperpolarization (Taber et al., 1986; Figure 12). Thus, taurine effects are complex, there being a number of actions on the membrane currents of cardiac cells, vascular smooth muscle cells, and neurons.

ACKNOWLEDGMENTS

This study was supported by grant HL-31942 from the National Institutes of Health (Sperelakis) and by grants MT-9819 and ME-9788 from the Medical Research Council of Canada and a grant from the Quebec Heart Foundation (Bkaily). Dr. Bkaily is a Scholar of the Canadian Heart Foundation.

The authors wish to acknowledge the able assistance of Rhonda S. Hentz and Judy McMahan in preparation of the manuscript, and Anthony Sperelakis in preparation of the figures.

REFERENCES

Azuma, J., Takihara, K., Awata, N., Ohta, H., Sawamura, A., Harada, H., and Kishimoto, S., 1984, Beneficial effect of taurine on congestive heart failure induced by chronic aortic regurgitation in rabbits, *Res. Commun. Chem. Pathol. Pharmacol.* 45:261-270.

Baskin, S.I., and Finney, C.M., 1979, Effects of taurine and taurine analogue on the cardiovascular system, *Gamyu Aminosan* (Sulfur-containing Amino Acids) 2:1-18.

Bigger, J.T., and Mandel, W.J., 1970, Effect of lidocaine on the electrophysiological properties of ventricular muscle and Purkinje fibers, *J. Clin. Invest.* 46:63-77.

Bigger, J.T., Basset, A.L. and Hoffman, B.F., 1968, Electrophysiological effects of diphenylhydantoin on canine Purkinje fibers, *Circ. Res.* 22:221-236.

Bkaily, G., and Sperelakis, N., 1984, Injection of protein kinase inhibitor into cultured heart cells blocks calcium slow channels, *Am. J. Physiol/Heart & Circ. Physiol.* 246:H630-H634.

Dietrich, J., and Diacono, J., 1971, Comparison between ouabain and taurine effects on isolated rat and guinea-pig hearts in low calcium medium, *Life Sci.* 10:499-507.

Dumaine, R., and Schanne, O.F., 1991, Depression of I_{Na} by taurine in isolated rabbit cardiac myocytes, *Biophys. J.* 59:739 (abst.).

Fabiato, A., and Fabiato, F., 1979, Calculator programs for computing the composition of the solutions containing multiple metals and ligands used for experiments in skinned muscle cells, *J. Physiol.* (Lond.) 75:463-505.

Hamill, O.P., Marty, A., Neher, E., Sakmann, B., and Sigworth, J., 1981, Improved patch-clamp techniques for high-resolution current recording from cells and cell-free membrane patches, *Pflugers Arch.* 391:88-100.

Kojima, M., and Sperelakis, N., 1984, Properties of oscillatory after-potentials in young embryonic chick hearts, *Circ. Res.* 55:497-503.

Li, T., and Sperelakis, N., 1983, Stimulation of slow action potentials in guinea pig papillary muscle cells by intracellular injection of cAMP, Gpp(NH)p, and cholera toxin, *Circ. Res.* 52:111-117.

Satoh, H., and Sperelakis, N., 1991a, Actions of taurine on fast Na+ current (I_{Na}) in embryonic chick ventricular myocytes, *Faseb J.* 5(6):A1742.

Satoh, H., and Sperelakis, N., 1991b, Identification of the hyperpolarization-activated inward current in young embryonic chick heart myocytes, *J. Develop. Physiol.* 15:247-252.

Sawamura, A., Azuma, J., Harada, H., Hasegawa, H., Ogura, K., Sperelakis, N., and Kishimoto, S., 1983, Protection by oral pretreatment with taurine against the negative inotropic effects of low calcium medium on isolated perfused chick heart, *Cardiovas. Res.* 17:620-626.

Sawamura, A., Sada, H., Azuma, J., Kishimoto, S., and Sperelakis, N., 1990, Taurine modulates ion influx through cardiac Ca^{2+} channels, *Cell Calcium* 11:251-259.

Sawamura, A., Sperelakis, N., and Azuma, J., 1986a, Protective effect of taurine against decline of cardiac slow action potentials during hypoxia, *Eur. J. Pharm.* 120:235-239.

Sawamura, A., Sperelakis, N., Azuma, J., and Kishimoto, S., 1986b, Effects of taurine on the electrical and mechanical activities of embryonic chick heart, *Can. J. Physiol. Pharmacol.* 64:649-655.

Shigenobu, K., and Sperelakis, N., 1972, Calcium current channels induced by catecholamines in chick embryonic hearts whose fast sodium channels are blocked by tetrodotoxin or elevated potassium, *Circ. Res.* 31:932-952.

Sperelakis, N., 1984, Hormonal and neurotransmitter regulation of Ca^{2+} influx through voltage-dependent slow channels in cardiac muscle membrane, Proceedings of the Membrane Biophysics Symposium on "Molecular Mechanism of Hormonal Regulation of Ion Movement" (A. Shamoo, ed), San Diego, February 13, 1983, *Membrane Biochem.* 5:131-166.

Sperelakis, N., and Schneider, J.A., 1976, A metabolic control mechanism for calcium ion influx that may protect the ventricular myocardial cell, *Am. J. Cardiol.* 37:1079-1085.

Sperelakis, N., Yamamoto, T., Bkaily, G., Sada, H., Sawamura, A., and Azuma, J., 1989, Taurine effects on action potentials and ionic currents in chick embryonic cells. In: Taurine and the Heart, edited by H. Iwata, J.B. Lombardini, and T. Segawa, Kluwer Academic Pub., Boston, pp. 1-20.

Taber, K.H., Lin, C.T., Liu, J.W., Thalmann, R.T., and Wu, J.Y., 1986, Taurine in hippocampus: Localization and postsynaptic action, *Brain Res.* 386:113-121.

Takihara, K., Azuma, J., Awata, N., Ohta, H., Hamaguchi, T., Sawamura, A., Tanaka, Y., Kishimoto, S., and Sperelakis, N., 1986, Beneficial effect of taurine in rabbits with chronic congestive heart failure, *Am. Heart J.* 112(6):1278-1284.

Takihara, K., Azuma, J., Awata, N., Ohta, H., Sawamura, A., Kishimoto, S., and Sperelakis, N., 1985, Taurine's possible protective role in age-dependent response to calcium paradox, *Life Sci.* 37:1705-1710.

Tsien, R.Y., and Rink, T.J., 1980, Neutral carrier ion-selective microelectrodes for measurements of intracellular-free calcium, *Biochim. Biophys. Acta* 559:623-638.

Vogel, S., and Sperelakis, N., 1981, Induction of slow action potentials by microiontophoresis of cyclic AMP into heart cells, *J. Mol. Cell Cardiol.* 13:51-64.

Wahler, G.M., and Sperelakis, N., 1985, Intracellular injection of cyclic GMP depresses cardiac slow action potentials, *J. Cyclic Nucleotide & Protein Phosphorylation Res.* 10:83-95.

TAURINE ATTENUATES CONTRACTURE INDUCED BY PERFUSION WITH LOW SODIUM, HIGH CALCIUM MEDIUM IN CHICK HEARTS

Yoshiji Ihara[1], Kyoko Takahashi[1], Hisato Harada[1],
Akihiko Sawamura[1], Stephen W. Schaffer[2] and Junichi Azuma[1]

[1]Department of Medicine III
Osaka University Medical School
Osaka, Japan

[2]Department of Pharmacology
University of South Alabama
School of Medicine
Mobile, AL USA

INTRODUCTION

Taurine is the most abundant free amino acid in heart. It has various effects on cardiac function, including improved cardiac performance in congestive heart failure[1] and regulation of calcium homeostasis[2]. Because taurine prevents calcium overload in several heart failure models, it was felt that it might improve cardiac function in hearts subjected to a low sodium, high calcium medium. It is known that a remarkable decrease in extracellular sodium concentration leads to contracture[3]. Thus, in this study we examined the effect of taurine on changes in intracellular calcium concentration and tissue high energy phosphate content of perfused chick heart exposed to buffer containing a low sodium and high calcium concentration.

MATERIALS AND METHODS

Perfusion Technique

Two or three day old chicks were killed by decapitation, their hearts cannulated immediately after excision and perfused according to the modified Langendorff method[4] with normal Tyrode solution containing (in mM): 130 NaCl, 5.4 KCl, 1.0 $MgCl_2$, 1.06 NaH_2PO_4, 20 $NaHCO_3$, 1.8 $CaCl_2$ and 5.55 glucose. After a 20 min equilibration period, one series of hearts was perfused for 30 min with Tyrode solution modified to contain only 65 mM NaCl and 3.6 mM $CaCl_2$; in this nontaurine group, total medium sodium was reduced from 151 to 85 mM. Osmolarity was adjusted with sucrose. Hearts from the other experimental group (taurine group) were perfused with normal and modified Tyrode which had been supplemented with

Taurine, Edited by J.B. Lombardini *et al.*
Plenum Press, New York, 1992

20 mM taurine. Hearts belonging to the control group were perfused with normal Tyrode for 50 min. The pH of every medium was maintained at 7.4 by saturation with 95% O_2,5% CO_2 gas at 37°C. All hearts were perfused under a constant pressure head of 80 cm H_2O and electrically stimulated at a constant rate of 200 beats/min. The voltage of the stimulating pulses was set at 30% above the threshold level for each individual experiment. Developed tension and its first derivative (dT/dt) were recorded using an isometric force-displacement transducer connected to the apex of the ventricle by a string.

Heart Cell Preparation and Fura-2 Loading

Chick cardiomyocytes were used in the measurement of intracellular free calcium ion concentration, $[Ca^{2+}]i$. To prepare the myocytes, chick hearts were perfused with normal Tyrode and then digested with 0.1% collagenase dissolved in Ca^{2+}-free Tyrode solution. Digested ventricular muscle was minced and washed in Ca^{2+}-free HEPES buffer containing (in mM) 150 NaCl, 5.4 KCl, 1.0 $MgCl_2$, 1.06 NaH_2PO_4 and 5.55 glucose bubbled with 100% O_2 gas at pH 7.4. The Ca^{2+} concentration of the HEPES buffer was gradually increased to 1.8 mM, leaving a field of isolated cardiomyocytes with about 80% viability and normal appearance according to analysis by light microscopy. The cardiomyocytes were loaded with the fluorescence indicator dye fura-2 for 30 min at 37°C in 1.8 mM Ca^{2+} containing HEPES buffer (normal HEPES) supplemented with 3 μM fura-2/AM.

Imaging Methods

Fura-2 loaded cells were incubated with normal HEPES buffer for 20 min and then incubated for an additional 30 min with modified HEPES buffer containing 85 mM Na^+, 3.6 mM Ca^{2+}, an appropriate amount of sucrose to adjust for osmolarity differences and the same concentration of other ions as normal HEPES (nontaurine group). Fura-2 loaded cells belonging to the taurine group were incubated with normal and modified HEPES media containing 20 mM taurine in addition to the other components. Fluorescence of fura-2 loaded cells and background measured at 510 nm after exciting at either 340 or 380 nm was monitored using a television equipped with a fluorescence imaging microscope. The 340/380 nm ratio of the fluorescence signal was calculated with a computer linked to the fluorescence imaging system. This ratio is proportional to $[Ca^{2+}]i$.

High Performance Liquid Chromatography

Some hearts perfused by the same procedure as described above were rapidly frozen using a clamp precooled in liquid nitrogen. All samples were freeze-dried for 20 hrs, weighed and then deproteinized with perchloric acid. After neutralization with KOH, the supernatant was used for determination of high energy phosphate content using high performance liquid chromatography[5]. The total adenine nucleotide pool is represented by ATP + ADP + AMP.

Data Analysis

The data are presented as mean $\pm$ S.E.M. in Figures 1-5 and mean $\pm$ S.D. in Table 1. Statistical analysis was performed by the Student's unpaired t-test. A probability of < 0.05 was considered statistically significant.

RESULTS

Effect of Low Na⁺, High Ca²⁺ Medium on Contraction of Isolated Chick Hearts

Perfusion of isolated chick hearts with low Na^+ (85 mM), high Ca^{2+} (3.6 mM) Tyrode solution (modified medium) induced changes in their contractile state. Figure 1 shows representative tracings of developed tension and its first derivative (dT/dt) of chick heart perfused initially with normal Tyrode solution, followed by modified medium. Shortly after exposure to the modified medium, there was an abrupt increase in developed tension, followed by a prolonged decline in developed tension and an increase in resting tension (Figure 1-a). These changes were attenuated in hearts exposed to 20 mM taurine throughout the protocol. Mean values of developed tension, resting tension and both positive and negative maximal dT/dt ($\pm$ dT/dt) are summarized in Figures 2-4, respectively.

The time course of alterations in developed tension is shown in Figure 2. Addition of 20 mM taurine to the buffer significantly attenuated the decline in developed tension. In the untreated control, developed tension declined to 39.1 $\pm$ 11.9% and 21.2 $\pm$ 7.8% by 20 min and 30 min, respectively, while tension in the taurine-treated group only fell to 93.1 $\pm$ 9.4% and 71.6 $\pm$ 7.6% over the same time period.

Figure 3 reveals the time course of resting tension changes during perfusion with modified medium. Untreated control hearts exposed to modified medium exhibited prolonged increases in resting tension after 5 min. Addition of 20 mM taurine to the medium significantly reduced the increase in resting tension noted 30 min after the change in medium (0.87 $\pm$ 0.16 g in control vs. 1.67 $\pm$ 0.27 g in the taurine group).

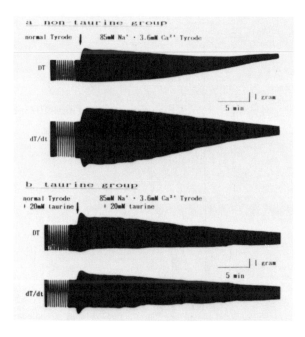

Figure 1. Representative tracings of developed tension (upper record) and its first derivative (dT/dt: lower record) taken from isolated chick heart perfused with modified medium either lacking (nontaurine group, Figure 1-a) or containing 20 mM taurine (taurine group, Figure 1-b).

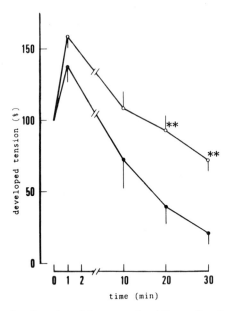

Figure 2. Time course of developed tension of the nontaurine (O , n=6) and taurine groups (O , n=5). At time 0, perfusion medium was modified as described in Figure 1. Each point is % of value at the end of perfusion with normal Tyrode buffer either containing or lacking taurine. Values are means ± S.E.M. ** denotes significant difference between the taurine and nontaurine groups (p < 0.01).

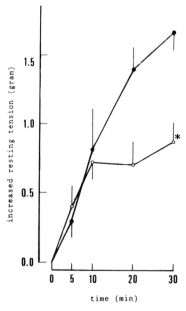

Figure 3. Time course of resting tension (g) of the nontaurine (O , n=6) and taurine groups (O , n=5). At time 0, perfusion medium was changed as described in Figure 1. Values are means ± S.E.M. * denotes significant difference between the taurine and nontaurine groups (p < 0.05).

Changes in -dT/dt often run parallel and proportional to changes in +dT/dt. In this study, the time course of +dT/dt mimicked that of developed tension (Figure 4). Taurine significantly attenuated the decrease in both +dT/dt and -dT/dt which accompanied the switch in buffer to the low Na$^+$, high Ca^{2+} medium. After 20 min and 30 min of exposure to modified buffer, +dT/dt was $50.9 \pm 17.5\%$ and $30.2 \pm 10.9\%$ in the nontaurine group, respectively, but 111 ± 11 and $94.8 \pm 9.8\%$ in the taurine group at the same time points. Changes in -dT/dt were similar (at 20 min it was $33.5 \pm 13.5\%$ in the nontaurine group vs. $81.1 \pm 13.6\%$ in the taurine group and at 30 min the value was $18.4 \pm 8.0\%$ for the nontaurine group vs. $66 \pm 10\%$ for the taurine group).

Measurements of [Ca^{2+}]$_i$

Changes in intracellular free Ca^{2+} concentration ([Ca^{2+}]$_i$) of isolated cardiomyocytes subjected to modified medium were examined using the fluorescence indicator dye, fura-2, and a fluorescence imaging microscope. The time course of the changes in the fluorescence 340/380nm ratio, a reflection of [Ca^{2+}]$_i$, is shown in Figure 5. In the absence of taurine, exposure of the myocytes to low Na$^+$, high Ca^{2+} HEPES buffer led to an immediate increase in [Ca^{2+}]$_i$ followed by a more gradual rise. Inclusion of 20 mM taurine in the buffer attenuated the increase in [Ca^{2+}]$_i$ of nonbeating isolated chick cardiomyocytes observed following exposure to the low Na$^+$, high Ca^{2+} buffer; the 340/380 nm ratio was 2.26 ± 0.09 in the untreated cells but 1.80 ± 0.06 in the taurine group 30 min after the change in buffer.

Measurement of Tissue High Energy Phosphate Content

The effect of taurine treatment on tissue high energy phosphate content was measured utilizing high performance liquid chromatography (Table 1). Tissue ATP and total adenine nucleotide content fell significantly following the change in perfusion medium to the low Na$^+$, high Ca^{2+} buffer. However, taurine treatment decreased the degree of ATP and the high energy phosphate loss seen 30 min after the change in medium (Table 1).

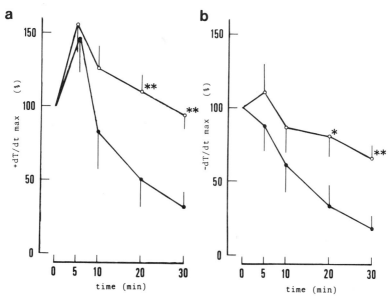

Figure 4. Time course of +dT/dt (Figure 4a) and -dT/dt (Figure 4b). Hearts were perfused with buffer as described in Figure 1. Each point is % of the value at the end of perfusion with normal Tyrode either lacking (O , nontaurine group, n=6) or containing taurine (O , taurine group, n=5).

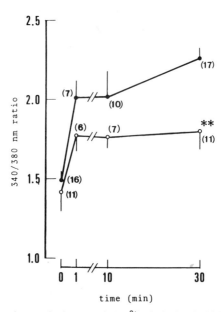

Figure 5. Effect of 20 mM taurine on the increase in $[Ca^{2+}]_i$ in isolated chick cardiomyocytes. At time 0, medium was switched from normal HEPES buffer to a low Na^+, high Ca^{2+} HEPES buffer lacking (O) or containing (O) 20 mM taurine. The vertical axis represents the emission ratio when excited at 340 nm and 380 nm. The ratio reflects $[Ca^{2+}]_i$. Each point is a mean $\pm$ S.E.M. Values in parentheses indicate the number of experiments.

Table 1. High energy phosphate content of control, nontaurine group and taurine group

Group	ATP	ADP	AMP	Total
Control	26.6 ± 4.0	4.92 ± 0.41	1.02 ± 0.41	32.6 ± 5.5
Nontaurine	0.7 ± 0.5	0.75 ± 0.10	1.61 ± 0.25	3.1 ± 0.5
Taurine	$3.9 \pm 1.2*$	0.80 ± 0.15	$0.80 \pm 0.55*$	$5.5 \pm 0.9*$

Each value represents the mean $\pm$ S.D. and refers to $\mu mol/g$ dry wt. * and ** denote significant difference between the nontaurine and taurine groups ($p<0.05$ and $p<0.01$, respectively). Total signifies total adenine nucleotide phosphate content (ATP + ADP + AMP).

DISCUSSION

In this study, perfusion with modified Tyrode solution containing 85 mM Na^+ and 3.6 mM Ca^{2+} was found to alter the contractile state of isolated chick heart (Figures 1 and 2). Exposure to the modified buffer caused an initial, rapid increase in developed tension, followed by a gradual decline in contraction. The amplitude of the transient increase was associated with the alteration in extracellular sodium ion concentration, $[Na^+]_o$, and has been previously shown to be linearly related to the $[Ca^{2+}]_o/[Na^+]_o^2$ ratio over a $[Na^+]_o$ range of 75 to 200 mM[6]. In this study, the ratio of $[Ca^{2+}]_o/[Na^+]_o^2$ was increased from 8×10^{-5} (normal Tyrode solution) to 4.98×10^{-4} mM^{-1} (modified Tyrode solution), accounting for the nearly 60% initial increase in developed tension. Interestingly, there was no significant difference in the transient increase in developed tension between the nontaurine and taurine groups.

The slow decline in developed tension following the initial increase was observed in both the nontaurine and taurine groups. This fall in developed tension was accompanied by an increase in resting tension, eventually resulting in a state of contracture (Figures 1-3). However, addition of taurine to the perfusion medium significantly decreased the degree of contracture. Also, no contracture occurred in chick hearts perfused with Tyrode solution containing 150 mM Na^+ and 3.6 mM Ca^{2+}, 85 mM Na^+ and 1.8 mM Ca^{2+} or 85 mM Na^+ and 0.9 mM Ca^{2+} (data not shown).

Niedergerke[7,8] has reported that ^{45}Ca uptake per heart beat was enhanced when either $[Na^+]_o$ was reduced or $[Ca^{2+}]_o$ was increased. Therefore, in our experiments in which low $[Na^+]_o$ and high $[Ca^{2+}]_o$ was used, one would expect excessive Ca^{2+} entry into the myocardium. Moreover, the decrease in the transarcolemmal Na^+ gradient should reduce efflux of Ca^{2+} from the cell via the Na^+, Ca^{2+} exchanger, suggesting that both voltage dependent slow channels and Ca^{2+} efflux via the Na^+, Ca^{2+} exchanger are involved in alterations in $[Ca^{2+}]_i$. The resulting calcium overload appears to be the primary basis underlying the decrease in developed tension, the increase in resting tension and the development of contracture.

Figure 5 shows the time course of $[Ca^{2+}]_i$ before and after shifting incubation medium from normal to modified HEPES buffer. Associated with the increase in medium Ca^{2+} and decrease in Na^+ was an increase in $[Ca^{2+}]_i$. According to Goldmann's formula, both lowering $[Na^+]_o$ and increasing $[Ca^{2+}]_o$ causes membrane hyperpolarization[9]. Therefore, Ca^{2+} entry into quiescent cells incubated in the presence of medium containing low Na^+ and high Ca^{2+} HEPES buffer would be slowed suggesting that Ca^{2+} entry does not occur solely via the Ca^{2+} channel. Another possibility is that Ca^{2+} could enter the cell via the Na^+, Ca^{2+} exchanger, although this is unlikely because the transarcolemmal Na^+ gradient and hyperpolarizing membrane potential would be unfavorable for promotion of Ca^{2+} entry in exchange for Na^+. The most likely scenario is that the increase in $[Ca^{2+}]_i$ shown in Figure 5 is caused by Ca^{2+} entry via the Ca^{2+} channel and passive diffusion, combined with inhibition of Ca^{2+} efflux via the Na^+, Ca^{2+} exchanger.

Addition of taurine to modified HEPES buffer decreased the elevation in $[Ca^{2+}]_i$, an effect possibly due to the promotion of Ca^{2+} extrusion via the Na^+, Ca^{2+} exchanger, modulation of Ca^{2+} entry via the Ca^{2+} channel, protection against passive Ca^{2+} diffusion and/or increases in Ca^{2+} binding to cardiac sarcolemma. According to Bers and Langer[10] cardiac sarcolemma contains Ca^{2+} binding sites which affect Ca^{2+} transport. By potentiating Ca^{2+} binding to the internal side of the cell membrane, taurine could affect Ca^{2+} transport[11,12]. Taurine is also an osmoregulator. It has been reported[3] that reducing $[Na^+]_o$ decreases $[Na^+]_i$ in sheep cardiac Purkinje fibers. Low Na^+ perfusion could alter intracellular osmolarity, an effect which might be minimized by taurine. Another important action of taurine is modulation of the slow inward Ca^{2+} current. Sawamura et al.[9] showed using the patch clamp technique that 20 mM taurine transiently decreased the slow inward Ca^{2+} current (I_{Ca}) in isolated guinea pig cardiomyocytes incubated in medium containing 3.6 mM Ca^{2+}. Therefore, it is possible that taurine diminishes Ca^{2+} entry through the slow Ca^{2+} channel of chick heart perfused with the low Na^+, high Ca^{2+} Tyrode solution.

Calcium overload occurs in chick heart perfused with modified medium. Although the sarcoplasmic reticulum plays an important role in redistributing intracellular Ca^{2+}, its capacity is limited and during calcium overload, the mitochondria becomes an important organelle in buffering Ca^{2+}. However, excessive Ca^{2+} entry into the mitochondria can lead to reduced ATP production[13]. Table 1 reveals that ATP and the total adenine nucleotide pool of the nontaurine and taurine groups were significantly lower than hearts not exposed to the low Na^+, high Ca^{2+} medium. Taurine attenuated the depression in ATP and total adenine nucleotide content. Because ATP is consumed by the sarcolemmal and sarcoplasmic reticular calcium pumps and these transporters are activated when $[Ca^{2+}]_i$ is elevated, taurine could reduce the rate of ATP degradation by reducing intracellular calcium overload. In this way, the effect is augmented because ATP depletion alone can cause Ca^{2+}

pump failure, resulting in further cytoplasmic Ca^{2+} accumulation.

In conclusion, isolated chick hearts perfused with moderately low Na^+ and high Ca^{2+} medium develop contracture. This is presumably caused by excessive Ca^{2+} entry through the slow Ca^{2+} channel and inhibition of Ca^{2+} efflux via the Na^+,Ca^{2+} exchanger. In addition to causing contracture, calcium overload also leads to depletion of tissue high energy phosphate content, which in turn leads to a worsening of the degree of calcium overload. Taurine attenuates these effects through modulation of $[Ca^{2+}]_i$.

REFERENCES

1. J. Azuma, H. Hasegawa, A. Sawamura, N. Awata, H. Harada, I. Ogura and S. Kishimoto, Taurine for treatment of congestive heart failure, *Int. J. Cardiol.* 2:303-309 (1982).
2. R.J. Huxtable and L. A. Sebring, Cardiovascular actions of taurine, *in*: "Sulfur Amino Acids: Biochemical and Clinical Aspects," K. Kuriyama, R.J. Huxtable and H. Iwata, eds., Alan R. Liss, New York, pp. 5-38 (1983).
3. G. Vassort, Influence of sodium ions on the regulation of frog myocardial contractility, *Pflugers Arch.* 339:225-240 (1973).
4. T. Hamaguchi, J. Azuma, H. Harada, K. Takahashi, S. Kishimoto and S.W. Schaffer, Protective effect of taurine against doxorubicin-induced cardiomyopathy in perfused chick hearts, *Pharmacol. Res.* 21:729-734 (1989).
5. H. Harada, J. Azuma, H. Hasegawa, H. Ohta, K. Yamauchi, K. Ogura, N. Awata, A. Sawamura, N. Sperelakis and S. Kishimoto, Enhanced suppression of myocardial slow action potentials during hypoxia by free fatty acids, *J. Mol. Cell Cardiol.* 16:261-276 (1984).
6. J.H. Tillisch, L.K. Fung, P.M. Hom and G.A. Langer, Transient and steady-state effects of sodium and calcium on myocardial contractile response, *J. Mol. Cell Cardiol.* 11:261-276 (1984).
7. R. Niedergerke, Movements of Ca^{2+} in frog ventricle at rest and during contractures, *J. Physiol.* (London) 167:515-550 (1963).
8. R. Niedergerke, Movements of Ca^{2+} in beating ventricles of the frog heart, *J. Physiol.* (London) 167:551-580 (1963).
9. A. Sawamura, H. Sada, J. Azuma, S. Kishimoto and N. Sperelakis, Taurine modulates ion influx through cardiac Ca^{2+} channels, *Cell Calcium* 11:251-259 (1990).
10. D. Bers and G. Langer, Uncoupling cation effects on cardiac contractility and sarcolemmal Ca^{2+} binding, *Am. J. Physiol.* 237:H332-H341 (1979).
11. J.P. Chovan, E.C. Kulakowski, B.W. Benson and S.W. Schaffer, Taurine enhancement of calcium binding to rat heart sarcolemma, *Biochim. Biophys. Acta* 551:129-136 (1979).
12. L.A. Sebring and R.J. Huxtable, Taurine modulation of calcium binding to cardiac sarcolemma, *J. Pharmac. Exp. Therap.* 232:445-451 (1984).
13. J.A. Hoerter, M.V. Miceli, D.G. Reulund, W.E. Jacobus, G. Gerstenblith and E.G. Lakatta, A phosphorus-31 nuclear magnetic resonance study of the metabolic, contractile and ionic consequences of induced calcium alterations in the isovolumic rat heart, *Circ. Res.* 58:539-551 (1986).

EFFECT OF TAURINE ON INTRACELLULAR CALCIUM
DYNAMICS OF CULTURED MYOCARDIAL CELLS
DURING THE CALCIUM PARADOX

Kyoko Takahashi,[1] Hisato Harada,[1] Stephen W. Schaffer[2]
and Junichi Azuma[1]

[1]Department of Medicine III
Osaka University Medical School
Osaka, Japan

[2]Department of Pharmacology
University of South Alabama School of Medicine
Mobile, AL, USA

INTRODUCTION

The "calcium paradox" phenomenon, first described by Zimmerman and Hulsman[1], occurs when hearts are reperfused with calcium after a short period of calcium-free perfusion. The Ca^{2+} repletion phase causes irreversible myocardial damage, characterized by reduced electrical activity, extensive ultrastructural damage, depletion of tissue high-energy phosphate content, massive release of intracellular constituents and an increase in cytosolic Na^+ and Ca^{2+}[2].

Takihara et al.[3] have reported that the response of hearts to the calcium paradox could be partially regulated by myocardial taurine content. In this report, a link between calcium and taurine is confirmed. However, the exact mechanism of this cardiac action of taurine has not been fully elucidated.

Fura-2, one of the calcium-sensitive fluorescent dyes, has provided a new technique to monitor fluctuations in intracellular Ca^{2+} during the contraction cycle and to quantify the concentration of intracellular free calcium, $[Ca^{2+}]_i$[4]. This technique has permitted the investigation of normal Ca^{2+} transients and the cellular mechanisms that regulate $[Ca^{2+}]_i$.

In the present study, we examined the effect of taurine on $[Ca^{2+}]_i$ dynamics of spontaneously beating cultured myocardial cells subjected to the calcium paradox.

Taurine, Edited by J.B. Lombardini *et al.*
Plenum Press, New York, 1992

MATERIALS AND METHODS

Mouse Myocardial Cell Culture

The method of preparing myocardial cell cultures has been described by Takahashi et al.[5,6]. Isolated cardiac cells (2-4 X 10^5) were seeded into Petri dishes (35 mm i.d.) containing a few glass coverslips. The cells were maintained for 24 hr at 37°C in Eagle's minimum essential medium (Eagle MEM) supplemented with 10% newborn calf serum in a humidified environment containing 95% air-5% CO_2.

Measurement of Intracellular Free Calcium Transients

Fura-2 loading was performed by the addition of fura-2/AM (3 μM dissolved in dimethyl sulfoxide with 0.2% cremophor EL) into a Petri dish containing glass coverslips on which cells were attached and 1 ml of culture medium (Eagle MEM with 10% serum). After mixing, the dishes were incubated in the dark for 1 hr in humidified 5% CO_2-95% air atmosphere at 37°C. The medium was removed, the cells were rinsed three times with phosphate buffered saline and then reincubated with culture medium. A coverslip with fura-2 loaded cells was placed in an experimental chamber of 37 $\pm$ 0.5°C. The cells were incubated in control medium [modified Eagle MEM without phenol red but supplemented with 5% serum containing 10 mM N-2-hydroxyethyl-piperazine-N'-2-ethanesulfonic acid (HEPES)(pH 7.4)]. The Ca^{2+}-free buffer had the following constituents (in mM): NaCl, 137; KCl, 2.7; Na_2HPO_4, 8; KH_2PO_4, 1.5; ethyleneglycol-bis-(aminoethylether)-N,N'-tetracetic acid (EGTA), 2; HEPES, 10 (pH 7.4). An INTERDEC #M-1000 fluorescent spectromicroscope system (Osaka, Japan) was used to evaluate $[Ca^{2+}]_i$. The results were generally presented as fura-2 ratios. Background was determined using nonlabelled cells and was subtracted from the data obtained with fura-2 loaded cells. No dye leakage from the cells could be detected 1 hr after initiating the experiments. All data were expressed as Max (peak $[Ca^{2+}]_i$ during systole), Min (diastolic $[Ca^{2+}]_i$) and Ca-T (difference between the Max and Min values; calcium transient) and represented the mean of 10 beats.

Morphological Evaluation

Morphological status of myocardial cells was monitored with an inverted phase-contrast microscope and videomonitor at magnifications of 150 to 400 in a chamber controlled at 37°C. Data were expressed as percent of cells exhibiting morphological changes.

Statistics

Statistical significance was determined by Student's t-test. Each value was expressed as mean $\pm$ S.E.M. Differences were considered significant when the calculated P value was less than 0.05.

RESULTS

Calcium Transients of Cultured Myocardial Cells

Figure 1 shows a typical recording of calcium transients from spontaneously beating cultured myocardial cells loaded with fura-2. These cells, superfused with control medium containing 2 mM Ca^{2+}, beat spontaneously and exhibited typical calcium transients. In accordance with previous studies, fluorescence intensity of the fura-2 signal increased at 340 nm and decreased at 380 nm upon cell contraction, indicative of an elevation in $[Ca^{2+}]_i$ {Figure 1(A)}. Also shown is the fluorescence intensity ratio at 340 nm divided by the 380 nm signal {Figure 1(B)} and $[Ca^{2+}]_i$ as calculated by the equation of Grynkiwcz et al.[7] {Figure 1(C)}.

During the course of one cycle of contraction and relaxation, the 340/380 fluorescence ratio rises to a maximum during systole {Max or "a" in Figure 1(B)} and falls to a minimum during diastole {Min or "b" in Figure 1(B)}. The calcium

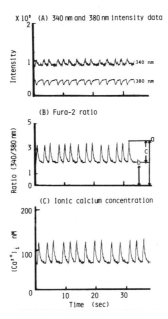

Figure 1. Calcium transients from control spontaneously beating cultured myocardial cells loaded with fura-2. All data in (A)-(C) were taken from the same cell. In (B), "a" refers to peak systolic value (Max), "b" represents diastolic value (Min) and "c" is the difference between "a" and "b" and is designated as $[Ca^{2+}]_i$ transient (Ca-T).

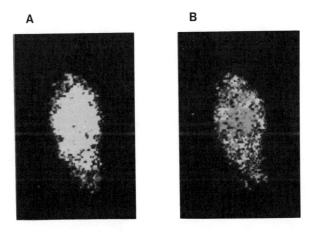

Figure 2. Fluorescence ratio image (340/380nm) of myocardial cells loaded with fura-2 in control medium. A. Peak systole, B. Diastole.

transient (Ca-T) depicted in Figure 1(C) represents the difference between the Max and Min values. Typically, Max is 3.1 ± 0.4, Min is 1.8 ± 0.2 and Ca-T is 1.3 ± 0.3 in normal myocytes for all coverslips utilized.

Figure 2 is a black and white image of beating myocardial cells loaded with fura-2 in control medium and demonstrates the spatial distribution of $[Ca^{2+}]_i$. During systole, white spots, which represent high $[Ca^{2+}]_i$, occupy large areas of the cell (Figure 2-A). In contrast, during diastole the fluorescence intensity of the area of the white spots are reduced (Figure 2B). The fluorescence ratio image reveals discrete, as well as clustered, white spots.

The Effect of Taurine on Intracellular Calcium Dynamics of Cultured Myocardial Cells during the Calcium Paradox

Figure 3 depicts an example of the types of changes in $[Ca^{2+}]_i$ which occur during the calcium paradox, the latter referring to the events and reactions which take place within isolated myocytes directly after reintroduction of calcium into incubation medium following a period of calcium-free exposure. Myocardial cells, which are incubated in control medium containing 2 mM Ca^{2+}, beat spontaneously and exhibit typical calcium transients. However, when these cells become exposed to Ca^{2+}-free buffer containing EGTA for 15 min, there is an immediate cessation of both spontaneous beating and calcium transients. Within a short time, $[Ca^{2+}]_i$ falls to a level of 1.5-1.7, which corresponds to an intracellular concentration of 3-6 X 10^{-8}M (Figure 3).

Restoration of medium calcium increased the values of Max and Min within 20 sec from 4.1 to 21.2 and from 2.2 to 15.2, respectively. Large increases in $[Ca^{2+}]_i$ are associated with contracture and blebbing of the cells; however, these changes were reversible. Major fluctuations in $[Ca^{2+}]_i$ were observed during the first few minutes

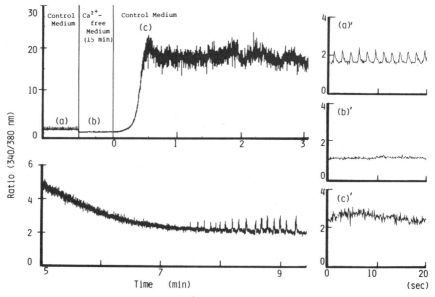

Figure 3. Time course of alterations in $[Ca^{2+}]_i$ during the calcium paradox. Myocardial cells were preincubated for 15 min in Ca^{2+}-free medium and then reexposed to medium containing normal Ca^{2+}. The data are expressed as ratio of 340nm/380nm. Inserts (a)'-(c)' represent the data using an expanded scale.

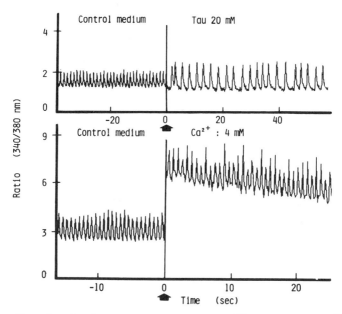

Figure 4. The effect of taurine (Tau) and elevated medium Ca^{2+} on changes in $[Ca^{2+}]_i$ of cultured myocardial cells. Following preincubation in control medium containing 2 mM Ca^{2+}, myocardial cells were exposed to medium containing either 20 mM taurine or 4 mM Ca^{2+}.

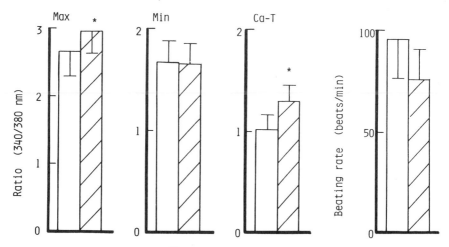

Figure 5. The effect of taurine on $[Ca^{2+}]_i$ and beating rate of cultured myocardial cells. Data shown represent measurements before and 15 min after exposure to 20 mM taurine and are expressed as means $\pm$ S.E.M. of 7 experiments. Asterisks indicate significant differences from control ($p < 0.05$). Control: □, Taurine: ▨ .

of the Ca^{2+} reintroduction step (Figure 3c). Subsequently, $[Ca^{2+}]_i$ decreased gradually over the next 4 min. Recovery of spontaneous beating and of calcium transients were only observed after an additional 7 min. Even after 14 min of Ca^{2+} restoration, the myocardial cells failed to beat rhythmically. These phenomena occurred independent of the beating rate over a range of 50-190 beats/min.

Since the pioneering study of Dolara et al.[8], several subsequent investigations have verified the biological interactions between taurine and calcium ion[9-11].

Nonetheless, the exact mechanisms underlying the cardiac actions of taurine have not been elucidated. As a result, we examined the action of taurine on $[Ca^{2+}]_i$ dynamics of the normally beating myocardial cell. Figure 4 shows that taurine caused an immediate increase in the Ca^{2+}-transient without significantly altering Min. It also mediated a decrease in beating rate. By contrast, elevations in $[Ca^{2+}]_o$ immediately increased all parameters (Max, Min and Ca-T), with the rise in Min being particularly noteworthy. After the initial increase, $[Ca^{2+}]_i$ gradually declined towards normal values.

We previously reported that at a dose of 20 mM, taurine uptake by myocardial cells reached a maximum by 15 min[6]. Based on these results, we decided to examine the effect of 15 min of taurine exposure on $[Ca^{2+}]_i$ and myocyte beating rate. It was found that taurine significantly increased Max and Ca-T from 2.7 to 3.0 and from 1.0 to 1.3, respectively, without significantly altering Min (Figure 5). At the same time, beating rate declined from 99 to 77 beats/min.

The effect of 20 mM taurine on $[Ca^{2+}]_i$ during the calcium paradox was also examined (Figure 6 and 7). Cells exposed to 20 mM taurine throughout the course of the experiment exhibited less accumulation of $[Ca^{2+}]_i$ during the Ca^{2+} restoration phase (Figure 6 vs Figure 3). The taurine-treated cells also recovered their characteristic beating pattern and calcium transients earlier than untreated cells, although during the Ca^{2+}-free phase, taurine caused no detectable effect on $[Ca^{2+}]_i$ (Figure 7, point b). Within 1 min after restoration of Ca^{2+}, $[Ca^{2+}]_i$ reached maximal levels in both taurine-treated and untreated cells (Figure 7, point c). Taurine reduced Max and Min from 21.2 to 11.4 and from 15.2 to 6.6, respectively, but had no effect on Ca-T at 1 min.

Morphological changes, such as formation of blebs or ballooning of the cell membrane, were observed in 57% of the cells upon restoration of Ca^{2+} (Figure 8). Taurine significantly decreased the number of morphologically abnormal cells from 57% to 35%. In both untreated and taurine-treated cells, the morphological changes were reversible and the cells gradually regained their normal shape and beating pattern.

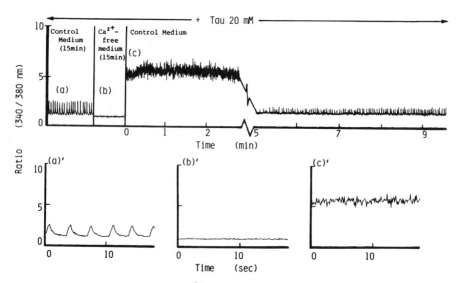

Figure 6. The effect of taurine (Tau) on $[Ca^{2+}]_i$ dynamics of cultured myocardial cells during the calcium paradox. Myocardial cells were preincubated for 15 min in medium containing 20 mM taurine prior to initiation of the experiment. The calcium paradox protocol described in Figure 3 was used, except all buffer contained 20 mM taurine.

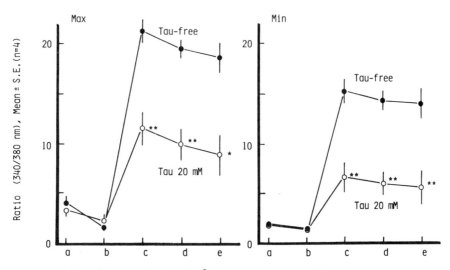

Figure 7. The effect of taurine (Tau) on $[Ca^{2+}]_i$ of cultured myocardial cells undergoing the calcium paradox. a: Control; b: Ca^{2+}-depletion phase; c-e: Ca^{2+}-repletion phase. "c" represents peak $[Ca^{2+}]_i$ reached less than 1 min after initiation of Ca^{2+} repletion. "d" and "e" refer to $[Ca^{2+}]_i$ 1 min and 2 min following introduction of Ca^{2+}. Open circles: taurine-free; closed circles: 20 mM taurine. Each value is the mean $\pm$ S.E.M. of 4 experiments. Asterisks indicate significant difference from taurine-free group (*:$p < 0.05$; **:$p < 0.01$).

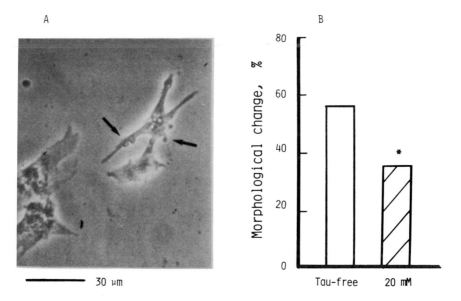

Figure 8. Improvement in myocardial cell morphology by exposure to taurine (Tau) during the calcium paradox. (A) characteristic morphological change induced by the calcium paradox during peak $[Ca^{2+}]_i$. (B) effect of taurine on the percentage of cells exhibiting altered morphology as a result of the calcium paradox. Asterisks represent significant difference from taurine-free group ($p < 0.05$).

DISCUSSION

In the present study, we measured $[Ca^{2+}]_i$ changes in isolated beating myocardial cells using fura-2. Although values obtained here for $[Ca^{2+}]_i$ were similar to those previously published[4,12] variations in the calibration and loading procedures between different laboratories are known to influence absolute levels of $[Ca^{2+}]_i$; therefore, our data was expressed as the fura-2 340nm/380nm fluorescence ratio. Generally, the most useful parameters were peak $[Ca^{2+}]_i$ during systole (Max), diastolic $[Ca^{2+}]_i$ (Min), and calcium transient (Ca-T) because they revealed the clearest changes in $[Ca^{2+}]_i$. A typical pattern of Ca^{2+} transients was observed using normal myocardial cells. Particularly revealing was the spatial distribution of Ca^{2+} within the cell. Figure 2 clearly demonstrates that the spatial distribution of $[Ca^{2+}]_i$ within beating cells was heterogeneous and that the degree of heterogeneity was enhanced during diastole.

It is well known that taurine has a positive inotropic action on the heart[13]. We have demonstrated in the present study that addition of taurine to normal medium increases the calcium transients of normal cells 30% while reducing beating rate and having no effect on Min. These results suggest that taurine modulates $[Ca^{2+}]_i$ during excitation-contraction coupling. By comparison, raising medium calcium concentration resulted in an elevation in all parameters examined (Max, Min, Ca-T). These differences in cellular response to taurine and elevated medium calcium concentration indicate that both agents increase $[Ca^{2+}]_i$ by a different mechanism.

The calcium paradox is characterized by a rapid and uncontrolled entry of Ca^{2+}. Although the gain in $[Ca^{2+}]_i$ that occurs during Ca^{2+} repletion is of critical importance, its route of entry has not been fully established. The known routes of Ca^{2+} entry include the voltage-dependent Ca^{2+} channel, the Na^+-Ca^{2+} exchanger, passive diffusion and abnormal sites of Ca^{2+} entry[2,14].

In the present study, it was shown that taurine treatment clearly improved the condition of myocardial cells subjected to the calcium paradox. Upon reintroduction of Ca^{2+} following a period of Ca^{2+}-free exposure, cells which had been pretreated for 15 min with 20 mM taurine recovered their spontaneous beating pattern and calcium transients earlier than untreated cells (5 min vs 7 min). The taurine-treated cells also exhibited less accumulation of $[Ca^{2+}]_i$, an effect which may account for the early recovery of beating. Taurine treatment also attenuated the genesis of morphological changes within myocardial cells induced by the calcium paradox. These results suggest that taurine plays an important role in stabilizing the membrane during the calcium paradox.

It is generally accepted that the initial event triggering the series of reactions leading to the calcium paradox is an alteration in Ca^{2+} permeability of the cell membrane during the Ca^{2+} free perfusion period[15]. The depletion of Ca^{2+} from the Ca^{2+} extracellular space leads to removal of Ca^{2+} from the membrane[1,15], resulting in increased membrane fluidity[16]. It has been suggested that taurine modulates Ca^{2+} homeostasis in hearts through its interaction with the sarcolemma[9,10]. Our data reveals that taurine causes no detectable effect on $[Ca^{2+}]_i$ during Ca^{2+}-free exposure. Therefore, it is likely that taurine provides protection against the calcium paradox by preventing non-specific permeability changes which lead to massive influx of Ca^{2+} upon Ca^{2+} repletion, although some other mechanism might be responsible for the protective action of taurine.

In summary: (a) Taurine increases calcium transients of normal myocardial cells by a different mechanism than the elevation in medium Ca^{2+}. (b) Taurine has no significant effect on $[Ca^{2+}]_i$ during Ca^{2+} depletion. (c) Taurine inhibits excessive accumulation of $[Ca^{2+}]_i$ during the Ca^{2+} repletion phase of the calcium paradox. (d) Taurine attenuates the development of morphological and beating abnormalities arising as a result of the calcium paradox. We suggest that important physiological and pharmacological roles of taurine include the regulation of calcium movement and membrane stabilization.

REFERENCES

1. A.N.E. Zimmerman and W.C. Hulsmann, Paradoxical influence of calcium ions on the permeability of the cell membranes of the rat heart, *Nature* 211:646-647 (1966).
2. R.A. Chapman and J. Tunstall, The calcium paradox of the heart, *Prog. Biophys. Molecular Biol.* 50:67-96 (1987).
3. K. Takihara, J. Azuma, S. Kishimoto, S. Onishi, and N. Sperelakis, Taurine prevention of calcium paradox-related damage in cardiac muscle, *Biochem. Pharmacol.* 37:2651-2658 (1988).
4. S. Bals, M. Bechem, W. Paffhausen, and L. Pott, Spontaneous and experimentally evoked $[Ca^{2+}]_i$-transients in cardiac myocytes measured by means of a fast fura-2 technique, *Cell Calcium* 11: 385-396 (1990).
5. K. Takahashi, Y. Fujita, T. Mayumi, T. Hama, and T. Kishi, Effect of adriamycin on cultured mouse embryo myocardial cells, *Chem. Pharm. Bull.* 35:326-334 (1987).
6. K. Takahashi, J. Azuma, N. Awata, A. Sawamura, S. Kishimoto, T. Yamagami, T. Kishi, H. Harada, and S.W. Schaffer, Protective effect of taurine on the irregular beating pattern of cultured myocardial cells induced by high and low extracellular calcium ion, *J. Mol. Cell. Cardiol.* 20: 397-403 (1988).
7. G. Gynkiewcz, M. Poenie, and R.Y. Tsien, A new generation of Ca^{2+} indicators with greatly improved fluorescence properties, *J. Biol. Chem.* 260:3440-3450 (1985).
8. P. Dolara, A. Agresti, A. Giotti, and G. Pasquini, Effect of taurine on calcium kinetics of guinea-pig heart, *Eur. J. Pharmacol.* 24:352-358 (1973).
9. R.J. Huxtable and L.A. Sebring, Cardiovascular actions of taurine, in: "Sulfur Amino Acids: Biological and Clinical Aspects," K. Kuriyama, R.J. Huxtable and H. Iwata, eds., Alan R. Liss, New York, 5-37 (1983).
10. S.W. Schaffer, J. Kramer, and J.P. Chovan, Does taurine have a function? Regulation of calcium homeostasis in the heart by taurine, *Fed. Proc.* 29:2691-2694 (1980).
11. F. Franconi, F. Martini, F. Stendardi, R. Matucci, L. Zilletti, and A. Giotti, Effect of taurine on calcium levels and contractility in guinea-pig ventricular strips, *Biochem. Pharmacol.* 31:3181-3185 (1982).
12. K.P. Burton, A.C. Morris, K.D. Massey, L.M. Buja, and H.K. Hagler, Free radicals alter ionic calcium levels and membrane phospholipid in cultured rat ventricular myocytes, *J. Mol. Cell. Cardiol.* 22:1035-1047 (1990).
13. R. Bandinelli, F. Franconi, A. Giotti, G. Martini, I. Stendardi, and L. Zilletti, The positive inotropic effect of taurine and calcium and the levels of taurine in ventricular strips, *Br. J. Pharmacol.* 72:115P-116P (1981).
14. T.J.C. Ruigrok, Is an increase of intracellular Na^+ during Ca^{2+} depletion essential for the occurrence of the calcium paradox? *J. Mol. Cell. Cardiol.* 22:499-501 (1990).
15. P.M. Grinwald and W.G. Nayler, Calcium entry in the calcium paradox, *J. Mol. Cell. Cardiol.* 13:867-880 (1981).
16. J. Campsi and C.J. Scandella, Calcium-induced decrease in membrane fluidity of sea urchin egg cortex after fertilization, *Nature Lond* 286:185-186 (1980).

INTRACELLULAR EFFECTS OF TAURINE:
STUDIES ON SKINNED CARDIAC PREPARATIONS

D.S. Steele and G.L. Smith

Institute of Physiology
Glasgow University
Scotland, G12 8QQ

INTRODUCTION

Taurine (2-aminoethane sulphonic acid) is the most abundant amino acid in the heart and contributes approximately 50% of the total amino acid pool. The intracellular concentration of taurine is species dependent, commonly about 5-20 mM[1]. Intracellular levels are maintained despite a much lower plasma concentration of about $60\,\mu M$[2]. The sarcoplasmic taurine concentration is linked to that of intracellular sodium suggesting the presence of sarcolemmal Na-taurine co-transport[3,4]. Elevated taurine levels have been found in heart tissue from patients with congestive heart failure and in experimental models of cardiac hypertrophy[5] while the taurine content of the heart decreases as a consequence of ischemia[6].

Taurine is reported to be an effective treatment for congestive heart failure[7] and protects against calcium Ca overload in a variety of conditions including cardiomyopathy[8], Ca paradox[9], hypoxic paradox[10], and isoprenaline toxicity[11]. When extracellular $[Ca^{2+}]$ exceeds the physiological range, taurine decreases contractility and cell calcium in guinea-pig ventricular muscle[12]. However, taurine increases contractility and cell calcium in the presence of low extracellular $[Ca^{2+}]$ and antagonises the negative inotropic effects of verapamil[13]. In addition, taurine acts antiarrythmically in a variety of experimental models[14,15].

While it is clear that taurine influences the Ca^{2+} homeostasis of the heart, its mechanism of action and physiological role remain unclear. Most studies have focused on the sarcolemma as a probable site of action. In isolated sarcolemmal preparations, taurine has been shown to affect Ca^{2+}-binding[13] and inhibit Na/Ca exchange[16]. In guinea-pig ventricular myocytes, introduction of taurine altered the inward Ca^{2+} current . The direction of this effect was dependent upon the extracellular $[Ca^{2+}]$, consistent with the inotropic action of taurine on intact tissue[17]. Taurine has been reported to inhibit α-adrenergic responses[18] but was without effect on ß-adrenergic responses or cAMP levels[7].

Reports of the effects of taurine on isolated intracellular membranes are contradictory. Taurine increased the rate of calcium accumulation and the maximum sequestering capacity in skeletal muscle SR[5]. Other studies found no effect of taurine on cardiac SR[19,20] but an increase in Ca^{2+}-binding was observed in mitochondrial

Taurine, Edited by J.B. Lombardini *et al.*
Plenum Press, New York, 1992

membrane fractions. However, taurine binding sites have been identified in isolated cardiac SR fragments from several species[21].

In this and previous studies,[22,23] we have investigated the intracellular effects of taurine in chemically skinned cardiac muscle. Taurine (1-40 mM) shifted the relationship between Ca^{2+} and tension towards lower $[Ca^{2+}]$ with little effect on maximum Ca^{2+}-activated force in Triton-treated rat ventricular trabeculae. This result is consistent with an increase in Ca^{2+} binding to troponin-C in the presence of taurine. In selectively saponin-skinned preparations, the amplitude of caffeine-induced contractures was used as an assay of the Ca^{2+}-content of the SR. Inclusion of taurine in a Ca^{2+}-loading solution potentiated the subsequent caffeine-induced contracture. The observed increase in Ca^{2+}-sensitivity was apparently insufficient to explain the potentiation of the caffeine response, so we concluded that taurine may increase Ca^{2+}-accumulation by the SR. However, this interpretation requires that the effect of taurine on steady-state force at a maintained $[Ca^{2+}]$ can be directly related to transient tension responses where Ca^{2+} is not in equilibrium with the myofilaments.

In this study we have employed a novel technique which allows measurement of Ca^{2+} within saponin-skinned trabeculae to study directly the effects of taurine on Ca^{2+} release by the SR. Furthermore, we will report that taurine affects rigor tension developed in the effective absence of Ca^{2+} and discuss the implications this may have for its mechanism of action on the contractile proteins.

METHODS

Sprague Dawley rats (200-250 g) were killed by a blow to the head and cervical dislocation. Hearts were removed rapidly and bathed in Tyrode's solution. Free running trabeculae (diameter 70-120 μm, length 2-3 mm) were dissected from the right ventricle. Experiments were done at room temperature (22-23°C).

Chemical skinning

The mounted preparation was exposed to a 'relaxing' solution (Table 1 solution B) including 50 μg/ml saponin (Sigma Chemicals Ltd.) for 30 mins. The skinning agent was then removed by bathing the preparation in solution C. In some experiments, measurements were made first after saponin-treatment and subsequently the SR membranes were destroyed by exposure to Triton X-100 (1% vol. 20 min) before the investigating the effects of taurine on the myofilaments.

Solution composition

The Ca^{2+} buffer EGTA was used to control $[Ca^{2+}]$. The solutions contained ATP and creatine phosphate to support contraction of the skinned muscle. Experiments were done at a pH(activity) of 7.00. Compositions of the solutions used are shown in Table 1. Solutions with a range of $[Ca^{2+}]$ were created by mixing solutions A and B in different proportions and used for myofilament Ca^{2+}-sensitivity determination. Solution D was used to provide the highest $[Ca^{2+}]$ and made by adding excess Ca^{2+} to solution A. Caffeine-induced contractures were evoked in 0.1 mM EGTA (solution C): concentrations higher than about 0.5 mM buffer the Ca^{2+} ions released from the SR and prevent the caffeine-induced contracture. Calcium chloride (1 M titration standard, BDH) was added to solution C to provide a range of $[Ca^{2+}]$ from pCa 7.2-6.0 and any necessary adjustment of pH was made. The equilibrium concentrations of metal ions were calculated using a computer program with the affinity constants for H^+, Ca^{2+} & Mg^{2+} and EGTA taken from Miller & Smith[24]. The affinity constants used for ATP and PCr were those quoted by Fabiato and Fabiato[25]. Corrections for ionic strength, details of pH measurement, allowance for EGTA purity and the

Table 1. Composition of solutions (in mM except where stated)

Solution	K	Na	Mg	Total [Ca²⁺]	[Ca²⁺] (μM)	ATP	PCr	EGTA	HDTA	HEPES
A	130	40	7	10.0	57.50	5	15	10.0	-	25
B	130	40	7	0.02	0.001	5	15	10.0	-	25
C	130	40	7	0.02	0.15	5	15	0.1	9.9	25
D	130	40	7	10.1	100.00	5	15	10.0	-	25

All solutions had a pH(activity) of 7.00. Free [Mg²⁺] was 2.1 mM in all solutions. Total chloride concentration varies from 110-120 mM. In some experiments $4\,\mu M$ Indo-1 (Calbiochem) was added to solution C.

principles of the calculations are detailed elsewhere[26,27].

Measurement of calcium in saponin-treated trabeculae.

Trabeculae were mounted between a force transducer (Akers AE875) and a fixed point by means of monofilament snares (Figure 1). The bath was placed on the stage of a Nikon Diaphot inverted microscope, and a perspex column was lowered close to the muscle to minimize the volume of solution above the preparation. The muscle was perfused by pumping solution (with $4\,\mu M$ Indo-1) down the central bore of the column at a rate of 1 ml/min. Caffeine (20 mM) was rapidly injected via the manifold at the base of the column. The preparation was illuminated with light of wavelength 360 nm, and [Ca²⁺] changes within a volume of solution containing the preparation were monitored by measuring the ratio of emitted light intensities at 405 nm and 495 nm. Light emitted from areas of the visual field not occupied by the muscle was reduced using a variable rectangular window on the side camera port before the TV camera and photomultiplier tubes. In some figures, a calibration bar showing the relationship between the Ca²⁺ and the measured ratio is provided. This applies only to *steady-state* [Ca²⁺]. During transient Ca²⁺-release from the SR, the fluorescence signal depends on the [Ca²⁺] within the preparation and that of the surrounding fluid.

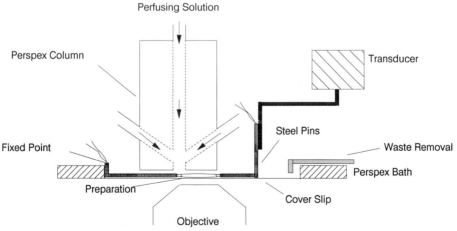

Figure 1. Diagrammatic representation of the apparatus used to measure Ca²⁺ within skinned preparations.

RESULTS

Caffeine-induced calcium transients in saponin-skinned muscle

The effect of bathing $[Ca^{2+}]$ on caffeine-induced Ca^{2+} and tension transients is shown in Figure 2. Injection of caffeine caused a transient increase in the fluorescence ratio and tension, as a consequence of Ca^{2+}-release from the SR followed by activation of the myofilaments. After 3 injections of caffeine, the $[Ca^{2+}]$ of the perfusing solution was increased from about 0.15 μM to 0.3 μM as indicated by a sustained rise in the fluorescence ratio. Subsequent applications of caffeine initiated larger Ca^{2+} and tension transients consistent with an increase in Ca^{2+}-accumulated by the SR during the period prior to release. After returning the $[Ca^{2+}]$ of the bathing solution to 0.15 μM the amplitude of both Ca^{2+} and tension responses gradually decreased towards control levels over 3-4 load and release cycles. Caffeine-induced calcium and tension transients were abolished by treatment with the non-ionic detergent Triton-X100 (1% vol for 10 min) which disrupts the SR (not shown).

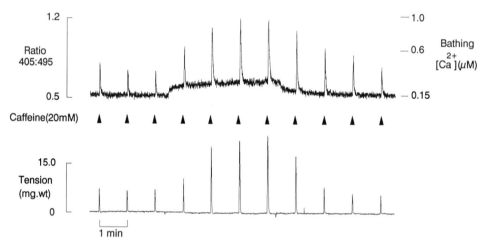

Figure 2. Simultaneous records of fluorescence ratio and isometric tension in a saponin-treated trabecula. The relationship between steady-state $[Ca^{2+}]$ and the fluorescence ratio is shown to the right of the diagram. Caffeine (20 mM) was injected for 50 msec at 1 min intervals. The basic solution composition as shown in Table 1 (c). $CaCl_2$ was added to produce the free $[Ca^{2+}]$ shown. All solutions contained Indo-1 (4 μM).

Figure 2 illustrates that the size of the Ca^{2+} and tension transients is dependent upon the $[Ca^{2+}]$ prior to addition of caffeine. In this study, the amplitude of the caffeine-induced calcium transient has been used as a indicator of the releasable Ca^{2+}-content of the SR. As caffeine is known to inhibit the accumulation of calcium by the SR, the relaxation phases were not analyzed in detail. In subsequent experiments the $[Ca^{2+}]$ of the bathing solution was routinely adjusted to result in responses to caffeine of 15-30% maximum Ca^{2+}-activated tension with a loading period of 1 minute. Such trains of contractures are reproducible over prolonged periods without significant deterioration.

Effect of taurine on caffeine-induced calcium release

We have previously demonstrated that the presence of taurine in both Ca^{2+}-loading and caffeine containing solutions increases the size of the caffeine-induced

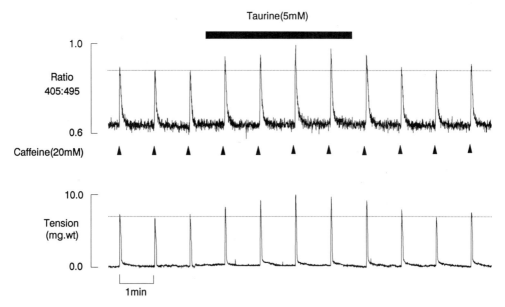

Figure 3. The effect of taurine (5 mM) on caffeine-induced Ca^{2+} and tension transients in a saponin-treated trabecula. Simultaneous records of fluorescence ratio (a) and tension (b) are shown. The basic solution composition was that shown in Table 1 (c) with added $CaCl_2$ to produce a final free $[Ca^{2+}]$ of $0.25 \mu M$. All solutions contained $4 \mu M$ Indo-1.

contractures[22]. However, concentrations of taurine within the physiological range also increase myofilament Ca^{2+}-sensitivity making the interpretation of this result difficult. Figure 3 shows that taurine also increases the amplitude the caffeine-induced Ca^{2+}-transient. In this experiment the $[Ca^{2+}]$ was maintained at $0.25 \mu M$. A train of regularly evoked contractures was produced in response to caffeine injected at 1 minute intervals. Once the responses had stabilized, taurine (5 mM) was introduced to both Ca^{2+}-loading and caffeine solutions. As indicated by the fluorescence ratio taurine had no effect on the free $[Ca^{2+}]$ of the solution. However, subsequent caffeine-induced Ca^{2+} and tension transients increased in size until a new steady-state was reached after 3-4 responses. After removal of taurine, the Ca^{2+} and tension responses gradually decreased to control levels over several load-release cycles. The average increase in Ca^{2+} and tension transients was $32 \pm 12\%$ (n=4) and $68 \pm 21\%$ (n=7) respectively in the presence of taurine.

Effects of taurine on spontaneous Ca^{2+} release

Saponin-treated trabeculae exhibited spontaneous oscillations of Ca^{2+} uptake and release at $[Ca^{2+}]$s of about 0.5-$1.0 \mu M$. Such oscillations have been reported in skinned[28] and intact cardiac preparations under conditions of Ca^{2+}-overload[29]. While asynchronous oscillations are occurring in different parts of the multicellular preparation, little or no tension may be produced due to the nonuniformity of activation, even if the $[Ca^{2+}]$ of the bathing solution is above the threshold for tension under steady-state conditions. Figure 4 shows spontaneous oscillations of calcium in the presence of $0.5 \mu M$ $[Ca^{2+}]$. In this case, no tension was generated. Introduction of taurine (5 mM) induced an increase in the amplitude of the spontaneous oscillations. Similar results were obtained in 4 other preparations.

Effects of taurine on Ca^{2+}-activated force and Ca^{2+}-independent rigor tension.

We have previously reported that taurine increases steady-state tension developed at a submaximal $[Ca^{2+}]$. Since this occurs in Triton-treated preparations where the SR is not functional it appears that taurine can act at the level of the myofilaments to increase Ca-sensitivity. This effect is shared with a variety of synthetic compounds including caffeine[30,31]. In Figure 5 we have compared the effects of 30 mM taurine (A) and 10 mM caffeine (B) on submaximal Ca^{2+}-activated force production. Taurine (30 mM) decreased the $[Ca^{2+}]$ required for half maximal activation by 0.051 ($\pm$ 0.015, n=4) pCa units while 10 mM caffeine typically produced a larger increase in Ca^{2+}-sensitivity (0.2 pCa units[31]). Taurine (30 mM) caused only a small increase in maximum Ca-activated force (0.8% $\pm$ 0.4%, n=4) while caffeine (10 mM) was without effect on this parameter (not shown).

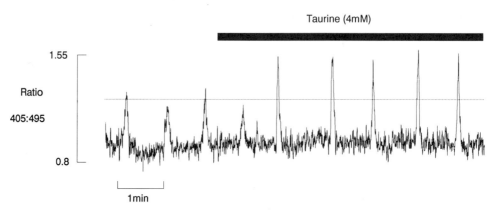

Figure 4. Effects of taurine (4 mM) on spontaneous Ca^{2+} and tension oscillations in a saponin-treated trabecula. Solution composition was that shown in Table 1(c) with added $CaCl_2$ to produce a final $[Ca^{2+}]$ of 0.5 μM. All solutions contained 4 μM Indo-1.

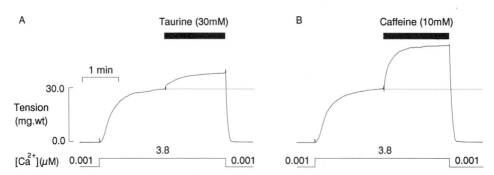

Figure 5. The effect of taurine and caffeine on submaximal Ca^{2+}-activated force in a Triton-treated trabecula. Panel (A), and (B), submaximal activation was achieved by increasing the $[Ca^{2+}]$ from 0.001 μM to 3.7 μM in solutions strongly Ca^{2+}-buffered with EGTA (10 mM). The steady-state tension achieved (dashed line) represents approximately half-maximal Ca^{2+}-activated tension. Addition of 30 mM taurine (A) and (10 mM) caffeine (B) caused a maintained increase in force.

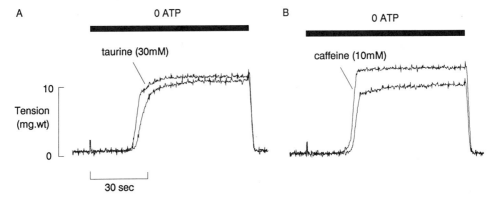

Figure 6. The effects of caffeine (10 mM) and taurine on rigor tension in a Triton-treated trabecula. A rigor contracture was induced by transferring the preparation from a solution containing 5 mM ATP (solution B) to an identical solution without ATP. $[Ca^{2+}]$ was 0.001 μM in all solutions.

It is generally assumed that such Ca-sensitizing compounds increase the affinity of troponin-C for calcium. However, as shown in Figure 6, taurine and caffeine increase the final level of rigor tension developed in the effective absence of Ca^{2+} 10^{-9}M. In this experiment, rigor was induced by transferring the muscle to an ATP-free solution. Figure 6 (A) shows superimposed rigor contractures developed in the absence (lower trace) and presence (upper trace) of taurine (30 mM). Taurine typically increased the final level of rigor tension by (11.4 $\pm$ 2.5 % n=6). Rigor contractures developed in the presence of caffeine (10 mM) were potentiated to a greater extent (Figure 6B) than those in the presence of taurine (25.3 $\pm$ 12.1 % n=7). This is consistent with the relative potencies of these compounds on Ca^{2+}-activated force.

DISCUSSION

Effect of taurine on the Ca^{2+}-content of the SR

Taurine increased the size of caffeine-induced Ca^{2+}-transients in saponin-skinned trabeculae (Figure 3). Under these conditions a maximum response to caffeine was achieved at approximately 10 mM; increasing the concentration of caffeine to 20 mM, as used in this study, caused no further Ca^{2+}-release (not shown). Thus, it seems likely that potentiation of the caffeine-induced Ca^{2+}-transient in presence of taurine reflects a increase in Ca^{2+}-accumulation by the SR, during the period prior to release, rather than facilitation of the caffeine-induced Ca^{2+}-release process. Consistent with this, taurine also increased the amplitude of spontaneous calcium release (Figure 4).

Mechanism of action of taurine on the SR

The simplest explanation of this phenomenon is that taurine stimulates directly the SR Ca^{2+}-pump protein, however, there are several other possible mechanisms. One theory which has received considerable attention is that taurine forms ionic bonds with phospholipids, thereby altering the structure of the sarcolemma and the affinity of Ca^{2+}-binding sites on the membrane[32] . This theory is equally applicable to the SR. Alternatively, taurine may act by inhibiting the action of calmodulin[33]. However, this should reduce calmodulin-stimulated phospholamban phosphorylation and decrease Ca^{2+}-accumulation by the SR. Clearly this is not consistent with the results presented in this study. However, calmodulin may be lost from the cytosol following saponin-treatment. Further work is required to establish whether the action of taurine on the SR is affected by introduction of exogenous calmodulin in saponin-treated trabeculae. Other possibilities are that taurine could decrease the affinity of the SR Ca^{2+}-pump for other ionic species such as magnesium, decrease the passive leak of Ca^{2+} from the SR via an action on the SR Ca^{2+}-channel or interact with structures within the SR such as calsequistrin.

Effect of taurine on the contractile proteins.

We have reported that taurine increases the apparent Ca^{2+}-sensitivity of the myofilaments in Triton-treated cardiac muscle[22]. This was subsequently confirmed in chemically skinned Crayfish skeletal muscle and pig heart preparations[34]. A similar Ca^{2+}-sensitizing action has been reported to occur with endogenous imidazole derivatives such as homocarnosine and a variety of synthetic compounds including caffeine[31,35].

It is generally assumed that such Ca^{2+}-sensitizers act by increasing the affinity of troponin-C for calcium. However, both caffeine and taurine potentiate the development of rigor tension in the effective absence of calcium[22]. As rigor is independent of $[Ca^{2+}]$ over range $10^{-10}M$ -$10^{-7}M$ Ca^{2+}, this suggests that caffeine and taurine may affect force production via a mechanism which is independent of Ca^{2+}-binding to troponin-C[22]. However, this does not exclude the possibility of an additional direct action on troponin-C. Indeed this is consistent with taurine's reported action as a inhibitor of calmodulin[33] as other such inhibitors increase myofilament Ca^{2+}-sensitivity by a direct action on troponin-C[36].

CONCLUSION

The effects of taurine on the SR and contractile proteins may contribute to the reported actions of taurine on intact tissue. For example, an increase in Ca^{2+}-accumulation by the SR could explain the positive inotropic action of taurine and its ability to antagonise the negative inotropic effects of low extracellular Ca^{2+}[12]. However, these results cannot explain the antiarrythmic action of taurine or the decrease in contractility and the fall in cell calcium observed in the presence of high extracellular Ca^{2+}. Indeed the potentiation of spontaneous Ca^{2+}-release (figure 4) would seem to predispose the muscle to arrythmia. However, such spontaneous activity requires that the $[Ca^{2+}]$ at the external surface of the SR is maintained at a relatively high level. Therefore, it seems likely that other well documented effects of taurine on the sarcolemma serve to decrease intracellular $[Ca^{2+}]$ under conditions which would normally lead to Ca^{2+}-overload and arrhythmia.

ACKNOWLEDGMENTS

This work was financially supported by the Wellcome Trust and the British Heart Foundation.

REFERENCES

1. J.G. Jacobson and J.R.L.H. Smith, Biochemistry and physiology of taurine and taurine derivatives, *Physiol. Rev.* 48:424-511, 1968.
2. T.L. Perry and S. Hansen, Technical pitfalls leading to errors in the quantification of plasma amino acids, *Clinica Chimica Acta* 25:53, 1969.
3. J. Bahl, C.J. Frangakis, B. Larsen, S. Cahng, D. Grosso, and R.A. Bressler, Accumulation of taurine

by isolated rat heart cells and rat heart slices, *in:* "The Effects of Taurine on Excitable Tissues," S.W. Schaffer, S.W. Baskin, and J. Kocsis, ed., Lancaster: Spectrum Publications Inc., 1981, p. 247-258.

4. R. A. Chapman and M.S. Suleiman, Na-dependent taurine uptake in isolated bovine cardiac sarcolemmal vesicles, *J. Physiol.* 430:72**P**, 1991.

5. R. Huxtable and R. Bressler, Effect of taurine on a muscle intracellular membrane, *Biochemica et Biophysica Acta* 323:573, 1974.

6. M.F. Crass, and J.B. Lombardini, Loss of cardiac muscle taurine after left ventricular ischaemia, *Life Sci.* 21:951-958, 1977.

7. J. Azuma, A. Sawamura, and N. Awata, Theraputic effect of taurine in conjestive heart failure; a double-blind cross-over trial, *Clin. Cardiol.* 8:276-282, 1985.

8. M.J. McBroom and J.D. Welty, Effect of taurine on heart calcium in the cardiomopathic hamster, *J. Mol. Cell. Cardiol.* 9:853-859, 1977.

9. J.H. Kramer, J.P. Chovan, and S.W. Schaffer, The effect of taurine on calcium paradox and ischaemic heart failure, *Amer. J. Physiol.* 240:H238-H246, 1981.

10. F. Franconi, I. Stenardi, and P. Failli, The protective effects of taurine on hypoxia and reoxygenation in guinea-pig heart, *Biochem. Pharmacol.* 34:2611-2615, 1985.

11. H. Ohta, A. Junichi, N. Awata, Mechanism of the protective effect of taurine against isoprenaline induced myocardial damage, *Cardiovas. Res.* 22:407-413, 1988.

12. F. Franconi, F. Martini, I. Stendardi, R. Matucci, L. Zilletti, and A. Giotti, Effect of taurine on calcium levels and contractility in guinea-pig ventricular strips, *Biochem. Pharmacol.* 31:3181--3185, 1982.

13. J.P. Chovan, E.C. Kulakowski, S. Sheakowski, and S.W. Schaffer, Calcium regulation by the low-affinity taurine binding sites of cardiac sarcolemma, *Mol. Pharmacol.* 17:295-300, 1980.

14. J. Hernandez, S. Artillo, M.I. Serrano, and J.S. Serrano, Further evidence for the antiarrhythmic efficacy of taurine in the heart, *Res. Commun. Chem. Pathol. Pharmacol.* 43:343-346, 1984.

15. K. Takahashi, J. Azuma, N. Awata, Protective effect of taurine on the irregular beating pattern of cultured myocardial cells induced by high and low extracellular calcium ion, *J. Mol. Cell. Cardiol.* 20:397-403, 1988.

16. T. Matsuda, T. Gemba, A. Baba, and H. Iwata, Inhibition by taurine of Na-Ca exchange in sarcolemmal membrane vesicles from bovine and guinea-pig hearts. *Comp. Biochem. Physiol.* 94C:335-339, 1989.

17. A. Sawamura, H. Sada, J. Azuma, S. Kishimoto, and N. Sperelakis, Taurine modulates ion influx through cardiac Ca^{2+} channels, *Cell Calcium* 11:251-259, 1990.

18. F. Franconi, F. Bennardini, R. Matucci, Functional and binding evidence of taurine inhibition of alpha-adrenoceptor effects on guinea-pig ventricle, *J. Mol. Cell. Cardiol.* 18:461-468, 1986.

19. J.D. Welty and C.M. Welty, Effects of taurine on subcellular calcium dynamics in normal and cardio myopathic hamster heart, *in: "Effects of Taurine on Excitable Tissues,"* S.W. Baskin, J. Kocsis, and S.W. Schaffer, eds., Lancaster: Spectrum Publications, 1991,

20. M.L. Entman and B.P. Bornet, Effect of calcium on cardiac sarcoplasmic reticulum, *Life Sci.* 21:543-550, 1977.

21. M.C. Quennedey, J. Velly, and J. Schwartz, [3H] Taurine binding on a cardiac sarcoplasmic fraction from rats, rabbits and pigs, *Eur. J. Pharmacol.* 38:73-76, 1986.

22. D.S. Steele, G.L. Smith, and D.J. Miller, The effects of taurine on Ca^{2+} uptake by the sarcoplasmic reticulum and Ca^{2+} sensitivity of chemically skinned rat heart, *J. Physiol.* 422:499-511, 1990.

23. D.J. Miller, G.L. Smith, and D.S. Steele, Taurine enhances sarcoplasmic reticulum function and myofilament calcium sensitivity in chemically skinned rat ventricular trabeculae, *J. Physiol.* 415:111-111, 1989.

24. G.L. Smith and D.J. Miller, Potentiometric measurements of stoichiometricand apparent affinity constants of EGTA for protons and divalent ions including calcium, *Biochim. Biophy. Acta* 839:287-299, 1985.

25. A. Fabiato and F. Fabiato, Calculator programs for computing the composition of the solutions containing multiple metals and ligands used for experiments in skinned muscle cells, *J. Physiol.* 75:463-505, 1979.

26. D.J. Miller and G.L. Smith, EGTA purity and the buffering of calcium ions in physiological solutions, *Amer.J. Physiol.* 246:C160-C166, 1984.

27. S.M. Harrison, C. Lamont, D.J. Miller, and D.S. Steele, Sulmazol (AR L 115BS) induces caffeine-like contractures in mammalian cardiac muscle selectively skinned with saponin. *J. Physiol.* 407:124-124, 1988.

28. A. Fabiato, Spontaneous versus triggered contractions of "calcium tolerant" cardiac cells from the adult rat ventricle, *Basic Res. in Cardiol.* 80:83-88, 1985.

29. C.J. Nieman and D.A. Eisner, Effects of caffeine, tetracaine and ryanodine on calcium-dependent oscillations in sheep cardiac purkinje fibers. *J. Gen. Physiol.* 86:877-889, 1985.

30. B. Wetzel and N. Hauel, New cardiotonic agents-a promising approach for treatment of heart failure, *TIPS* 9:166-170, 1988.

31. I.R. Wendt and D.G. Stephenson, Effects of caffeine on Ca-activated force production in skinned cardiac and skeletal muscle fibres of the rat, *Pflugers. Archive.* 398:210-216, 1983.
32. R. Huxtable and L.A. Sebring, "Cardiovascular Actions of Taurine," New York:Sulfur amino acids:Biochemical and Clinical Aspects, 1983. pp. 5-37.
33. S.W. Schaffer, S. Allo, H. Harada, and J. Azuma, "Regulation of Calcium Homeostasis by Taurine: Role of Calmodulin, Wiley-Liss Inc., New York, 1990. pp. 217-225.
34. S. Galler, C. Hutzler, and T. Haller, Effects of taurine on Ca^{2+}-dependent force development of skinned muscle fibre preparations, *J. Exp. Biol.* 152:255-264, 1990.
35. D.J. Miller, J. Campbell, J.J. O'Dowd, and D.J. Robins, Novel endogenous imidazoles calcium-sensitize chemically skinned rat heart muscle, *J. Physiol.* 427:54P, 1990.
36. R.J. Solaro, P. Bousquet, and J.D. Johnson, Stimulation of cardiac myofilament force,ATPase activity and troponin-C Ca^{2+} binding by Bepridil, *J. Pharm. Exp. Ther.* 238:502-507, 1986.

EFFECTS OF TAURINE DEFICIENCY ON ARRHYTHMOGENESIS AND EXCITATION-CONTRACTION COUPLING IN CARDIAC TISSUE

Norma Lake

Departments of Physiology and Ophthalmology
McGill University, 3655 Drummond Street
Montreal, Quebec, Canada H3G 1Y6

INTRODUCTION

Taurine is the major constituent of the free amino acid pool of mammalian myocardium, but its precise function is little understood. There is little biosynthesis of taurine within the heart; instead its mM levels are determined by active transmembrane transport by a specific carrier. It is not found in proteins, neither is it a substrate for metabolism. Taurine deficiency, however, produces cardiac electrophysiological and mechanical abnormalities (Lake et al., 1987; 1990) and life-threatening dilated cardiomyopathy documented by echocardiography in cats and foxes (Pion et al., 1987; Moise et al., 1989). Novotny et al. (1991) have recently described deficits in left ventricular systolic pressure development in such cats. Taurine supplements can reverse these conditions, and are also beneficial in the treatment of human congestive heart failure of many etiologies (Azuma et al., 1985) and a rabbit model of this condition (Azuma et al., 1984). The sites of action of most of these effects are unknown. The majority of studies have described the pharmacological effects of exogenous taurine in doses which far exceed the micromolar levels normally found in the plasma. Interactions with calcium (Ca) are suggested by the observations that taurine administration has positive inotropic effects (Huxtable, 1976), prevents Ca paradox (Kramer et al., 1981) and delays the onset of Ca accumulation which causes necrotic lesions in genetic cardiomyopathic hamsters (McBroom and Welty, 1977).

Some groups, including our own, have put more effort into attempts to understand the role played by taurine in the intracellular compartment, since that is where the millimolar amounts are located. Experiments with sarcolemmal (SL) vesicles suggest that through effects on phospholipids taurine modulates binding of a pool of Ca at the cytosolic face rather than externally (Huxtable and Sebring, 1986). Harada et al. (1988; 1990) have proposed that depression of the SL Ca-ATPase (Ca extrusion pump) underlies the deleterious effects of taurine depletion and the enhanced toxicity (Ca overload) of doxorubicin they have observed. These conclusions were based in part on SL vesicle preparations. Although the SL Ca pump may be depressed, we think that its overall contribution to Ca homeostasis may be rather minor, in view of its low capacity, and the likelihood that any deficit could be compensated for by increased extrusion via the high capacity Na/Ca exchanger, and/or reuptake by the sarcoplasmic reticulum (SR).

Action Potential Characteristics of Papillary Muscle Fibres

Our initial studies (Lake et al., 1987) described QT prolongation in rats followed with EKG recordings during 20 weeks of treatment with guanidinoethane sulfonate (GES), an inhibitor of taurine uptake by the heart which leads to taurine depletion, since local biosynthesis is inadequate or absent (Huxtable et al., 1979). At the end of that time papillary muscles were used to study action potential characteristics. These studies showed significant prolongations of APD_{75} and APD_{95} (Action Potential Duration to 75% or 95% repolarization) in taurine-deficient animals. The prolongations (QT and APD) were reversible by taurine supplements.

Treatment with GES for as little as 6 weeks produces considerable taurine depletion, not so for ß-alanine, which *in vitro* was a more potent taurine uptake inhibitor than GES (Huxtable et al., 1979). We have shown that ß-alanine *in vivo* had a transient effect on taurine levels (Lake and DeMarte, 1988) which declined, presumably since it can be metabolized (and with treatment the liver is induced to do so), whereas GES is more metabolically resistant.

The APD alterations produced by 4 or 8 wk GES are very similar and both greater than the peak effects of ß-alanine, which parallels the treatments' efficacies in reducing taurine levels: GES-treated animals have ventricles depleted to about 30% of control, compared to 70% of control following ß-alanine. In ß-alanine treated animals the slow response action potentials recorded from cells showed no differences from control cells in the upstroke velocity or amplitude, which processes reflect the underlying current inflow through the calcium channels, and suggest *it* did not differ between the groups. These slow response action potentials are recorded when the fast sodium channels are inactivated by a high potassium medium.

With GES or ß-alanine treatment, prolongation of the late phase of repolarization of the fast response action potentials is seen in taurine-deficient muscles compared to controls (see Table 1 for the values from short periods of GES or ß-alanine treatment). We propose that decrements in outward K currents and/or increments in inward current from Ca extrusion via the Na/Ca exchanger, thought to be active in this late phase of the action potential (Schouten and ter Keurs, 1985; Noble, 1986) can underlie this prolongation. Since many membrane channels are modulated by Ca transients, it could be that the depression of the SL Ca pump reported by Harada et al., (reviewed above) combined with depressed SR function suggested from our studies (discussed below) alter internal Ca homeostasis sufficiently to have consequences for the membrane ionic channel properties (Lake, 1990). Alternatively, the outcome of these two effects, depression of the SL and SR Ca pumps, could redirect more of the Ca extrusion to the Na/Ca exchanger which being electrogenic would thus generate more inward current which would prolong APD.

Taurine Deficiency and Arrhythmogenesis

Because action potential prolongation via increased late inward current could predispose to triggered activity, we have explored the idea that taurine deficiency may enhance vulnerability to arrhythmias. The paradigm was left coronary artery occlusion in control and taurine-depleted rats (GES-treated for 6 weeks). Just prior to occlusion the rats were given a 0.5 ml bolus of saline or 1 mM taurine in saline, and infusion was continued at 0.5 ml per hour, higher rates being detrimental (Curtis et al., 1987). Thus there were four groups of rats: control or taurine-depleted, treated with saline or taurine infusion. Occlusion of the coronary vessel leads in about 4

Table 1. Effects of taurine depletion on papillary muscle action potentials.

	Control n = 50	GES 4 wk 19	Control 18	ß-alanine 28
Parameter				
Ampl (mV)	102.4 ± 8.7	89.2 ± 14*	105.7 ± 4.7	103.4 ± 3.4
Vm (mV)	-83 ± 7	-76 ± 8**	-81.5 ± 2.3	-80.2 ± 2.3
Vmax (V/s)	241 ± 67	163 ± 46**	251 ± 62	224 ± 23*
Action potential duration (ms)				
APD_5	1.5 ± 0.6	2.2 ± 1.0	1.4 ± 0.6	1.4 ± 0.3
APD_{10}	2.3 ± 0.9	3.3 ± 1.2	2.1 ± 0.8	2.2 ± 0.6
APD_{25}	4.9 ± 1.4	6.5 ± 1.9	4.5 ± 1.3	5.3 ± 1.7
APD_{40}	7.5 ± 2.1	9.9 ± 2.7	7.2 ± 1.4	8.4 ± 2.7
APD_{50}	9.5 ± 2.4	13.4 ± 4.3	9.2 ± 1.8	10.7 ± 3.5
APD_{60}	12.5 ± 3.0	20.7 ± 8.0*	12.1 ± 2.4	14.4 ± 4.9
APD_{75}	20.6 ± 6.0	45.9 ± 16.3**	20.6 ± 4.6	26.5 ± 9.0*
APD_{90}	44.1 ± 14.4	83.5 ± 19.1**	44.7 ± 9.4	58.1 ± 13.9**
APD_{95}	61.8 ± 19.0	103.0 ± 18.9**	65.6 ± 6.8	79.1 ± 10.1*

Values are mean ± S.D. n = number of cells; Ampl = amplitude; Vm = resting membrane potential; Vmax = maximal upstroke velocity. Stimulation frequency was 1 Hz. Significance of difference between treated and control means given by *$p < 0.05$; **$p < 0.01$ using post ANOVA t-tests.

minutes to arrhythmias easily seen on the EKG record. The animals were monitored for four hours (if they survived that long). Death was usually a result of several periods of severe ventricular fibrillation. At the conclusion of the experiment indocyanine green dye was perfused to reveal the occluded zone (dye excluded) which was excised, weighed and recorded as percentage of total ventricular weight. Subsequently, the ventricular tissue was sliced and incubated in tetrazolium dye which gives an estimate of the infarcted zone (white) while the viable tissue is purple (Botting et al., 1983).

The size of the occluded zone (30-35% ventricular weight) varied little amongst the rats. Compared to the dog for example, the rat is advantageous for this type of study because its uniform coronary anatomy between animals gives a high reproducibility of occlusion effects. The infarct size (approximately 25%) also showed little variation between groups. Curtis et al. (1987) have shown that there is an important correlation between arrhythmias and size of the ischemic region, presumably because the site of arrhythmogenesis is at the border between the ischemic and normal tissue. Although the infarct sizes were not different, taurine-depleted animals had lower survival rates and more PVC's (premature ventricular contractions) and more and longer episodes of ventricular tachycardia (VT) and fibrillation (VF). Taurine infusion increased survival time and decreased PVCs, VT and VF in both control and depleted rats. Figure 1 shows results obtained from 15 rats in each of the four groups: control rats either with saline or taurine infusion, or taurine-deficient rats with either saline or taurine infusion. These experiments indicate that (1) taurine-deficient animals appear to be more susceptible to ischemia-induced arrhythmias and (2) taurine may be a useful anti-arrhythmic compound.

Effect of Taurine Deficiency on Excitation-Contraction Coupling

Our current understanding of excitation-contraction coupling, e.g. in rat ventricle, is that a relatively small amount of "trigger" Ca which enters through Ca channels during the action potential, releases a much larger amount of Ca from the sarcoplasmic reticulum (SR), which subsequently activates the myofibrils to contract. Trans-

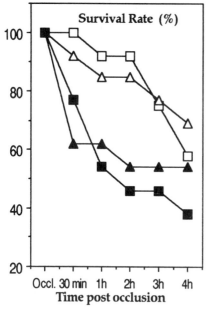 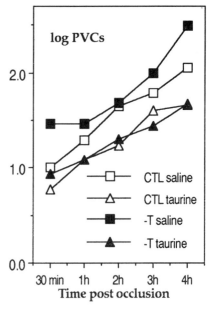

Figure 1. Effects of acute left coronary occlusion in control (CTL, open symbols) and taurine-depleted (-T, closed symbols) rats, receiving infusions of saline (squares) or taurine (triangles). Survival rates are shown for 15 rats in each of these four groups. The premature ventricular contraction (PVC) data are shown only for the animals which survived for 4 h in order to show time course of effects.

sarcolemmal extrusion and resequestration of Ca by the SR allows relaxation to proceed and re-primes the SR for the next release. The SR release channels recover rapidly, so the amount released by the SR is a positive function of the Ca accumulation which can occur in the interval between stimuli, and the force trace is a good monitor of the Ca released.

Our initial studies (Lake et al., 1990) were carried out on papillary muscles from control rats and others treated for 6 weeks with the taurine uptake antagonist, GES (resulting in 75% depletion of myocardial taurine). There were no changes in DNA, i.e. no cell death occurred. All studies were conducted with the muscles at initial length, Lmax, the length that produced maximal active tension. Taurine-depleted muscles did not differ in dimensions or preload required for Lmax, but generated tensions only two thirds of control. From measurements of after- and unloaded contractions, the velocity-tension relationship and the derived maximal velocity of shortening (Vmax) appeared not different between the groups. This suggests that the interaction of Ca with contractile proteins and their inhibitors was unchanged by taurine depletion. Since peak tension depends on the *number* of activated contractile elements which is a function of the level of the systolic Ca transient, the tension deficits we observed could be due to lower levels of calcium or fewer responsive elements. While the velocity of shortening was unchanged, relaxation times with stimulation at 0.1 Hz were significantly prolonged in taurine-deficient muscles and contraction durations were prolonged. These studies suggested that taurine deficiency may lead to reductions in action potential triggered Ca release from internal stores, and/or deficits in Ca resequestration. This might result from disfacilitation of Ca binding to the SR and other storage sites during taurine deficiency, in view of the *in vitro* observations of taurine enhancement of Ca binding to phospholipids (Huxtable and Sebring, 1986).

The papillary muscle preparations we used were about 1 mm in diameter which may be close to the upper limit that can be supported metabolically by superfusion. In fact, work by Schouten and ter Keurs indicates that *in vitro* the reduction of peak force and action potential duration observed at high stimulation frequencies is a function of muscle diameter owing to metabolic considerations (Schouten and ter Keurs, 1986). In order to pursue more detailed and sophisticated studies of the effects of taurine deficiency on excitation-contraction coupling it appeared prudent to move to the use of smaller papillary muscles and free running ventricular trabeculae. In 250-350 g rats thin (100-150 μ) flat ribbon-like or cylindrical trabeculae can be found which are also optimal for utilizing laser diffraction techniques to monitor sarcomere length during contractility experiments (ter Keurs et al., 1980).

Using such trabeculae, we have found that taurine-deficient muscles (20-30% of control taurine levels) showed significant decrements in maximal calcium-activated force (Fmax), confirming our previous findings with papillary muscles (see Table 2). To examine SR function, a short train of extra systoles (e.g. 10 @ 2 Hz) is inserted into the basic rate (0.5 Hz) which tends to load the muscle and the SR with Ca. The subsequent beat is of enhanced amplitude due to the higher SR release, and the fraction of Ca recirculated through the SR (rather than being extruded from the cell) can be calculated from the beat-to-beat decay of this post extra systolic potentiation. Our results from this kind of protocol have suggested that the fraction of Ca which recirculates through the SR (B_f) is reduced from control values in taurine-depleted trabeculae (see Table 2).

While no differences from control were found in force-interval relationships (except that maximal force production was less), taurine-depleted muscles required significantly higher Ca levels for half maximal force generation, determined from computer fits to the Hill equation describing the force-Ca relation. While doing these protocols it became obvious that taurine-depleted muscles showed after-contractions and spontaneous activity (an index of SR Ca overload) more frequently and at lower Ca concentrations than did control trabeculae. This by itself would tend to reduce muscle performance and is also consistent with our hypothesis of decreased Ca sequestering capacity in taurine-depleted muscles.

We have carried out preliminary screening of the calsequestrin (the primary calcium-binding protein of SR) content of control and taurine-depleted hearts with Dr. M. Michalak (University of Alberta, Edmonton). The quantity of calsequestrin is not different between the groups. Other experiments are required to investigate if the calcium binding capacity or affinity of the calsequestrin is altered.

Since taurine depletion is achieved through treating the rats with GES, a taurine transport inhibitor in their drinking water, I have tried to rule out direct effects of GES. In papillary muscle studies restoration of tissue taurine levels in the continued presence of GES was associated with recovery of QT, APD and Fmax values back to control. One of the most convincing pieces of evidence against direct effects of GES on excitation-contraction coupling has come from studies by Robert Godt, Medical College of Georgia, using chemically skinned trabeculae and GES which I synthesized

Table 2. Effects of taurine depletion on cardiac trabeculae.

Parameter	Control	Depleted (% of control)	p (signif)
Ventricular taurine (μmol/mg DNA)	39.1 $\pm$ 2.4	38	<0.001
Stress (mN/mm^2) in 0.7 mM Ca	57.8 $\pm$ 5.4	55	0.01
mN/mm^2 (Fmax) Ca activated	80.0 $\pm$ 11.4	62	0.03
mN/mm^2 (Fmax) Strontium	109.8 $\pm$ 19.1	56	0.05
Recirculation fraction (B_f)	0.82 $\pm$ 0.03	81	<0.01

and provided to him. While 24 mM taurine significantly increased both Ca sensitivity and maximal Ca activated force (normalized to cross sectional area), 24 mM GES was entirely without effect (personal communication).

The overall conclusions from the contractility studies I have carried out are that during taurine deficiency SR Ca handling function is depressed *and/or* the force-generating machinery is somehow reduced in efficacy. The latter suggestion comes from considering the outcomes of various potentiation protocols such as twin pulses, force-frequency, rest potentiation, etc., that we have tested in trabeculae and papillary muscles (e.g., Lake et al., 1990). In all cases the extent of potentiation was highly similar in control and depleted muscles such that the differences remained between them in the force generated (depleted muscles gave only 60% of control). Recent studies using strontium-supplemented media to bypass the SR and activate the myofibrils directly have shown a significant deficit compared to controls in the maximal strontium-activated force produced by taurine-depleted trabeculae (Table 2). This persisting deficit suggests that whether or not there is a problem with SR calcium resequestration and release, the number of responsive contractile elements appears subnormal in taurine-depleted muscles.

ACKNOWLEDGEMENTS

These studies were supported in part by grants from the Medical Research Council of Canada and the Canadian Heart and Stroke Foundation. I am indebted to my collaborators, Drs. H.E.D.J. ter Keurs and D. Eley, for advice and assistance with studies using the trabeculae. I am grateful for the skillful work of Sue Cocker and Dr. Shimin Wang in some of the electrophysiological studies, Luisa DeMarte for the biochemical assays, and Carole Verdone-Smith for preparation of the manuscript.

REFERENCES

Azuma, J., Sawamura, A., Awata, N., Ohta, H., Hamaguchi, T., Harada, H., Takihara, K., Hasegawa, H., Yamagami, T., Ishiyama, T., Iwata, H., and Kishimoto, S., 1985, Therapeutic effect of taurine in congestive heart failure: a double-blind crossover trial, *Clin. Cardiol.* 8:276-282.

Azuma, J., Takihara, K., Awata, N., Ohta, H., Sawamura, A., Harada, H., and Kishimoto, S., 1984, Beneficial effect of taurine on congestive heart failure induced by chronic aortic regurgitation in rabbits, *Res. Commun. Chem. Pathol. Pharmacol.* 45:261-270.

Botting, J.H., Johnston, K.M., MacLeod, B.A., and Walter, M.J.A., 1983, The effect of modification of sympathetic activity on responses to ligation of a coronary artery in the conscious rat, *Br. J. Pharmacol.* 79:265-271.

Curtis, M.J., Macleod, B.A., and Walter, M.J.A., 1987, Models for the study of arrhythmias in myocardial ischemia and infarction: the use of the rat, *J. Mol. Cell Cardiol.* 19:399-419.

Harada, H., Allo, S., Viyuoh, N., Azuma, J., Takahashi, K., and Schaffer, S.W., 1988, Regulation of calcium transport in drug-induced taurine-depleted hearts, *Biochim. Biophys. Acta* 944:273-278.

Harada, H., Cusak, B.J., Olson, R.D., Stroo, W., Azuma, J., Hamaguchi, T., and Schaffer, S.W., 1990, Taurine deficiency and doxorubicin: interaction with the cardiac sarcolemmal calcium pump, *Biochem. Pharmacol.* 39:745-751.

Huxtable, R.J., 1976, Metabolism and function of taurine in the heart, *in*: "Taurine," R.J. Huxtable and A. Barbeau, eds., Raven Press, New York, pp. 99-119.

Huxtable, R.J., Laird, H.E., and Lippincott, S., 1979, The transport of taurine in the heart and the rapid depletion of tissue taurine content by guanidinoethyl sulfonate, *J. Pharmacol. Exp. Ther.* 211:465-472.

Huxtable, R.J. and Sebring, L.A., 1986, Towards a unifying theory for the action of taurine, *Trends Pharm. Sci.* 7:4851-485.

Kramer, J.H., Chovan, J.P., and Schaffer, S.W., 1981, Effect of taurine on calcium paradox and ischemic heart failure, *Amer. J. Physiol.* 240:H238-H246.

Lake, N., 1990, Effects of taurine depletion on excitable tissues: recent studies, *in*: "Taurine: Functional

Neurochemistry, Physiology and Cardiology", H. Pasantes-Morales, D.L. Martin, W. Shain, and R. Martin del Rio, eds., A.R. Liss, New York.

Lake, N. and De Marte, L., 1988, Effect of ß-alanine treatment on the taurine and DNA content of the rat heart and retina, *Neurochem. Res.* 13:1003-1006.

Lake, N., De Roode, M., and Nattel, S., 1987, Effects of taurine depletion on rat cardiac electrophysiology: *in vivo* and *in vitro* studies, *Life Sci.* 40:997-1005.

Lake, N., Eley, D.W., and ter Keurs, H.E.D.J., 1991a) The effects of taurine depletion on excitation-contraction coupling in rat cardiac trabeculae, *Biophysical J.* 59:64A.

Lake, N., Eley, D.W., and ter Keurs, H.E.D.J., 1991b, Taurine depletion and excitation-contraction coupling in rat myocardium, *Canadian Cardiovascular Society* in press.

Lake, N., Splawinski, J.B., Juneau, C., and Rouleau, J.L., 1990, Effects of taurine depletion on intrinsic contractility of rat ventricular papillary muscles, *Can. J. Physiol. Pharmacol.* 68:800-806.

McBroom, M.J. and Welty, J.D., 1977, Effects of taurine on heat calcium in the cardiomyopathic hamster, *J. Molec. Cell Cardiol.* 9:853-858.

Moise, N.S., Pacioretty, L.M., Kallfeiz, F.A., and Gilmour, R.E., 1989, Association between low plasma taurine and dilatative cardiomyopathy in silver foxes, *FASEB J.* 3:6040A.

Noble, D., 1986, Sodium-calcium exchange and its role in generating electric current, *in*: "Cardiac muscle: the regulation of excitation and contraction," R.D. Nathan, ed., Academic Press, Orlando, pp. 171-200.

Novotny, M.J., Hogan, P.M., Paley, D.M., and Adams, R.H., 1991, Systolic and diastolic dysfunction of the left ventricle induced by dietary taurine deficiency in cats, *Amer. J. Physiol.* 261:H121-H127.

Pion, P.D., Kittleson, M.D., Rogers, Q.R., and Morris, J.G., 1987, Myocardial failure in cats associated with low plasma taurine: a reversible cardiomyopathy, *Science* 237:764-768.

Schouten, V.J.A. and ter Keurs, H.E.D.J., 1985, The slow repolarization phase of the action potential in rat heart, *J. Physiol.* 360:13-25.

Schouten, V.J.A. and ter Keurs, H.E.D.J., 1986, The force-frequency relationship in rat myocardium. The influence of muscle dimensions, *Pflügers Arch.* 407:14-17.

Sturman, J.A., Gargano, A.D., Messing, J.M., and Imaki, H., 1986, Feline maternal taurine deficiency: effect on mother and offspring, *J. Nutr.* 166:655-667.

ter Keurs, H.E.D.J., Rijnsburger, W.H., van Heuningen, R., and Nagelsnit, M.J., 1980, Tension development and sarcomere length in rat cardiac trabeculae. Evidence of length-dependent activation, *Circ. Res.* 46:703-714.

TAURINE POTENTIATES THE ANTIAGGREGATORY
ACTION OF ASPIRIN AND INDOMETHACIN

Flavia Franconi[1], Mauro Miceli[2], Federico Bennardini[1],
Antonella Mattana[1], Jesus Covarrubias[4], and Giuseppe Seghieri[3]

[1]Istituto di Chimica Biologica, University of Sassari, Via Muroni
    23/A - 07100, Sassari, Italy
[2]Clin. Lab. Ospedale Annunziata, Florence, Italy
[3]Diabetes Unit, Spedali Riuniti Pistoia, Italy
[4]Istituto Farmochimico Nativelle, Florence, Italy

INTRODUCTION

In plasma, the physiological levels of taurine have been reported to be 0.05-0.22 mM depending on the species[1] while platelet levels are a hundred times higher than the plasma[2]. These cellular fragments possess active transport sites[2]. However, the physiological significance of taurine in the platelets is still obscure although it has been shown that taurine stabilizes platelets against platelet activating factor (PAF) in guinea-pigs and ADP in man[3,4].

Taurine supplementation (400, 1600 mg/day for 14 days) decreased the platelet sensitivity when collagen was the aggregating agent. This decrease in aggregability was accomplished with an alteration in thromboxane A_2 release and an increase in glutathione (GSH) levels[5]. Also, in isolated rabbit heart coronary artery, taurine dose-dependently reduced the synthesis of thromboxane A_2 and PGI_2[6]. The questions we posed were whether taurine influences platelet aggregability in vitro and whether taurine affects the antiaggregating effect of aspirin and indomethacin.

METHODS

Forty-nine healthy volunteers (24-45 years of age) completely free of any medications for at least 15 days were studied. Blood samples were drawn from the anticubital vein and were anticoagulated with 3.8% trisodium citrate (9:1 v/v). Platelet rich plasma (PRP) was prepared by centrifugation at 800 x g for 10 min. The residual plasma was further centrifuged at 3,000 x g for 10 min to obtain platelet-poor plasma (PPP). The platelet count of PRP was adjusted to about 280,000 cells/μl.

Taurine, Edited by J.B. Lombardini *et al.*
Plenum Press, New York, 1992

Measurement of Platelet Aggregation

Platelet aggregation was measured turbidimetrically according to Born and Cross[7] using an aggregometer (Daichii model PA-3220). ADP (1μM), epinephrine (5μM), arachidonic acid (1 mM), collagen (5μg/ml) and PAF (1μM) were used as aggregating agents (final concentrations are in the parentheses). Experiments designed to evaluate the effects of taurine were conducted by the following procedures. Taurine at different concentrations was added to PRP and incubated for 10 min at 37°C before further processing. For experiments involving aspirin + taurine- or indomethacin-treated platelets, aspirin or indomethacin were added after 10 min preincubation with taurine and 5 min before the aggregating agents. To express the aggregation of platelets and the inhibitory action exerted by the drugs, the absorption of PRP itself was set at 0%, while that of PPP was set at 100%. The aggregation rate was also evaluated.

Measurement of Malondialdehyde (MDA)

MDA was measured spectrophotometrically according to De Gaetano et al.[8] with slight modifications.

Measurement of Lactate Dehydrogenase (LDH) Activity

The intactness of the platelets was ensured by measuring released LDH activity in the plasma. This activity was measured spectrophotometrically according to Wroblewski and La Due[9].

RESULTS

Preincubation with taurine up to a concentration of 40 mM did not increase LDH release from platelets. Moreover, it did not produce any change in aggregability induced by 5μg/ml collagen and 1 mM arachidonic acid, although small decreases were observed with 5μM adrenaline (data not shown), 1μM ADP and 1μM PAF. ADP, adrenaline and PAF were used at a concentration that caused reversible aggregation.

Aspirin and indomethacin, as is well known, dose-dependently reduce aggregation induced by arachidonic acid and collagen. Taurine-pretreated platelets showed increased sensitivity to aspirin and indomethacin; in fact, the dose-response curve of aspirin was shifted to the left (Table 1 and 2). The taurine effect was dose-dependent (Table 3).

MDA production was proposed as an indicator of cycloxygenase and thromboxane synthetase activities and also of the release of arachidonate in human platelets. Taurine preincubation per se did not modify MDA production induced by arachidonic acid but it reduced MDA production in taurine + aspirin- treated platelets (Table 4).

Also, using collagen as an aggregating agent, taurine shifted to the left the dose-response curves of aspirin in a dose-dependent manner (Tables 5 and 6). A simultaneous increase in latency-time was observed in taurine + aspirin-treated platelets (Table 6).

DISCUSSION

These experiments show that the sulphonic amino acid taurine markedly potentiates the inhibition of platelet aggregation induced by aspirin and indomethacin. Although in a previous study performed in human platelets taurine antagonized the

Table 1. The inhibitory action of aspirin (ASA) on taurine(40 mM)-pretreated platelets using 1 mM arachidonic acid as the aggregating agent.

ASA	% Rate	% Inhibition
	Control	
1 x 10^{-6} M	43.8 ± 6.1	2.5 ± 1.4
5 x 10^{-6} M	23.0 ± 6.1	39.7 ± 10.0
1 x 10^{-5} M	21.8 ± 3.6	44.0 ± 9.6
2.5 x 10^{-5} M	23.0 ± 2.8	43.1 ± 2.5
	40 mM Taurine	
1 x 10^{-6} M	30.6 ± 3.4	22.2 ± 4.3*
5 x 10^{-6} M	12.7 ± 3.0*	63.2 ± 9.7*
1 x 10^{-5} M	13.4 ± 3.1*	78.5 ± 9.7*
2.5 x 10^{-5} M	0.0 ± 0.0*	100.0 ± 0.0*

The values are means ± S.E. of at least 4 experiments. Analysis of variance gives *$P < 0.001$.

Table 2. The inhibitory action of indomethacin on taurine(20 mM)-pretreated platelets using 1 mM arachidonic acid as the aggregating agent.

Indomethacin	% Rate	% Inhibition
	Control	
10^{-8} M	82.6 ± 2.7	0
10^{-7} M	67.3 ± 1.8	0
10^{-6} M	31.9 ± 1.5	33.7 ± 5.8
1.25 x 10^{-6} M	5.7 ± 2.7	100.0 ± 0.0
	20 mM Taurine	
10^{-8} M	81.0 ± 1.0	1.7 ± 1.6
10^{-7} M	24.3 ± 2.0*	44.7 ± 4.7*
10^{-6} M	2.6 ± 2.6*	100.0 ± 0.0*
1.25 x 10^{-6} M	1.8 ± 1.5	100.0 ± 0.0

The values are means ± S.E. of at least 3 experiments. Analysis of variance gives *$P < 0.001$.

Table 3. Effect of different concentrations of taurine on the inhibitory action of aspirin (ASA; 2.5 10^{-5} M) and indomethacin (10^{6} M) on arachidonic acid (1 mM)-induced platelet aggregation.

Taurine Concentration	% Inhibition	
	ASA	Indomethacin
0	45.3 ± 4.2	33.7 ± 5.8
10^{-6}	45.2 ± 4.5	34.0 ± 0.6
10^{-5}	68.0 ± 5.1	73.2 ± 12.0
10^{-4}	97.0 ± 1.9	100.0 ± 0.0
10^{-3}	100.0 ± 0.0	100.0 ± 0.0

Values are means ± S.E. of at least 4 experiments.

Table 4. MDA production in aspirin(2.5×10^{-5} M)- and taurine+aspirin-treated platelets stimulated by 1 mM arachidonic acid.

Treatments	MDA (nmoles 10^8 · cells)		P
1 mM AA	0.44 ± 0.02	(11)	--
4 x 10^{-2} M TAU+1 mM AA	0.45 ± 0.04	(8)	--
2.5 x 10^{-5} M ASA+1 mM AA	0.24 ± 0.01	(11)	--
4 x 10^{-2} M TAU+ASA+1 mM AA	0.03 ± 0.005	(5)	0.001
1 x 10^{-4} M TAU+ASA+1 mM AA	0.04 ± 0.01	(3)	0.001
1 x 10^{-5} M TAU+ASA+1 mM AA	0.16 ± 0.01	(3)	0.001
1 x 10^{-6} M TAU+ASA+1 mM AA	0.38 ± 0.08	(3)	--

Values are means ± S.E. The number of experiments is designated in the parentheses. TAU = taurine, ASA = aspirin, AA = arachidonic acid.

Table 5. The inhibitory action of aspirin in taurine-pretreated platelets stimulated by collagen (5 μg/ml).

	% Rate	
ASA (M)		ASA + 40 mM Taurine
No ASA	76.5 ± 5.0	80.2 ± 6.0
1 x 10^{-5}	59.0 ± 6.4	63.7 ± 5.8
2.5 x 10^{-5}	59.0 ± 8.5	$26.3 \pm 7.7*$
5 x 10^{-5}	41.7 ± 5.5	$11.9 \pm 4.4*$
7.5 x 10^{-5}	30.0 ± 2.9	$4.3 \pm 2.1*$

	% Inhibition	
ASA (M)		ASA + 40 mM Taurine
1 x 10^{-5}	11.7 ± 4.3	17.0 ± 4.3
2.5 x 10^{-5}	12.6 ± 3.3	$45.9 \pm 11.5*$
5 x 10^{-5}	28.3 ± 6.2	$71.4 \pm 10.5*$
7.5 x 10^{-5}	46.6 ± 3.5	$88.6 \pm 4.0*$

	Latency Time (sec)	
ASA (M)		ASA + 40 mM Taurine
No ASA	50.1 ± 2.9	---
1 x 10^{-5}	61.3 ± 3.0	$75.4 \pm 3.2*$
2.5 x 10^{-5}	57.4 ± 3.4	$131.6 \pm 43 *$
5 x 10^{-5}	71.4 ± 3.4	$143.0 \pm 39 *$
7.5 x 10^{-5}	77.0 ± 3.6	> 180

Values are means ± S.E. of at least 4 experiments. Variance analysis gives *P < 0.001. ASA = aspirin.

Table 6. Dose-dependency of taurine on inhibition and latency time in aspirin(7.5×10^{-5} M)-treated platelets aggregated by 5 μg/ml collagen.

Taurine Concentration	% Inhibition	Latency Time (sec)
0	45.6 ± 3.4 (6)	74.1 ± 5.5 (6)
10^{-5} M	43.0 ± 2.5 (6)	82.3 ± 2.3 (5)
5 x 10^{-5} M	51.0 ± 4.2* (4)	93 ± 3.4* (4)
10^{-4} M	73.0 ± 3.5* (6)	117 ± 8.1* (6)
4 x 10^{-2} M	89.2 ± 5.0* (4)	> 180 (4)

The number of experiments is designated in the parentheses. Analysis of variance gives a *$P < 0.001$. Values are means ± S.E.

aggregation induced by 3.5 μM ADP[4], in our laboratory taurine does not significantly change the aggregation induced by 1 μM ADP in the absence of antagonists (data not shown).

Due to its sensitivity and simplicity, the MDA method is the most common assay used to determine lipid peroxidation *in vitro*. However, due to the potential non-specificity of the MDA reaction, the assay can be utilized only as a qualitative indicator of lipid peroxidation. Nevertheless, the samples exposed to arachidonic acid accumulated substantial amounts of a MDA reactive substance indicating the peroxidative degradation of lipids. This increase is prevented by aspirin and taurine pretreatments; taurine effectively potentiates this aspirin effect but taurine alone is not able to inhibit lipid peroxidation at least in our experimental conditions.

Although the antioxidant effect of taurine has been described *in vitro* and *in vivo*[10,11], in aqueous solutions taurine does not react with O_2^- and H_2O_2. However, taurine scavenges hydroxyl radical but about two orders of magnitude less than mannitol[12].

Taurine might contribute to the inhibition of platelet activation in the presence of aspirin and indomethacin by stabilizing the platelet membrane.

Taurine is effective in further inhibiting platelet aggregation in the presence of aspirin or indomethacin. However, it must be remembered that thromboxane A_2 synthesis is inhibited *in vivo* by taurine supplementation[5], but this effect appears not to be important in our experimental model because taurine *per se* is without effect on aggregation and MDA production.

In many tissues taurine modulates calcium homeostasis and although no data are available on the interaction of taurine and calcium in platelets, a possible interaction at this level cannot be excluded. However, our data suggest that taurine does not act alone but only in the presence of aspirin.

Taurine administration together with aspirin could be of some benefit in antithrombotic therapy. With few exceptions it has been shown in many animal species[1,13] and clinical studies[14] that taurine administration, even in high doses, is generally free of serious side-effects. Moreover, it has been demonstrated that taurine does not change the ulcerogenic effect of indomethacin in the rat[15]. Thus, it is also possible that subjects which are non-responders to aspirin could become responders with taurine administration.

In conclusion, the data presented here demonstrate that taurine potentiates the aspirin and indomethacin inhibition of platelet aggregation in humans.

REFERENCES

1. J.C. Jacobsen and L.H. Smith, Jr., Biochemistry and physiology of taurine and taurine derivatives, *Physiol. Rev.* 48:424 (1968).
2. L. Althee, S.J. Boullin, and M.K. Paasonen, Transport of taurine by normal human platelets, *Br. J. Pharmacol.* 52:245 (1974).

3.   M. Kuracki, K. Hongoh, A. Watanabe, and H. Aihara, Suppression of bronchial response to platelet activating factor following taurine administration, *in*: "The Biology of Taurine, Methods and Mechanisms", R.J. Huxtable, F. Franconi, and A. Giotti, eds., Plenum Press, New York, pp. 189 (1987).

4.   V.A. Alzanov, V.S. Gurevich, I.A. Mikhailova, and E.N. Streltsova, The influence of taurine on Ca^{++}, Mg^{++} ATPase activity and aggregation of human platelets, *Bull. Eksp. Fisiol. Med.* 100:398 (1985).

5.   K.C. Hayes, A. Pronezuk, A.E. Addesa, and Z.F. Stephan, Taurine modulates platelets aggregation in cats and humans, *Am. J. Clin. Nutr.* 49:1211 (1989).

6.   A. Pham Huu Chanh, R. Chahine, S. Pham Huu Chanh, V. Dossou-Obete, and C.H. Navarro Delmasure, Taurine and eicosanoids in the heart, *Prostaglandin Leukotrienes and Med.* 28:243 (1987).

7.   G.U.R. Born and M.J. Cross, The aggregation of blood platelets, *J. Physiol.* 168:178 (1963).

8.   G. De Gaetano, G. Reytor, M. Livio, and Y. Merino, Arachidonic acid-induced MDA-formation in rat diabetes, *Naunym-Schmiedeberg Arch. Pharmacol.* 312:85 (1980).

9.   F. Wroblewski and G.S. La Due, Lactate dehydrogenase activity in blood, *Proc. Soc. Exp. Med.* 90:210 (1955).

10. H. Pasantes-Morales and C. Cruz, Protective effect of taurine and tocopherol on retinal induced damage in human lymphoblastoid cells, *J. Nutr.* 114:2256 (1984).

11. M. Abe, M. Takahashi, K. Takenchi, and M. Fukuda, Studies on the significance of taurine in radiation injury, *Radiat. Res.* 33:563 (1968).

12. O.I. Arouma, B. Halliwell, B.M. Hoey, and J. Butler, The antioxidant action of taurine, hypotaurine and their metabolic precursors, *Biochem. J.* 256:251 (1988).

13. D.P. Earle, K. Small, and J. Victor, Effect of excess dietary cysteic acid, D-L methionine and taurine on the rat liver, *J. Exp. Med.* 76:317 (1942).

14. J. Azuma, A. Sawamura, N. Awata, H. Hasegawa, K. Ogura, H. Harada, H. Ohta, K. Yamauchi, S. Kishimoto, T. Yamagami, E. Ueda, and T. Ishyiama, Double blind randomized crossover trial of taurine in congestive heart failure, *Curr. Ther. Res.* 34:543 (1983).

15. S. Hara, Yamagihara, K. Satoh, and S. Morioka, Exacerbation of indomethacin induced gastric ulceration by systemically administered GABA in rats: possible involvement of peripheral GABA receptor, *J. Pharmacol.* 47:333 (1988).

ANTIARRHYTHMIC ACTION OF TAURINE

G.X. Wang, J. Duan, S. Zhou, P. Li and Y. Kang

Department of Pharmacology, Tianjin Medical College,
Tianjin, 300070, People's Republic of China

INTRODUCTION

Taurine (2-aminoethane sulfonic acid), a ß-amino acid present in human diet, is one of the most abundant sulfur containing amino acids in mammalian tissues, found in excess of 30 μmol/g tissue in rat heart[1,2]. Taurine exhibits a variety of cardiovascular pharmacological activity, including membrane stabilizing effects, inotropic actions and hypotensive activity. Recently, evidence has been obtained indicating that taurine also attenuates isoprenaline-induced free radical production. Although some investigators have reported that taurine is an antiarrhythmic agent[3,4], the precise mechanism underlying this action has not been established. In the present study, we examined the effects of taurine on arrhythmias caused by adrenaline infusion, ischemia-reperfusion injury and veratrine administration in anesthetized rats. We also have explored the relationship between the antiarrhythmic activity of taurine and changes in myocardial cAMP and free radical production.

METHODS

Ischemia-reperfusion Arrhythmias

Wistar rats (220-280 g) were anesthetized with sodium pentobarbital (50 mg/kg, i.p.). Their chests were opened and the rats were placed on artificial respiration with room air. A suture was placed around the left anterior descending coronary artery and ischemia was induced by passing both ends of the ligature through a small plastic tube which was then pressed against the surface of the vessel. When desired, the coronary artery was occluded for 15 min and subsequently reperfused. Lead II of the ECG was analyzed for the number of premature ventricular contractions, incidence and duration of ventricular tachycardia and fibrillation, time required to restore sinus rhythm and mean score defined according to Mest and Foerster[5]. Myocardial malondialdehyde content was measured using the method of Ohkawa and Tanizawa[6]. Superoxide dismutase activity was determined by Marklund and Marklund's method[7].

Adrenalin-induced Arrhythmias

Wistar rats (200-300 g) were anesthetized with urethane (1 g/kg, i.p.). Arrhythmias were induced by adrenalin infusion via the left external jugular vein at the rate of 250 μg/min for 10 min. In the taurine-treated group, taurine (either 70 or 140 mg/kg) was injected 10 min prior to adrenalin infusion. Myocardial function was evaluated by continuously monitoring lead II of the ECG. Some hearts were examined for changes in myocardial cAMP content. In these experiments, hearts were rapidly frozen in dry ice-ethanol. Approximately 100 mg of frozen ventricular tissue was homogenized with 2 ml of 5% perchloric acid and then centrifuged for 10 min at 5000 rpm. The supernatant was washed three times by ether saturated with water and then evaporated. One ml of tris buffer was then added and the samples centrifuged for 10 min at 10000 rpm. cAMP in the extract was assayed using the cAMP assay kit produced by the Beijin Atomic Energy Research Institute.

Veratrine-induced Arrhythmias

Wistar rats (180-220 g) were anesthetized with urethane (1 g/kg, i.p.). The femoral vein was cannulated and arrhythmias were induced by administration of veratrine (1 mg/kg, i.v.). Lead II of the ECG was recorded via limb leads and monitored with an oscilloscope. In treated rats, taurine (70 or 140 mg/kg, i.v.) was administered 3 min before the veratrine injection. Antiarrhythmic activity was evaluated by determining decreased incidence, time required to recover sinus rhythm and mean scores according to the method of Mest and Foerster[5].

RESULTS

Ischemia-reperfusion Arrhythmias

Table 1 shows the effects of taurine on arrhythmias produced by subjecting the heart to ischemia-reperfusion injury. In control rats, the incidence of premature ventricular contractions, ventricular fibrillation and ventricular tachycardia were 94%, 56% and 88%, respectively. The mortality rate was 44%; surviving rats recovered sinus rhythm within 367 $\pm$ 160 sec. In rats treated with 70 mg/kg taurine, the duration of ventricular tachycardia and time required to recover sinus rhythm were significantly shortened; ventricular tachycardia time decreased from 78 $\pm$ 33 in the untreated group to 23 $\pm$ 13 sec in the taurine group while sinus rhythm recovery time fell from 367 $\pm$ 160 to 206 $\pm$ 140 sec in these same groups. Taurine administration at the higher dose of 140 mg/kg also protected against the number of premature ventricular contractions (80.7 $\pm$ 73 in control group vs. 19.4 $\pm$ 14.7 in taurine group), the duration of ventricular tachycardia (78 $\pm$ 32 vs. 9.3 $\pm$ 3.8 sec), the incidence of ventricular tachycardia (88% vs. 36%) and the incidence of ventricular fibrillation (56% vs. 9%). None of the taurine-treated animals died as a result of ischemia-reperfusion injury.

Reperfusion of the ischemic heart was associated with an elevation in myocardial malondialdehyde content from 213 $\pm$ 75 to 313 $\pm$ 64 nmol/g wet wt. This change in myocardial malondialdehyde content was accompanied by a decrease in myocardial superoxide dismutase activity (1.6 $\pm$ 0.3 in control vs. 1.2 $\pm$ 0.3 U/mg protein in ischemic heart). In rats treated with 70 mg/kg taurine, myocardial malondialdehyde content was lowered significantly compared with that in the untreated rats (313 $\pm$ 63 vs. 230 $\pm$ 50 nmol/g wet wt), although there was no significant difference in myocardial superoxide dismutase activity.

Table 1. Effect of taurine on ischemia-reperfusion arrhythmias

Parameter	Control (N = 16)	Taurine (70 mg/kg) (N = 10)	Taurine (140 mg/kg) (N = 11)
Premature ventricular contractions			
Number (10 min)	80.7 ± 73.0*	47.4 ± 24.8	19.4 ± 14.7**
Incidence (%)	94	80	82
Ventricular tachycardia			
Duration (sec)	78.4 ± 32.5	22.9 ± 12.8**	9.25 ± 3.77**
Incidence (%)	88	70	36
Ventricular fibrillation			
Duration (sec)	50.8 ± 36.1	20.7 ± 8.3**	2.0
Incidence (%)	56	20	9
Time to recover sinus rhythm (sec)	367 ± 160	206 ± 140*	224 ± 188**
Mean score	161 ± 67	106 ± 68	82 ± 57**
Mortality (%)	44	0	0

* Values shown represent means ± S.D.
** Represents significant difference between control and taurine groups (p < 0.05).

Adrenalin-induced Arrhythmias

Toxic doses of adrenalin were shown to induce both mono- and multifocal ventricular tachycardia, an occasional ventricular premature contraction and A-V block. These arrhythmias were accompanied by increased myocardial cAMP content. Taurine treatment (70 mg/kg) postponed the onset of ventricular tachycardia from 241 ± 62 to 152 ± 57 sec and increased the dose of adrenalin required to induce arrhythmias from 63 ± 24 to 104 ± 66 μg/kg. At higher doses of taurine (140 mg/kg), ventricular tachycardia occurred in 64% of the animals (92% in control), mean time of onset of arrhythmias was increased from 241 ± 62 to 385 ± 90 sec and the dose of adrenalin required to induce arrhythmias increased from 63 ± 24 to 164 ± 38 μg/kg.

Data shown in Table 2 indicate that taurine (70 mg/kg and 140 mg/kg) significantly inhibited adrenalin-mediated increases in myocardial cAMP.

Veratrine-induced Arrhythmias

Veratrine (1 mg/kg, i.v.) was shown to cause premature ventricular contractions and ventricular tachycardia in all rats tested (Table 3). The mean duration of these ventricular arrhythmias was 3.53 ± 0.48 min. Partial A-V block was seen in some rats even in the absence of ventricular arrhythmias. The mean score of arrhythmias and sinus rhythm recovery time was 201 ± 16 and 15.8 ± 3.3 min, respectively.

A significant decrease in the incidence of ventricular tachycardia, ventricular fibrillation and A-V block was observed in rats pretreated with taurine, particularly at the higher taurine doses. At 140 mg/kg taurine, none of the ten rats examined developed ventricular tachycardia, ventricular fibrillation or A-V block. The duration of ventricular arrhythmias and sinus rhythm recovery time were shortened by taurine pretreatment from 3.53 ± 0.48 to 0.86 ± 0.12 min and from 15.8 ± 3.3 to 1.35 ±

Table 2. Attenuation by taurine of adrenalin-mediated increase in heart cAMP content

Condition	Myocardial cAMP (nmol/g wet wt)
Control group	
Without taurine	$1.62 \pm .21^{*}$
With taurine (70 mg/kg)	$1.61 \pm .20$
With taurine (140 mg/kg)	$1.72 \pm .25$
Adrenalin-treated group	
Without taurine	$2.15 \pm .28$
With taurine (70 mg/kg)	$1.67 \pm .11^{**}$
With taurine (140 mg/kg)	$1.52 \pm .18^{**}$

* Values shown represent means ± S.D. of 5 rats.
** Denotes significant difference between taurine-treated and untreated groups ($p < 0.05$).

Table 3. Effect of taurine on veratrine-induced arrhythmias in anesthetized rats

Parameter	Control (N = 10)	Taurine (70 mg/kg) (N = 18)	Taurine (140 mg/kg) (N = 10)
Incidence of Arrhythmias (%)			
Premature ventricular contractions	100	70	20
Ventricular tachycardia	100	50	0
Ventricular fibrillation	30	10	0
A-V nodal block	60	10	0
Duration of Arrhythmias (min)	$3.53 \pm 0.48^{*}$	$1.17 \pm 0.33^{**}$	$0.86 \pm 0.12^{**}$
Mean score	201 ± 16	$100 \pm 29^{**}$	$26 \pm 20^{**}$
Sinus rhythm recovery time (min)	15.8 ± 3.3	$3.73 \pm 0.7^{**}$	$1.35 \pm 1.0^{**}$

* Values shown represent means ± S.D. of N rats.
** Denotes significant difference between control and taurine groups.

1.00 min, respectively. Somewhat surprising was the observation that veratrine injection was associated with enhanced nasal discharge and oral secretion. Because both veratrine-induced arrhythmias and nasal discharge were antagonized by atropine, it seems likely that vagus nerve activity contributes to the development of veratrine-induced arrhythmias.

DISCUSSION

Some investigators[3,4] have shown that taurine exerts antiarrhythmic activity against adrenalin and digoxin-induced arrhythmias. In the present experiments, we also observed antiarrhythmic effects of taurine against adrenalin, veratrine and ischemia-reperfusion injury. Although not presented here, we have also shown that taurine prevents arrhythmias induced by aconitine, $CaCl_2$, CsCl and ouabain.

The mechanism responsible for the generation of ischemia-reperfusion arrhythmias has not been established, although it is thought by many investigators that free radical formation and calcium overload may play important roles in arrhythmogenesis. It has been established that abrupt reperfusion following ischemia results in an increase in free radical formation, resulting in damage to the cell membrane and enhancing calcium permeability. Inhibition of calcium uptake by isolated sarcoplasmic reticulum has also been reported. Both of these free radical-mediated effects would be expected to increase intracellular calcium, causing delayed afterdepolarizations and promoting ischemia-reperfusion arrhythmias[8,9]. In accordance with this hypothesis, we found a good correlation between the increase in malondialdehyde content, a measure of lipid peroxidation, and the development of arrhythmias. This relationship also could be important in the antiarrhythmic actions of taurine. By inhibiting ischemia-mediated elevations in lipid peroxidation, taurine would be expected to prevent disturbances in membrane structure and function. This action, combined with the direct modulation of calcium transport, could account for taurine's ability to prevent calcium overload and arrhythmias.

Adrenalin-induced arrhythmias have been used to examine the effectiveness of antiarrhythmic drugs. It has been proposed that the mechanism underlying adrenalin-induced arrhythmias is the elevation in intracellular calcium secondary to enhanced calcium influx through the slow calcium channels. The activity of these L-type slow channels are directly regulated by channel phosphorylation catalyzed by cAMP-dependent protein kinase. Because taurine attenuates adrenalin-mediated increases in cAMP content, it should decrease the level of active cAMP-dependent protein kinase, thereby reducing membrane phosphorylation and the opening of the calcium channel.

The mechanism by which taurine reduces tissue cAMP content has received some attention in the literature. Schaffer et al.[10] initially observed an effect of taurine on myocardial cAMP content, an effect they later attributed to inhibition of adenylate cyclase[11]. A subsequent study by Malchikova et al.[12] revealed that taurine blocked stress-induced increases in cAMP levels. However, it was their feeling that taurine's effect was related to activation of phosphodiesterase activity[13]. Further studies are required to explain the effects of taurine on adrenalin-induced arrhythmias.

The arrhythmogenic mechanism of veratrine also remains unclear. One possibility is that it relates to alterations in parasympathetic nervous activity; we found that veratrine activated the parasympathetic system, producing increased nasal and oral secretion and A-V block. In this regard, it is important that taurine exhibits anticholinergic activity. Kuriyama et al.[14] and Hue et al.[15] have shown that taurine decreases the release of acetylcholine from adrenals and superior cervical ganglia. Moreover, taurine inhibits carbamylcholine-induced contractions of frog gastrocnemius muscle[16]. Therefore, taurine could reduce veratrine-induced arrhythmias by inhibiting parasympathetic activity.

REFERENCES

1. S.W. Schaffer, E.C. Kulakowski, and J.H. Kramer, Taurine transport by reconstituted membrane vesicles, in: "Taurine in Nutrition and Neurology," R.J. Huxtable and H. Pasantes-Morales, eds., Plenum Press, New York, pp 143-160 (1982).
2. R. Huxtable, Metabolism and functions of taurine in the heart, in: "Taurine," R. Huxtable and A. Barbeau, eds., Raven Press, New York, pp 99-120 (1976).
3. W.O. Read and J.D. Welty, Effect of taurine on epinephrine and digoxin-induced irregularities of the dog heart, J. Pharmacol. Exp. Therap. 139:283-289 (1963).
4. E.T. Chazov, L.S. Malchikova, N.V. Lipina, G.B. Asafov, and V.M. Smirnov, Taurine and electrical activity of the heart, Circ. Res. 34 and 35 (Supp. III):11-12 (1974).
5. H.J. Mest and W. Foerster, The central nervous effect of prostaglandins I_2, E_2 and $F_{2\alpha}$ on aconitine-

induced cardiac arrhythmias in rats, *Prostaglandins* 18:235-238 (1979).

6. H. Ohkawa, N. Ohnishi, and K. Yagi, Assay for lipid peroxides in animal tissues by thiobarbituric acid reaction, *Anal. Biochem.* 95:351-358 (1979).

7. S. Marklund and G. Marklund, Involvement of the superoxide anion radical in the autoxidation of pyrogallol and a convenient assay for superoxide dismutase, *Eur. J. Biochem.* 47:469-474 (1974).

8. A.S. Manning and D.J. Hearse, Reperfusion-induced arrhythmias: Mechanisms and prevention, *J. Mol. Cell. Cardiol.* 16:497-518 (1984).

9. B. Woodward, N. Mohamed, and M. Zakaria, Effect of some free radical scavengers on reperfusion-induced arrhythmias in the isolated rat heart, *J. Mol. Cell. Cardiol.* 17:485-493 (1985).

10. S.W. Schaffer, J.P Chovan, and R.F. Werkman, Dissociation of cAMP changes and myocardial contractility in perfused rat heart, *Biochem. Biophys. Res. Commun.* 81:248-253 (1978).

11. S.W. Schaffer, J.H. Kramer, W.G. Lampson, E. Kulakowski, and Y. Sakane, Effect of taurine on myocardial metabolism: Role of calmodulin, *in*: "Sulfur Amino Acids: Biochemical and Clinical Aspects," K. Kuriyama, R.J. Huxtable and H. Iwata, eds., Alan R. Liss, Inc., New York, pp 39-50 (1983).

12. L.S. Malchikova, N.V. Speranskaya, and E.P. Elizarova, Effect of taurine on the cAMP and cGMP content in rat hearts during stress, *Medicina* (Moscow) 20:133-137 (1979).

13. L.S. Malchikova and E.P. Elizarova, Taurine and cAMP content in the heart, *Medicina* (Moscow) 21:85-89 (1981).

14. K. Kuriyama and E. Ueno, Functional roles of taurine in central nervous system: Possible role as a neuromodulator, *Proc 9th Symposium Act. and Mesh* (Osaka) S32 (1980).

15. B. Hue and T. Chanelet, Modulation of acetylcholine release by taurine in the central nervous system of the cockroach, *Periplaneta americana, L. J. Pharmac.* Paris 15:65-78 (1981).

16. A. Lehmann and A. Hamberger, Inhibition of cholinergic response by taurine in frog isolated skeletal muscle, *J. Pharm. and Pharmacol.* 36:59-61 (1984).

MECHANISM UNDERLYING PHYSIOLOGICAL MODULATION OF MYOCARDIAL CONTRACTION BY TAURINE

S.W. Schaffer[1], S. Punna[1], J. Duan[1], H. Harada[1,2],
T. Hamaguchi[1,2] and J. Azuma[2]

[1]Department of Pharmacology
University of South Alabama School of Medicine
Mobile, AL 36688 USA and

[2]Department of Medicine III
Osaka University Medical School
Osaka, Japan

INTRODUCTION

One of the established actions of taurine is the modulation of myocardial contraction. Initial studies demonstrating this effect employed pharmacological doses of taurine to alter contractile function[1-3]. Although it was assumed that this action of taurine might represent a physiological function, only recently has evidence been obtained confirming this theory. Pion et al.[4] reported in 1987 that cats fed a taurine deficient diet developed a cardiomyopathy. More recently, Steele et al.[5] and Galler et al.[6] found that taurine altered contractile function of skinned muscle fibers, a preparation in which the cell membrane is permeabilized, allowing changes in the intracellular milieu.

The modulation of contraction by taurine has largely been attributed to alterations in calcium transport. In an early study, Chovan et al.[7] suggested that taurine stimulated contraction by enhancing Ca^{2+} binding to the cell membrane. However, as more information became available regarding the regulation of calcium homeostasis in the heart, it became apparent that the actions of taurine must be linked to a specific calcium transporter. Attempts to clarify the site of taurine action have been thus far inconclusive. The aim of this paper is to examine the possibility that the physiological actions of taurine involve the modulation of specific calcium transporters.

METHODS

Junctional and longitudinal sarcoplasmic reticula were isolated according to the

Taurine, Edited by J.B. Lombardini *et al.*
Plenum Press, New York, 1992

procedure described by Schaffer et al.[8] The rate of ATP-dependent Ca^{2+} uptake by longitudinal vesicles was assayed at a free Ca^{2+} concentration of 2.0 μM using the method described by Harada et al.[9] Junctional sarcoplasmic reticulum was loaded with $^{45}Ca^{2+}$ by preincubating the membranes for 2 hrs at 30°C with 20 mM MOPS buffer (pH 7.0) containing 160 mM KCl, 5 mM ATP, 5 mM $MgCl_2$, 0.1 mM EGTA and 2 μM free $^{45}Ca^{2+}$ (2 X 10^5 cpm/50 μl). After achieving a steady state of Ca^{2+} loading, $^{45}Ca^{2+}$-loaded vesicles were placed in 40 mM histidine buffer (pH 7.0, 30°C) containing 110 mM KCl, 100 μM EGTA and 10 μM free Ca^{2+}. This triggered the release of $^{45}Ca^{2+}$ from the membrane vesicles. The release reaction was stopped by filtration over Millipore filters and washing with ice-cold 20 mM MOPS buffer (pH 6.8) containing 160 mM KCl, 10 mM $MgCl_2$, 20 mM EGTA and 15 μM ruthenium red. The filters were washed three times with the stop buffer and then dried and counted for radioactivity.

Sarcolemma from rat heart was isolated according to the method described by Harada et al.[9] Both ATP-dependent sarcolemmal Ca^{2+} uptake and sarcolemmal Na^+-Ca^{2+} exchange were measured according to procedures described previously[9].

RESULTS

The studies of Steele et al.[5] and Galler et al.[6] using skinned muscle fibers suggest that the modulation of myocardial contraction by taurine must involve an intracellular site of taurine action. Steele et al.[5] proposed on the basis of his contractile studies that taurine's actions were related to modulation of sarcoplasmic reticular calcium pool size. However, this is in apparent contradiction to earlier studies showing no effect of taurine on calcium uptake by isolated sarcoplasmic reticulum[10,11]. Since those earlier studies were completed, improvements have been made in the preparation of sarcoplasmic reticulum, both in terms of its purity as it relates to other organelles, as well as separation of different fractions of sarcoplasmic reticulum, namely, ryanodine-sensitive and insensitive membrane fractions. Therefore, we decided to reexamine the effect of taurine on calcium movement by enriched longitudinal and junctional sarcoplasmic reticulum. Figure 1 reveals that neither

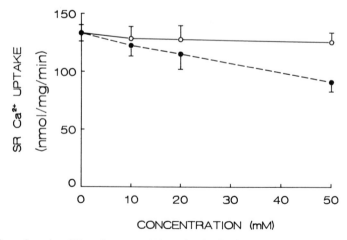

Figure 1. Effect of taurine (●) and sucrose (O) on longitudinal sarcoplasmic reticular Ca^{2+} uptake. Uptake of $^{45}Ca^{2+}$ by isolated sarcoplasmic reticulum in medium containing 2 μM $^{45}Ca^{2+}$ and the appropriate concentration of either taurine or sucrose was examined. A significant difference between taurine and sucrose was observed at 50 mM.

taurine nor sucrose over a concentration range of 10-25 mM influenced ATP-dependent Ca^{2+} uptake by longitudinal sarcoplasmic reticulum. However, at 50 mM, while sucrose had no effect on Ca^{2+} transport, 50 mM taurine caused a modest, but significant reduction in the rate of Ca^{2+} transport.

In contrast to the rather small changes in longitudinal ATP-dependent Ca^{2+} transport evoked by taurine, substantial changes in the function of the ryanodine-sensitive Ca^{2+} channel were observed in the presence of taurine. In agreement with previous studies[12], we found that isolated junctional sarcoplasmic reticulum exposed to extravesicular calcium, immediately released Ca^{2+} into the medium (Figure 2). Within 2 sec following introduction of $^{45}Ca^{2+}$-loaded junctional sarcoplasmic reticulum into medium containing $10\,\mu M\ Ca^{2+}$, there was a 25% drop in the amount of $^{45}Ca^{2+}$ found within the vesicles of the sarcoplasmic reticulum. The amount of vesicular $^{45}Ca^{2+}$ loss was enhanced in a concentration-dependent manner by taurine. The maximal effect of taurine was observed at the highest taurine concentration examined (50 mM) and represented a 70% increase in the initial rate of $^{45}Ca^{2+}$ release (Figure 2A). By 30 sec the taurine effect was still apparent. Approximately 40% of initial $^{45}Ca^{2+}$ was lost from the vesicles in the control but nearly 50% more was lost in the presence of 50 mM taurine (Figure 2B). Although lower concentrations of taurine had smaller effects, the promotion of calcium release was statistically significant at all concentrations tested except 10 mM taurine. Sucrose over the same concentration range had no effect on release indicating that the taurine effect was unrelated to osmotic forces.

In the normally beating myocardium, intracellular calcium homeostasis is modulated not only by the sarcoplasmic reticular Ca^{2+} transporters but also sarcolemmal transporters. One of the sarcolemmal transporters which influences both the rate of pressure development, as well as the amplitude of contraction, is the Na^+-Ca^{2+} exchanger. Nevertheless, over a concentration range of 10-50 mM, taurine had no effect on Na^+-Ca^{2+} exchanger activity (Table 1).

A less important sarcolemmal Ca^{2+} transporter, but one which can influence diastolic function of the heart is the ATP-dependent Ca^{2+} pump. Figure 3 reveals that 20 mM taurine significantly decreases Ca^{2+} transport by this carrier, particularly

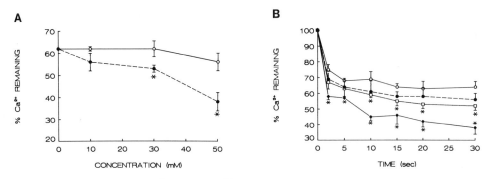

Figure 2. Effect of taurine on Ca^{2+}-induced, Ca^{2+} release from isolated junctional sarcoplasmic reticulum (SR). In Panel A, the rate of Ca^{2+} release from $^{45}Ca^{2+}$-loaded SR vesicles was examined by determining the amount of $^{45}Ca^{2+}$ remaining within the vesicles. The release reaction was initiated by placing the vesicles in medium containing $10\,\mu M\ Ca^{2+}$ and varying the concentration of taurine: 0 mM (O), 10 mM (●), 30 mM (□) and 50 mM (♦). Panel B illustrates the amount of $^{45}Ca^{2+}$ remaining associated with the SR vesicles following 30 sec of the release reaction taking place in medium containing sucrose (O) or taurine (●). Data for 30 and 50 mM taurine were significantly different from the taurine and sucrose groups ($p < 0.05$). Asterisks denote significant differences between the taurine and sucrose groups.

Table 1. Effect of taurine on the rate of sarcolemmal Na^+-Ca^{2+} exchange

Concentration	Sucrose (nmol Ca^{2+}/mg/sec)	Taurine (nmol Ca^{2+}/mg/sec)
10 mM	2.67	2.58
30 mM	2.67 ± 0.04	2.43 ± 0.16
50 mM	2.24 ± 0.19	2.42 ± 0.22

Sarcolemmal vesicles were loaded with 160 mM Na^+ before being placed in buffer supplemented with $^{45}Ca^{2+}$ and containing both 160 mM K^+ and the appropriate concentration of either taurine or sucrose. The rate of Na^+-Ca^{2+} exchange was monitored by observing the rate of $^{45}Ca^{2+}$ uptake. No significant difference was observed between the various groups. Values represent the means ± S.E.M. of 4 preparations.

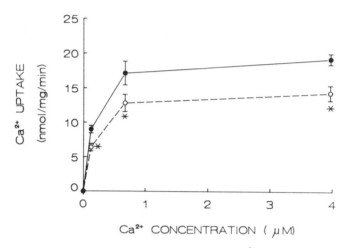

Figure 3. Effect of taurine on sarcolemmal ATP-dependent Ca^{2+} transport. Uptake into isolated sarcolemmal vesicles was initiated by addition of $^{45}Ca^{2+}$ and ATP. The rate of transport in the presence (●) or absence (O) of 10 mM taurine was monitored as a function of extravesicular Ca^{2+} concentration. Values shown represent means ± S.E.M. of 5-7 samples. Asterisks denote significant difference between the control and taurine groups (p < 0.05).

at the lower Ca^{2+} concentrations. Reducing medium taurine concentration to 1 mM eliminated the taurine effect at all Ca^{2+} concentrations examined.

DISCUSSION

This study indicates that taurine modulates the function of several Ca^{2+} transporters, with the largest and potentially most significant action relating to the enhancement of Ca^{2+}-induced, Ca^{2+} release from intracellular Ca^{2+} stores. This process presumably occurs in junctional sarcoplasmic reticulum and is believed to play a major role in the development of contractile force. During stimulation of the heart, Ca^{2+} enters the myocyte through the L-type Ca^{2+} channel. This source of Ca^{2+} is insufficient to mediate maximal contraction. However, it elevates cytoplasmic Ca^{2+} levels enough to cause the ryanodine-sensitive Ca^{2+} channel located on the junctional sarcoplasmic reticulum to open. The release of Ca^{2+} from this internal store is referred to as Ca^{2+}-induced, Ca^{2+} release and is responsible for significantly increasing intracellular calcium levels and stimulating myocardial contraction. Thus, the promotion of Ca^{2+} release from junctional sarcoplasmic reticulum appears to contribute to the taurine-mediated positive inotropic effect observed using the skinned fiber preparation[15,16]. Although the potential involvement of this channel in the positive inotropic effect of taurine was not discussed by Steele et al.[15], they did indicate that skinned fibers exposed to medium containing $0.47 \, \mu M \, Ca^{2+}$ and 30 mM taurine underwent large fluctuations in caffeine-induced contracture, an effect attributed by the authors to release of Ca^{2+} from the sarcoplasmic reticulum.

In addition to altering Ca^{2+}-induced, Ca^{2+} release from the sarcoplasmic reticulum, taurine also inhibits sarcolemmal ATP-dependent Ca^{2+} transport. Normally this transporter functions to pump Ca^{2+} from the cell during diastole. Because it has a low capacity to pump Ca^{2+}, it is thought to significantly affect intracellular Ca^{2+} levels only when the intracellular Ca^{2+} concentration falls below the K_m of the sarcoplasmic reticular Ca^{2+} pump, the latter having a high capacity to remove Ca^{2+} from the cytoplasm only during the early phases of diastole. Thus, as intracellular Ca^{2+} levels fall, the sarcolemmal Ca^{2+} pump assumes a more central role in maintaining low cytoplasmic Ca^{2+} levels. Inhibition of this transporter by taurine might contribute to a slight elevation in cytoplasmic Ca^{2+} levels during diastole, although in the compromised myocardium the effect may be magnified. Harada et al.[13] have shown that loss of sarcolemmal Ca^{2+} pump activity in Adriamycin-treated animals may contribute to Ca^{2+} overload and injury to the myocardium. Nevertheless, in the normal heart this effect should be inconsequential in relation to the change in Ca^{2+}-induced, Ca^{2+} release.

The present study agrees with previous reports indicating that taurine at concentrations less than 20 mM exhibits little, if any effect on ATP-dependent Ca^{2+} uptake by isolated sarcoplasmic reticulum[10,11]. However, higher doses of taurine appear to significantly inhibit Ca^{2+} uptake. This conclusion is supported by the work of Steele et al.[5], who reported that higher doses of taurine reduced the amplitude of caffeine-induced contractures of skinned rat trabeculae. Two factors appeared to contribute to the effect observed by Steele et al.[5]: (1) the osmotic effect of high doses of taurine and (2) the promotion of taurine-mediated diminution in the size of the sarcoplasmic reticular Ca^{2+} pool. Our data suggest that the second factor is more important. While sucrose was found incapable of influencing sarcoplasmic reticular Ca^{2+} transport, higher doses of taurine modestly inhibited ATP-dependent Ca^{2+} uptake.

No stimulatory effect of taurine on ATP-dependent Ca^{2+} uptake by isolated longitudinal sarcoplasmic reticulum was observed in the present study. This is in apparent conflict with the conclusions of Steele et al.[5] One potential explanation for this discrepancy relates to the system used to monitor Ca^{2+} transport. The procedure commonly used to measure the rate of Ca^{2+} accumulation by isolated sarcoplasmic reticulum, utilizes intravesicular medium containing oxalate, which binds Ca^{2+}, concentrating it within the membrane vesicles. This artificial system presumably leads to a highly calcium-loaded vesicle, a condition in which only the inhibiting effects of taurine may be observed.

Earlier studies examining the effects of taurine on the heart focused on the sarcolemma as a site of taurine action. According to Chovan et al.[7] stimulation of myocardial contraction by taurine was linked to taurine-mediated changes in sarcolemmal Ca^{2+} binding. This theory assumed that increases in the size of the sarcolemmal Ca^{2+} pool promoted contraction by enhancing Ca^{2+} uptake during excitation. Nevertheless, the theory failed to address the mechanism by which sarcolemmal Ca^{2+} loading stimulated cellular Ca^{2+} uptake. In spite of a considerable effort to clarify this point, the issue still remains unresolved. Thus, more recent studies have focused on the effects of taurine on the activity of various sarcolemmal Ca^{2+} transporters. Sawamura et al.[14] found that taurine exerted a positive inotropic effect in embryonic chick heart, but this effect was not mediated through an increase in flux through the L-type Ca^{2+} channel or by altering the fast Na^+ channels. The present data also rules out the Na^+-Ca^{2+} exchanger as a factor in the positive inotropic effect of taurine. However, the possibility that taurine indirectly regulates intracellular Ca^{2+} levels through the modulation of K^+ or Na^+ permeability deserves further consideration[15,16].

REFERENCES

1. J. Dietrich and J. Diacona, Comparison between ouabain and taurine effects on isolated rat and guinea pig hearts in low calcium medium, *Life Sci.* 10:499-507 (1971).
2. S.W. Schaffer, J.P. Chovan, and R.F. Werkman, Dissocation of cAMP changes and myocardial contractility in taurine perfused rat heart, *Biochem. Biophys. Res. Commun.* 81:248-253 (1978).
3. J.C. Khatter, P.L. Soni, R.J. Hoeschen, L.E. Alto, and N.S. Dhalla, Subcellular effects of taurine on guinea pig heart, *in*: "The Effects of Taurine on Excitable Tissues," S.W. Schaffer, S.I. Baskin, and J.J. Kocsis, eds., Spectrum Publications, New York, pp. 281-293 (1981).
4. P.D. Pion, M.D. Kittleson, Q.R. Rogers, and J.G. Morris, Myocardial failure in cats associated with low plasma taurine: A reversible cardiomyopathy, *Science* 237:764-768 (1987).
5. D.S. Steele, G.L. Smith, and D.J. Miller, The effects of taurine on Ca^{2+} uptake by the sarcoplasmic reticulum and Ca^{2+} sensitivity of chemically skinned rat heart, *J. Physiol.* 422:499-511 (1990).
6. S. Galler, C. Hutzler, and T. Haller, Effects of taurine on Ca^{2+}-dependent force development of skinned muscle fibre preparations, *J. Exp. Biol.* 152:255-264 (1990).
7. J.P. Chovan, E.C. Kulakowski, B.W. Benson, and S.W. Schaffer, Taurine enhancement of calcium binding to rat heart sarcolemma, *Biochim. Biophys. Acta* 551:129-136 (1979).
8. S.W. Schaffer, S. Allo, S. Punna, and T. White, Defective response to cAMP-dependent protein kinase in non-insulin-dependent diabetic heart, *Am. J. Physiol.* 261:E369-E376 (1991).
9. H. Harada, S. Allo, N. Viyuoh, J. Azuma, K. Takahashi, and S.W. Schaffer, Regulation of calcium transport in drug-induced taurine-depleted hearts, *Biochim. Biophys. Acta* 944:273-278 (1988).
10. J.D. Welty and M.C. Welty, Effects of taurine on subcellular calcium dynamics in the normal and cardiomyopathic hamster heart, *in*: "The Effects of Taurine on Excitable Tissues," S.W. Schaffer, S.I. Baskin, and J.J. Kocsis, eds., Spectrum Publications, New York, pp. 295-312 (1981).
11. M.L. Entman, E.P. Bornet, and R. Bressler, The effect of taurine on cardiac sarcoplasmic reticulum, *Life Sci.* 21:543-550 (1977).
12. M. Michalak, Identification of the Ca^{2+}-release activity and ryanodine receptor in sarcoplasmic reticulum membranes during cardiac myogenesis, *Biochem. J.* 253:631-636 (1988).
13. H. Harada, B.J. Cusack, R.D. Olson, W. Stroo, J. Azuma, T. Hamaguchi, and S.W. Schaffer, Taurine deficiency and doxorubicin: Interaction with the cardiac sarcolemmal calcium pump, *Biochem. Pharmacol.* 39:745-751 (1990).
14. A. Sawamura, N. Sperelakis, J. Azuma, and S. Kishimoto, Effects of taurine on the electrical and mechanical activities of embryonic chick heart, *Can. J. Physiol. Pharmacol.* 64:649-655 (1986).
15. N. Lake, M. de Roode, and S. Nattel, Effects of taurine depletion on rat cardiac electrophysiology: In vivo and in vitro studies, *Life Sci.* 40:997-1005 (1987).
16. N. Sperelakis, T. Yamamoto, G. Bkaily, H. Sada, A. Sawamura, and J. Azuma, Taurine effects on action potentials and ionic currents in chick myocardial cells, *in*: "Taurine and the Heart," H. Iwata, J.B. Lombardini, and T. Segawa, eds., Kluwer Academic Publishers, Boston, pp. 1-19 (1989).

EFFECTS OF TAURINE DEPLETION ON MEMBRANE ELECTRICAL
PROPERTIES OF RAT SKELETAL MUSCLE

Annamaria De Luca[1], Sabata Pierno[1], Ryan J. Huxtable[2]
Paola Falli[3], Flavia Franconi[4], Alberto Giotti[3], and
Diana Conte Camerino[1]

[1]Unità di Farmacologia, Dipartimento Farmacobiologico
   Università di Barí, Italy 61801
[2]Department of Pharmacology, College of Medicine
   University of Arizona, Tucson, USA
[3]Dipartimento di Farmacologia, M. Aiazzi Mancini,
   Università di Firenze, Italy
[4]Istituto di Chimica Biologica, Università di Sassari, Italy

INTRODUCTION

Taurine is abundantly present in mammalian skeletal muscle but its physiological role is not yet completely understood. Findings from our laboratory support the proposed hypothesis of a membrane-stabilizer action exerted by the amino acid in this tissue (Huxtable and Sebring, 1986). Indeed, in vitro application of taurine can markedly reduce membrane excitability of rat skeletal muscle as a result of a specific increase of membrane chloride conductance (GCl) (Conte Camerino et al., 1987). It is well known that a large membrane GCl ensures the electrical stability of sarcolemma, since an abnormally low GCl triggers the hyperexcitability characteristic of myotonic muscles (Bryant and Morales-Aguilera, 1971; Rüdel and Lehmann-Horn, 1985). Therefore, our finding has given a rationale for the observed relief of myotonic signs in human patients after taurine administration (Durelli et al., 1983). To better understand the action of taurine on skeletal muscle chloride channels, we have further investigated the ability of the amino acid to reverse or prevent the decrease of GCl and the myotonic hyperexcitability in different experimental models of myotonia induced pharmacologically in rats (Conte Camerino et al., 1989). The finding that taurine fails to restore GCl when the channels have been sterically blocked by anthracene-9-carboxylic acid (Bryant and Morales-Aguilera, 1971; Palade and Barchi, 1977), whereas it normalizes GCl when it is lowered indirectly by 20-25 diazacholesterol which mainly changes the lipid environment of the membrane (Furman and Barchi, 1981), suggests that taurine is not a chloride channel ionophore. Instead, it might allosterically modulate chloride channel kinetics by binding to low-affinity sites (Conte Camerino et al., 1989). An analogous mechanism has been proposed to explain the taurine-dependent increase of Ca^{2+} binding to acidic

Taurine, Edited by J.B. Lombardini *et al.*
Plenum Press, New York, 1992

phospholipids in cardiac sarcolemma (Huxtable and Sebring, 1987). In the present study we have investigated in more detail the long-term modulatory role of taurine on membrane ionic conductances and in particular on GCl by evaluating the effect of taurine depletion in skeletal muscle. For depleting the muscles of taurine, we used a chronic treatment with guanidinoethane sulfonate (GES) which is known to be a specific inhibitor of taurine transport in several tissues (Huxtable et al., 1979). Whether or not changes in excitation-contraction coupling occur as a consequence of long-term taurine depletion of striated fibers was also evaluated.

METHODS

Adult male Wistar rats were used for all the experiments. For in vivo treatment four rats received GES at the concentration of 1% in drinking water ad libitum for four weeks (Huxtable et al., 1979).

For in vitro studies, extensor digitorum longus (EDL) muscles, removed under urethane anaesthesia from untreated and GES treated rats, were placed in a muscle bath at 30 $\pm$ 1°C and perfused with either chloride-containing or chloride-free physiological solutions. A standard two microelectrode technique was used for making intracellular measurements of membrane potentials and cable properties (Conte Camerino et al., 1989). The current pulse generation, acquisition of the voltage records, and calculation of fiber cable constants by standard cable analysis were done in real-time under computer control as described in detail elsewhere (Bryant and Conte Camerino, 1991). The reciprocal of membrane resistance in normal physiological solution was assumed to be the total membrane conductance (Gm) and the same parameter measured in chloride-free solution was considered largely the potassium conductance (GK). The mean chloride conductance (GCl) of each preparation was estimated as the mean Gm minus the mean GK. The excitability characteristics were determined by recording the intracellular membrane potential response to square-wave constant current pulses. In each fiber the membrane potential was set by a steady holding current to -80 mV, before passing the depolarizing pulses.

Mechanical threshold of EDL muscle fibers was determined by using a two-microelectrode point voltage clamp method (De Luca and Conte Camerino, in press), in the presence of $3 \mu M$ TTX. The holding current was set at -90 mV. To determine the rheobase mechanical threshold, depolarizing command pulses of 500 ms duration were given repetitively at a rate of 0.3 Hz, while the impaled fibers were viewed continuously with a stereomicroscope. The command voltage was increased until just visible contractions were observed; the threshold membrane potential at this point could be read from a digital sample-and-hold voltmeter (De Luca and Conte Camerino, in press).

The intracellular concentrations of taurine were determined using a Perkin-Elmer 3B Liquid Chromatograph according to the method of Krusz et al. (1978). After perchloric acid homogenization of the tissue and neutralization, the supernatant was used for HPLC taurine determination.

Normal and chloride-free physiological solutions, prepared as previously detailed (Conte Camerino et al., 1989) were continuously gassed with 95% O_2 and 5% CO_2. Taurine (Merck, Darmstadt, Germany) at a concentration of 60 mM was dissolved in either normal or chloride-free medium. The pH of all solutions was carefully maintained between 7.2 - 7.3. The data are expressed as mean $\pm$ SEM. Significance of differences between group means was calculated by the Student's t-test. The estimates of the SEM of GCl were obtained from the variance of Gm and GK, assuming no covariance, using standard methods (Green and Margerison, 1978).

RESULTS

After four-weeks treatment with GES, the intracellular taurine content (determined by HPLC) of EDL muscle was 50% lower than that normally found in untreated control muscles (Table 1).

Table 1. Effect of guanidinoethane sulfonate (GES) chronic treatment on taurine content of extensor digitorum longus muscle from adult rats.

Treatment	n	Taurine Content (nmol/mg w.w.)
Control	9	24.1 ± 1.7
GES	2	13.6 ± 0.3*

The columns from left to right are as follows: Treatment: GES was administered ad libitum in drinking water at the dose of 1% for 4 weeks; n: number of muscles sampled; Taurine Content: intracellular level of taurine (mean $\pm$ S.E.) determined by the HPLC method.
* Significantly different with respect to control value ($p < 0.025$).

The electrophysiological recordings performed in vitro in normal physiological solution showed different values of membrane resistance (Rm) in the GES treated muscle ($446 \pm 34 \Omega \cdot cm^2$; n = 24) with respect to the untreated controls ($320 \pm 18 \Omega \cdot cm^2$; n = 10). A significant difference in the Rm value was also found in the chloride-free physiological solution, being Rm $2207 \pm 124 \Omega \cdot cm^2$ (n = 38) and $4147 \pm 465 \cdot cm^2$ (n = 15) in GES-treated and control rats, respectively. The conversion of Rm values to conductances showed that the GES treatment produced a marked decrease of Gm, mostly due to a significant 32% decrease of GCl; interestingly an 80% increase of GK was also observed (Table 2). The in vitro application of taurine on the GES treated fibers markedly increased membrane conductance. Indeed Gm was $4116 \pm 358 \mu S/cm^2$ (n = 9) after application of 30 mM and 60 mM taurine, respectively. These values were highly significant not only with respect to the Gm value ($2480 \pm 147 \mu S/cm^2$, n = 24) of GES treated fibers but also versus Gm ($3197 \pm 135 \mu S/cm^2$, n = 15) of untreated controls, suggesting an over sensitivity of taurine depleted muscles to the amino acid. The chronic GES treatment also modified membrane excitability characteristics (Table 3). According to the observed decrease

Table 2. Effects of guanidinoethane sulfonate (GES) chronic treatment on membrane component conductances of extensor digitorum longus muscle fibers from adult rats.

Treatment	n	Gm ($\mu S/cm^2$)	n^1	GK ($\mu S/cm^2$)	GCl ($\mu S/cm^2$)
Control	15	3197 ± 135	15	280 ± 27	2917 ± 37
GES	24	2480 ± 147*	38	508 ± 28*	1973 ± 94

The columns from left to right are as follows: Treatment: GES was administered ad libitum in drinking water at the dose of 1% for 4 weeks; n, number of fibers for GK, the component potassium conductance, Gm, the total membrane conductance, and GCl, the component chloride conductance. Each value is expressed as mean $\pm$ S.E.
* Significantly different with respect to control values ($p < 0.005$ or less).

Table 3. Effects of guanidinoethane sulfonate (GES) chronic treatment of membrane excitability parameters of extensor digitorum longus muscle fibers from adult rats.

Treatment	n	RP (-mV)	AP (mV)	Ith (nA)	Ith/I2nd	Lat (msec)	Dur (msec)	N Spikes
Control	7	67 ± 2	102 ± 7	129 ± 5.6	0.70 ± 0.03	5.3 ± 0.1	9.4 ± 0.4	3.4 ± 0.5
GES	7	70 ± 2	98 ± 2	131 ± 11	0.84 ± 0.03	7.7 ± 0.5*	13.2 ± 0.7*	5.2 ± 0.8

The columns from left to right are as follows: GES was administered ad libitum in drinking water at the dose of 1% for 4 weeks; n: number of fibers; RP: resting membrane potential; AP: amplitude of the action potential; Ith: threshold current; Ith/I2nd: ratio betwen threshold current and the current necessary to elicit the second action potential; Lat (latency): time interval between the onset of the current pulse and the beginning of an action potential; Dur: duration of the action potential; N spikes: maximum number of action potential elicited. Each value is expressed as mean ± S.E.
* Significantly different with respect to control value (p < 0.01 or less).

of GCl we found that the time interval between the onset of the current pulse and the beginning of an action potential (latency, Lat) as well as the duration of the action potential were significantly prolonged in the GES treated fibers. Although the rheobasic current (Ith) did not change, the ratio between Ith and the current necessary to elicit two action potentials (Ith/I2nd) was significantly increased by the GES treatment. Accordingly, the firing capability of the GES treated fibers showed a tendency to increase; such an increase was not significant probably because of the large increase of GK which in turn has a stabilizing effect on membrane excitability (Table 3).

The effect of GES treatment on the rheobase voltage for mechanical activation of striated fibers is shown in Table 4. The rheobase value of the GES treated fibers was significantly lower with respect to the untreated EDL, i.e. the GES treated fibers contracted at more negative potentials.

Again, the in vitro application of 60 mM taurine shifted the rheobase towards the control value (Table 4).

Table 4. Effect of guanidinoethane sulfonate (GES) chronic treatment on the rheobase mechanical threshold of rat extensor digitorum longus muscle fibers.

Treatment	N/n	Rheobase voltage (mV)
Control	4/22	-61.4 ± 2.2
GES	2/11	-71.1 ± 2.3*
TAU 60 mM on GES Treated	2/9	-64.8 ± 2.9

The columns from left to right are as follows: Treatment: GES was administered ad libitum in drinking water at the dose of 1% for 4 weeks; Taurine (TAU) at the concentration of 60 mM was applied in vitro on the EDL muscle from GES treated rats; N/n number of rats/number of fibers sampled; rheobase voltage expressed as mean $\pm$ S.E.
* Significantly different with respect to untreated controls ($p < 0.01$).

DISCUSSION

In the present study we found that a 50% reduction of taurine content in rat skeletal muscle by chronic treatment with GES (Huxtable et al., 1979) significantly changes the membrane electrical properties. Indeed, we found a marked decrease of GCl along with a significant increase of GK. The excitability characteristics were also changed along the line of the observed effect on GCl and GK. The changes of membrane component conductances were really mediated by the depletion of taurine, since in vitro application of the amino acid to the GES treated fibers completely overcame the decrease of Gm and the amount of the effect produced by taurine was so large to suggest a taurine rebound effect in such depleted muscles.

The present data again corroborate the modulatory role of taurine in the control of chloride channel functions (Conte Camerino et al., 1989). Also, the finding that GK significantly increased after taurine depletion, whereas this parameter is not significantly affected when taurine is applied in vitro to normal muscles (Conte Camerino et al., 1987) suggests other interesting long-term functions of this amino acid in skeletal muscle. An attractive hypothesis to explain the observed increase of

GK is the opening of Ca^{2+}-dependent potassium channels, so called big K^+ channels, which have been clearly identified in skeletal muscle membrane (Latorre et al., 1989). Although we do not have direct pharmacological proof that this hypothesis is correct, several pieces of evidence suggest that taurine modulates the availability of Ca^{2+} in various tissues. Indeed studies on cardiac fibers suggest that taurine by binding to low affinity sites enhances the amount of Ca^{2+} bound to acidic phospholipids (Huxtable and Sebring, 1986; 1987) and that the inotropic effect of taurine depends upon the extracellular Ca^{2+} concentration (Franconi et al., 1984). It has also been proposed that taurine may increase Ca^{2+} uptake by sarcoplasmic reticulum in both cardiac (Lake et al., 1991) and skeletal muscle (Huxtable and Bressler, 1973).

In support of this view, in the present study we found that the mechanical threshold of taurine depleted muscle is significantly changed, since the GES-treated fibers contracted at significantly more negative potential with respect to the untreated controls. This effect is again reversed by the in vitro application of taurine.

Finally, these findings show that even in skeletal muscle taurine has a long term control on the availability of Ca^{2+} which in turn may modulate the open/shut probability of Ca^{2+}-dependent K^+ channels. Such effects together with the modulatory role on chloride channels demonstrate that taurine has an important function on membrane excitability and contractile properties of mammalian skeletal muscle.

ACKNOWLEDGMENT

The financial support of Italian MURST 40%, 1990 and of Telethon, Italy for the project "Abnormal ion channel function of skeletal muscle membrane in genetically occurring diseases: electrophysiology and pharmacology" are gratefully acknowledged.

REFERENCES

Bryant, S.H. and Morales-Aguilera, A., 1971, Chloride conductance in normal and myotonic muscle fibers and the actions of monocarboxylic aromatic acids, *J. Physiol.* 219:367-383.

Bryant, S.H. and Conte Camerino, D., 1991, Chloride channel regulation in the skeletal muscle of normal and myotonic goats, *Pflüger Arch.* 417: 605-610.

Conte Camerino, D., Franconi, F., Mambrini, M., Bennardini, F., Failli, P., Bryant, S.H., and Giotti, A., 1987, The action of taurine on chloride conductance and excitability characteristics of rat striated muscle fibers, *Pharmacol. Res. Commun.* 19:685-701.

Conte Camerino, D., De Luca, A., Mambrini, M., Ferrannini, E., Franconi, F., Giotti, A., and Bryant, S.H., 1989, The effects of taurine on pharmacologically induced myotonia, Muscle & Nerve, 12:898-904.

De Luca, A. and Conte Camerino, D., Effect of aging on the mechanical threshold of a rat skeletal muscle, *Pflügers Arch.* in press.

Durelli, L., Mutani, R., and Fassio, F, 1983, The treatment of myotonia:evaluation of chronic oral taurine therapy, *Neurol.* 33:559-603.

Franconi, F., Stendardi, I., Matucci, R., Failli, P., Bennardini, F., Antonini, G., and Giotti, A., 1984, Inotropic effect of taurine in guinea-pig ventricular strips, *Eur. J. Pharmacol.* 102:511.

Furman, R.E. and Barchi, R.L., 1981, 20,25-Diazacholesterol myotonia: an electrophysiological study, *Ann. Neurol.* 10:251-260.

Green, J.R. and Margerison, D., 1978, "Statistical treatment of experimental data", New York, Elsevier pp. 86-88.

Huxtable, R. and Bressler, R., 1973, Effect of taurine on a muscle intracellular membrane, *Biochim. Biophys. Acta.* 323:573-583.

Huxtable, R.J., Laird, H.E., and Lippincott, S.E., 1979, The transport of taurine in the heart and the rapid depletion of tissue taurine content by guanidinoethyl sulfonate, *J. Pharmacol. Exp. Ther.* 211:465-471.

Huxtable, R.J. and Sebring, L.A., 1986, Towards a unifying theory for the action of taurine, *TIPS* 7:481-485.

Huxtable, R.J. and Sebring, L.A., 1987, Modulation by taurine of calcium binding to phospholipid vesicles and cardiac sarcolemma, *Proc. West Pharmacol.* 30:153-155.

Krusz, J., Kendrick, Z.V., and Baskin, S.I., 1978, Taurine content in the pineal of aging monkey, In "Aging in the Nonhuman Primate", Raven Press, new York, pp. 106-111.

Lake, N., Eley, D.W., ter Keurs, H.E.D.J., 1991, The effects of taurine depletion on excitation-contraction coupling in rat cardiac trabeculae, *Biophys. J.* 59:64a.

Latorre, R., Oberhanser, A., Labarca, D., and Alvarez, O., 1989, Varieties of calcium-activated potassium channels, *Ann. Rev. Physiol.* 51:385-399.

Palade, P.T. and Barchi, R.L., 1977, On the inhibition of muscle membrane chloride conductance by aromatic carboxylic acids, J. Gen. Physiol. 69:879-896.

Rüdel, R. and Lehmann-Horn, F., 1985, Membrane change in cells from myotonia patients, *Physiol. Rev.* 65:310-356.

REGRESSION OF TAURINE DEPLETION IN RHESUS
MONKEYS DEPRIVED OF DIETARY TAURINE
THROUGHOUT THE FIRST YEAR

Martha Neuringer[1,2] John A. Sturman[3], Humi Imaki[3],
and Thomas Palackal[3]

[1]Departments of Medicine & Ophthalmology,
Oregon Health Sciences University,
Portland, OR 97201,

[2]Division of Neuroscience,
Oregon Regional Primate Research Center,
Beaverton, OR 97006, and

[3]Department of Developmental Biochemistry,
Institute for Basic Research in Developmental Disabilities,
Staten Island, NY 10314

INTRODUCTION

We are studying dietary taurine deficiency in developing rhesus monkeys in order to define its biochemical, morphological and functional effects and their timing and reversibility. Given the similarity between human and rhesus monkey infants in their nutritional requirements, sensory capacities and nervous system structure and function, these studies provide the best available model for assessing the long-term importance of dietary taurine for the development of the human retina and brain.

In a recent series of studies, infant rhesus monkeys were fed a taurine-free, soy-protein based human infant formula from birth until 3 months of age. Compared to monkeys fed the same formula supplemented with taurine, they showed reductions in plasma taurine levels similar to those reported for human infants fed low-taurine formulas (Sturman et al., 1988). At 1 - 3 months of age, the taurine-deprived infants' visual acuity was below normal (Neuringer & Sturman, 1987), and they showed alterations in the cortical visual evoked potential to patterned stimuli (Neuringer et al., 1990). Morphological studies demonstrated degenerative changes in retinal photoreceptor cells in the foveal region (Imaki et al., 1987) and alterations in the primary visual area of the cerebral cortex (Palackal et al., 1991). These studies demonstrated that rhesus monkeys require a dietary source of taurine during early infancy.

Unlike infants, adult rhesus monkeys require no dietary taurine in order to

Taurine, Edited by J.B. Lombardini *et al.*
Plenum Press, New York, 1992

maintain normal levels of taurine in plasma and tissues (Neuringer et al., 1985). In these studies, the monkeys were fed taurine-free casein-based diets for several years. They also showed no impairment in the electroretinogram or in visual acuity compared to monkeys fed a standard stock diet containing taurine. Thus, at some point during development, primates lose their dependence on dietary taurine.

We now have extended our studies of early taurine deprivation through the first year of life in order to determine the duration of dependence on dietary taurine and to evaluate the longer-term consequences of postnatal taurine deprivation.

METHODS

Rhesus monkeys were hand-reared from birth and were fed one of two diets: a taurine-free, soy protein-based human infant formula (Isomil®, Ross Laboratories), or the same formula supplemented with 70 μmoles/100 ml taurine, approximately the level in rhesus milk (Rassin et al., 1978). Within each diet group, 5 monkeys continued on these diets until 3 months, 5 until 6 months, and 5 until 12 months of age. At the completion of the assigned feeding duration, the animals were killed for biochemical and morphological studies.

Concentrations of taurine in plasma and tissues were determined at the completion of the study in each group, as described previously (Sturman et al., 1988, 1991). Additional plasma samples from the 3-month groups were analyzed at 1 and 2 months of age. A variety of tissues were examined, including 20 neural and ocular tissues, 5 muscle samples and 5 internal organs.

Bile samples were obtained at necropsy and analyzed by HPLC (Rossi et al., 1987; Sturman et al., 1988) for bile acid composition and for the proportion of each type of bile acid conjugated with taurine.

The activities of several enzymes involved in taurine biosynthesis were determined in liver and brain (Sturman et al., 1988, 1991). The activity of cysteinesulfinic acid decarboxylase (more correctly, 3-sulfinoalanine decarboxylase) was determined by collecting $^{14}CO_2$ produced from [1-^{14}C]cysteinesulfinic acid. In brain, measurements were carried out in the presence of saturating levels (100 mmol/L) of unlabeled L-glutamic acid to prevent decarboxylation of cysteinesulfinic acid by glutamic acid decarboxylase. Cystathionine synthase and cystathionase were measured by a modification of the method of Gaull et al. (1969) and cysteine dioxygenase by Misra and Olney's (1975) modification of the method of Yamaguchi et al. (1973).

One retina from each monkey was processed for light and electron microscopy using standard procedures (Imaki et al., 1987).

RESULTS

Taurine concentrations in plasma were significantly lower in the taurine-deprived monkeys at 2 through 6 months, but by 12 months of age levels were no longer different from those in taurine-supplemented monkeys (Figure 1). Similarly, in most tissues, the taurine-deprived groups had significantly lower taurine concentrations at 3 and 6 months but not at 12 months of age. In the retina, which retained taurine more tenaciously than most other tissues, levels were reduced by approximately 20% at 3 and 6 months (Figure 2). In occipital cortex, the primary visual area of the cerebral cortex, levels were reduced by 25% at 3 months and by 46% at 6 months (Figure 3). The liver showed the greatest taurine depletion, with reductions of 62% at 3 months and 54% at 6 months (Figure 4). At both 3 and 6 months, the taurine-deprived group had significantly lower taurine concentrations in 14 out of 20 neural and ocular tissue samples and in all 5 internal organs (liver, kidney, spleen, lung and adrenal) which were analyzed; among the 5 muscle samples (heart, diaphragm, biceps, triceps and gastrocnemius), all had lower taurine levels at 3 months and all but one (heart) at 6 months (Sturman et al., 1988, 1991). In agreement with previous findings that taurine levels in brain are highest early in development (Sturman & Gaull, 1975), all neural tissues in the supplemented group showed the expected age-related decline in taurine concentrations between 3 and 12 months of age.

The proportion of bile acids conjugated with taurine was reduced significantly at 3 and 6 months in each of the bile acids analyzed--cholic acid, chenodeoxycholic acid, deoxycholic acid and lithocholic acid (Sturman et al., 1988, 1991). For all bile acids

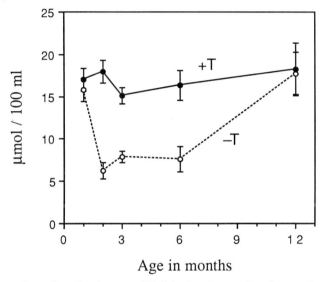

Figure 1. Concentrations of taurine (mean ± SEM) in the plasma of taurine-supplemented (+T) and taurine-deprived (-T) rhesus monkeys during the first year. Values for the deprived group were significantly lower (p<0.01) at 2, 3 and 6 months but not at 1 or 12 months of age.

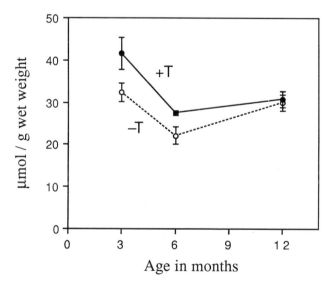

Figure 2. Concentrations of taurine (mean ± SEM) in the retina of taurine-supplemented (+T) and taurine-deprived (-T) rhesus monkeys at 3 months (p<0.15), 6 months (p<0.04) and 12 months of age.

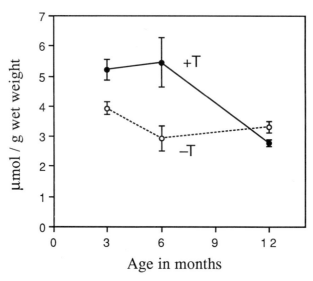

Figure 3. Concentrations of taurine (mean ± SEM) in the occipital cortex of taurine-supplemented (+T) and taurine-deprived (-T) rhesus monkeys. Values for the deprived group were significantly lower at 3 months (p<0.01) and 6 months (p<0.025) but not at 12 months.

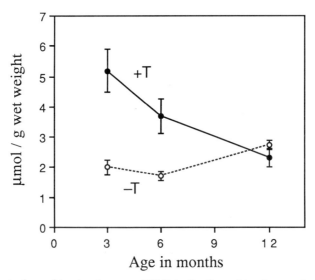

Figure 4. Concentrations of taurine (mean ± SEM) in the liver of taurine-supplemented (+T) and taurine-deprived (-T) rhesus monkeys. As in most tissues, values for the deprived group were significantly lower at 3 months (p<0.0025) and 6 months (p<0.01) but not at 12 months.

combined, the proportion conjugated with taurine was reduced by 59% in the deprived group at both ages--from 34% to 14% at 3 months and from 44 to 18% at six months (Figure 5). This difference also disappeared by 12 months of age.

The activity of liver cysteinesulfinic acid decarboxylase, a key enzyme in taurine biosynthesis, was significantly greater in the taurine-deprived group at 3 months of age but not at 6 or 12 months (Figure 6). Differences in liver cystathionase and cysteine dioxygenase at 3 months only approached significance (p = 0.09 and 0.13, respective-

ly), while liver and brain cystathionine synthase, brain cysteinesulfinic acid decarboxylase and brain cysteine dioxygenase showed no effect of diet at any age. None
of these enzymes showed significant developmental changes except for brain
cystathionine synthase, which increased nearly 2-fold in the deprived group and nearly
3-fold in the supplemented group from 3 to 6 months of age (Figure 7).

Degenerative morphological changes were found in the retinas of taurine-
deprived infants at 3 months, as previously reported (Imaki et al., 1987). Particularly
in cones in the foveal region, the orientation of photoreceptor outer segments was
disturbed and many disk membranes were fragmented, disorganized and vesiculated.
Inner segments and synaptic endings also showed alterations, including swollen

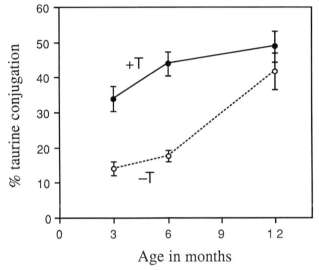

Figure 5. The proportion of all bile acids conjugated with taurine (mean $\pm$ SEM) in taurine-supplemented (+T) and taurine-deprived (-T) rhesus monkeys. Values were significantly lower for the
deprived group at 3 months (p<0.02) and 6 months (p<0.001) but not at 12 months.

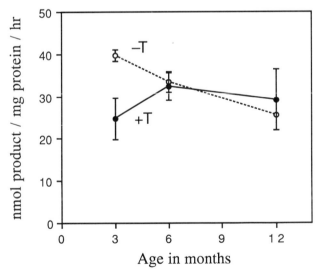

Figure 6. Activity of cysteinesulfinic acid decarboxylase (mean $\pm$ SEM) in the liver of taurine-supplemented (+T) and taurine-deprived (-T) rhesus monkeys. Activity was higher in the deprived group at 3
months (p<0.02).

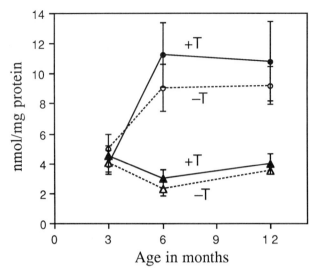

Figure 7. Activity of cystathionine synthase (mean ± SEM) in the brain (circles) and liver (triangles, values divided by 10) of taurine-supplemented (+T) and taurine-deprived (-T) rhesus monkeys. Activity was not affected by diet but increased in the brain of both groups between 3 and 6 months.

mitochondria and large vacuoles. These changes became progressively less severe with age, but some abnormalities in outer segment and synaptic morphology persisted at 12 months of age (Imaki et al., 1991).

DISCUSSION

These results indicate that in monkey infants, dependence on dietary sources of taurine continues for at least the first 6 months of life but declines thereafter. In monkeys fed a taurine-free soy protein-based human infant formula from birth, plasma and most tissues showed substantial taurine depletion at 3 and 6 months of age. However, by 12 months of age all levels had normalized despite continued dietary taurine deprivation. This spontaneous regression of taurine depletion could be due in part to an increase in synthetic ability, but our results show only a modest increase from 3 to 6 months in the maximal activity of brain cystathionine synthase, and this enzyme is not likely to play a major role in determining overall synthetic capacity. A more important factor probably is the decrease in taurine incorporation into developing tissue, particularly muscle, as the growth rate slows progressively during the first year.

Taurine deprivation elicited only minor compensatory metabolic adjustments, including a 61% increase in the activity of liver cysteinesulfinic acid decarboxylase at 3 months. A 59% reduction in the proportion of bile acids conjugated with taurine provided a mechanism to conserve taurine, a mechanism not available to the cat (Rabin et al., 1976). However, these relatively small adaptations were insufficient to balance the capacity for taurine biosynthesis with demand.

Although biochemical indices of taurine depletion normalized by 12 months of age, the morphological consequences were slower to recover. The degree of disruption of photoreceptor outer segment membranes appeared to decline with age, but changes were still present at 12 months (Imaki et al., 1991). This finding provides confirmation of an earlier study in which monkeys fed a taurine-free casein hydrolysate formula from birth showed widespread degenerative changes in cone photoreceptors even at 26 months of age (Sturman et al., 1984). Thus, taurine depletion during the early postnatal period may produce persistent effects on photoreceptor structure. In addition, we previously found quantitative and qualitative changes in the morphology of the primary visual cortex in 3-month-old taurine-deprived monkeys (Palackal et al., 1991), and some of the qualitative changes also appear to persist through the first year (unpublished results). Current studies are following taurine-deprived monkeys for longer periods to determine if these morphological changes eventually disappear or if they are irreversible.

A 6-month period of vulnerability to taurine depletion in monkeys corresponds to a 2-year period in human infants, based on the widely accepted generalization that development proceeds four times more rapidly in monkeys (Boothe et al., 1985). As a rough estimate, then, our findings suggest that human infants should be provided with dietary sources of taurine throughout this period. However, this comparison may underestimate the potential vulnerability of human infants to morphological or functional deficits. The maturation of the retina and nervous system is considerably more advanced in monkeys than in human infants at the time of birth. For example, in the monkey retina only the foveal region remains immature at birth, a fact which may be related to this area's specific vulnerability to taurine-deprivation-induced damage. In human newborns, and especially in premature infants, the impact of postnatal taurine deprivation could be greater and more widespread.

ACKNOWLEDGEMENTS

This research was supported by National Institutes of Health grants HD-18678 and RR-00163 and by grants from the National Retinitis Pigmentosa Foundation and the Medical Research Foundation of Oregon. This is publication 1830 of the Oregon Regional Primate Research Center.

REFERENCES

Boothe, R.G., Dobson, V., and Teller, D.Y., 1985, Postnatal development of vision in human and non-human primates, *Ann. Rev. Neurosci.* 8:495-545.

Gaull, G.E., Rassin, D.K., and Sturman, J.A., 1969, Enzymatic and metabolic studies of homocystinuria: effects of pyridoxine, *Neuropadiatrie* 1:199-226.

Imaki, H., Moretz, R., Wisniewski, H., Neuringer, M., and Sturman, J., 1987, Retinal degeneration in 3-month-old rhesus monkey infants fed a taurine-free human infant formula, *J. Neurosci. Res.* 18:602-614.

Imaki, H., Sturman, J., and Neuringer, M., 1991, Retinal morphology in rhesus monkeys fed formula with and without taurine for 6 and 12 months, *Invest. Ophthalmol. Vis. Sci.* 32:1231 (abstract).

Misra, C.H., and Olney, J.W., 1975, Cysteine oxidase in brain, *Brain Res.* 97:117-126.

Neuringer, M., Palackal, T., Kujawa, M., Moretz, R.C., and Sturman, J.A., 1990, Visual cortex development in rhesus monkeys deprived of dietary taurine, *in:* "Taurine: Functional Neurochemistry, Physiology and Cardiology," H. Pasantes-Morales, D.L. Martin, W. Shain, and R. Martin del Rio, eds., Wiley-Liss, New York, pp. 415-422.

Neuringer, M., and Sturman, J.A., 1987, Visual acuity loss in rhesus monkey infants fed a taurine-free human infant formula, *J. Neurosci. Res.* 18:597-601.

Neuringer, M., Sturman, J.A., Wen, G.Y., and Wisniewski, H.M., 1985, Dietary taurine is necessary for normal retinal development in monkeys, *in:* "Taurine: Biological Actions and Clinical Perspectives," S.S. Oja, L. Ahtee, P. Kontro, and M.K. Paasonen, eds., Alan R. Liss, New York, pp. 53-62.

Palackal, T., Kujawa, M., Moretz, R.C., Neuringer, M., and Sturman, J.A., 1991, Laminar analysis of the number of neurons, astrocytes, oligodendrocytes and microglia in the visual cortex (area 17) of 3-month-old rhesus monkeys fed a human infant soy-protein formula with or without taurine supplementation from birth, *Dev. Neurosci.* 13:20-33.

Rabin, B., Nicolosi, R.J., and Hayes, K.C., 1976, Dietary influence on bile acid conjugation in the cat, *J. Nutr.* 106:1241-1246.

Rassin, D.K., Sturman, J.A., and Gaull, G.E., 1978, Taurine and other free amino acids in milk of man and other mammals, *Early Hum. Dev.* 2:1-13.

Rossi, S.S., Converse, J.L., and Hofmann, A.E., 1987, High pressure liquid chromatographic analysis of conjugated bile acids in human bile: simultaneous resolution of sulfated and unsulfated lithocholyl amidates and the common conjugated bile acids, *J. Lipid Res.* 28:589-595.

Sturman, J.A., Gargano, A.D., Messing, J.M., and Imaki, H., 1986, Feline maternal taurine deficiency: effect on mother and offspring., *J. Nutr.* 116:655-667.

Sturman, J.A., and Gaull, G.E., 1975, Taurine in the brain and liver of the developing human and monkey, *J. Neurochem.* 25:831-835.

Sturman, J.A., Messing, J.M., Rossi, S.S., Hofmann, A.F., and Neuringer, M., 1988, Tissue taurine content and conjugated bile acid composition of rhesus monkey infants fed a human infant soy-protein formula with or without taurine supplementation for three months, *Neurochem. Res.* 13:311-316.

Sturman, J.A., Messing, J.M., Rossi, S.S., Hofmann, A.F., and Neuringer, M., 1991, Tissue taurine content, activity of taurine synthesis enzymes and conjugated bile acid composition of taurine-deprived and taurine-supplemented rhesus monkey infants at 6 and 12 months of age, *J. Nutr.* 121:854-862.

Sturman, J.A., Wen, G.Y., Wisniewski, H., and Neuringer, M., 1984, Retinal degeneration in primates raised on a synthetic human infant formula, *Int. J. Devel. Neurosci.* 2:121-129.

Yamaguchi, K., Sakakibara, S., Asamizu, J., and Ueda, I., 1973, Induction and activation of cysteine oxidase of rat liver. II. The measurement of cysteine metabolism in vivo and the activation of in vivo activity of cysteine oxidase, *Biochim. Biophys. Acta* 297:48-59.

TAURINE TRANSPORT IN THE MOUSE CEREBRAL CORTEX DURING DEVELOPMENT AND AGEING

Pirjo Saransaari and Simo S. Oja

Tampere Brain Research Center
Department of Biomedical Sciences
University of Tampere, Box 607
SF-33101 Tampere, Finland

INTRODUCTION

The importance of taurine in the developing brain has already been recognized. A deficiency of taurine in early development has been shown in kittens to interfere with the normal cell migration from the cerebellar granule cell layer, manifesting itself in a nervous system dysfunction.[1] In immature brain tissue taurine levels are also very high,[2,3] and depolarizing concentrations of K^+ evoke a strikingly large release of both endogenous and exogenous preloaded taurine.[4,5,6] On the other hand, very little is known of the possible functions of taurine in the ageing brain. We have now studied the uptake and the release of endogenous and labeled preloaded taurine in the cerebral cortex of mice during the whole life-span, focusing mainly on changes during ageing.

MATERIALS AND METHODS

Young, adult and ageing NMRI mice were used in the experiments. [^{3}H]Taurine (1.07 PBq/mol) was obtained from Amersham International, U.K. The concentrations of taurine were measured by high-performance liquid chromatography as described by Kontro and Oja.[7] The uptake of [^{3}H]taurine by cerebral cortical synaptosomal preparations was determined as in Kontro.[8] The release of endogenous and preloaded labeled taurine from cerebral cortical slices was assessed in a superfusion system and the efflux rate constants for taurine were computed.[4,6]

RESULTS AND DISCUSSION

The concentration of taurine was very high in the developing mouse cerebral cortex, decreasing then gradually with age and remaining at a constant level from 12 to 24 months,

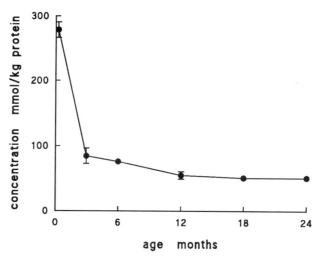

Figure 1. Concentrations of taurine during development and ageing in mouse cerebral cortical slices. The results are mean values (± SEM) of 3-6 determinations.

Table 1. Kinetic parameters of taurine uptake by synaptosomal preparations from the mouse cerebral cortex

Age (months)	K_{m1} (mM)	V_1 (μmol/s/kg)	K_{m2} (μM)	V_2 (nmol/s/kg)
3	4.4	16.8	61.3	38.3
6	4.8	23.9	75.6	66.7
12	4.7	18.9	43.0	71.7
18	4.8	22.7	70.0	79.5
24	4.0	16.0	45.4	137.2

The results were calculated per kg protein in the synaptosomal preparations by fitting the original data by a simplex algorithm. The nonsaturable uptake subtracted varied from 0.15 to 0.21 x 10^{-3} s^{-1}.

the oldest age studied (Fig. 1). The developmental changes in the level of taurine in the mouse cerebral cortex were similar to those in other animal species.[9,10] The fairly constant concentration of taurine during ageing may indicate that there are no great changes in biosynthetic capacity or in the dietary supply of taurine in old age.

The uptake of taurine by the synaptosomal preparations consisted of two saturable transport components, low and high affinity, and nonsaturable uptake (diffusion). The transport constant of the low-affinity uptake did not significantly change during ageing between 3 and 24 months (Table 1). The transport constant of the high-affinity uptake was lower in the oldest animals than in the young adults, whereas the maximal velocity of transport increased during ageing. Both the maximal velocity and the affinity of the low-affinity taurine uptake component remained apparently unchanged during ageing. The high-affinity uptake was, however, increased in senescent mice, which could be of importance for the synaptic functions of taurine, since the high-affinity uptake system has generally been considered to terminate the synaptic actions of neuromodulators. On the other hand, in synaptosomes prepared from the immature mouse brain the high-affinity uptake of taurine has been found to be similar to that in synaptosomes from adults but the affinity of low-affinity uptake higher.[8]

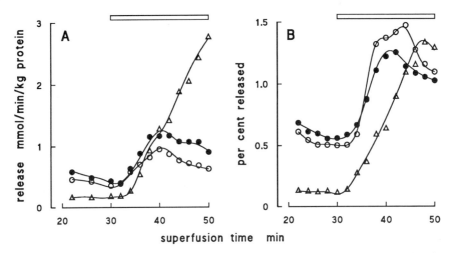

Figure 2. Release of endogenous (A) and preloaded labeled (B) taurine from cerebral cortical slices from 3-day-old (-Δ-), 3-month-old (-o-) and 24-month-old (●) mice. The slices were superfused for 50 min with Krebs-Ringer-Hepes medium which contained 50 mM K^+ from 30 min onwards as indicated by the bar. The results are mean values of 3 separate experiments with SEMs less than 10 %.

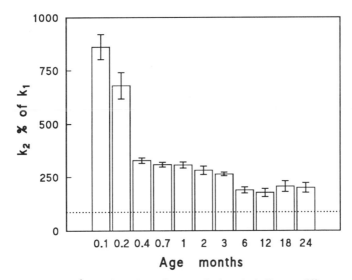

Figure 3. K^+ stimulation of [^{3}H]taurine release from cerebral cortical slices at different ages in mice. The efflux rate constants k_2 (34-50 min) are given in per cent (± SEM) of the basal efflux constant k_1 (20-30 min). The magnitude of k_2 for unstimulated efflux is indicated by the dotted line.

The basal unstimulated release of endogenous taurine was very low in the immature cerebral cortex, but K^+ stimulation (50 mM) enhanced the release almost 20-fold. The release steadily increased during the whole stimulation period (Fig. 2A). It was stimulated only about 1.5-fold by K^+ in 3-month-olds and somewhat more in 24-month-olds (Fig. 2A). Also the release of preloaded labeled taurine was highest in the cerebral cortex of 3-day-old mice (Fig. 2B), but in this case the release was somewhat greater at 3 months than at 24 months. K^+ stimulation also evoked a strikingly large release of exogenous taurine from the immature cerebral cortex, as shown by the calculated efflux rate constants, k_2, for the superfusion interval of 34 to 50 min (Fig. 3). The magnitude of K^+ stimulation decreased during maturation, remaining then at the same level from 6 months onwards.

A consistent finding in every brain area has been a strikingly large release of both endogenous and preloaded labeled taurine from immature brain tissue caused by K^+ stimulation.[4,5,6,11,12] We have earlier suggested that taurine could act as a modulator that switches off any prolonged excitation in neuronal networks,[13] being particularly active in the developing brain. Another salient feature is the slow time course of the evoked release of taurine,[4,11,13,14,15] which was here also evident in the ageing brain. The time course of taurine release was closely similar in 3- and 24-month-old mice. The parallelism in the release of exogenous and endogenous taurine was marked at each age. The only exception was that in senescent mice the release of endogenous taurine was relatively somewhat larger than that of exogenous labeled taurine. The releasable taurine pool(s) may thus change during ageing.

The basal unstimulated release of [³H]taurine from cerebral cortical slices was greatly enhanced when Ca^{2+} was omitted and EDTA (2.0 mM) added to superfusion medium in immature, adult and ageing mice (Fig. 4). The K^+ stimulation was still discernible in slices from 3-day-old and 3-month-old mice, whereas there was no K^+ stimulation in the absence of Ca^{2+} in 18-month-olds. Depending on the preparation studied and on the experimental design, the release of taurine in the adult brain has been claimed to be more or less Ca^{2+}-dependent, unaffected by extracellular Ca^{2+} or even enhanced in the absence of Ca^{2+}.[16] Evaluation as to whether or not the K^+-stimulated release is dependent on Ca^{2+} is hampered by the great enhancement of the basal efflux in Ca^{2+}-free media, which enhancement

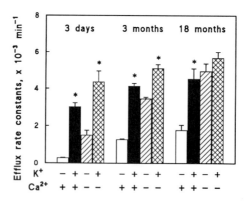

Figure 4. Release of [³H]taurine from cerebral cortical slices from mice of different ages. The graph shows efflux rate constants $k_2 \pm$ SEM (34-50 min) in the absence and presence of K^+ (50 mM) and Ca^{2+} as indicated. The asterisk indicates a statistically significant (P<0.01) K^+ stimulation.

apparently increases with age (Fig. 4). It has earlier been inferred that only a part of the release of taurine is Ca^{2+}-dependent in both adult and developing brain.[4] The present results show that the stimulated release of taurine from cerebral cortical slices from old mice is completely Ca^{2+}-dependent.

The glutamate agonists NMDA, kainate and quisqualate (all 0.1 mM) potentiated the release of taurine from the developing cerebral cortex, but in the adults only NMDA and kainate were effective (Fig. 5). The agonists had no significant effects on the release of

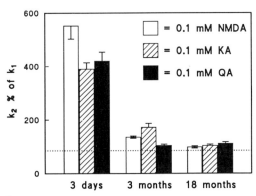

Figure 5. Stimulation of [³H]taurine release from mouse cerebral cortex slices by glutamate agonists. The graph shows efflux rate constants k_2 (34-50 min) ± SEM. The magnitude of k_2 for unstimulated efflux is indicated by the dotted line.

taurine from cerebral cortical slices from 18-month-old mice. The excitatory amino acid receptor agonists seem to be more effective in the developing than in the adult brain in this respect. The NMDA and AMPA subtypes of the glutamate receptor have been implicated in the release of taurine in the immature cerebral cortex.[17] This kind of modulation seems to be lacking in ageing mice. In keeping with this, the maximal capacity for glutamate binding decreases during ageing; particularly the number of NMDA receptors diminishes in the cerebral cortex.[18]

In conclusion, the results corroborate the importance of taurine as a regulator of neuronal activity in immature cerebral cortex, while showing that the tissue levels and transport systems of taurine do not in general exhibit any great changes during ageing.

ACKNOWLEDGEMENTS

The skillful technical assistance of Ms. Paula Kosonen and Ms. Oili Pääkkönen and the financial support of the Emil Aaltonen Foundation, Finland, are gratefully acknowledged.

REFERENCES

1. J.A. Sturman, R.C. Moretz, J.H. French, and H.M. Wisniewski, Taurine deficiency in the developing cat: persistence of the cerebellar external granule cell layer, *J. Neurosci. Res.* 13:405 (1985).
2. S.S. Oja and P. Kontro, Taurine, *in*: "Handbook of Neurochemistry," 2nd edn, vol. 3, p. 501, A. Lajtha, ed., Plenum, New York (1983).
3. R.J. Huxtable, Taurine in the central nervous system and the mammalian actions of taurine, *Prog. Neurobiol.* 32:471 (1989).
4. P. Kontro and S.S. Oja, Taurine and GABA release from mouse cerebral cortex slices: potassium stimulation releases more taurine than GABA from developing brain, *Devl Brain Res.* 37:277 (1987).
5. P. Kontro and S.S. Oja, Release of taurine and GABA from cerebellar slices from developing and adult mice, *Neuroscience* 29:413 (1989).
6. S.S. Oja and P. Kontro, Release of endogenous taurine and γ-aminobutyric acid from brain slices from the adult and developing mouse, *J. Neurochem.* 52:1018 (1989).
7. P. Kontro and S.S. Oja, Taurine and GABA binding in mouse brain: effects of freezing, washing and Triton X-100 treatment on membranes, *Int. J. Neurosci.* 32:881 (1987).
8. P. Kontro, Comparison of taurine, hypotaurine and ß-alanine uptake in brain synaptosomal preparations from developing and adult mouse, *Int. J. Devl Neurosci.* 2:465 (1984).
9. S.S. Oja, A.J. Uusitalo, M.-L. Vahvelainen, and R.S. Piha, Changes in cerebral and hepatic amino acids in the rat and guinea pig during development, *Brain Res.* 11:655 (1968).

10. J.A. Sturman and G.E. Gaull, Taurine in the brain and liver of the developing human and monkey, *J. Neurochem.* 25:831 (1975).
11. P. Kontro and S.S. Oja, Taurine efflux from brain slices: potassium-evoked release is greater from immature than mature brain tissue, *in*: "The Biology of Taurine: Methods and Mechanisms," p. 79, R.J. Huxtable, F. Franconi and A. Giotti, eds, Plenum, New York (1987).
12. P. Kontro and S.S. Oja, Release of taurine, GABA and dopamine from rat striatal slices: mutual interactions and developmental aspects, *Neuroscience* 24:49 (1988).
13. P. Kontro and S.S. Oja, Taurine and GABA release from mouse cerebral cortex slices: effects of structural analogues and drugs, *Neurochem. Res.* 12:475 (1987).
14. E.R. Korpi, P. Kontro, K. Nieminen, K.-M. Marnela, and S.S. Oja, Spontaneous and depolarization-induced efflux of hypotaurine from mouse cerebral cortex slices: comparison with taurine and GABA, *Life Sci.* 29:811 (1981).
15. J.-P. Pin, S. Weiss, M. Sebben, D.E. Kemp, and J. Bockaert, Release of endogenous amino acids from striatal neurons in primary culture, *J. Neurochem.* 47:594 (1986).
16. P. Saransaari and S.S. Oja, Release of GABA and taurine from brain slices, *Prog. Neurobiol.* 38:455 (1992).
17. P. Saransaari and S.S. Oja, Excitatory amino acids evoke taurine release from cerebral cortex slices from adult and developing mice, *Neuroscience* 45:451 (1991).
18. G.L. Wenk, L.C. Walker, D.L. Price, and L.C. Cork, Loss of NMDA, but not GABA-A, binding in the brains of aged rats and monkeys, *Neurobiol. Aging* 12:93 (1991).

PHOSPHOLIPIDS, PHOSPHOLIPID METHYLATION AND TAURINE
CONTENT IN SYNAPTOSOMES OF DEVELOPING RAT BRAIN

P.-L. Lleu, S. Croswell, and R.J. Huxtable

Department of Pharmacology
College of Medicine
University of Arizona
Tucson, Arizona 85724

INTRODUCTION

Many phenomena involving taurine appear to be based on a membrane-modifying action, although the nature of this action has remained largely undefined. Taurine stabilizes membranes under a variety of conditions, reducing malondialdehyde release induced by carbon tetrachloride (Nakashima et al., 1983), and antagonizing the loss of membrane function produced by phospholipase C (Huxtable and Bressler, 1973) among other actions. Taurine also modifies the calcium binding characteristics of membranes, increasing the affinity of calcium binding while decreasing the maximum binding capacity (Huxtable and Peterson, 1988; Sebring and Huxtable, 1985). Perhaps associated with this is a taurine-induced shift in the membrane transition temperature (Lombardini, 1985), and a stimulation of high affinity, energy-dependent calcium transport in the presence of bicarbonate (reviewed in Huxtable, 1989).

It has been suggested that the effects of taurine on the binding of calcium is secondary to an interaction of taurine with membrane phospholipids (Huxtable and Sebring, 1986; Huxtable, 1990). In support of this suggestion, characteristics of the low affinity binding of taurine to biological membranes can be reproduced in artificial phospholipid membranes, and the effects of taurine on calcium binding parameters can likewise be mimicked in artificial phospholipid membranes (Sebring and Huxtable, 1986). The phospholipid hypothesis for the action of taurine indicates that the phospholipid composition of a membrane and the taurine content of the enclosed cytoplasm mutually affect calcium binding and calcium movements. If true, this implies a possible biochemical relationship between taurine concentration and phospholipid composition of the membrane. We have examined for such a relationship using synaptosomes from the developing rat brain as a model system.

ADVANTAGES OF THE DEVELOPING SYNAPTOSOME

Enriched synaptosomes, commonly known as the P_2B fraction (Gray and

Taurine, Edited by J.B. Lombardini *et al.*
Plenum Press, New York, 1992

Whittaker, 1962), and the synaptic plasma membranes which are prepared from them (Cotman, 1974; Cotman and Matthews, 1971), have proven to be useful in the study of taurine and membrane interactions. Taurine modifies calcium binding characteristics to the synaptosomes (Lazarewicz et al., 1985; Huxtable and Peterson, 1988), inhibits protein phosphorylation (Li and Lombardini, 1991) and stimulates calcium transport [although this may, in fact, be occurring in contaminating mitochondria (Huxtable, 1989)]. The binding and transport of taurine itself have also been studied in this preparation (Meiners et al., 1980; Allen et al., 1986; Kontro and Oja, 1983; Korpi and Oja, 1984; Lombardini, 1977; Hruska et al., 1978; Huxtable and Peterson, 1989).

During postnatal development in the rat, there is a rapid increase in synaptosomal density in the brain (Huxtable et al., 1989). The developmental process in the synaptosome is also associated with changes in phospholipid composition, such as the phosphatidylcholine:phosphatidylethanolamine ratio (Figure 1), numerous biochemical changes, including alterations in S-adenosylmethionine and S-adenosylhomocysteine concentrations (Gharib et al., 1982; Sarda et al., 1989), and developmental changes in amino acid content. The most dramatic change among the amino acids is a profound fall in taurine concentration (Figure 2). Over the first eight weeks of life, these decreased to 1/6 of their original levels. Other amino acids, such as glutamate, increase slightly in concentration, and others remain unchanged, as is the case with GABA (Figure 2).

In examining the relationship between these various developmental changes, we found a strong correlation between synaptosomal taurine concentration and the ratio of phosphatidylcholine to phosphatidylethanolamine in synaptosomal membranes (Figure 3) (Huxtable et al., 1989).

Phosphatidylcholine and phosphatidylethanolamine are neutral phospholipids, the charged headgroups of which are closely related in structure to taurine (Figure 4), that exhibit low affinity binding sites for taurine in keeping with the high cellular

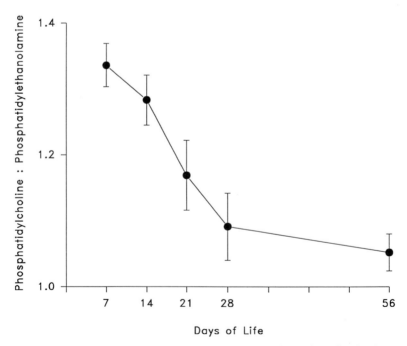

Figure 1. Changes in the phosphatidylcholine:phosphatidylethanolamine ratio during development in rat brain synaptosomes.

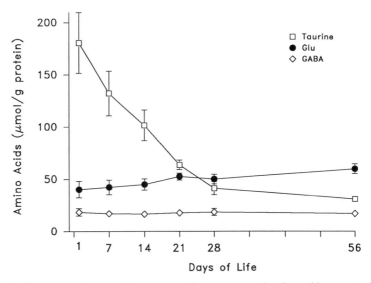

Figure 2. Patterns of change and development in synaptosomal amino acid concentrations.

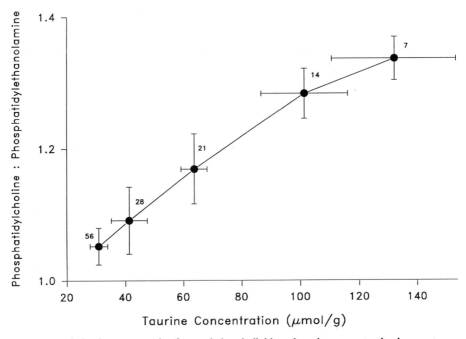

Figure 3. Correlation between a ratio of neutral phospholipids and taurine concentration in synaptosomes from developing rat brain. The day of life is indicated by the number above each datum point.

concentration of this substance (Sebring and Huxtable, 1986). In phospholipid vesicles prepared from single classes of phospholipids, phosphatidylcholine has a binding affinity (K_D) of 60 mM, phosphatidylethanolamine of 70 mM, phosphatidylserine of 107 mM, while binding is undetectable to phosphatidylinositol (Huxtable and Sebring, 1986). Taurine binds to mixed phospholipid vesicles containing the various classes of phospholipids in the same proportion as present in cardiac sarcolemma with an affinity of 64 mM. If cholesterol is included in the liposome to reduce membrane fluidity, the affinity increases to 29 mM, which may be compared to the K_D of 19 mM found for the binding of taurine to sarcolemma itself. It thus appears that the low affinity binding of taurine observed in biological membranes is a result of the binding of taurine to neutral phospholipids in the membrane. This low affinity binding site has been equated with the Ca^{2+}-modulatory site for taurine (Huxtable and Sebring, 1986; Sebring and Huxtable, 1986; Huxtable, 1990).

PHOSPHOLIPID METHYLATION

The neutral phospholipids have opposing effects on membranes, phosphatidylcholine stabilizing lipid bilayers, and phosphatidylethanolamine destabilizing (Sun and Sun, 1985). A direct metabolic relationship also subsists between them. Although probably most phosphatidylcholine is biosynthesized via the CDP choline route, phosphatidylcholine can also be biosynthesized directly from phosphatidylethanolamine by the phospholipid methylation route.

The phospholipid methylation route starts with the adenosylation of methionine to S-adenosylmethionine. The enzyme catalyzing this conversion is methionine adenosyltransferase (EC 2.5.1.6). S-Adenosylmethionine is a sulfonium compound, serving as a source of methyl groups in numerous methylation reactions. The second enzyme is phospholipid methyltransferase I. This enzyme transfers a methyl group from S-adenosylmethionine to the membrane phospholipid, phosphatidylethanolamine, yielding monomethylphosphatidylethanolamine (Crews et al., 1980). Methyltransferase I activity is high in brain synaptosomes, where it is localized on the inner (cytoplasmic) aspect of the membrane. Further transfer of methyl groups from two additional molecules of S-adenosylmethionine yield phosphatidylcholine. The enzyme carrying out these methylations is phospholipid methyltransferase II. This enzyme is

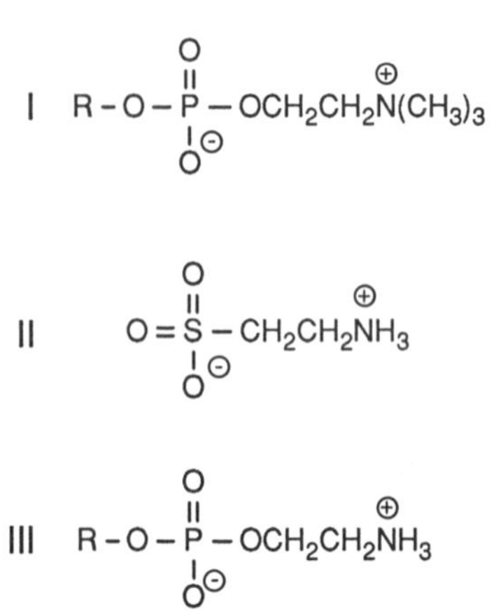

Figure 4. The structural similarities of taurine (II) and the charged head group portion of the neutral phospholipids phosphatidylcholine (I) and phosphatidylethanolamine (III).
R = Diacylglycerol.

localized in synaptosomes on the outer aspect of the membrane. These sequential methylation reactions exhibit different affinities for S-adenosylmethionine, the first methylation proceeding at an S-adenosylmethionine concentration of 0.1 μM, the second at 10 μM, and the third at 150 μM (Hirata and Axelrod, 1978; Panagia et al., 1984). Methyltransferase I and methyltransferase II have also been found in adrenals, heart, and erythrocytes (Hirata and Axelrod, 1978). In all cases, the enzymes are asymmetrically distributed on the membrane, methyltransferase I being on the inside and methyltransferase II on the outside. These enzymes may be one factor controlling the topological distribution of membrane phospholipids, phosphatidylethanolamine being typically present in greater amounts on the cytoplasmic aspect of the membrane, and phosphatidylcholine being present in greater amounts on the external aspect of the membrane.

The methylation route is being increasingly recognized as an important regulator of numerous membrane functions (Crews, 1985). These functions include Ca^{++} channel opening (McGivney et al., 1981), altered Ca^{++} binding and Ca^{++} ATPase activities (Panagia et al., 1987a), inhibition of Na^+ and Ca^{++} exchange in the heart (Panagia et al., 1987b; Dyer and Greenwood, 1988), histamine release and various receptor activities (Okumura et al., 1987). Thus, N-methylation unmasks cryptic ß-adrenergic receptors (Strittmatter et al., 1979; Tallman et al., 1979).

The two methyltransferases seem to have independent actions on membrane function. Methyltransferase I enhances the coupling of ß-adrenergic receptors with adenylate cyclase (Hirata et al., 1979), while methyltransferase II increases the number of ß-adrenergic receptors (Strittmatter et al., 1979).

Phospholipid methylation, in turn, regulated in a number of ways, including by the S-adenosylmethionine: S-adenosylhomocysteine ratio in the cell, S-adenosylmethionine stimulating and S-adenosylhomocysteine inhibiting methylation (Fonlupt et al., 1981). This ratio decreases in the brain with age (Gharib et al., 1982; Sarda et al., 1989). S-Adenosylmethionine levels, in turn, are strongly influenced by methionine levels (Dyer and Greenwood, 1988). Methionine adenosyltransferase, the enzyme converting methionine to S-adenosylmethionine, has a K_m of 10 μM (Mitsui et al., 1988) while methionine levels in the brain average 1 μM. The enzyme is, therefore, unsaturated, its activity responding sensitively to changes in methionine concentrations. Methionine adenosyltransferase is also influenced by adenosine levels, these increasing in the brain with age (Sarda et al., 1989; Gharib et al., 1982). Methylation is also increased by both α-adrenergic (Hirata et al., 1979) and ß-adrenergic stimulation (Okumura et al., 1987; Hirata et al., 1979).

The influence of phospholipid methylation on β-adrenergic receptor activity is intriguing inasmuch as taurine transport and efflux is altered by β-adrenergic activation in the heart, salivary glands, pineal, and other systems (Huxtable and Chubb, 1977; Azari and Huxtable, 1980; Wheler and Klein, 1980; Madelian et al., 1985; Shain and Martin, 1984; Lleu and Rebel, 1989). This may form part of a feed-back system, therefore, regulating membrane structure and function.

In keeping with the correlation between taurine concentration and neutral phospholipid ratio (Figure 3), we also observe a correlation between taurine concentration and the rate of phospholipid methylation in synaptosomes from developing brain (Figure 5). In this experiment, [^{3}H-methyl]methionine was injected intraperitoneally, and cortical cerebral synaptosomes isolated 9 h later. Synaptosomal methionine concentrations were measured, and the specific activity calculated for methionine in the synaptosome used to calculate methylation rates. For this purpose, the contribution for [^{3}H-methyl]-S-adenosylmethionine to the soluble radioactivity was ignored. Methionine concentrations in the brain are an order of magnitude greater than S-adenosylmethionine concentrations (Gharib et al., 1982; Dyer and Greenwood, 1988; Tudball and Griffiths, 1976; Kontro et al., 1980). As the enzyme synthesizing S-adenosylmethionine, methionine adenosyltransferase, is unsaturated, its activity responds sensitively to changes in methionine concentration. The specific activities of precursor and product, therefore, rapidly equilibrate. No corresponding

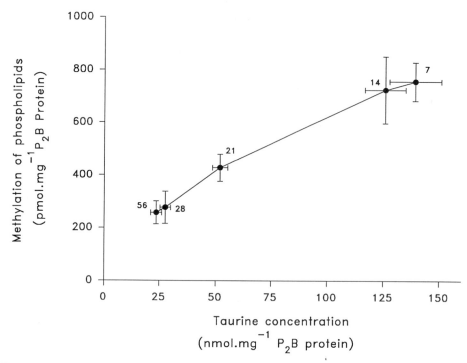

Figure 5. Correlation of taurine concentration and phospholipid methylation rate in synaptosomes from developing rat brain. The day of life is indicated by each point. [³H-methyl]Methionine was injected i.p. and animals killed 9 h later. The specific activity of methionine in isolated synaptosomes was determined, and this used to calculate the rate of incorporation into synaptosomal phospholipids.

correlation was found between methionine incorporation into protein and taurine concentration.

TAURINE-MEMBRANE INTERACTIONS

These results suggest there are three levels at which interactions between taurine and cell membranes occur: taurine may modify membrane phospholipid composition by affecting phospholipid methylation, taurine modifies the properties of a membrane of a given structure, and the phospholipid composition modifies the interaction of the membrane with taurine. Examples of taurine modifying membrane properties include the alteration in calcium binding characteristics, stimulation of high-affinity bicarbonate-dependent calcium transport, alteration of the transition temperature for calcium transport, and an increase in chloride conductance. Examples of membrane composition modifying the interaction with taurine includes the variation in taurine binding affinity or modifying phospholipid composition or cholesterol content, the increase in V_{max} of taurine transport with increase in membrane polyunsaturated fatty acid content (Yorek et al., 1984), and the developmental changes in transport and diffusion of taurine that occur.

If taurine is a regulator of the phospholipid methylation pathway, this would provide a prime example of the enantiostatic function of taurine which has been recently proposed (Huxtable, 1992). In enantiostasis, the effects of a change in the chemical and physical properties of the internal milieu of a cell is opposed by a further change (Mangum and Towle, 1977). As a consequence, although the milieu is unstable, cell functions are stable. Alterations in membrane composition and the concentration of taurine to which the membrane is exposed vary in a coordinated way such that the functions of the membrane which maintain cell viability are preserved intact.

REFERENCES

Allen, I.C., Schousboe, A. and Griffiths, R., 1986, Effect of L-homocysteine and derivatives on the high affinity uptake of taurine and GABA into synaptosomes and cultured neurons and astrocytes, *Neurochem. Res.* 11:1487-1496.

Azari, J. and Huxtable, R.J., 1980, The mechanism of the adrenergic stimulation of taurine influx in the heart, *Eur. J. Pharmacol.* 61:217-223.

Cotman, C.W., 1974, Isolation of synaptosomal and synaptic plasma membrane fraction, *Meth. Enzymol.* 31:445-452.

Cotman, C.W. and Matthews, D.A., 1971, Synaptic plasma membranes from rat brain synaptosomes: Isolation and partial characterization, *Biochim. Biophy. Acta* 249:380-394.

Crews, F., 1985, Phospholipid methylation and membrane function, *in*: "Phospholipids and Cellular Regulation, Vol. I," K. F. Kuo, ed., CRC Press, Boca Raton, Florida, pp. 131-158.

Crews, F.T., Hirata, F. and Axelrod, J., 1980, Identification and properties of methyltransferases that synthesize phosphatidylcholine in rat brain synaptosomes, *J. Neurochem.* 34:1491-1498.

Dyer, J.R. and Greenwood, C.E., 1988, Evidence for altered methionine methyl-group utilization in the diabetic rat's brain, *Neurochem. Res.* 13:517-523.

Fonlupt, P., Rey, C. and Pacheco, H., 1981, Phosphatidylethanolamine methylation in membranes from rat cerebral cortex: effect of exogenous phospholipids and S-adenosylhomocysteine, *Biochem. Biophys. Res. Comm.* 100:1720-1726.

Gharib, A., Sarda, N., Chabannes, B., Cronenberger, L. and Pacheco, H., 1982, The regional concentrations of *S*-adenosyl-L-methionine, *S*-adenosyl-L-homocysteine, and adenosine in rat brain, *J. Neurochem.* 38:810-815.

Gray, E.G. and Whittaker, V.P., 1962, The isolation of nerve endings from brain: an electron microscopic study of cell fragments derived by homogenization and centrifugation, *J. Anat.* 96:431-435.

Hirata, F., Strittmatter, W.J. and Axelrod, J., 1979, β-Adrenergic receptor agonists increase phospholipid methylation, membrane fluidity and β-adrenergic receptor-adenylate cyclase coupling, *Proc. Natl. Acad. Sci. USA* 76:368-372.

Hirata, F. and Axelrod, J., 1978, Enzymatic synthesis and rapid translocation of phosphatidylcholine by two methyltransferases in erythrocyte membranes, *Proc. Natl. Acad. Sci. USA* 75:2348-2352.

Hruska, R.E., Huxtable, R.J. and Yamamura, H.I., 1978, High affinity, temperature-sensitive and sodium-dependent transport of taurine in the rat brain, *in*: "Taurine and Neurological Disorders," A. Barbeau and R. J. Huxtable, eds., Raven Press, New York, pp. 109-117.

Huxtable, R. and Chubb, J., 1977, Adrenergic stimulation of taurine transport by the heart, *Science* 198:409-411.

Huxtable, R.J., 1989, Taurine in the central nervous system and the mammalian actions of taurine, *Prog. Neurobiol.* 32:471-533.

Huxtable, R.J., Crosswell, S. and Parker, D., 1989, Phospholipid composition and taurine content of synaptosomes in developing rat brain, *Neurochem. Int.* 15:233-238.

Huxtable, R.J., 1990, The interaction between taurine, calcium and phospholipids: Further investigations of a trinitarian hypothesis, *in*: "Taurine: Functional Neurochemistry, Physiology, and Cardiology," H. Pasantes-Morales, D. L. Martin, W. Shain and R. M. del Río, eds., Wiley-Liss, New York, pp. 185-196.

Huxtable, R.J., 1992, The physiological actions of taurine, *Physiol. Rev.* 72:101-163.

Huxtable, R.J. and Bressler, R., 1973, Effect of taurine on a muscle intracellular membrane, *Biochim. Biophys. Acta* 323:573-583.

Huxtable, R.J. and Peterson, A., 1988, The effect of taurine on calcium binding to brain synaptosomes, *Pharmacologist* 30:A86.

Huxtable, R.J. and Peterson, A., 1989, Sodium-dependent and sodium-independent binding of taurine, *Neurochem. Int.* 140:79-84.

Huxtable, R.J. and Sebring, L.A., 1986, Towards a unifying theory for the action of taurine, *TIPS* 7:481-485.

Kontro, P., Marnela, K.-M. and Oja, S.S., 1980, Free amino acids in the synaptosome and synaptic vesicle. Fractions of different bovine brain areas, *Brain Res.* 184:129-141.

Kontro, P. and Oja, S.S., 1983, Binding of taurine to brain synaptic membranes, in: "CNS Receptors - From Molecular Pharmacology to Behavior: Advances in Biochemical Psychopharmacology, Vol. 37," P. Mandel and F. V. DeFeudis, eds., Raven Press , New York, pp. 23-34.

Korpi, E.R. and Oja, S.S., 1984, Calcium chelators enhance the efflux of taurine from brain slices, *Neuropharmacology* 23:377-380.

Lazarewicz, J.W., Noremberg, K., Lehmann, A. and Hamberger, A., 1985, Effects of taurine on calcium binding and accumulation in rabbit hippocampal and cortical synaptosomes, *Neurochem. Int.* 7:421-428.

Li, Y.-P. and Lombardini, J.B., 1991, Inhibition by taurine of the phosphorylation of specific synaptosomal proteins in the rat cortex: Effects of taurine on the stimulation of calcium uptake in mitochondria and inhibition of phosphoinositide turnover, *Brain Res.* 553:89-96.

Lleu, P.L. and Rebel, G., 1989, Effects of HEPES on the taurine uptake by cultured glial cells, *J. Neurosci. Res.* 23:78-86.

Lombardini, J.B., 1977, High affinity uptake systems for taurine in tissue slices and synaptosomal fractions prepared from various regions of the rat central nervous system. Correction of transport data by different experimental procedures, *J. Neurochem.* 29:305-312.

Lombardini, J.B., 1985, Taurine effects on the transition temperature in Arrhenius plots of ATP-dependent calcium ion uptake in rat retinal membrane preparation, *Biochem. Pharmacol.* 34:3741-3745.

Madelian, V., Martin, D.L., Lepore, R., Perrone, M. and Shain, W., 1985, β-Receptor stimulated and cyclic adenosine 3',5'-monophosphate-mediated taurine release from LRM55 glial cells, *J. Neurosci.* 5:3154-3160.

Mangum, C. and Towle, D., 1977, Physiological adaptation to unstable environments, *Am. Scient.* 65:67-75.

McGivney, A., Crews, F.T., Hirata, F., Axelrod, J. and Siraganian, R.R., 1981, Rat basophilic leukemia cells defective in phospholipid methyltransferase enzyme, Ca^{2+} influx and histamine release: Reconstruction by hybridization, *Proc. Natl. Acad. Sci. USA* 78:6176-6180.

Meiners, B.A., Speth, R.C., Bresolin, N., Huxtable, R.J. and Yamamura, H.I., 1980, Sodium-dependent, high-affinity taurine transport into rat brain synaptosomes, *Fed. Proc.* 39:2695-2700.

Mitsui, K., Teraoka, H. and Tsukada, K., 1988, Complete purification and immunochemical analysis of S-adenosylmethionine synthetase from rat brain, *J. Biol. Chem.* 263:11211-11216.

Nakashima, T., Takino, T. and Kuriyama, K., 1983, Therapeutic and prophylactic effects of taurine administration on experimental liver injury, *in*: "Sulfur Amino Acids: Biochemical and Clinical Aspects," K. Kuriyama, R. J. Huxtable and H. Iwata, eds., Alan R. Liss, Inc., New York, pp. 449-460.

Okumura, K., Panagia, V., Beamish, R.E. and Dhalla, N.S., 1987, Biphasic changes in the sarcolemmal phosphatidylethanolamine N-methylation activity in catecholamine-induced cardiomyopathy, *J. Mol. Cell. Cardiol.* 19:356-366.

Panagia, V., Ganguly, P. and Dhalla, N.S., 1984, Characterization of heart sarcolemmal phospholipid methylation, *Biochim. Biophys. Acta* 792:245-253.

Panagia, V., Elimban, V., Ganguly, P.K. and Dhalla, N.S., 1987a, Decreased Ca2+ binding and Ca2+ ATPase activities in heart sarcolemma upon phospholipid methylation, *Mol. Cell. Biochem.* 78:65-71.

Panagia, V., Makino, N., Ganguly, P.K. and Dhalla, N.S., 1987b, Inhibition of Na^+-Ca^{2+} exchange in heart sarcolemmal vesicles by phosphatidylethanolamine N-methylation, *Eur. J. Biochem.* 166:597-603.

Sarda, N., Reynaud, D. and Gharib, A., 1989, S-adenosylmethionine, S-adenosylhomocysteine and adenosine system. Age-dependent availability in rat brain, *Dev. Pharmacol. Ther.* 13:104-112.

Sebring, L. and Huxtable, R.J., 1985, Taurine modulation of calcium binding to cardiac sarcolemma, *J. Pharmacol. Exptl. Therap.* 232:445-451.

Sebring, L.A. and Huxtable, R.J., 1986, Low affinity binding of taurine to phospholiposomes and cardiac sarcolemma, *Biochim. Biophys. Acta* 884:559-566.

Shain, W.G. and Martin, D.L., 1984, Activation of beta-adrenergic receptors stimulates taurine release from glial cells, *Cell. Molec. Neurol.* 4:191-196.

Strittmatter, W.J., Hirata, F. and Axelrod, J., 1979, Phospholipid methylation unmasks cryptic β-adrenergic receptors in rat reticulocytes, *Science* 204:1207-1209.

Sun, G.Y. and Sun, A.Y., 1985, Ethanol and membrane lipids, *Alcohol. Clin. Exp. Res.* 9:164-180.

Tallman, J.F.J., Henneberry, R.C., Hirata, F. and Axelrod, J., 1979, Control of β-adrenergic receptors in Hela cells, *in*: "Catecholamines: Basic and Clinical Frontiers, Vol. I," E. Usdin, I. J. Kopin and J. Barchas, eds., Pergamon Press, pp. 489-491.

Tudball, N. and Griffiths, R., 1976, Biochemical changes in the brain of experimental animals in response to elevated plasma homocystine and methionine, *J. Neurochem.* 26:1149-1154.

Wheler, G.H.T. and Klein, D.C., 1980, Taurine release from the pineal gland is stimulated via a beta-adrenergic mechanism, *Brain Res.* 187:155-164.

Yorek, M.A., Strom, D.K. and Spector, A.A., 1984, Effect of membrane polyunsaturation on carrier-mediated transport in cultured retinoblastoma cells: alterations in taurine uptake, *J. Neurochem.* 42:254-261.

THE EFFECT OF TAURINE ON THE AGE-RELATED
DECLINE OF THE IMMUNE RESPONSE IN MICE:
THE RESTORATIVE EFFECT ON THE T CELL
PROLIFERATIVE RESPONSE TO COSTIMULATION WITH
IONOMYCIN AND PHORBOL MYRISTATE ACETATE

Shigeru Negoro and Hideki Hara

Department of Medicine III
Osaka University Medical School
Osaka, Japan

INTRODUCTION

The proliferative responses of lymphocytes to mitogen in humans and experimental animals decrease with age[1-5]. In previous experiments, we showed that the ability of highly purified T and B cells to repeat replication under mitogenic stimulation was significantly depressed in the aged person compared to the young person[6-8]. We used highly purified T and B cells as responding cells and costimulation of both ionomycin and phorbol-12-myristate-13-acetate (PMA) as mitogen to analyze the proliferative response. Miller et al.[9,10] suggested the diminished lymphocyte reactivity to the stimulation by the mitogen might reflect an age-related impairment in the ability to raise the concentration of intracellular free calcium ion ($[Ca^{2+}]_i$) in response to the stimulation.

Many investigators have tried to prevent or improve the age-related immune decline[11-15]. Taurine, a sulfur containing amino acid, is known to have several physiological actions. It has been reported that taurine modulates calcium movement through the sarcolemma and consequently exerts beneficial effects in modifying susceptibility to heart and aorta cell injury due to the disturbance of calcium homeostasis[16-18]. A relationship between taurine and Ca^{2+} has also been reported in the retina[19-21], erythrocytes[22] and in cerebellar astrocytes[23]. Furthermore, other investigators reported that taurine augments the immune response as an adjuvant[24,25]. In the present experiments, we examined whether taurine could modulate the levels of $[Ca^{2+}]_i$ and restore the declined proliferative activity in the lymphocytes of aged mice, using ionomycin and PMA as mitotic stimulators.

MATERIALS AND METHODS

Mice

BALB/cCrSlc (BALD/c) mice were purchased from Japan SLC Co. (Hamamatsu Shizuoka). The mice were aged at our animal facility under conventional conditions

Taurine, Edited by J.B. Lombardini *et al.*
Plenum Press, New York, 1992

2-3-month-old mice and 21-28-month-old mice were used as the young and the aged donors, respectively.

Cell Preparations and Cell Culture

Single cell suspension of purified splenic T and B cells were prepared as described previously[26,27]. Cells were cultured with PMA and ionomycin in the presence or absence of 79.9 mM taurine in a humidified atmosphere of 5% CO_2 in air at 37°C. The proliferative response of the cells were measured by using tritium thymidine incorporation.

Measurement of Taurine Uptake by Spleen Cells

After washing 3 times with Hanks balanced salt solution (HBSS) without sodium bicarbonate (Nissui Seiyaku Co., Tokyo), the purified T and B cells were suspended with the same medium; 5 x 10^5 cells were placed in each well of 96-well culture plate. Each well received 0.033 nmole of [^{3}H]taurine and then cold taurine was added to achieve a final taurine concentration of 79.9 mM with a final volume of 200 μl/well. After incubation in a humidified atmosphere of 5% CO_2 in air at 37°C, cells were placed in a small centrifuge tube containing 100 μl HBSS and centrifuged for 60 s at 8000 rpm (1700 g) to terminate the uptake of taurine by the cells. After washing 3 times with ice-cold HBSS, the cells were resuspended in 100 μl ice-cold HBSS and adsorbed on a glass fiber filter. Radioactivity on each filter was measured by a liquid scintillation spectrometer.

Measurement of Intracellular Free Ca^{2+}

Purified T or B cells (2 x 10^7 cells/ml) were loaded with the acetoxymethyl ester of Fura-2 (Fura-2 AM). The fluorescence of the cell suspension (2 x 10^7 cells/2 ml) was monitored with a Hitachi F-3000 spectrofluorimeter (Hitachi, Tokyo) utilizing a 2-ml quartz cuvette. The cell suspension was excited at 335 nm and emission of fluorescence was measured at 500 nm. During measurements, the cell suspension was maintained at 37°C utilizing a thermostatically controlled cuvette holder and stirring apparatus. $[Ca^{2+}]_i$ was calculated by the method of Grynkiewicz et al.[28] with slight modification. In brief, the intensities of the maximum (F_{max}) and the minimum (F_{min}) fluorescence for each sample were measured and were substituted along with the measured fluorescence (F) into the following equation: $[Ca^{2+}]_i = K_d(F-F_{min})/(F_{max}-F)$. The value used for K_d was 224 mM, which assumes an internal concentration of salt of approximately 140-160 mM[28-30]. Measurements were made in a solution (pH 6.80) containing 110 mM NaCl, 25 mM $NaHCO_3$, 5 mM KCl, 1 mM Na_2HP_4O·- NaH_2PO_4, 1 mM sodium acetate, 1.8 mM $CaCl_2$, 1 mM $MgCl_2$, 8.3 mM glucose, 0.1% BSA (KRB). Cells were washed in RPMI 1640 (culture medium) and resuspended to 2 x 10^7 cells/ml in the same medium. Fura-2 AM was added (final concentration 2 μM) and cells were incubated for 20 min at 37°C in the dark. The cells were then washed once with culture medium and resuspended to 2 x 10^7 cells/ml in KRB. After 1 ml of cell suspension was transferred to a cuvette, either 1 ml of KRB (without taurine) or 1 ml of 159.8 mM taurine solution containing the same salts as KRB was added. PMA was added to the cuvette to achieve a final concentration of 1 ng/ml, then each of the various doses of ionomycin was added. To obtain the F_{max}, excess ionomycin (final concentration 3 μM) was added to the cuvette. To obtain the F_{min}, $MnCl_2$ (final concentration 0.5 mM) was added to the cuvette after cell lysis with Triton X-100 (final 0.25%).

Statistical Analysis

Statistical analysis of data was performed by using the Student's t-test; a P value of less than 0.05 was considered statistically significant in the present experiments.

RESULTS

Taurine Uptake by Spleen Cells

We measured the uptake of taurine in young and old purified lymphocytes to examine whether taurine uptake in lymphocyte changes with aging. In preliminary experiments, we examined the time course of taurine uptake at 0°C and at 37°C by spleen cells depleted of adherent cells (non-adherent cells) in the presence of 165 μM [^{3}H]taurine. To stop the uptake of taurine, 0.1% NaN_3 was added to the cell suspension which was incubated at 0°C. Uptake of taurine at 0°C during the incubation for 40 min was negligible, while the uptake at 37°C was significant. The amount of taurine uptake reached a maximum at 6 min after the start of the incubation at 37°C. This is a more rapid uptake compared to cerebral cortical neurons (about 40 min)[31] or hepatocytes (about 90 min)[32]. We next examined the uptake of taurine by purified T and B cells from young and old mice. As shown in Figure 1A, the pattern of uptake of taurine by purified T cells in the old mice was the same as that in the young mice. The maximum taurine uptake was attained within 6 min after the initiation of incubation in the young and the old mice. The maximum taurine uptake was attained within 10 min in purified B cells in the old and the young mice (Figure 1B).

Proliferative Responses of Purified Lymphocyte Subsets to the Combined Stimulation of Ionomycin and PMA

Lymphocytes were prepared from aged or young mice and purified preparations of T cells and B cells were obtained. Each of the cell preparations was stimulated with the combination 1 ng/ml PMA and various doses of ionomycin. As shown in Table 1, proliferative responses of purified T cells from the old mice were significantly lower than that from the young mice under stimulation with ionomycin at the concentration of 0.25 μM, 0.5 μM and 1 μM. However, the ratio of [^{3}H]thymidine uptake by old T cells to that by young ones (O/Y) gradually increased to 1.0 with increasing ionomycin concentration. The proliferative responses between purified B cells in the old and in the young mice were, in general, not as statistically different than those observed between the young and old T cells (Table 1 and Table 2).

The Effect of Taurine on the Proliferative Responses of Purified T Cells and B Cells Induced by Costimulation with Ionomycin and PMA

In the presence or absence of taurine, purified T cells and purified B cells from young and old mice were cultured with 1 ng/ml of PMA and various doses of ionomycin, and the effect of 79.9 mM taurine on the proliferative responses was tested. As shown in Figure 2, an augmenting effect of taurine on the T cell proliferative response was observed at suboptimal doses of ionomycin (0.125 μM in the young, 0.125 μM and 0.25 μM in the old). Taurine also tended to augment the B cell proliferative response at 0.25 μM of ionomycin in the young and at 0.25 μM and 0.5 μM in the old (Figure 3). In order to confirm the significance of the

augmenting effects of taurine, we repeated experiments at suboptimal doses of
ionomycin. Results from three experiments were pooled and listed in Tables 3 and
4. The augmenting effect of taurine was found to be significant at 0.125 μM of
ionomycin (P<0.025) in young T cells, at 0.125 μM (P<0.025) and 0.25 μM (P<0.05)

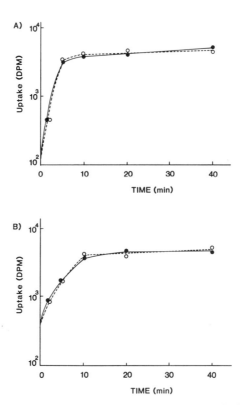

Figure 1. Time course of taurine uptake by purified T (A) and B (B) cells from young and old mice.
The cells were incubated at 37°C in HBSS containing 165 μM [³H]taurine and the radioactivity of
[³H]taurine transported into the purified T cells was measured as described in detail in Materials and
Methods. Each value represents the mean value of two separate experiments. (●-●), young; (O-O), old.

Table 1. Proliferative responses of purified T cells from old or young mice to the
combined stimulation with ionomycin and PMA.

IM(μM)[a]	[³H]Thymidine incorporation (Mean ± S.E.dpm)		O/Y[b]
	Young (n = 9)	Old (n = 6)	
0	316.0 ± 60.1	321.1 ± 94.9	1.00 (N.S)[c]
0.125	921.7 ± 261.9	543.5 ± 134.9	0.45 (N.S)
0.25	75844.0 ± 13949.1	19221.6 ± 8362.7	0.25 (P<0.01)
0.5	122944.0 ± 11676.2	61304.5 ± 6140.6	0.50 (P<0.005)
1.0	136431.0 ± 5941.6	85636.8 ± 8104.0	0.62 (P<0.005)

[a] Purified T cells were cultured at a cell density of 1 x 10⁵/ml under
stimulation with 1 ng/ml of PMA and various doses of ionomycin for 3 days.
[b] O/Y is the ratio of [³H]thymidine uptake by old T cells to that of young
T cells.
[c] Amount of [³H]thymidine uptake by old T cells and that of young T cells
at each ionomycin dose was compared. N.S is not significant.

Table 2. Proliferative responses of purified B cells from old or young mice to the combined stimulation with ionomycin and PMA.

IM(μM)	[³H]Thymidine incorporation (Mean ± S.E.dpm)		O/Y
	Young (n = 9)	Old (n = 8)	
0	280.8 ± 75555.5	318.6 ± 81.2	1.10 (N.S)
0.125	396.5 ± 77.8	386.8 ± 67.8	0.98 (N.S)
0.25	753.9 ± 266.6	1219.9 ± 170.9	1.62 (N.S)
0.5	11012.6 ± 2543.8	3875.7 ± 674.2	0.35 (P<0.025)
1.0	15540.5 ± 2279.1	13473.4 ± 1593.7	0.87 (N.S)

Purified B cells were cultured as in the experiment in Table 1. See footnotes for Table 1.

PURIFIED T CELL

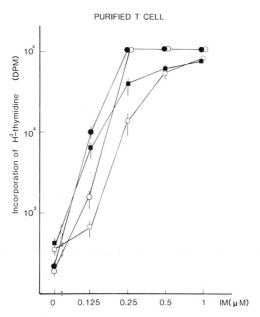

Figure 2. The effect of taurine on the proliferative response of purified spleen T cells from young and old mice. Purified T cells from old ($\square$ and $\blacksquare$) or young (O and ●) mice were cultured at a cell density of 1 x 10⁵/ml under stimulation with 1 ng/ml of PMA and various doses of ionomycin (abscissa) in the presence (closed symbol) or absence (open symbol) of 79.9 mM taurine for 3 days. Each point and vertical bar in the figure denote the arithmetic mean value of 4 mice and standard error of the mean, respectively. The bar is omitted when smaller than the symbol.

in old T cells, at 0.25 μM (P<0.05) in young B cells, and at 0.25 μM (P<0.05) and 0.5 μM (P<0.05) in old B cells.

The Effect of Taurine on the Concentration of $[Ca^{2+}]_i$ in Purified T Cells Under Costimulation of PMA and Ionomycin

Purified T cells from young and old mice were stimulated with 1 ng/ml of PMA and various doses of ionomycin in the presence or absence of taurine. In the absence of taurine, the influx of Ca^{2+} was significantly lower in the old cells than that

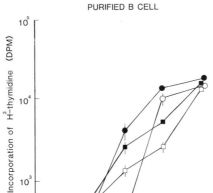

PURIFIED B CELL

Figure 3. The effect of taurine on the proliferative response of purified spleen B cells from young and old mice. See legend for Figure 2.

Table 3. The effect of taurine on the purified T cell proliferative response under costimulation with ionomycin and PMA.

Donor	IM(μM)[a]	[³H]Thymidine incorporation (Mean ± S.E.dpm)		(+)/(-)[b]
		Tau(-)	Tau(+)	
Young (n=7)	0.125	1042.9 ± 379.3	8652.3 ± 2274.1	8.3**c
Old (n=7)	0.125	557.9 ± 140.9	6218.3 ± 1549.3	11.1**
	0.25	10809.4 ± 3902.4	47703.4 ± 12516.8	4.4*

[a] Purified T cells were cultured at cell density of 1×10^5/ml under stimulation with 1 ng/ml of PMA and 0.125 or 0.25 μM of ionomycin and in the presence [Tau(+)] or absence [Tau(-)] of 79.9 mM taurine for 3 days.
[b] The ratio of [³H]thymidine uptake by Tau(+) T cells to that by Tau(-) T cells.
[c] The significant differences between the amount of [³H]thymidine uptake by Tau(+) T cells and that by Tau(-) T cells were *P<0.05 and **P<0.025.

observed in the young cells at 0.25 and 0.5 μM of ionomycin. In the presence of taurine, the influx of Ca^{2+} was significantly increased only in the T cells from the old mice. There is no difference between $[Ca^{2+}]_i$ in the old cells and that in the young cells at 0.25 and 0.5 μM of ionomycin in the presence of taurine (Table 5).

The Effect of Taurine on the concentration of $[Ca^{2+}]_i$ in Purified B Cells Under Costimulation of PMA and Ionomycin

In the absence of taurine, the influx of Ca^{2+} was similar in young and old groups at any doses of ionomycin. In the presence of taurine, the influx of Ca^{2+} was

Table 4. The effect of taurine on the purified B cell proliferative response under costimulation with ionomycin and PMA.

Donor group	IM(μM)[a]	[³H]Thymidine incorporation (Mean ± S.E.dpm) Tau(-)	Tau(+)	(+)/(-)[b]
Young (n=11)	0.25	648.4 ± 202.2	5854.2 ± 376.5	9.0*c
Old (n=8)	0.25	982.9 ± 376.5	4523.0 ± 1140.5	4.6*
	0.5	3875.7 ± 674.2	6681.9 ± 1184.2	1.7*

[a] Purified B cells were cultured at a cell density of 1 x 10⁵/ml under stimulation with 1 ng/ml of PMA and 0.125 or 0.25 μM of ionomycin and in the presence [Tau(+)] or absence [Tau(-)] of 79.9 mM of taurine for 3 days.
[b] The ratio of [³H]thymidine uptake by Tau(+) B cells to that by Tau(-) B cells.
[c] The significant difference between the amount of [³H]thymidine uptake by Tau(+) cells and that by Tau(-) cells was *P<0.05.

Table 5. The effect of taurine on the calcium influx of purified T cells.

	IM(μM)	Intracellular free Ca²⁺ concentration[a] (Mean ± S.E.nM) Young	Old	O/Y[b]
Tau(-)	0	249.7 ± 5.2	290.1 ± 26.0	1.16NS
	0.125	794.6 ± 123.0	625.8 ± 52.7	0.79NS
	0.25	1590.0 ± 104.1	1144.5 ± 73.2	0.72*
	0.5	3342.2 ± 183.8	1700.3 ± 185.6	0.51***
Tau(+)	0	384.5 ± 32.8	490.8 ± 79.2	1.28NS
	0.125	1327.5 ± 190.4	1485.1 ± 212.6	1.12NS
	0.25	2452.6 ± 522.7	2312.3 ± 127.8	0.94NS
	0.5	3477.3 ± 378.2	3132.5 ± 213.2	0.90NS
(+)/(-)[c]	0	1.54NS	1.69*	
	0.125	1.67NS	2.40*	
	0.25	1.54NS	2.02****	
	0.5	1.04NS	1.84*	

[a] [Ca²⁺]$_i$ of purified T cells was measured at a cell density of 1 x 10⁷/ml under stimulation with 1 ng/ml of PMA, and various doses of ionomycin in the presence [Tau(+)] or absence [Tau(-)] of 79.9 mM. Each value represents the mean of four separate (at 0 and 0.125 μM concentrations of ionomycin) or three separate (at 0.25 μM and 0.5 μM concentrations of ionomycin) experiments. In each experiment, purified cell preparations were prepared from pooled spleen cells of 4-6 young or old mice.
[b] The ratio of [Ca²⁺]$_i$ of old T cells to that of young T cells.
[c] The ratio of [Ca²⁺]$_i$ of Tau(+) T cells to that of Tau(-) T cells.
*P<0.05, **P<0.025, ***P<0.01, ****P<0.005, NS = not significant.

significantly increased at most doses of ionomycin in both age groups. That is, the effect of taurine on the concentration of [Ca²⁺]$_i$ seemed to be similar in young and old B cells (Table 6).

Table 6. The effect of taurine on the calcium influx of purified B cells.

	IM(μM)	Intracellular free Ca^{2+} concentration[a] (Mean $\pm$ S.E.nM)		O/Y[b]
		Young	Old	
Tau(-)	0	290.7 $\pm$ 10.6	365.2 $\pm$ 27.0	1.26[NS]
	0.125	458.4 $\pm$ 30.7	559.3 $\pm$ 37.1	1.22[NS]
	0.25	555.3 $\pm$ 84.0	649.3 $\pm$ 38.8	1.17[NS]
	0.5	1390.3 $\pm$ 67.1	1264.1 $\pm$ 97.7	0.91[NS]
Tau(+)	0	477.3 $\pm$ 23.8	442.9 $\pm$ 38.7	0.93[NS]
	0.125	1049.1 $\pm$ 97.8	989.3 $\pm$ 93.4	0.94[NS]
	0.25	1153.1 $\pm$ 32.7	1316.7 $\pm$ 209.4	1.14[NS]
	0.5	2711.2 $\pm$ 188.6	2228.0 $\pm$ 417.7	0.82[NS]
(+)/(-)	0	1.64**	1.21[NS]	
	0.125	2.29***	1.77**	
	0.25	1.63**	2.03**	
	0.5	2.27**	1.76*	

$[Ca^{2+}]_i$ of purified B cells was measured as in the experiments of Table 5. See footnote for Table 5 for details.
*P<0.05, **P<0.025, ***P<0.01, NS = not significant.

DISCUSSION

In the present experiments, we demonstrated that taurine improved the proliferative response of old T cells under costimulation with PMA and suboptimal doses of ionomycin (Figure 2 and Table 3). Age-related decline of the *in vitro* proliferative response of lymphocytes to specific antigens or lectins is well known[1-5]. However, the proliferative responses to antigens or lectins involve complex cell-to-cell interactions among macrophages/monocytes and various lymphocytes. To exclude these cell interactions and to make the analysis easier and clearer, in the present experiments, we used purified T cell and purified B cell preparations as responding cells and ionomycin and PMA as combined mitogenic stimulators, which induce cell proliferation at a single-cell state[33]. Thus, the observed effects of taurine in the present experiments are attributable to the direct effect of taurine on T or B cells.

We also stimulated purified T cells with Con A and purified B cells with lipopolysaccharide in the presence or absence of taurine (data not shown). Similar augmenting effects of taurine were also observed at suboptimal doses of the mitogens. Furthermore, we used not only BALB/c mice but also C57BL/6CrSlc (B6) mice which have shown relatively low age-associated decline in antibody response to the TI-2 antigen as compared to other strains of mice[34]. The augmenting effect of taurine on the proliferative response of old lymphocytes in B6 mice was not as significant as in BALB/c mice, but the similar effect of taurine was observed in B6 mice. Thus, the augmenting effect of taurine on the proliferative response of lymphocytes as shown here may be an ubiquitous phenomenon.

Taurine is a sulfur containing amino acid that is most abundant in mammalian mononuclear leukocytes and polymorphonuclear leukocytes. The concentration of taurine occupies 45-75% of the total free amino acids of the leukocytes[35]. Nevertheless, the biochemical role of taurine in leukocytes or immune cells has not been fully elucidated. Rosenberg and Scriver[36] speculated that taurine might have some role on leukocyte mobility and phagocytosis. Zglicznski et al.[37] indicated a role for taurine in regulation of myeloperoxidase activity. Some investigators reported taurine augmented antibody production[24,25]. However, the mechanism by which taurine augments the immune response has not been defined. Recently, taurine was found to have an important regulatory role for Ca^{2+} influx through Ca^{2+} channels in myocardial cells[38]. Taurine is positively inotropic at low extracellular Ca^{2+} concentrations[39,40], whereas negatively inotropic at high extracellular Ca^{2+} concentrations[39]. Furthermore, in previous reports, it has been demonstrated that taurine has a protective effect on the irregular beating pattern of cultured myocardial cells induced by high and low extracellular calcium ion[41].

Miller[9,10] reported that the age-related decline of the purified T cell proliferative response to a mitogenic stimulation is caused by age-related impairment in the ability to raise $[Ca^{2+}]_i$ in response to the stimulation. We confirmed these results. As shown in Table 5, the level of $[Ca^{2+}]_i$ in old T cells was significantly lower than that in young T cells after costimulation with PMA and either 0.25 μM or 0.5 μM of ionomycin. We also examined the effects of taurine on the movement of Ca^{2+} in lymphocytes. Taurine enhanced the influx of Ca^{2+} under costimulation of PMA and ionomycin in T cells and B cells (Tables 5 and 6). The enhanced influx of Ca^{2+} by taurine seems to result in the augmentation of the proliferation of T and B cells. Thus, the augmenting effect of taurine on the proliferative response was significant under suboptimal doses of ionomycin, where taurine also enhanced markedly the influx of Ca^{2+}. Furthermore, in old T cells, not only the influx of Ca^{2+} but the proliferation induced by costimulation with PMA and ionomycin were restored to the levels of young T cells after the addition of taurine. The marked augmenting effects of taurine observed in old T cells are not due to an age-dependent increase in taurine uptake, as is evidenced from Figure 1A.

In contrast, such preferential effects of taurine on old T cells were not observed in old B cells, which showed no age-related decline of Ca^{2+} influx under costimulation with PMA and ionomycin as shown in Table 6. Taurine demonstrated a similar degree of the augmenting effect on Ca^{2+} influx under costimulation with PMA and ionomycin in both young and old B cells.

We have already shown in heart cells that taurine stimulated Ca^{2+} influx through slow Ca^{2+} channels which depends on cyclic nucleotide metabolism and the phosphorylation process. The analysis of the precise mechanism of taurine on T and B cells is under progress.

In the present experiments, we used a high concentration of taurine in *in vitro* experiments. Taurine is a physiological substance and can be administered *in vivo*. Experiments are now in progress to examine the effect of taurine on the age related decline of T cell function *in vivo*.

SUMMARY

1.   Proliferative responses to the costimulation with phorbol-12-myristate-13-acetate (PMA) and suboptimal doses of ionomycin in the purified T and B cells from old mice were lower than those from young mice.
2.   The degree of the age-related decline was more significant in T cells than in B cells.
3.   Taurine, a sulfur containing amino acid, augmented the proliferative responses of T cells from both young and old mice.
4.   The augmentation of the proliferative response by taurine was more marked in old T cells than in young T cells.
5.   The concentration of intracellular free calcium ion ($[Ca^{2+}]_i$) was significantly lower in old T cells when stimulated with PMA and ionomycin than observed in young T cells.
6.   In the presence of taurine, the concentration of $[Ca^{2+}]_i$ in the old T cells significantly increased under stimulation by PMA and ionomycin..
7.   The results indicate that taurine improved the proliferative response in old T

cells by restoration of the increment of the concentration of $[Ca^{2+}]_i$ under the stimulation by PMA and ionomycin.

REFERENCES

1. A.V. Pisciotta, D.W. Westing, C. DePrey, and B. Walsh, Mitogenic effect of phytohemagglutinin at different ages, *Nature* 215:193-194 (1967).
2. H.M. Hallgren, C.E. Buckley, V.A., Gilbertsein, and E.J. Yunis, Lymphocytes phytohemagglutinin responsiveness, immunoglobulins and autoantibodies in aging humans, *J. Immunol.* 111:1101-1107 (1973).
3. C.S. Walters and H.N. Claman, Age-related changes in cell-mediated immunity in BALB/c mice, *J. Immunol.* 115:1438-1443 (1975).
4. B. Inkeles, J.B. Innes, M.M. Kuntz, A.S. Kadish, and M.E. Weksler, Immunological studies of aging. III. Cytokinetic basis for the impaired response of lymphocytes from aged humans to plant lectines, *J. Exp. Med.* 145:1176-1187 (1977).
5. R.E. Callard and A. Basten, Immune function in aged mice. I. T-cell responsiveness using phytohemagglutinin as a function probe, *Cell. Immunol.* 31:13-25 (1977).
6. S. Negoro, H. Hara, S. Miyata, O. Saiki, T. Tanaka, K. Yoshizaki, T. Igarashi, and S. Kishimoto, Mechanism of age-related decline in antigen specific T cell proliferative response: IL-2 receptor expression and recombinant IL-2 induced proliferative response of purified TAC-positive T cells, *Mech. Ageing Dev.* 36:223-241 (1986).
7. H. Hara, S. Negoro, S. Miyata, O. Saiki, K. Yoshizaki, T. Tanaka, T. Igarashi, and S. Kishimoto, Age-associated changes in proliferative and differentiative response of human B cells and production of T cell-derived factors regulating B cell function, *Mech. Ageing Dev.* 38:245-258 (1987).
8. S. Negoro, H. Hara, S. Miyata, O. Saiki, T. Tanaka, K. Yoshizaki, N. Nishimoto, and S. Kishimoto, Age-related changes of the function of T cell subsets: Predominant defect of the proliferative response in CD8 positive T cell subset in aged persons, *Mech. Ageing Dev.* 39:263-279 (1987).
9. R.A. Miller, Immunodeficiency of aging: Restorative effects of phorbol ester combined with calcium ionophore, *J. Immunol.* 137:805-808 (1986).
10. R.A. Miller, B. Jacobson, G. Weil, and E.R. Simons, Diminished calcium influx in lectin-stimulated T cells from old mice, *J. Cell. Physiol.* 132:337-342 (1987).
11. R.L. Walford, "The Immunologic Theory of Aging," Munksgaard, Copenhagen (1969).
12. K. Hirokawa, J.W. Albright, and T. Makinodan, Restoration of impaired immune functions in aging animals. I. Effect of syngenic thymus and bone marrow grafts, *Clin. Immunol. Immunopathol.* 5:371-376 (1976).
13. D. Frasca, L. Adorini, and G. Doria, Enhancement of helper and suppressor T cell activities by thymosin alpha 1 injection in old mice, *Immunopharmacology* 10:41-49 (1985).
14. M.L. Heidrick, J.W. Albright, and T. Makinodan, Restoration of impaired immune functions in aging animals. IV. Action of 2-mercaptoethanol in enhancing age-reduced immune responsiveness, *Mech. Ageing Dev.* 13:367-378 (1980).
15. T. Furukawa, S.N. Meydani, and J.B. Blumberg, Reversal of age-associated decline in immune responsiveness by dietary glutathione supplementation in mice, *Mech. Ageing Dev.* 38:107-117 (1987).
16. K. Takihara, J. Azuma, N. Awata, H. Ohta, A. Sawamura, S. Kishimoto, and N. Sperelakis, Taurine's possible protective role in age-dependent response to calcium paradox, *Life Sci.* 37:1705-1710 (1985).
17. H. Ohta, J. Azuma, S. Onishi, N. Awata, K. Takihara, and S. Kishimoto, Proliferative effect of taurine against isoprenaline-induced myocardial damage, *Basic Res. Cardiol.* 18:473-481 (1986).
18. T. Hamaguchi, J. Azuma, N. Awata, H. Ohta, K. Takihara, H. Harada, S. Kishimoto, and N. Sperelakis, Reduction of doxorubicin-induced cardioxicity in mice by taurine, *Res. Commum. Chem. Pathol. Pharmacol.* 59:21-23 (1988).
19. H. Pasantes-Morales, R.M. Ademe, and A.M. Lopez-Colome, Taurine effects on $^{45}Ca^{2+}$ transport in retinal subcellular fractions, *Brain Res.* 17:131-138 (1979).
20. H. Pasantes-Morales, Taurine-calcium interactions in frog rod outer segments: Taurine effects on an ATP-dependent calcium translocation process, *Vision Res.* 22:1487-1493 (1982).
21. J.B. Lombardini, Effects of ATP and taurine on calcium uptake by membrane preparations of the rat retina, *J. Neuro. Chem.* 40:402-406 (1983).
22. M.V. Leite and L. Goldstein, Ca^{2+} ionophore and phorbol ester stimulate taurine efflux from skate erythrocyte, *J. Exp. Zool.* 242:95-97 (1987).
23. R.A. Philibert and G.R. Dutton, Phorbol ester and dibutyryl cyclic AMP reduce content and efflux of taurine in primary cerebellar astrocytes in culture, *Neuro. Sci. Lett.* 95: 323-328 (1988).
24. G.E. Gaull, Taurine in the nutrition of the human infant, *Acta. Paediatr. Scand.* 296:38-40 (1982).

25. M. Masuda, K. Horisaka, and T. Koeda, Influences of taurine on functions of rat neutrophils, *Jpn. J. Pharmacol.* 34: 116-118 (1984).

26. M.J. Julius, E. Simpson, and L.A. Herzenberg, A rapid method for the isolation of functional thymus-derived murine lymphocytes, *Eur. J. Immunol.* 3:645-649 (1973).

27. S. Nishio, S. Negoro, T. Hosokawa, H. Hara, T. Tanaka, Y. Deguchi, J. Ling, N. Awata, J. Azuma, A. Aoike, K. Kawai, and S. Kishimoto, The effect of taurine on age-related immune decline in mice, *Mech. Ageing Dev.* 52:125-139 (1990).

28. G. Grynkiewicz, M. Poenie, and R.Y. Tsien, A new generation of Ca^{2+} indicators with greatly improved fluorescence properties, *J. Biol. Chem.* 260:3440-3450 (1985).

29. J. LaBaer, R.Y. Tsien, K.A. Fahey, and A.L. DeFranco, Stimulation on the antigen receptor on WEHI-231 B lymphoma cells results in a voltage-independent increase in cytoplasmic calcium, *J. Immunol.* 137:1836-1844 (1986).

30. G. Pantaleo, D. Olieve, A. Poggi, W.J. Kozuumbo, L. Moretta, and A. Moretta, Transmembrane signaling via the T11-dependent pathway of human T cell activation. Evidence for the involvement of 1,2-diacylglycerol and inositol phosphates, *Eur. J. Immunol.* 17:55-60 (1987).

31. M. Kishi, S. Ohkuma, M. Kimori, and K. Kuriyama, Characteristics of taurine transport system and its developmental pattern in mouse cerebral cortical neurons in primary culture, *Biochim. Biophys. Acta.* 939:615-623 (1988).

32. S. Ohkuma, J. Tamura, K. Kuriyama, and T. Takino, Characteristics of taurine transport in freshly isolated rat hepato cytes, *Jpn. J. Pharmacol.* 31:1061-1070 (1981).

33. G.A. Koretszky, R.P. Daniele, W.C. Greene, and P.C. Nowell, Evidence for an interleukine-independent pathway for human lymphocyte activation, *Proc. Natl, Acad. Sci. U.S.A.* 80:3444-3447 (1983).

34. T. Hosokawa, A. Aoike, M. Hosono, K. Kawai, and B. Cinader, Strain differences of age-dependent changes in the responsiveness to a T-independent type-2 antigen in mice, *Mech. Ageing Dev.* 45:9-21 (1988).

35. K. Fukuda, Y. Hirai, H. Yoshida, T. Nakajima, and T. Usui, Free amino acid content of lymphocytes and granulocytes compared, *Clin. Chem.* 28:1758-1761 (1982).

36. L.E. Rosenberg and C.R. Scriver, Disorders of amino acid metabolism: Taurinuria, *in:* "Duncan's Disease of Metabolism," P.K. Bondy and L.E. Rosenberg, eds., W.B. Saunders Co., Philadelphia, pp. 555-556 (1974).

37. J.M. Zgliczynski, T. Stelmaszynska, W. Ostrowski, J. Naskalski, and J. Sznajd, Myeloperoxidase of human leukemic leukocytes. Oxidation of amino acids in the presence of hydrogen peroxide, *Eur. J. Biochem.* 4:540-547 (1968).

38. N. Sperelakis, T. Yamoto, G. Bkaily, H. Sada, A. Sawamura, and J. Azuma, Taurine effects on action potentials and ionic currents in chick heart, *Can. J. Physiol. Pharmacol.* 64:1-19 (1988).

39. F. Franconi, F. Martini, I. Stendardi, R. Matsucci, L. Zilletti, and A. Giotti, Effect of taurine on calcium level and contractility in guinea-pig ventricular strips, *Biochem. Pharmacol.* 31:3181-3186 (1982).

40. A. Sawamura, J. Azuma, H. Harada, H. Hasegawa, K. Ogura, N. Sperelakis, and S. S. Kisimoto, Protection by oral pretreatment with taurine against the negative inotropic effects of low calcium medium on isolated perfused chick heart, *Cardiovasc. Res.* 17:620-626 (1983).

41. K. Takahashi, J. Azuma, N. Awata, A. Sawamura, S. Kishimoto, T. Yamagami, T. Kishi, H. Harada, and W. Schaffer, Protective effect of taurine on the irregular beating pattern of cultured myocardial cells induced by high and low extracellular calcium ion, *J. Mol. Cell Cardiol.* 20:397-403 (1988).

EFFECTS OF TAURINE DEFICIENCY ON IMMUNE FUNCTION IN MICE

Norma Lake, Erin D. Wright, and Wayne S. Lapp

Department of Physiology
McGill University, 3655 Drummond Street
Montreal, Quebec, Canada H3G 1Y6

INTRODUCTION

The role of the sulfur-containing amino acid, taurine, in immune function has received scant attention until recently. In the past few years reports have described its adjuvant effect for viral antigens in mice and human subjects (Kuriyama et al., 1988; Ishizaka et al., 1990), its reversal of age-related immune decline in mice (Nishio et al., 1990) and a decline in polymorphonuclear cell function in severely taurine-deficient cats (Schuller-Levis et al., 1990).

Taurine Depletion

We have found rapid taurine depletion in the lymphoid organs of 3 to 6 month old CBA mice by treating them for 11 days *in vivo* by addition to their drinking water of ß-alanine or GES (guanidinoethane sulfonate), which are antagonists of the taurine transport system (Huxtable et al., 1979). Taurine concentrations in the spleen and thymus (assayed as described in Lake and De Marte, 1988) were reduced by 35-50%, compared to controls that were housed in adjacent cages and consumed water containing no additives.

Plaque-forming Cell Assays

We compared control and treated groups for humoral antibody responses to sheep red blood cell antigens using a modification of previously described techniques for the plaque-forming cell assay (Cunningham and Szenberg, 1968; Kongshavn and Lapp, 1972). Animals were immunized intravenously with sheep red blood cell four days before the plaque-forming cells assay. Spleen cell suspensions were made by gently tamping the cells through a stainless steel screen and collecting the cells in Hanks Balanced Salt solution. The final volume was made to 15 ml and the total number of spleen cells harvested was estimated by counting aliquots. The test mixture consisted of complement solution, sheep red blood cells and spleen cell suspension, which was put into chambers and incubated for 1 hr at 37°C. The numbers of plaque-forming cells were counted macroscopically, and the results expressed as per spleen, and per 10^6 spleen cells.

We did a series of five separate experiments with 5-6 mice per control or depleted group. For three sets the treated animals received ß-alanine, the remaining two sets were treated with GES. Animals drawn from the same or adjacent cages (receiving the same drinking solution) were used for the biochemical assays of taurine content. Similar effects were observed with ß-alanine and GES. In taurine-deficient animals the total number of viable spleen cells harvested was unchanged from control. However, the number of plaque-forming cells was significantly reduced with a mean reduction of $40 \pm 5\%$ (See Table 1).

Table 1. Plaque forming cell (PFC) assays of taurine-depleted spleens

	PFCs per Spleen (X 10^5)		
Treatment	Trial 1	Trial 2	Trial 3
Control	2.40 ± 0.57	2.31 ± 0.37	1.35 ± 0.22
3% ß-alanine	$1.37 \pm 0.47**$	$1.73 \pm 0.23**$	$0.72 \pm 0.21**$
Control	1.93 ± 0.40	2.73 ± 0.55	--
1% GES	$1.22 \pm 0.24*$	$1.32 \pm 0.20**$	--

Note: The data are expressed as means $\pm$ S.D. Five or six spleens were used per group per trial. Differences between the means were evaluated with t-tests. * p <0.025; **p < 0.01.

Table 2. Percent of lymphocytes positive for given markers

Organ	Lyt2+	L3T4+	Lyt2+/L3T4+	Lyt2-/L3T4	μ-HC+
SPLEEN					
control	12.1 ± 2.6	24.9 ± 3.8	0.06 ± 0.05	62.9 ± 5.7	43.8 ± 5.4
taurine-depleted	12.0 ± 1.5	27.0 ± 1.3	0.06 ± 0.04	61.0 ± 2.4	40.8 ± 4.5
THYMUS					
control	6.0 ± 0.7	15.2 ± 4.5	74.3 ± 5.6	4.6 ± 1.7	--
taurine-depleted	6.3 ± 0.4	17.2 ± 3.5	71.3 ± 4.6	5.3 ± 0.9	--
LYMPH NODE					
control	23.8 ± 2.3	54.2 ± 4.8	0.46 ± 0.5	21.5 ± 4.9	16.7 ± 2.5
taurine-depleted	22.5 ± 1.6	58.1 ± 6.5	0.29 ± 0.1	19.1 ± 6.4	18.5 ± 3.9

Note: Results are shown as means $\pm$ S.D. Ten organs were examined per group. 5000 cells were examined per organ sample. Taurine-depleted animals were given ß-alanine treatment.

Fluorescence Flow Cytometry Studies

In order to quantitate the proportion of lymphocytes bearing specific surface markers in the spleen, thymus and lymph nodes, cell suspensions were made from these organs, labelled with primary or primary plus secondary antibodies, and then processed with a fluorescence-activated cell scanning instrument (Becton Dickinson) interfaced to an HP microcomputer.

The markers we used were the T cell antigens Lyt 2, L3T4, and the B cell μ heavy chain antigen (μ-HC). Table 2 shows the frequency distribution of various lymphocyte subtypes in the spleen, thymus and lymph nodes from control or taurine-depleted (ß-alanine treated) mice. While the frequency distribution of subtypes differed among the organs in the expected manner, there were no differences in the profiles from control or depleted animals.

The reductions in the functional immune response associated with taurine deficiency that we have described here may arise through suppression of cell function without loss of viability in the lymphocyte subtypes we have identified. Another possibility is that the responsiveness of a separate cell sub-population that we have not evaluated, such as antigen-presenting cells for example, has been altered.

ACKNOWLEDGEMENTS

These studies were supported in part by funds from the Medical Research Council of Canada. We are grateful to Ms. A Lee Loy for excellent technical assistance and to Dr. M.M. Frojmovic for the use of the flow cytometer.

REFERENCES

Cunningham, A.J., Szenberg, A., 1968, Further improvements in the plaque technique for detecting single antibody-forming cells, *Immunology* 14:599-600.

Huxtable, R.J., Laird, H.E., and Lippincott, S., 1979, The transport of taurine in the heart and rapid depletion of tissue taurine content by guanidinoethyl sulfonate, *J. Pharmacol. Exp. Ther.* 211:465-471.

Kongshavn, P.A.L. and Lapp, W.S., 1972, Immunosuppressive effect of male mouse submandibular gland extracts on plaque-forming cells in mice: abolition by orchiectomy, *Immunology* 22:227-230.

Ishizaka, S., Yoshiwaka, M., Kitagami, K., and Tsujii, T., 1990, Oral adjuvants for viral vaccines in humans, *Vaccine* 8:337-341.

Kuriyama, S., Tsujii, T., Ishizaka, S., Kikuchi, E., Kinoshita, K., Nishimura, K., Kitagami, K., Yoshikawa, M., and Matsumoto, M., 1988, Enhancing effects of oral adjuvants on anti-HBs response induced by hepatitis B vaccine, *Clin. Exp. Immunol.* 72:383-389.

Lake, N. and De Marte, L., 1988, Effects of ß-alanine treatment on the taurine and DNA content of the rat heart and retina, *Neurochem. Res.* 13:1003-1006.

Nishio, S.-I., Negoro, S., Hosokawa, T., Hara, H., Tanaka, T., Deguchi, Y., Ling, J., Awata, N., Azuma, J., Aoike, A., Kawai, K., and Kishimoto, S., 1990, The effect of taurine on age-related immune decline in mice: the effect of taurine on T cell and B cell proliferative response under costimulation with ionomycin and phorbol myristate acetate, *Mech. Age. Dev.* 52:125-139.

Schuller-Levis, G., Mehta, P.D., Rudelli, R., and Sturman, J.A., 1990, Immunologic consequences of taurine deficiency in cats, *J. Leukocyte Biol.* 47:321-331.

REVIEW: RECENT STUDIES ON TAURINE IN THE CENTRAL NERVOUS SYSTEM

John B. Lombardini

Departments of Pharmacology and Ophthalmology &
    Visual Sciences
Texas Tech University Health Sciences Center
Lubbock, Texas 79430

INTRODUCTION

The functions of taurine in the central nervous system are ambiguous at best and any review concerning this ubiquitous compound, unless truly exhaustive, suffers from the problem of where to start. With this problem in mind I have opted to review select papers dealing with taurine and the brain that have been published since 1989 when the very elegant and comprehensive review entitled "Taurine in the Central Nervous System and the Mammalian Actions of Taurine" by Ryan Huxtable was published[1]. Articles referring to taurine and the retina will also be omitted since this subject has been recently reviewed elsewhere[2].

The papers that will be reviewed deal with levels of taurine and the release of taurine from brain tissues monitored after some insult and in some instances the administration of taurine as a therapeutic agent. The general categories of these papers are: 1) ischemia-hypoxia; 2) seizures due to vitamin B_6 deficiency, hypoxia, and audiogenic stimulation; 3) potpourri that includes trauma, Alzheimer's Disease, ageing, pregnancy, hypertension and irradiation; and 4) taurine as a therapeutic agent.

ISCHEMIA-HYPOXIA

It is well-known that taurine and other amino acids and even low molecular weight peptides are released from various brain regions and the spinal cord when the animal is subjected to ischemia or hypoxia. A powerful tool for studying the release mechanism of these compounds is *in vivo* microdialysis coupled with high pressure liquid chromatography to analyze the composition of the perfusate. In these experiments various types of dialysis probes are inserted into the test area of the brain and then the probes are perfused with either artificial cerebral spinal fluid or Ringer's solution.

In the spinal cord of rabbits that had total occlusion of the distal aorta, taurine concentrations in the dialysate increased immediately during the ischemic period and

Taurine, Edited by J.B. Lombardini *et al*.
Plenum Press, New York, 1992

continued to increase during the reperfusion period[3]. However, by 20 minutes post-ischemia the taurine levels had returned to control values. The taurine levels increased by approximately a factor of 2 from 3 to 8 μmol/liter. In contrast, glycine levels increased by a factor of 4 (6 to 23 μmol/liter) during the period immediately following ischemia.

Taurine concentrations were also elevated in striatal dialysates of postnatal day 7 rats after a hypoxic-ischemic insult brought about by ligation of the right carotid artery and exposure to 8% oxygen for 2.5 hours[4]. These conditions are used as a model of perinatal stroke. In these experiments the levels of taurine increased greater than 7 fold over baseline only when both procedures were performed on the rats, i.e., ligation of the carotid artery and exposure to reduced oxygen. Interestingly, there was no changes in the efflux of glycine in this animal model. However, there appeared to be a correlation between the elevations in efflux between glutamate and taurine. Elevated taurine effluxes were observed in 7 out of 8 hypoxic-ischemic rats in which elevated glutamate effluxes were observed. It was thus speculated that the increase in taurine effluxes could be either a compensatory response to local edema (an osmoregulatory agent) or a direct response to glutamate on specific receptors as originally suggested by Menéndez et al.[5].

The possibility that taurine release from the rat hippocampus in vivo is not solely due to cellular swelling has also been suggested by additional microdialysis studies of Menéndez et al.[6]. In these experiments taurine release was measured after perfusion with quisqualic acid, kainic acid and N-methyl-D-aspartate (NMDA). The increase in the extracellular taurine levels that was prompted by quisqualic acid and kainic acid was for the most part suppressed when the osmolarity of the perfusate was increased. These data thus suggested that taurine release in these circumstances was due to cellular swelling since when the cells were prevented from swelling taurine release did not occur. However, taurine release induced with NMDA was not affected by hyperosmotic medium and thus it was concluded that taurine may have functions other than osmoregulation. A similar conclusion, i.e., that taurine release may not necessarily be only a function of cellular swelling, has been presented by Oja and Saransaari in this symposium.

Matsumoto and colleagues[7] have also used the in vivo microdialysis technique to assay the evoked release of both excitatory amino acids, glutamate and aspartate, and inhibitory amino acids, GABA and taurine, from the rat hippocampus in a 4-vessel occlusion model. Potassium (100 mM)-evoked release of these amino acids was measured on the 5th day after 20 minutes of ischemia. They found that the release of glutamate and aspartate was reduced in the ischemic animal while the release of GABA and taurine was unchanged and increased (50%), respectively. The decreases in glutamate and aspartate release were attributed to damage of the hippocampal CA_1 pyramidal cell layer containing glutamatergic and aspartatergic neurons and thus it was concluded these neurons are more vulnerable to ischemia than the neurons that contain GABA and taurine.

The lowering of taurine levels measured by immunocytochemical techniques in the rat hippocampus following an ischemic episode[8] support the data of Matusmoto and colleagues[7] demonstrating an increased taurine release observed in the dialysates. In the former study by Torp and colleagues[8] taurine levels were studied in the same animal model (4 vessel occlusion model) and found to be decreased in the pyramidal cell bodies. However, in a different animal model using spontaneously hypertensive rats in which forebrain ischemia was induced by bilateral carotid artery ligation for 60 minutes the taurine content of various ischemic brain regions including the hippocampus was not decreased[9]. On the contrary the taurine content of two (parietal cortex and nucleus accumbens) of nine regions assayed were actually increased by 11% which was significant although certainly not remarkable.

The taurine dipeptide, γ-glutamyltaurine, an endogenous constituent of brain tissue[10] is considered to be an excitatory amino acid antagonist[11]. γ-Glutamyltaurine was found to be elevated 1.6 fold in the rat CA_1 pyramidal cells in the hippocampus during ischemia (4-vessel occlusion model)[12]. A function of γ-glutamyltaurine may be to protect against neuronal toxicity and subsequent neuronal death caused by the excitatory amino acids which are released during ischemia or immediately postischemia. It is postulated that the excitatory amino acids cause excessive activation of their receptors leading to the cellular influx of toxic levels of calcium[13].

Finally, there is an interesting paper on the release of inhibitory neurotransmitters in response to anoxia in the turtle brain striatum[14]. The turtle brain is quite different in its response to anoxia compared to the mammalian brain in that anoxia is tolerated, depolarization does not occur, and there is no massive release of the excitatory neurotransmitter glutamate. However, hypoxia caused taurine to be released into the dialysate approximately 24 times over control values (0.5 μM to 12 μM)[14] while the intracellular content of taurine rose by 1.5 times (2.1 to 3.2 mM)[15]. The rise in taurine and other inhibitory amino acids (GABA and glycine) may account for the decrease in brain activity and energy consumption that is observed in the anoxic turtle brain allowing the turtle to survive long periods of total anoxia[14].

SEIZURES DUE TO VITAMIN B$_6$ DEFICIENCY

The levels and release of taurine in various brain regions has also been studied in a number of animal models in which seizures are induced. In the vitamin B$_6$ deficient neonatal rat the levels of taurine along with other amino acids such as glutamate and GABA are reduced while glycine is increased[16]. Decreases in brain taurine levels of cortex, hippocampus, caudate/putamen, substantia nigra, and pons/medulla were observed in the 14 day old rat which is the age when spontaneous seizures due to vitamin B$_6$ deficiency also occur. Taurine levels in only two of the brain regions (hippocampus and pons/medulla) were also reduced in the 28 day old rat while at 56 days there were no changes. The unanswered question was posed as to whether the changes in taurine (and the other amino acids) was a cause or a consequence of the seizures.

In a follow-up study, the same investigator[17] measured the basal and potassium-evoked release of taurine from the hippocampus and cortex of the vitamin B$_6$ deficient 14 day old rat. In this series of experiments the basal release of taurine was decreased in both the hippocampus and cortex while the potassium-evoked release was unchanged. It was suggested that these data may be explained by alterations in the efflux mechanisms due to vitamin B$_6$ deficiency or a reduced synthesis of taurine since vitamin B$_6$ is a required cofactor in the decarboxylation of cysteinesulfinic acid, the precursor of taurine.

SEIZURES DUE TO HYPOXIA

The protective effect of taurine on induced seizures due to hypoxia was tested by Malcangio et al.[18]. In this study the intracerebroventricular administration of taurine was effective as a protectant against convulsions induced by nitrogen but was without effect on convulsions induced by pentylenetetrazol and hyperbaric oxygen. Survival time of the mice exposed to hypoxia was increased in the taurine-treated animals compared to controls. However, while taurine had a protective effect against hypoxia-induced amnesia, convulsions and death, taurine also appeared to cause a learning impairment in control animals.

SEIZURES DUE TO ACOUSTIC STIMULUS

The levels of taurine and other amino acids in 17 brain areas of three sublines of RB mice with different susceptibilities to audiogenic seizures have been reported[19]. While tissue levels of taurine appear not to be a marker for seizure susceptibility in the RB mice glutamine (or GABA[20]) may provide such a role. On the contrary, the inhibitory amino acids including taurine may be involved in the severity of the seizure in this particular animal model for experimental epilepsy. However, the extent of the changes in the taurine levels appears at best marginal even though statistically significant.

TRAUMA

Amino acid patterns have also been measured after experimental injury to the spinal cord[21]. Laminectomy at segment T9 was performed on rats and then the spinal cord was injured by dropping a 5 gram weight from different heights (5, 10, or 20 cm) onto the exposed dura mater. Only moderate reversible injury occurred when the weight was dropped at 5 and 10 cm while irreversible spastic paresis occurred when the weight was dropped at the 20 cm height. Taurine levels declined only slightly (but statistically significant) at 5 min and 4 hours after severe spinal injury in contrast to the levels of the excitatory amino acids, aspartate and glutamate, which decreased in greater quantities and for longer duration (up to 24 hours) at the trauma site. Interestingly, taurine levels increased after moderate injury and then declined after severe injury. Again as in damage due to hypoxia-ischemia it appears that the excitatory amino acids glutamate and aspartate have a role in exacerbating the secondary injury while the changes in the tissue taurine levels are still an enigma.

ALZHEIMER'S DISEASE

In Alzheimer's Disease a major characteristic in the neurons is the formation of neurofibrillary tangles. However, there is little information concerning the types and quantities of neurotransmitters in these pathological structures. In a study of 6 patients with Alzheimer's Disease, it was reported that severely degenerating neurons located in the hippocampal pyramidal neurons in the cornu ammonis (CA) fields contained neurofibrillary tangles along with taurine immunoreactivity[22]. Decreases in glutamate- and glutaminase-immunoreactivity were observed in the CA fields of the hippocampus of these Alzheimer's Disease patients.

In a subsequent study by Alom and colleagues[23], it was observed that the levels of taurine were decreased by 25% in the cerebrospinal fluid of patients with Alzheimer's Disease. The levels of other amino acids (aspartate, serine, glutamine, alanine, and homocysteic acid) were not changed. The source of the loss of taurine, whether from a glial or neuronal taurine pool, was not determined. It was suggested that the role of the taurine-containing neurons should be investigated in Alzheimer's Disease patients.

AGING

While it has long been known that many amino acids, and in particular taurine, change in their concentrations during the development of the CNS only recently has this phenomenon been examined in great detail in the rat brain and spinal cord. Banay-Schwartz and colleagues[24] have analyzed the taurine content along with the distribution of 4 other amino acids (glycine, serine, threonine and alanine) in 53 specific microdissected regions of the CNS of Fischer 344 male rats. Of the two age groups examined (3 months and 29 months) taurine decreased with age in 21 areas. The decreases ranged from -45 to -11%. Only one area, the nucleus cuneatus increased (+11%) in taurine content. While there is no immediate correlation of these values with pathology the data base have been recorded and may be useful in the future to evaluate age-related changes.

PREGNANCY

The levels in the cerebral cortex of taurine and other ninhydrin-positive substances taken as a group were measured in 19-day old pregnant rats[25]. Taurine levels decreased by 48% while the ninhydrin-positive substances decreased by 22%. However, while the total brain water did not significantly change in the pregnant rats compared to non-pregnant rats the plasma osmolality decreased by 10%. It was thus suggested that the loss of taurine (along with other ninhydrin-positive substances) in the pregnant animal is a possible mechanism by which brain cells are prevented from swelling during the induced hypoosmolality.

HYPERTENSION

Hypertension produces an increase (30-40%) in taurine content in various brain regions of the Brattleboro rat[26]. Since this strain of hypertensive rat is known to be deficient in arginine vasopressin the rise in brain taurine levels is most likely due to a response to the hypernatremia that develops.

IRRADIATION

This study reports on the levels of taurine and other amino acids obtained after focused microwave irradiation, a technique that is currently used to rapidly inactivate enzymes in brain tissue and thus prevent the metabolism of various *in vivo* substances[27]. The advantages of this procedure appeared to be that it prevented post-mortem biochemical changes within seconds of death and thus was initially considered to be a more accurate method for obtaining *in vivo* amino acid levels than various freezing techniques. However, as determined in this study the microwave irradiation technique produces a rapid diffusion of amino acids and other small molecules from areas of high concentrations to areas of low concentrations in adjacent regions. It was thus suggested that microwave irradiation causes mechanical injury of the tissue and damages the membrane barriers thereby permitting diffusion of low molecular weight compounds down concentration gradients.

TAURINE AS A THERAPEUTIC AGENT

A relatively new and reproducible model for focal status epilepticus has been described which involves chronic infusion of GABA into a specific brain area for a defined period (6 hours to 14 days) and then abruptly stopping the infusion[28]. Paroxysmal discharges appear within approximately 20 minutes and consequently the new model has been named the GABA-withdrawal syndrome[29]. Chronic intracortical infusion of taurine was unable to prevent the paroxysmal discharges due to GABA withdrawal.

Finally, taurine has been tested as a treatment of dyskinesias in 14 patients[30]. Unfortunately, while oral treatment with taurine (3 g/day) appeared to initially improve the dyskinesias in these patients the effect was not lasting and at the end of a 6-week treatment the clinical symptoms reappeared.

CONCLUSIONS

In general, taurine levels in the brain decrease and extracellular levels increase when an animal is subjected to pathologic conditions such as ischemia-anoxia and seizures. In addition, taurine content in the brain decreases during pregnancy. Taurine levels tend to increase in certain regions of the brain in hypertension. The mechanism of the changes in taurine levels in the above conditions appears to be related to cell swelling and/or osmoregulation and perhaps in certain instances a response to neurotoxic cellular edema induced by the release of excitatory amino acids such as glutamate and aspartate (see article by Dr. Nico van Gelder in this Proceedings). However, there is a component of the taurine release that may be independent of osmoregulation. Finally, it has been attempted to use taurine as a therapeutic agent in various types of dyskinesias and seizures due to GABA-withdrawal. Unfortunately, taurine was without effect in these pathologic conditions. On the contrary, taurine does appear to protect against seizures due to hypoxia induced by nitrogen.

REFERENCES

1. R.J. Huxtable, Taurine in the central nervous system and the mammalian actions of taurine, *Prog. Neurobiol.* 32:471-533 (1989).
2. J.B. Lombardini, Taurine: retinal function, *Brain Res. Rev.* 16:151-169 (1991).
3. R.K. Simpson, Jr., C.S. Robertson, and J.C. Goodman, Spinal cord ischemia-induced elevation of amino acids: extracellular measurement with microdialysis, *Neurochem. Res.* 215:635-639 (1990).
4. K.E. Gordon, J. Simpson, D. Statman, and F.S. Silverstein, Effects of perinatal stroke on striatal amino acid efflux in rats studied with *in vivo* microdialysis, *Stroke* 22:928-932 (1991).
5. N. Menéndez, O. Herreras, J.M. Solís, A.S. Herranz, and R. Martín del Río, Extracellular taurine increases in rat hippocampus evoked by specific glutamate receptor activation is related to the excitatory potency of glutamate agonists, *Neurosci. Lett.* 102:64-69 (1989).
6. N. Menéndez, J.M. Solís, O. Herreras, A.S. Herranz, and R. Martín del Río, Role of endogenous taurine on the glutamate analogue-induced neurotoxicity in the rat hippocampus *in vivo*, *J. Neurochem.* 55:714-717 (1990).
7. K. Matsumoto, S. Ueda, T. Hashimoto, and K. Kuriyama, Ischemic neuronal injury in the rat hippocampus following transient forebrain ischemia: evaluation using *in vivo* microdialysis. *Brain Res.* 543:236-242 (1991).
8. R. Torp, P. Andiné, H. Hagberg, T. Karagülle, T.W. Blackstad, and O.P. Ottersen, Cellular and subcellular redistribution of glutamate-, glutamine- and taurine-like immunoreactivities during forebrain ischemia: a semiquantitative electron microscopic study in rat hippocampus, *Neurosci.* 41:433-447 (1991).
9. H. Ooboshi, H. Yao, T. Matsumoto, H. Hirano, H. Uchimura, S. Sadoshima, and M. Fujishima, Excitatory and inhibitory amino acid changes in ischemic brain regions in spontaneously hypertensive rats, *Neurochem. Res.* 16:51-56 (1991).
10. K.-M. Marnel, H.R. Morris, M. Panico, M. Timonen and P. Lähdesmäki, Glutamyl-taurine is the predominant synaptic taurine peptide, *J. Neurochem.* 44:752-754 (1985).
11. V. Varga, R. Janáky, K.-M. Marnela, J. Gulyás, P. Kontro, and S.S. Oja, Displacement of excitatory amino acid receptor ligands by acidic oligopeptides, *Neurochem. Res.* 14: 1223-1227 (1989).
12. P. Andiné, O. Orwar, I. Jacobson, M. Sandberg, and H. Hagberg, Extracellular acidic sulfur-containing amino acids and γ-glutamyl peptides in global ischemia: postischemic recovery of neuronal activity is paralleled by a tetrodotoxin-sensitive increase in cysteinesulfinate in the CA1 of the rat hippocampus, *J. Neurochem.* 57:230-236 (1991).
13. D.W. Choi, Glutamate neurotoxicity and diseases of the nervous system, *Neuron* 1:623-634 (1988).
14. G.E. Nilsson and P.L. Lutz, Release of inhibitory neurotransmitters in response to anoxia in turtle brain, *Am. J. Physiol.* 261:R32-R37 (1991).
15. G.E. Nilsson, A.A. Alfaro, and P.L. Lutz, Changes in turtle brain neurotransmitters and related substances during anoxia, *Am. J. Physiol.* 259:R376-R384 (1990).

16. T.R. Guilarte, Regional changes in the concentrations of glutamate, glycine, taurine, and GABA in the vitamin B-6 deficient developing rat brain: association with neonatal seizures, *Neurochem. Res.* 14:889-897 (1989).

17. T.R. Guilarte, Abnormal endogenous amino acid release in brain slices from vitamin B-6 restricted neonatal rats, *Neurosci. Lett.* 121:203-206 (1991).

18. M. Malcangio, A. Bartolini, C. Ghelardini, F. Bennardini, P. Malmberg-Aiello, F. Franconi, and A. Giotti, Effect of ICV taurine on the impairment of learning, convulsions and death caused by hypoxia, *Psychopharmacol.* 98:316-320.

19. S. Simler, L. Ciesielski, J. Clement, and P. Mandel, Amino acid neurotransmitter alterations in three sublines of Rb mice differing by their susceptibility to audiogenic seizures, *Neurochem. Res.* 15:687-693 (1990).

20. L. Ciesielski, S. Simler, J. Clement, and P. Mandel, Effect of repeated convulsive seizures on brain γ-aminobutyric acid metabolism in three sublines of mice differing by their response to acoustic stimulation, *J. Neurochem.* 49:220-226 (1987).

21. P. Demediuk, M.P. Daly, and A.I. Faden, Effect of impact trauma on neurotransmitter and non-neurotransmitter amino acids in rat spinal cord, *J. Neurochem.* 52:1529-1536 (1989).

22. N.W. Kowall and M.F. Beal, Glutamate-, glutaminase-, and taurine-immunoreactive neurons develop neurofibrillary tangles in Alzheimer's Disease, *Ann. Neurol.* 29:162-167 (1991).

23. J. Alom, J.N. Mahy, N. Brandi, and E. Tolosa, Cerebrospinal fluid taurine in Alzheimer's Disease, *Ann. Neruol.* 30:735 (1991).

24. M. Banay-Schwartz, A. Lajtha, and M. Palkovits, Changes with ageing in the levels of amino acids in rat CNS structural elements II. Taurine and small neutral amino acids, *Neurochem. Res.* 14:563-570 (1989).

25. R.O. Law, Effects of pregnancy on the contents of water, taurine, and total amino nitrogen in rat cerebral cortex, *J. Neurochem.* 53:300-302 (1989).

26. R. Dawson, Jr., D.R. Wallace, and M.J. King, Monoamine and amino acid content in brain regions of Brattleboro rats, *Neurochem. Res.* 15:755-761 (1990).

27. C.F. Baxter, J.E. Parson, C.C. Oh, C.G. Wasterlain, and R.A. Baldwin, Changes of amino acid gradients in brain tissues induced by microwave irradiation and other means, *Neurochem. Res.* 124:909-913 (1989).

28. S. Brailowsky, M. Kunimoto, C. Silva-Barrat, C. Menini, and R. Naquet, Electroencephalographic study of the GABA-withdrawal syndrome in rats, *Epilepsia* 31:369-377 (1990).

29. S. Brailowsky, M. Kunimoto, C. Menini, and R. Naquet, The GABA-withdrawl syndrome: a new model of focal epileptogenesis, *Brain Res.* 442:175-179 (1988).

30. H. Nyland, B.A. Engelsen, and H. Blom, Taurine treatment of dyskinesias: an attempt, *Eur. Neurol.* 29:121-123 (1989).

PATHOLOGIES OF THE CNS AND ASSOCIATED
TAURINE CHANGES

Nico M. van Gelder

Centre de recherche en sciences neurologiques
Départemente de physiologie
Université de Montréal
C.P. 6128, succ. A
Montréal, Québec
Canada H3C 3J7

INTRODUCTION

Investigations into the action(s) of taurine over the past 20 years, have hinted at an important role of this amino acid in maintaining normal CNS function. Thus, changes of taurine content have been observed in association with certain forms of epilepsy (van Gelder et al., 1972), retinal blindness (Hayes et al., 1975a), abnormal neural (sensory) development (Sturman, 1990) and cardiac myopathies (Huxtable, 1989). These findings seem to suggest that taurine is needed in developing, as well as adult, tissues which cells are characterized by their ability to undergo rapid voltage discharges. What precisely, however, that function may be is not easy to uncover.

Unlike many other molecules implicated in neural signal transmission, the physicochemical properties of taurine mitigate against its direct participation in the process of electrical conduction or in the transsynaptic chemical messenger process. As far as can be determined, taurine is not incorporated into proteins or the membrane structure. Its release upon depolarization is slower than for most small molecular weight substances acting at the synapse; the release is usually observed only after the depolarizing process (in terminals) is well advanced or unless the discharge frequency is rather high and prolonged. Finally, although extracellular taurine will oppose Ca^{2+} influx, in the normal CSF or neural tissue superfusates the concentrations are low, well below those reported to exert a physiological action.

In view of the reputed metabolic inertness of the molecule, the essential need for taurine during development must be connected to the large (1-10 mM) amounts present within cells. As recently reviewed (Pasantes-Morales and del Río, 1990) an osmotic role for the amino acid would demand a quantitatively important, but fluctuating, pool of intracellular free taurine which does not interfere with either metabolic or physiological properties of electrically excitable cells; the fluctuating size of this pool, i.e. the sequestered to cytoplasmic ratio, would assure (i) a dynamic osmotic equilibration of the intracellular environment with the extracellular milieu and (ii) a constant cell volume by release of the free form and, with it, cell water.

Taurine, Edited by J.B. Lombardini *et al.*
Plenum Press, New York, 1992

Taurine release mostly seems to *follow* glutamate efflux but, nevertheless, is linked to the same physiological triggers which release the excitatory amino acid from cells. Since, however, the two amino acids are metabolically entirely unrelated, their tissue release must be cued by one or more intervening factors which are responsible, on depolarization, for the immediate discharge of glutamate, and the delayed but more prolonged taurine efflux. These factors have been proposed to represent the rise in intracellular ionic Ca^{2+} and the increase in water production from enhanced glucose metabolism upon neuronal discharge, or other osmotic cues (van Gelder, 1990).

Assuming, for a moment, that the postulated role of taurine is correct in that it regulates ionic Ca^{2+} concentrations in the cells as well as cell water volume, then certain *pathological* conditions associated with changed cellular taurine levels should be compatible with this quantative concept of taurine function. Here we review three such conditions: An excess of taurine, a deficit of taurine and, finally, an alteration in taurine distribution among diverse cells of the CNS. These conditions are found, respectively, in *Friedreich's Ataxia*, a *Kwashiorkor* model, and the *Hypersynchrony* of epilepsy and migraine.

Friedreich's Ataxia

The disease belongs to a group of ataxias marked by, among others, spino-cerebellar pathway degeneration and cardiomyopathy, and is acquired (as far as is known) almost exclusively by autosomal-recessive inheritance; sensory afferent pathways are particularly affected. The disease often is associated with cardiac failure in mid-life. It has been extensively studied, in terms of biochemical, physiological and clinical parameters, by a large group of investigators coordinated by the late A. Barbeau. Results of the many studies have been published in a series of volumes of the Can. J. Neurol. Sci. (Quebec Cooperative Study of Friedreich's Ataxia, 1978-1984).

This investigator's modest contribution (van Gelder et al., 1987) consisted of challenging patients and an age, sex matched control group with an oral 1 g dose of taurine, and to follow the blood taurine concentrations over 5 hrs. The only abnormality found, if any, was that F.A. patients clearly tended to exhibit higher peak taurine levels in blood following such challenge; no clinical signs or other discomfort were noted by either patients or attending physicians. Combined with other data on the rate of disappearance of taurine from blood (identical) and rates of urinary excretion (proportionally enhanced), the only conclusion to be reached was that F.A. individuals may temporarily exhibit higher blood taurine levels than unaffected individuals, when the diet contains the amino acid. Since neither the challenging dose in young adult F.A. patients nor oral feeding of taurine to many different individuals in single doses of up to 4 g have given rise to any reports of adverse reactions, no data seemed to suggest at the time that this (inherited) high taurine absorption "trait" in F.A. families was of any consequence. However, one can assume that during normal pregnancy, such a trait might occasionally expose the foetus transiently to perhaps unusually high taurine levels in the umbilical circulation.

In order to investigate a possible effect of high taurine on embryonic development, an experimental model was chosen where the nutritional supply of taurine during development represented an initial constant amount which could not be replenished during the embryonic period (summary ref.: van Gelder and Bélanger, 1988). For this purpose the chick embryo was studied, since the yolk essentially contains the entire pool of (non-renewable) amino acids needed for development. Because maturation of different organs, and cell number (e.g. DNA, RNA) as well as cell differentiation (e.g. protein), follows an individual time course during embryogenesis, for comparison purposes the content of each free amino acid in different tissues at different growth phases (1-14 days) was expressed relative to the valine level in the free amino acid pools. This amino acid, classified as essential and not derived from metabolism, appears to be freely accessible to all developing organs, and its tissue level directly and proportionally reflects yolk content, both with respect to circulating concentrations as well as in the organs. The method to compare amino acid content in different tissues obviated the need to take into account the variations between organs at the same age (but not state) of development.

While in the yolk, in the circulation, and in the various embryonic organs, the essential amino acids as a group varied within 20% of the valine levels, taurine (and, also, phosphoethanolamine) represents only a very small fraction of the total free amino acid supply in the yolk. Yet, already on day 1-2 after fertilization, the embryo has extracted this amino acid more than any of the others present in the nutritional reservoir. It should be noted that phosphoethanolamine, as highly concentrated from yolk as taurine, is an integral constituent of all membranes; taurine on the other hand has not been demonstrated to be incorporated into any lipoidal or proteinaceous structure of cells.

From Figure 1A it is apparent that the taurine level in the embryo CNS plateaus around 8-10 days but that subsequently, a second almost linear increase in taurine occurs; the secondary rapid concentration of taurine from the circulation appears specific for only neural tissue, since neither the blood nor the heart or retina exhibits this secondary steady rise in taurine during the same time span. The period coincides with the glial development phase of CNS maturation and this delayed accumulation of taurine in neural tissue may therefore represent uptake of the amino acid by these maturing cells (many other non-essential amino acids, especially glutamate and glutamine, exhibit a similar but less important increase; glutamine synthesis is the exclusive domain of glial cells in the CNS).

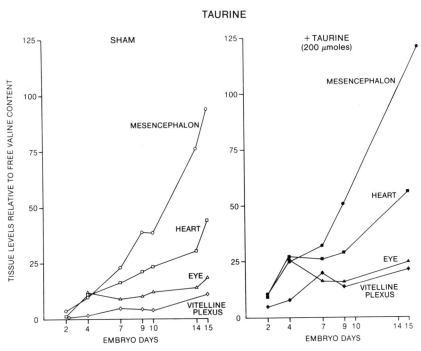

TAURINE

Figure 1. Tissue taurine content in chick embryos relative to valine ($= 1$) following injection into the yolk of 100 μmoles NaCl (Sham) or 200 μmoles of taurine. With increased taurine levels in the vitelline circulation, taurine levels are higher than usual during the first week of development, but after day 7, the *rates* of tissue taurine accumulation in the CNS are equal, despite differing blood taurine levels. This second phase of taurine acquisition coincides with glial proliferation and these cells may therefore be primarily responsible for the homeostasis of neural tissue taurine content in the face of varying nutritional taurine supplies or blood levels (van Gelder and Bélanger, 1988).

When up to 200 μmoles of taurine are injected into the yolk between 0-2 days after fertilization (Figure 1B), it can be seen that the taurine content in all organs plateaus much earlier, around day 4. However, the second phase of taurine accumulation, which starts around day 7, demonstrates the same *rate* of increase in the CNS as is seen in the nutritionally non-manipulated eggs. These data suggest that during the first week or so of embryonic growth the CNS (and other organs) acquires taurine at a rate dependent on the amount available to the embryo. The amount which has accumulated in the tissue at the end of this period is nevertheless limited, by possibly either retention and/or exchange mechanisms within the organ tissues. Taurine accumulation during the first week is thus governed by its availability in the nutritional supply until a tissue determined limit is reached. In contrast, the second phase of taurine acquisition in the CNS appears independent of this taurine supply, with a rate that now becomes almost entirely governed by the tissue itself, and which is constant.

From these observations one can conclude that any effects of fluctuating high taurine levels in the nutritional supply, whether positive or negative, would occur during early embryogenesis. In the chick CNS that phase represents primarily neuronal multiplication and functional differentiation. At a later period of embryogenesis, the CNS seems to prevent a further abnormal increase in taurine even when the levels in the supply are high. On the other hand, the early acquisition of extra taurine, in the chick at least, is not rectified at a later stage of development, with the result that at birth the chicks demonstrate above normal taurine levels (approximately 20%). However at 7-10 days post-hatching, the elevation is no longer observed, indicating that post-hatched chick CNS can eventually rectify the early embryonic exposure to a high taurine supply.

When batches of eggs so injected with taurine are allowed to develop to hatching, many of the chicks were incapable of extricating themselves from the egg. (Equimolar doses of valine or saline had no such effect; GABA at 20% amounts, or glycine at 50% of taurine doses were lethal to most early embryos). Among those taurine exposed chicks which were able to hatch, 80% of individuals exhibited a weakened stance and a staggering gate. Especially the staggering gate did neither worsen nor ameliorate in the first week following hatching.

Finally, because of the possible although very tenuous analogy between a F.A. foetus which may have been exposed occasionally to higher than usual blood taurine levels, and the chicks exposed to high taurine in the nutritional supply, the rare results of amino acid determinations in autopsied F.A. brain were recalculated from data published by others (Huxtable et al., 1979), to express the taurine levels relative to free valine content. While statistically without relevance, the two F.A. individuals tended to have up to 5 times more taurine relative to valine (or other free amino acids) in certain areas of the brain than individuals not carrying the F.A. trait. Table 1 shows the high taurine levels relative to other amino acids such as valine in the cerebellum, whereas another neural structure, the degenerating dorsal root ganglion shows this to a much lesser extent.

As intriguing as these data may appear, it is evident that few conclusions can be reached here. Unfortunately, with the demise of A. Barbeau a number of projects in progress had to be discontinued and there appears little chance that more metabolic data of clinical relevance will appear soon. What these studies have shown, however, is that if any pathology is associated with abnormally high circulating taurine levels, such pathology "most likely" may find its origin in the early phase of embryo development i.e. the neurogenic phase. This is further substantiated by the work of Lima et al. (1988) who have shown that taurine stimulates neurite outgrowth in goldfish retinal explants, both with respect to numbers as well as length, but this growth promoting action of taurine only occurs in the first week of culture. Subsequently, taurine has no further effect. Recent work to be published soon from this same group (B. D. Drujan and L. Lima, personal communications), not only

Table 1. Free amino acid abundance relative to valine ($=1$) in four control and two Friedreich's Ataxic cerebellar and brainstem structures

	TAU		GLU		GLN		GABA	
	Con.	F.A.	Con.	F.A.	Con.	F.A.	Con.	F.A.
PVC[a]	2.55	4.10	16.25	7.84	8.26	8.84	2.58	1.64
AVC	2.77	4.27	13.29	6.87	6.19	9.33	2.67	1.62
ION	1.75	1.98	9.22	2.61	9.87	2.64	1.89	0.54
DN	2.05	3.36	10.23	3.96	9.38	8.91	4.81	1.46
DRG	1.27	1.70	2.95	2.62	0.57	3.43	n.d.	0.19
CH	2.95	9.95	13.52	11.63	7.17	13.11	1.88	1.92

[a]PVC, AVC = posterior and anterior cerebellar vermis; ION = inferior olivary nucleus; DN-dentate nucleus; DRG = dorsal root ganglion; CH = cerebellar hemispheres. Data from Huxtable et al., 1979.

confirms this observation but also suggests that the effect is mediated via a Ca^{2+} mechanism. No action of taurine on retinal explants was observed in the absence of (external) Ca^{2+} in the medium.

Trenkner and Sturman (1990) using cerebellar cell cultures, report that taurine in the medium at physiologically relevant concentrations promotes morphological differentiation of neonatal mouse granule cells but that in kitten cerebellar cultures, taurine is toxic. Aside from (suggested) species differences, the degree of cell differentiation at the time the culture was established might also have influenced the outcome, especially if Ca^{2+} mechanisms are implicated (above).

Protein Malnutrition (Kwashiorkor)

Some time ago Hayes and coworkers (1975b) as well as, more recently, Pion and coworkers (1987) described in cats, respectively, retinal degeneration and a genetically determined cardiac myopathy, as a result of dietary taurine deficiency. The group of Sturman (1990) showed moreover that in the cat, embryological development as well as, subsequently, neonatal growth and CNS maturation was highly abnormal. Most of the kittens died either during foetal development or soon after birth. Addition of taurine to the diet reversed all the effects of a taurine deficient diet. All these findings strongly point to an important role of taurine during development. The data indicate that through unknown mechanisms taurine serves as an essential amino acid during maturation of excitable tissues.

In a different set of experiments (van Gelder and Parent, 1981) mice were allowed to bear normal litters, but one day after birth, the mothers were placed on diets containing either an excessive protein (68%) or a deficient protein content (8%), compared to a normal protein diet (57%). In accordance with the above cited results, the high mortality of pups nursed by the dams on a protein deficient diet, especially, but those in the excessive protein group as well, demonstrated a decreased mortality if taurine was supplied in the drinking water. When the quantity as well as the quality of the milk was analyzed, it was found that both abnormal protein diets caused large reductions in milk quantities available to the pups (stomach content). In addition, in the low protein group but not in the excess protein group, milk was deficient in taurine and protein. Supplementation of the mother's drinking water with 0.02% taurine restored both milk taurine and protein content; in the protein deficient and, to a lesser extent, the protein excessive dietary groups, the quantity of milk also increased. This then indicated that taurine during nursing (in mice) may be essential for the nutritional quality of the milk, as well as the quantity. Armstrong et al. (1973) already reported that soon after onset of pregnancy and extending into the lactating period, women will reduce urinary taurine excretion to negligible amounts. Assuming that little dietary modifications had occurred during these studies, one can infer that in humans, taurine is similarly needed to assure normal foetal development and adequate milk production of good quality.

One reassuring aspect of the post-natal feeding experiments in mice was that the milk taurine and protein composition are regulated by some form(s) of interdependent feedback mechanism(s) so that the protein concentration as well as that of taurine do not reach excessive levels in the milk. This would imply that, barring genetic dietary absorption anomalies (see section above), a dietary supplement of taurine + protein mixture in moderate quantities, to undernourished mothers may assure rectification of these constituents in milk, without a danger of excessive amounts appearing in the milk.

Despite the apparent tight regulation of blood taurine levels under normal dietary circumstance, primarily by the kidney (van Gelder et al., 1975), it nevertheless seems prudent not to assume that an excess of dietary taurine during pregnancy will not affect the foetus, nor on the other hand, that the apparent beneficial effects of taurine supplements following or during malnutrition and undernutrition are entirely attributable to taurine itself. Some reports would suggest that an interaction occurs between taurine and certain hormone systems (see Huxtable, 1989); in the case of pregnancy, the many reports of the cardiovascular actions of taurine (see other chapters) suggest that it may affect the foetus indirectly, through improvements in the foetal or umbilical circulation and through nutrient delivery to (some?) foetal organs.

Hypersynchrony, Seizures, and CNS Damage

All three above mentioned phenomena are associated with a release of tissue glutamate into the extracellular fluids. It is the extracellular presence (in excess) and regional diffusion of this excitatory amino acid which, in all probability, are responsible for many of the resulting neurological dysfunctions (review: van Gelder, 1990). The cause for the extracellular accumulation of glutamate has not yet been precisely elucidated; one interpretation (van Gelder, 1978; 1990) of the data proposes a dissociation or, alternatively, an overwhelming of the metabolic processes responsible for the detoxification of neuronally released glutamate by enveloping glial elements, to form glutamine (see Figure 2). This would result in increased extracellular glutamate levels and, also, in the case of a neuronal-glial metabolic dissociation, should cause an extracellular accumulation of other neuronal products released under similar circumstances, such as, notably, taurine. The consequence of a glial failure to take up and to neutralize the physiological actions of free, extracellular, glutamate and taurine not only will affect neuronal excitability and hypersynchrony in the region where this phenomenon occurs. In addition, it is likely to also disturb the water redistribution between the neuronal-glial network and the CSF (van Gelder, 1990). The resulting cytotoxic (glial) oedema and CSF disturbance may well contribute to the permanent damage which can occur following repeated and/or grouped neuronal discharges at maximal frequencies, e.g. epilepsy. Data reflecting both the partial redistribution of neuronal glutamate and taurine to their glial compartment have been obtained in several experimental epilepsy models as well as following rearrangement of the neural tissue cytoarchitecture (van Gelder and

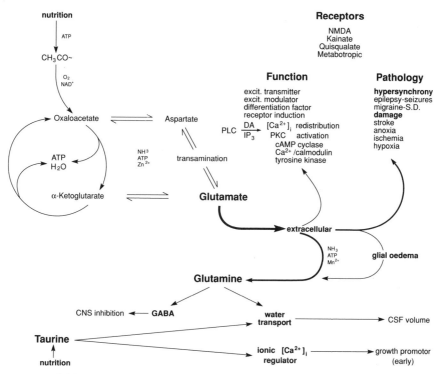

Figure 2. Superficial synopsis of the functions of taurine and of glutamate and its metabolites in the CNS. The "metabolic" interactions between the two amino acids are postulated to occur principally at the level of water control and osmotic homeostasis within and between diverse cells of neural tissue, and via internal Ca^{2+} redistribution among the cytoplasmic and sequestered fractions in cells.

The role of extracellular glutamate in maintaining mildly synchronized excitability in groups of neurons progresses towards pathological conditions as (increased) tissue release can no longer be compensated for by glial uptake and transamination to glutamine. The simultaneous perturbation in neuronal-glial uptake and release mechanisms, responsible for recycling of taurine between neurons and glia, contributes further to the development of cytotoxic oedema and increased intracellular Ca^{2+} mobility (see van Gelder, 1990).

Drujan, 1980; van Gelder, 1982). In addition, in familial tendencies for migraine and epilepsy, analogous disturbances in the glutamate-taurine ratios are detectable, with the most predictable change representing a rise in blood glutamate; in the case of epilepsy, such changed ratios occur independently of whether or not the predisposed individuals show clinical symptoms, aside from an increased tendency for some EEG abnormalities (van Gelder et al., 1980).

Unlike the preceding two CSN pathologies mentioned, where taurine, respectively, represents a greater or lesser percentage of the total intracellular pools of free amino acids, the group of dysfunctions cited in this section appears predominantly associated with a disturbance in the normal distribution of taurine between neuronal and apposing glial elements, without necessarily a net change in the total taurine content of neural tissue. Within the CNS, the similarly disturbed glutamate redistribution, on the other hand, is almost invariably reflected in a lowering of neural tissue glutamate, because glial glutamate is rapidly metabolized to glutamine, which subsequently in part, also represents the precursor for GABA (Figure 2). A careful analysis of many data bases will reveal evidence that the sum of neural tissue glutamate plus glutamine plus GABA may remain quite stable during moderate neural excitation but the glutamate to glutamine plus GABA ratio changes in favour of the metabolites (van Gelder, 1978; van Gelder and Drujan, 1980). During pathological conditions of hypersynchrony, such as the beginning of migraine (Spreading Depression) or seizures (summated, high amplitude EEG), the metabolic ratio of glutamate to its metabolites first is altered and then demonstrates a net loss of tissue glutamate (van Gelder et al., 1983). The cause appears to be a saturation of the transport of glutamate into glia as well as an impairment in these cells to further accelerate the amidation of the amino acid to glutamine; subsequently glutamine formation no longer can compensate for the additional (neuronal) release of glutamate. This results in an eventual loss of glutamate via the extracellular spaces by diffusion, interstitial water flow, and other "dilution" mechanisms. In the process the electrophysiological action of glutamate becomes apparent in the form of an increase in intracellular ionic Ca^{2+}, and enhanced excitation of a (hyper) synchronized group of neurons. If the extracellular glutamate dispersion is slow, it may cause Spreading Depression, especially in CNS areas where groups of neurons are densely packed.

Hence, a disturbance of the taurine to glutamate ratio in neural tissues, or in the CSF, or in the blood (refs. above) and/or a closer correlation of neural tissue taurine content with the levels of glutamate metabolites, rather than with a glutamate itself, will indicate that in these transient or chronic changes of CNS excitability a breakdown has occurred in the metabolic links between neurons and their closely associated glial envelope. While the resulting CNS dysfunctions can be primarily ascribed to changes in the extracellular glutamate levels (Figure 2), the disturbance in the normal distribution of taurine between these two cell types will cause, in addition, irregularities in the control of intracellular Ca^{2+} ionization, and a breakdown of water volume control of neurons and glia (van Gelder, 1990). The latter phenomenon, especially, can lead to permanent damage to the cytoarchitecture. This possibility brings, perhaps, in sharper focus the present discussion on the potentially adverse effects of deviating CNS taurine levels during embryonic and neonatal development.

Both as an early growth promotor, via Ca^{2+} ionization control, and as a regulator of osmotic equilibration in developing cells, an excess as well as a deficit of taurine during critical periods of neurogenesis could be expected to have a negative effect on CNS formation. In addition, even following this critical phase and into adulthood, a dynamic redistribution process of taurine during a cycle of neuronal excitation and neuronal quiescence, might compensate for water volume changes and altered ionic Ca^{2+} levels. Such a "permanent" function of taurine would indeed remain compatible with the proposal that it is the *quantity* of taurine inside cells and within the CNS which accounts for its essential function in certain tissues, rather than the somewhat weak electrophysiological actions of the amino acid. Tissues undergoing rapid depolarization-repolarization cycles and contraction, exhibit the greatest fluctuations in water volume and in the free to sequestered Ca^{2+} ratio. This, too, would explain why taurine is present in high, but nevertheless very constant and regulated quantities in certain types of organs, such as various neural tissues, the heart, and skeletal muscle.

REFERENCES

Armstrong, M.D., 1973, Decreased taurine excretion in relation to childbirth, lactation, and progestin-estrogen therapy, *Clin. Chim. Acta*, 46:253-256.

Hayes, K.C., Carey, R.E., and Schmidt, S.Y., 1975a, Retinal degeneration associated with taurine deficiency in the cat, *Science*, 188:949-951.

Hayes, K.C., Rabin, A.R., and Berson, E.L., 1975b, An ultrastructural study of nutritionally induced and reversed retinal degeneration in cats, *Am. J. Pathol.*, 78:505-524.

Huxtable, R.I., Azari, J., Reisine, T., Johnson, P., Yamamura, H., and Barbeau, A., 1979, Regional distribution of amino acids in Friedreich's Ataxia brains, *Can. J. Neurol. Sci.*, 6:255-258.

Huxtable, R.I., 1989, Taurine in the central nervous system and the mammalian actions of taurine, *Progr. Neurobiol.*, 32:471-533.

Lima, L., Matus, P., and Drujan, B., 1988, Taurine effect on neuritic growth from goldfish retinal explants, *Int. J. Devl. Neurosci.* 6:417-424.

Pasantes-Morales, H. and Martín del Río, R., 1990, Taurine and mechanisms of cell volume regulation, *Progr. Clin. Biol. Res.* 351:317-328.

Pion, P.D., Kittleson, M.D., Rogers, Q.R., and Morris, J.G., 1987, Myocardial failure in cats associated with low plasma taurine: a reversible cardiomyopathy, *Science* 237:764-768.

Sturman, J.A., 1990, Taurine deficiency, *in*: "Taurine: Functional Neurochemistry, Physiology, and Cardiology", H. Pasantes-Morales, D.L. Martin, W. Shain, R. Martín del Río, eds., Wiley-Liss, New York, pp. 385-395.

Trenker, E. and Sturman, J.A., 1990, The role of taurine in the survival and function of cerebellar cells in cultures of early postnatal cat, *Int. J. Devl. Neurosci.* 9:77-88.

van Gelder, N.M., Sherwin, A.L. and Rasmussen, T., 1972, Amino acid content of human epileptogenic brain: focal versus surrounding regions, *Brain Res.* 40:385-393.

van Gelder, N.M., Sherwin, A.L., Sacks, C., and Andermann, F., 1975, Biochemical observations following administration of taurine to patients with epilepsy, *Brain Res.*, 94:297-306.

van Gelder, N.M., 1978, Taurine, the compartmentalized metabolism of glutamic acid and the epilepsies, *Can. J. Physiol. Pharm.* 56:362-374.

van Gelder, N.M., Janjua, N.A., Metrakos, K., MacGibbon, B., and Metrakos, J.D., 1980, Plasma amino acids in 3/sec spike-wave epilepsy, *Neurochem. Res.* 5:659-671.

van Gelder, N.M. and Drujan, B.D., 1980, Alterations in the compartmentalized metabolism of glutamic acid with changed cerebral conditions, *Brain Res.* 200:443-455.

van Gelder, N.M. and Parent, M., 1981, Effect of protein and taurine content of maternal diet on the physical development of neonates, *Neurochem. Res.* 200:443-455.

van Gelder, N.M., 1982, Changed taurine-glutamic acid content and altered nervous tissue cytoarchitecture, *Adv. Expt. Med. Biol.* 139:239-256.

van Gelder, N.M., Siatitsas, I., Ménini, C., and Gloor, P., 1983, Feline generalized penicillin epilepsy: changes of glutamic acid and taurine parallel the progressive increase in excitability of the cortex, *Epilepsia*, 24:200-213.

van Gelder, N.M., Roy, M., Bélanger, Paris, S., and Barbeau, A., 1987, Subtle defects in the regulation of the free amino acid balance in Friedreich Ataxia: a relative deficiency of histidine combined with a mild but chronic hyperammonemia, *in*: "Basic and Clinical Aspects of Nutrition and Brain Development", D.K. Rassin, B. Haber, B. Drujan, eds., Alan R. Liss Inc., New york.

van Gelder, N.M. and Bélanger, F., 1988, Nutrition and selective accumulation processes establish the free amino acid pools of developing chick embryo brain, *in*: "Amino Acid Availability and Brain Function in Health and Disease", G. Heuther, ed., Springer-Verlag, Berlin.

van Gelder, N.M., 1990, Neuronal discharge hypersynchrony and the intracranial water balance in relation to glutamic acid and taurine redistribution: migraine and epilepsy, *Progr. Clin. Biol. Res.* 351:1-20.

NOTE: The disproportionate number of references of author, cited, merely reflects a greater familiarity with the work reported in these papers. It should not be construed, nor is it intended, to suggest claims to priority or to originality.

TAURINE RECEPTOR: KINETIC ANALYSIS AND PHARMACOLOGICAL STUDIES

J.-Y. Wu[1,2], X.W. Tang[1] and W.H. Tsai[2]

[1]Department of Physiology and Cell Biology
University of Kansas
Lawrence, KS 66045 USA

[2]Academia Sinica
Taipei, Taiwan ROC

INTRODUCTION

Taurine is one of the most abundant amino acids in the vertebrate CNS and is believed to be involved in many important physiological functions such as in the development of the CNS at least in some species[1,2], in maintaining the structural integrity of the membrane[3,4], in regulating calcium binding and transport[5,6], and serving as an osmoregulator[7,8] and inhibitory neurotransmitter[9-11]. Although many of these effects may be mediated through a specific taurine receptor, little is known with certainty about the detailed properties of the taurine receptor. Previously, we have reported the basic characterizations of the taurine receptor in brain membrane preparations, including binding kinetics, effect of divalent cations and specificity of the taurine receptor binding[12]. Other related studies dealing with taurine receptor binding in the brain, as well as other tissues, have also been reported[13-15]. In this communication, we further characterize the taurine receptor in the brain, describe a new procedure for the preparation of the taurine receptor and present further evidence supporting the presence of a specific taurine receptor in the brain. In addition, the effects of amino acids, nucleotides, taurine analogues, various ions and second messengers on the taurine receptor are included.

METHODS AND MATERIALS

Preparation of the Taurine Receptor

The membrane-bound taurine receptor was prepared as previously described[12] with the exception that the P_2 membrane was first subjected to freezing-thawing, followed by an extensive wash with Tris-citrate buffer, pH 7.4 and finally briefly washed three times with 0.1% Triton X-100.

Taurine Receptor Binding Assay.

For total binding (TB), the binding assay consisted of 800 μl crude taurine receptor (about 0.15 mg/ml), 100 μl of 5 mM Tris-Cl (pH 7.2), and 100 μl of ^{3}H-taurine (300 nM). For nonspecific binding (NB), the conditions were the same as TB except that 100 μl of 1 mM unlabeled taurine was included.

The binding mixture was incubated at 4°C for 45 min, and the reaction was terminated by centrifugation to separate the receptor-^{3}H-taurine complex from free ^{3}H-taurine. The pellet thus obtained was briefly rinsed twice with Tris-Cl buffer and then resuspended in the same buffer before adding scintillation fluid for counting.

RESULTS AND DISCUSSION

Kinetic Analysis

The taurine receptor prepared by the modified procedure described here has a much higher specific binding compared to that obtained by the original procedure[12]. The ratio of TB to NB is typically between 10-20 whereas the original procedure usually gives a ratio between 3-5.

The binding of ^{3}H-taurine to the receptor appears to be saturable. Scatchard analysis gives a K_d of 92 nM and B_{max} of 6.0 pmol/mg protein. The plot of Log (Bound/Bmax-Bound) vs Log (Free) shows a linear line with a Hill coefficient of 0.90, suggesting a single site model for the binding of ^{3}H-taurine to the receptor.

Effect of Agonists and Antagonists of Other Transmitter Receptors

The binding of ^{3}H-taurine to the receptor is not affected by agonists and antagonists of glutamate receptors (e.g., glutamate, quisqualic acid, kainic acid, and NMDA), glycine receptor (e.g., glycine, strychnine), benzodiazepine (BZ) receptor (e.g., FNZP), $GABA_A$ receptors (e.g., pictrotoxin, bicuculline) and $GABA_B$ receptor agonists and antagonists (e.g., GABA, baclofen, and phaclofen) (Table 1), suggesting that taurine receptor binding is highly specific.

Table 1. Effect of various agonists and antagonists of other amino acid receptors on taurine receptor binding.

Receptor	Agonists/Antagonists	Conc. (M)	Inh. (%)
Glu	Glutamate	10^{-3}	15
	Kainic Acid	10^{-3}	2
	NMDA	10^{-3}	8
	Quisqualic Acid	5×10^{-4}	0
	Kynurenic Acid	5×10^{-4}	25
Gly	Glycine	10^{-3}	0
	Strychnine	10^{-4}	2
BZ	FNZP	10^{-5}	0
$GABA_A$	Picrotoxin	10^{-4}	10
		10^{-6}	5
	Bicuculline	10^{-4}	0
		10^{-6}	5
$GABA_B$	GABA	10^{-3}	50
		10^{-4}	17
		10^{-6}	12
	Baclofen	10^{-4}	26
		10^{-6}	9
	Phaclofen	10^{-4}	7

Effect of Taurine Analogues

Taurine analogues such as hypotaurine and homotaurine were found to be quite potent inhibiting taurine receptor binding at 10^{-6} M to an extent of 85 and 78%, respectively (Table 2). On the other hand, GABA and ß-alanine are rather weak analogues inhibiting only 17 and 25% at 10^{-4} M. If hypotaurine and homotaurine do not elicit similar physiological responses as taurine, they certainly can serve as specific taurine receptor blockers, thus providing a new tool for pharmacological and physiological studies of the mechanism of taurine receptor mediated processes.

Table 2. Effect of taurine analogues on ^{3}H-taurine binding.

Conc. of	Inhibition (%)		
Analogues (M)	Taurine	Hypotaurine	Homotaurine
10^{-4}	100	100	92
10^{-5}	98	97	85
10^{-6}	93	85	78
10^{-7}	78	68	59
10^{-8}	31	42	21

Effect of Amino Acids

We have examined 23 amino acids on taurine binding. Among these amino acids, cysteic acid is the most potent inhibitor (inhibiting taurine binding by 84% at 1 mM) followed by ß-alanine, valine, tyrosine and cysteine as shown in Table 3.

Table 3. Effect of amino acids on ^{3}H-taurine binding.

Amino Acids	Conc. (mM)	Inh. (%)	Conc. (mM)	Inh. (%)
L-Alanine	1.0	40	0.10	14
ß-Alanine	1.0	61	0.10	25
L-Valine	1.0	49	0.10	8
L-Leucine	1.0	25		
L-Isoleucine	1.0	37		
L-Proline	1.0	24		
L-Phenylalanine	1.0	39	0.10	-2
L-Tryptophan	1.0	20		
L-Methionine	1.0	41	0.10	13
L-Glycine	1.0	0		
L-Serine	1.0	25		
L-Threonine	1.0	51	0.10	7
L-Tyrosine			0.01	12
L-Asparagine	1.0	9		
L-Glutamine	1.0	20		
L-Aspartate	1.0	52	0.10	-3
L-Glutamate	1.0	15		
L-Lysine			0.10	-5
L-Arginine	1.0	23		
L-Histidine	1.0	37		
L-Cysteine	1.0	58	0.10	17
L-Cystine	0.1	45	0.01	32
L-Cysteic Acid	1.0	84	0.01	41
L-Cystine Sulfinic Acid	1.0	60	0.10	17

Effect of Divalent and Monovalent Cations

It is known that some amino acid receptors, such as the NMDA receptor, are affected by Mg^{2+}[16]. It is also known that the uptake of amino acid transmitters is Na^+ dependent[17]. Hence, we have tested various divalent and monovalent cations in order to determine whether they have any effect on the taurine receptor. The results show that taurine receptor binding is not affected by monovalent cations at 1 mM concentration. No significant effect by Mg^{2+}, Ca^{2+}, Ba^{2+}, and Mn^{2+} at either 10^{-4} M or 10^{-6} M was observed. However, the binding was abolished by Co^{2+}, Zn^{2+}, and Hg^{2+} at 10^{-4} M (Table 4). These results suggest that the taurine receptor is probably not regulated by either divalent or monovalent cations. Furthermore, it also suggests that the taurine receptor probably contains essential free sulfhydryl groups since its binding activity is completely abolished by heavy metal ions (e.g., Hg^{2+}) at low concentrations.

Table 4. Effect of monovalent and divalent cations on ^{3}H-taurine binding.

Cations	Conc. (M)	Activity (%)
Na^+	10^{-3}	98
K^+	10^{-3}	113
Li^+	10^{-3}	93
NH_4^+	10^{-3}	96
Mg^{2+}	10^{-4}	87
	10^{-6}	93
Ca^{2+}	10^{-4}	90
	10^{-6}	100
Cu^{2+}	10^{-4}	115
	10^{-6}	101
Co^{2+}	10^{-4}	0
	10^{-6}	86
Zn^{2+}	10^{-4}	0
	10^{-6}	100
Ba^{2+}	10^{-4}	94
	10^{-6}	105
Mn^{2+}	10^{-4}	94
	10^{-6}	104
Hg^{2+}	10^{-4}	0
	10^{-6}	112

Effect of Nucleotides

Among the nucleotides tested, ATP is the most potent (inhibiting 53% at 1 mM) followed by GTP and ADP (inhibiting 28% at 1 mM). GDP, cAMP and cGMP do not have a significant effect (<17% inhibition) on taurine receptor binding at 1 mM (Table 5).

SUMMARY

A new procedure for the preparation of the taurine receptor from mammalian brain is described. The taurine receptor thus prepared shows a K_d of 92 nM, B_{max} of 6.0 pmol/mg protein and Hill coefficient of 0.90 suggesting a single site model for the binding of ^{3}H-taurine to the receptor. The binding of ^{3}H-taurine to the receptor is

Table 5. Effect of nucleotides on ^{3}H-taurine binding.

Nucleotides	Conc. (M)	Inhibition (%)
ATP	10^{-3}	53
	10^{-5}	12
ADP	10^{-3}	28
	10^{-5}	0
cAMP	10^{-3}	9
	10^{-5}	2
GTP	10^{-3}	28
	10^{-5}	13
GDP	10^{-3}	10
	10^{-5}	3
cGMP	10^{-3}	17
	10^{-5}	3

highly specific and is not affected by agonists and antagonists of other receptors such as glutamate, quisqualic acid, kainate and NMDA for the glutamate receptor; glycine and strychnine for the glycine receptor; FNZP for the benzodiazepine receptor; picrotoxin and bicuculline for the GABA$_B$ receptor. However, analogues of taurine (e.g., homotaurine and hypotaurine) are potent inhibitors inhibiting more than 50% of ^{3}H-taurine binding at $0.1\,\mu$M. Taurine receptor binding is not significantly affected by monovalent cations (e.g., Na$^+$, K$^+$, Li$^+$ and NH$_4^+$) at 1 mM or divalent cations (e.g., Mg^{2+}, Ca^{2+}, Ba^{2+} and Mn^{2+}) at 0.1 mM. However, the binding was completely abolished by Co^{2+}, Zn^{2+} and Hg^{2+} at 0.1 mM, suggesting the presence of free sulfhydryl groups near or at the ligand binding site. Among the amino acids tested, cysteic acid was the most potent inhibitor, followed by ß-alanine, valine, tyrosine and cysteine inhibiting ^{3}H-taurine to an extent of 84, 66, 63, 62, and 58% of 1 mM, respectively. Nucleotides and second messengers (e.g., ATP, ADP, cAMP, GTP, cGMP and diacyl glycerol) do not inhibit ^{3}H-taurine binding significantly at 0.1 mM. From above studies, it seems that the taurine receptor is not up- or down-regulated by ions or second messengers at the taurine binding site. Whether the taurine receptor is coupled to a G-protein mediated second messenger system is currently under investigation.

ACKNOWLEDGMENT

This study was supported in part from grants NS20978 (NIH) and BNS-8820581 (NSF) USA and a grant from the National Science Council, Taiwan, ROC. The expert typing of the manuscript by Jan Elder and Judy Wiglesworth is gratefully acknowledged.

REFERENCES

1. J.A. Sturman, R.C. Moretz, J.H. French, and H.M. Wisniewski, Postnatal taurine deficiency in the kitten results in a persistence of the cerebellar external granule cell layer: correction by taurine feeding, *J. Neuro. Res.* 13:521 (1985).
2. J.A. Sturman, A.D. Gargano, J.M. Messing, and H. Imaki, Feline maternal taurine deficiency: effect on mother and offspring, *J. Nutr.* 116:655 (1986).
3. H. Pasantes-Morales and C. Cruz, Taurine and hypotaurine inhibit light-induced lipid peroxidation and protect rod outer segment structures, *Brain Res.* 330:154 (1985).

4. J. Moran, P. Salazar, and H. Pasantes-Morales, Effect of tocopherol and taurine on membrane fluidity of retinal rod outer segments, *Experimental Eye Research* 45:769 (1988).

5. J.W. Lazarewicz, K. Noremberg, A. Lehmann, and A. Hamberger, Effects of taurine on calcium binding and accumulation in rabbit hippocampal and cortical synaptosomes, *Neurochem. Int.* 7:421 (1985).

6. J.B. Lombardini, Effects of taurine on calcium ion uptake and protein phosphorylation in rat retinal membrane preparations, *J. Neurochem.* 45:268 (1985).

7. J.M. Solis, A.S. Herranz, O. Herreras, J. Lerma, and R.M. Del Rio, Does taurine act as an osmoregulatory substance in the rat brain, *Neurosci. Lett.* 91:53 (1988).

8. J.V. Wade, J.P. Olson, F.E. Samson, S.R. Nelson, and T.L. Pazdernik, A possible role for taurine in osmoregulation within the brain, *J. Neurochem.* 51:740 (1988).

9. K. Okamoto, H. Kimura, and Y. Sakai, Evidence for taurine as an inhibitory neurotransmitter in cerebellar stellate interneurons: Selective antagonism by TAG (6-aminomethyl-3-methyl-4H,1,2,4-benzothiadiazine-1,1-dioxide), *Brain Res.* 265(1):163 (1983).

10. C.-T. Lin, Y.Y.T. Su, G.-X. Song, and J.-Y. Wu, Is taurine a neurotransmitter in rabbit retina?, *Brain Res.* 337:293 (1985).

11. T.C. Taber, C.-T. Lin, G.-X. Song, R.H. Thalman, and J.-Y. Wu, Taurine in the rat hippocampus-localization and postsynaptic action, *Brain Res.* 386:113 (1986).

12. J.-Y. Wu, C. Liao, C.-J. Lin, Y.H. Lee, J.-Y. Ho, and W.H. Tsai, Taurine receptors in the mammalian brain, *Prog. Clin. Biol. Res.* 351:147 (1990).

13. E.C. Kulakowski, J. Maturo, and S.W. Schaffer, The identification of taurine receptors from rat heart sarcolemma, *Biochem. Biophys. Res. Comm.* 80(4):936 (1978).

14. E.C. Kulakowski and S.W. Schaffer, Partial purification of a high affinity taurine binding protein by affinity chromatography, *Biochem. Pharmacol.* 32:753 (1983).

15. E. Kumpulainen, Purification and characterization of taurine-binding glycoprotein in calf brain, *in* "Metabolism and Function of Brain Amino Acids and Proteins in Health and Disease", pp. 129-134, P. Lahdesmaki and S.S. Oja, eds., University of Oulu, Oulu (1980).

16. A.C. Foster and G.E. Fagg, Taking apart NMDA receptors, *Nature* 329:395 (1987).

17. B.I. Kanner and S. Schuldiner, Mechanism of transport and storage of neurotransmitters, *CRC Crit. Rev. Biochem.* 22:1 (1987).

EVOKED ENDOGENOUS TAURINE RELEASE FROM
CULTURED CEREBELLAR NEURONS

Gary R. Dutton[1] and Keith L. Rogers[2]

Department of Pharmacology[1]
Department of Psychiatry[2]
University of Iowa College of Medicine
Iowa City, Iowa 52242

INTRODUCTION

The functional role(s) of taurine in the CNS has not yet been determined unambiguously. However, its absence is known to have a negative impact on brain development, and it has also been implicated in osmotic regulation, neuromodulation and neurotransmission (Huxtable, 1989). Many of these studies have utilized a variety of *in vitro* preparations from which the release or efflux of preloaded radiolabeled taurine has been monitored in response to a variety of stimuli. The focus of this chapter is on work done in this laboratory following HPLC analysis of the excitatory amino acid agonist evoked release of endogenous taurine from cultured cerebellar neurons containing a mixture of approximately 90% excitatory granule cells and 5-7% inhibitory (GABAergic) interneurons.

In order to approach the question of the neuronal origin of the stimulus-evoked release of endogenous taurine from these mixed cell cultures, we recently completed a study involving the use of kainic acid's (KA) selective neurotoxicity to remove the inhibitory neurons from these cultures (Simmons and Dutton, in press). The data demonstrate that following such treatment an approximate 50-75% permanent decrease in taurine release is produced in response to isosmotically elevating K^+. Thus, it is our current bias that, while all cells in these cultures probably take up taurine from the serum in which they are grown, there is a disproportionately larger K^+-releasable pool of this amino acid in the inhibitory interneurons compared to the granule neurons.

In earlier studies we and others demonstrated that there is a dose-dependent efflux of taurine from neuronal cultures following the isosmotic elevation of K^+ (Philibert et al., 1989). The general characteristics of this release are a delayed onset and slow rise to maximum following stimulation compared to the known transmitters glutamate and GABA (Philibert and Dutton, 1989; Schousboe et al., 1990). In addition, the per cent increase above basal release levels is around 5-fold compared to a 10 to 30-fold increase for glutamate and GABA, when stimulated under identical conditions (Rogers et al., 1991).

The Ca^{++}-dependency of K^{+}-evoked taurine efflux from these neuronal cultures has been reported to be partly dependent (Philibert et al., 1989), independent (Schousboe and Pasantes-Morales, 1989) or sensitive to its absence in that Ca^{++}-free conditions result in a marked increase in basal taurine release levels above which no further increase is seen (Rogers et al., 1991). This has lead to further study of the role of osmolarity changes in inducing taurine release, since it is known that KCl causes increases in intracellular volume due to swelling. Thus, it has been established recently that taurine is likely to play a role as an osmotically active substance in these neurons (Schousboe et al., 1990), since these investigators showed that K^{+}-stimulated taurine release was not caused by the K^{+} depolarization, but was Cl^{-}-dependent, and directly related to cell volume changes. Furthermore, they demonstrated that decreasing osmolarity resulted in an ever increasing release of taurine from these cells.

While it has now been established that changes in cell volume are directly related to taurine release, it is not at all clear that this is the only mechanism by which taurine is released from neurons. Therefore, we evaluated other conditions which might produce endogenous taurine efflux from these cultured neurons. For example, depolarization using the Na^{+}-channel agonist veratridine elevated efflux levels greater than 300% above basal, and was inhibitable by tetrodotoxin (Philibert et al., 1989). The experiments described in this chapter represent a continuation of these investigations, and specifically describe results obtained when cultured cerebellar neurons are exposed to excitatory amino acid agonists and their antagonists.

METHODS

Cerebellar Cultures

Primary, dissociated cerebellar neuronal cultures were prepared and maintained by the method of Dutton, 1990. Briefly, cerebella were removed from 7-9 day old rats and cells isolated by trypsinization, trituration and centrifugation, and grown on poly-D-lysine coated 35 mm diameter 6-well plastic culture dishes. The culture medium consisted of Eagles MEM supplemented with 10% fetal calf serum and 2.5% chick embryo extract. Proliferating cells were inhibited after 1 day in culture by the addition of fluorodeoxyuridine. Cells were maintained at 37°C in a humidified atmosphere of 6% CO$_2$ until use after 7-9 days in culture.

Static Stimulation Paradigm

Cultures were washed 3 times for 1 min each at 37°C with 3 mls of HEPES-containing buffer, given three 10 min incubation periods in 2 mls of the standard buffer (3rd incubation=pre-stimulation sample), then 10 min incubation in buffer containing the appropriate stimulus substance (stimulation sample), followed by a final 10 min incubation in standard washing buffer (post-stimulation sample) (Rogers et al., 1991). Aliquots of pre-stimulation, stimulation and post-stimulation (not shown) samples were collected for HPLC analysis as previously described (Rogers et al., 1987).

RESULTS

The first experiments were designed to test the dose-response of effects of KA on the efflux of endogenous taurine from neuronal cultures after a 10 min exposure over the concentration range of 1-1000 μM, and to determine the Ca^{++}-dependence of this phenomenon (Figure 1). In the presence of Ca^{++} it can be seen (upper panel,

Figure 1) that there was a clear dose-dependent increase in taurine produced by KA, which resulted in maximal increases of approximately 6-fold. However, in the absence of Ca^{++} (lower panel, Figure 1), the basal levels of taurine release were seen to be increased by approximately 4-fold, above which no further release could be seen in the presence of increasing amounts of KA, with the exception of the highest concentration used where an approximate doubling in Ca^{++}-independent release was seen.

In order to determine if the KA-evoked taurine release was mediated through its receptor, the non-NMDA antagonist 6-cyano-7-nitroquinoxaline-2,3-dione (CNQX) was tested over the concentration range of 1-100 μM after stimulation of the cultures with 200 μM KA. The selectivity of this effect was further tested using the competitive NMDA receptor antagonist 3-(($\pm$)-2-carboxypiperazine-4-yl)-propyl-1-phosphonic acid (CPP, 10 μM) and non-competitive antagonist ($\pm$)-5-methyl-10,11-dihydro-5H-dibenzo[a,d]cyclohepten-5,10-imine maleate (MK-801, 0.5 μM). Figure 2 shows that there was a dose-dependent decrease in taurine release produced by

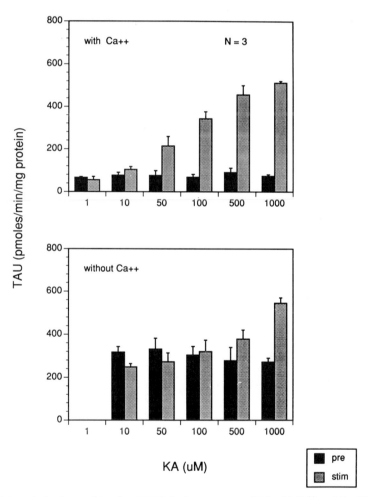

Figure 1. KA-evoked release of taurine (TAU) in the presence of 2.5 mM Ca^{++} and 1 mM Mg^{++} (top panel), or 0 Ca^{++} and 10 mM Mg^{++} (lower panel). Solid bars are pre-stimulation (pre) values and grey bars are stimulated (stim) values. N=number of independent experiments. These symbols are used throughout the following figures.

CNQX in the presence of 200 μM KA stimulation. However, no effect was seen with either CPP or MK-801, suggesting that this phenomenon was indeed mediated through KA receptors.

The effects on neuronal taurine release of increasing doses of the agonist N-methyl-D-aspartate (NMDA) over the range 1-100 μM under Mg^{++}-free conditions are shown in Figure 3. There was a concentration-dependent increase in endogenous taurine release from these cells, and the stimulatory effect of 100 μM NMDA was completely inhibited by 0.5 μM MK-801, suggesting that NMDA also produced taurine efflux by a receptor-mediated mechanism.

In contrast, neither the agonist quisqualate (QA) over the concentration range 0.01-100 μM (Figure 4) nor RS-α-amino-3-hydroxy-5-methyl-4-isoxazolopropionate (AMPA) over the concentration range 0.01-100 μM elicited taurine release above basal levels (Figure 5).

In order to determine if the KA-stimulated release of taurine might be influenced by the addition of other non-NMDA receptor agonists, increasing amounts of QA over the concentration range 1-100 μM were added to test its effect on the amount of taurine released by 100 μM KA (Figure 6). As is clearly demonstrated here, there is a dose-dependent decrease in KA-stimulus evoked taurine release with increasing concentrations of QA.

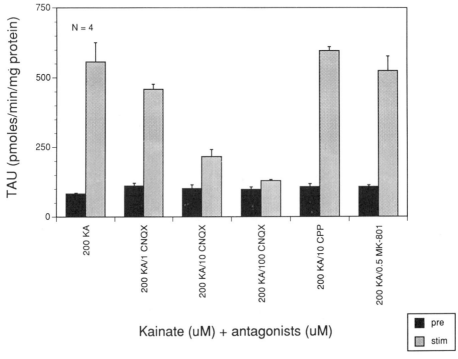

Figure 2. KA (200 μM)-stimulated release of taurine (TAU) from cultured cerebellar neurons in the presence of CNQX (1-100 μM), CPP (10 μM) and MK-801 (0.5 μM). For symbol definitions see Figure 1.

DISCUSSION

The main question at which the above work is aimed is, what can the characteristics of the stimulus-evoked release of taurine tell us about this amino acid's function in the brain. Much of the evidence previously reported is consistent with a

272

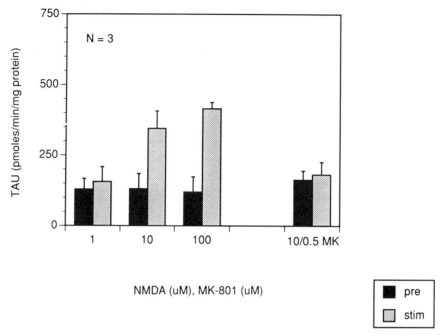

Figure 3. Dose-dependent effect of NMDA (1-100 μM) on evoked endogenous taurine (TAU) release from cultured cerebellar neurons performed in the absence of Mg^{++}. Far right - bars demonstrate the effect of combining 10 μM NMDA with 0.5 mM MK-801. For symbol definitions see Figure 1.

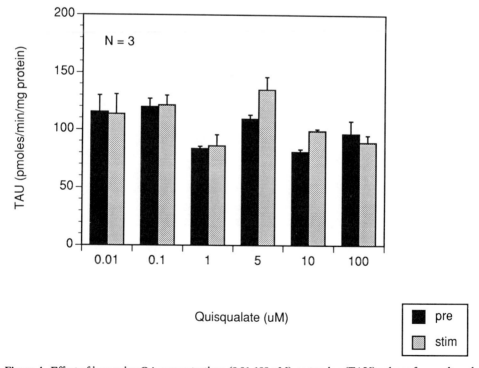

Figure 4. Effect of increasing QA concentrations (0.01-100 μM) on taurine (TAU) release from cultured cerebellar neurons. For symbol definitions see Figure 1.

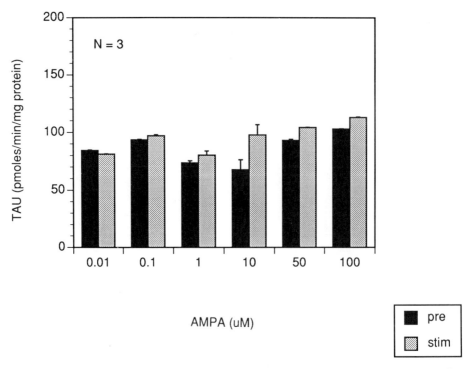

Figure 5. Effect of AMPA over the concentration range 0.01-100 μM on taurine (TAU) release from cultured cerebellar neurons. For symbol definitions see Figure 1.

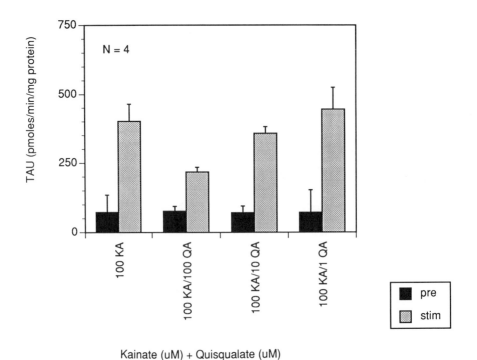

Kainate (uM) + Quisqualate (uM)

Figure 6. The effect of increasing QA concentrations (1-100 μM) on 100 μM KA-evoked release of endogenous taurine (TAU) from cultured cerebellar neurons. For symbol definitions see Figure 1.

role for taurine in the osmoregulatory mechanisms of both neurons and glial cells in response to swelling phenomena. However, an osmoregulatory role for taurine in cultured neurons is probably not its only function. This speculation is, in part, derived from our earlier experiments involving the evoked release of taurine from cultured cerebellar neurons seen in the presence of the voltage-regulated Na^+-channel agonist veratridine (Philibert et al., 1989). Veratridine elicits a dose-related increase in taurine efflux from these cells over the concentration range 0.5-50 μM (Miyazaki and Dutton, unpublished observations), which is Ca^{++}-independent at high concentrations (25-50 μM), but Ca^{++}-dependent at low concentrations (5 μM). This release is inhibited by tetrodotoxin at all veratridine concentrations. Thus, in the lower concentration range (~1-15 μM), depolarization induced Ca^{++}-dependent taurine release may require activation of voltage-regulated Na^+-channels, but at higher concentrations of veratridine (≥ 25 μM) a transport reversal process may underlie taurine efflux.

In the studies cited above using excitatory amino acid agonists and antagonists, it is clear that both KA and NMDA evoke the release of endogenous taurine from these cultured neurons through a receptor-mediated mechanism. This fact is further substantiated by the observation that QA may act by competing with KA at its receptor in a dose-dependent fashion, an observation made by other workers looking at cGMP formation using a similar culture system (Gallo et al., 1989 and 1990). Even though these responses appear to be receptor-mediated, a clear Ca^{++}-dependence for taurine release could not be established. This question is further complicated by our earlier reported observations that, using a perfusion (dynamic) stimulation paradigm, KA-evoked taurine release was seen to be Ca^{++}-dependent over the range 200-1000 μM (Rogers et al., 1990). Thus, even though these conflicting results may be explained as having been influenced by the differences in experimental methods employed (static vs. dynamic [perfusion] stimulation), the Ca^{++}-dependency question remains open, and the case for presynaptic vesicular taurine release remains unsubstantiated. Furthermore, it is possible that exposure to these excitatory agonists may have produced taurine release as a result of general damage to these neurons has previously been suggested (Magnusson et al., 1991).

In summary, the above results point to the ability of glutamate agonists to produce evoked endogenous taurine release from cultured cerebellar neurons via KA and NMDA receptors. The mechanism by which actual release from the cell is brought about, however, remains unclear. Questions also remain as to the role(s) of the released taurine in the central nervous system (CNS) function. In the absence of a clear Ca^{++}-dependence for both K^+- and KA-evoked taurine release, combined with the aberrant (compared to glutamate and GABA) temporal and quantitative dynamics of efflux following stimulation, a strong argument can be marshalled against a neurotransmitter role for this amino acid in the cerebellum. Determination of its possible function as a neuromodulator also awaits further study of the question of whether or not an adequate amount of taurine is released from these cells to produce changes in nerve cell activity.

ACKNOWLEDGMENTS

Work reported from this laboratory was supported by NIH grant NS 20632 (GRD). The authors wish to thank Kathy Andrews for assistance in preparation of this manuscript.

REFERENCES

Dutton, G.R., 1990, Isolation, culture and use of viable CNS perikarya, *in*: "Methods in Neurosciences," P.M., Conn, ed., Academic Press, San Diego.

Gallo, V., Biovannini, C., and Levi, G., 1989, Quisqualic acid modulates kainate responses in cultured cerebellar granule cells, *J. Neurochem.* 52:10.

Gallo, V., Giovannini, C., and Levi, G., 1990, Modulation of non-*N*-methyl-D-aspartate receptors in cultured cerebellar granule cells, *J. Neurochem.* 54:1619.

Huxtable, R.J., 1989, Taurine in the central nervous system and the mammalian actions of taurine, *Prog. Neurobiol.* 32:471.

Magnusson, K.R., Koerner, J.F., Larson, A.A., Smullin, D.H., Skilling, S.R., and Beitz, A.J., 1991, NMDA-, kainate- and quisqualate-stimulated release of taurine from electrophysiologically monitored rat hippocampal slices, *Brain Research* 549:1.

Philibert, R.A., Rogers, K.L., and Dutton, G.R., 1989, Stimulus-coupled taurine efflux from cerebellar neuronal cultures: On the roles of Ca^{++} and Na^{+}, *J. Neurosci. Res* 22:167.

Philibert, R.A., and Dutton, G.R., 1989, Dihydropyridines modulate K^+-evoked amino acid and adenosine release from cerebellar neuronal cultures, *Neurosci. Lett.* 102:97.

Rogers, K.L., Philibert, R.A., Allen, A.J., Molitor, J., Wilson, E.J., and Dutton, G.R., 1987, HPLC analysis of putative amino acid neurotransmitters released from primary cerebellar cultures, *J. Neurosci. Methods* 22:173.

Rogers, K.L., Philibert, R.A., and Dutton, G.R., 1990, Glutamate receptor agonists cause efflux of endogenous neuroactive amino acids from cerebellar neurons in culture, *Eur. J. Pharmacol.* 177:195.

Rogers, K.L., Philibert, R.A., and Dutton, G.R., 1991, K^+-stimulated amino acid release from cultured cerebellar neurons: Comparison of static and dynamic stimulation paradigms, *Neurochem. Res.* 16:899.

Schousboe, A., and Pasantes-Morales, H., 1989, Potassium-stimulated release of [^{3}H]taurine from cultured GABAergic and glutamate neurons, *J. Neurochem.* 53:1309.

Schousboe, A., Olea, R.S. and Pasantes-Morales, H., 1990, Depolarization induced neuronal release of taurine in relation to synaptic transmission: Comparison with GABA and glutamate, *in*: "Taurine: Functional Neurochemistry, Physiology, and Cardiology," H. Pasantes-Morales, D.L. Martin, W. Shain, and R. Martin del Rio, eds., Wiley-Liss, Inc., New York.

Schousboe, A., Morán, J., Pasantes-Morales, H., 1990, Potassium-stimulated release of taurine from cultured cerebellar granule neurons is associated with cell swelling, *J. Neurosci. Res.* 27:71.

Simmons, M.L., and Dutton, G.R., Chronic kainate treatment decreases K^+-stimulated release of endogenous amino acids from cultured cerebellar neurons, *J. Neurosci. Res.* in press.

EFFECT OF HEPES ON THE UPTAKE OF TAURINE BY CULTURED NERVOUS CELLS

G. Rebel, P.L. Lleu, V. Petegnief, M. Frauli-Meischner,
P. Guerin, and I.H. Lelong

Centre de Neurochimie
67084 Strasbourg, France

The ß amino acid taurine has many functions. Some functions involve many tissues of vertebrates such as regulation of cellular osmolarity[1], while other functions are restricted to a particular cell type such as biliary acid detoxification by hepatocytes[2]. Uptake and release of taurine has been studied in a variety of cell types. Uptake is linked to the existence of a transport mechanism specific for ß amino acids. This transporter recognizes taurine, ß-alanine[3,4] and probably also hypotaurine[5].

Among the numerous analogs of taurine which have been tested as inhibitors of taurine uptake, GES (guanidinoethanesulfonic acid) is the most potent inhibitor yet to be found[6]. In this context, Hepes (N hydroxyethyl piperazine N'ethanesulfonic acid) is the buffer system commonly used in incubation media for studying taurine uptake. Hepes, alone or in combination with bicarbonate, is also frequently used to buffer cell culture media. The structure of Hepes shows that this compound can be considered a N substituted taurine. Therefore, we have tested the effects of this buffer on the uptake of taurine in cultured glial cells. Glial cells have been chosen as they have a highly efficient uptake for taurine allowing millimolar quantities of taurine accumulation.

TAURINE UPTAKE DETERMINATION

Cells cultured in Petri dishes as described below, were preincubated for 5 min in either a Krebs-Ringer phosphate buffer supplemented with 20 mM Hepes (KRH) or without Hepes (KR) (final pH 7.4 adjusted with KOH). Non-radioactive taurine (100 μl of 40 μM) containing 0.4 μCi [³H]taurine was added to each dish. Cells were incubated for 10 min. The medium was then removed and the cells were washed with cold saline. The cells were frozen, homogenized in 0.1 N NaOH and aliquots were taken for activity and protein[7] determination.

EFFECT OF HEPES ON GLIAL CELL CULTURES

Glial Cell Cultures

Primary glial cell cultures were established by mechanical dissociation of brain hemispheres from 15 day old embryonic chickens or new born rats[8]. Cells were grown in 35 mm Petri dishes with Dulbecco modified Eagle medium (DMEM 74-1600 from Gibco) supplemented with selected heat inactivated fetal calf serum. Figure 1 depicts the experimental culture conditions.

Effect of Hepes on Glial Cell Morphology

Cultures of rat glial cells in a Hepes buffered medium cannot be distinguished from their sister cultures grown in the same medium buffered with bicarbonate. By contrast, a high vacuolization is observed when chicken glial cells are cultured in a Hepes buffered medium. This is a fast phenomenon which is completely reversible when cultures are returned to a bicarbonate buffered medium.

Modulation of the Uptake by Hepes

Figure 2 shows that taurine uptake is notably decreased when glial cells are grown and incubated in media containing 20 mM Hepes. A similar inhibition is observed when the glial cells are grown in other culture media as well as in the serum free medium (SFM) of Kim et al.[9] (Figure 3). When cells are grown in a medium buffered with 24 mM bicarbonate and 10 mM Hepes, the inhibition of the taurine uptake becomes detectable only after a week[10].

The observed inhibition of taurine varies widely from one experiment to the other (Figure 4). However, Hepes similarly inhibits growing and resting glial cells (7 day and 30 day old cultures)[11].

Hepes affects taurine uptake through two different mechanisms as shown in Figure 5. Cultures of chicken glial cells grown for two days in DMEM-Hepes does not change the taurine uptake. However, incubation of the same cells in the presence of Hepes induces a 50 percent decrease of the uptake. By contrast, a similar inhibition of the uptake is observed in rat glial cells grown in DMEM-Hepes for two days or incubated in Krebs-Ringer medium containing Hepes.

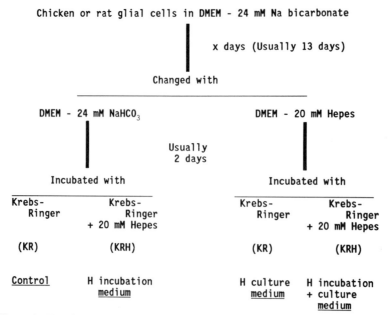

Figure 1. Experimental conditions used to test the effects of Hepes on cultured cells.

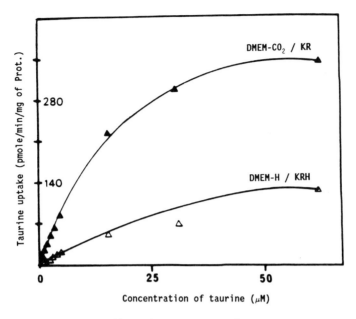

Effect of Hepes on rat glial cells.

Figure 2. Effect of Hepes on taurine uptake in rat glial cells. Cells are growth either in DMEM-CO$_2$ and incubated in KR or in DMEM-H and incubated in KRH.

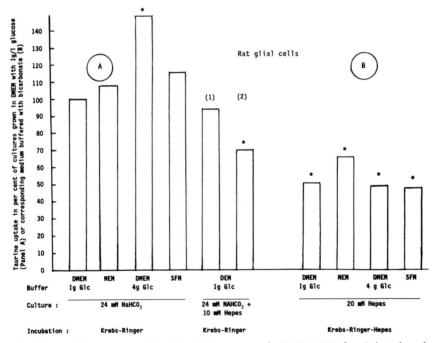

Figure 3. Effect of Hepes on rat glial cells. Cells are grown in DMEM-CO$_2$ for 13 days then changed to the indicated medium for 2 days. In the case of DMEM buffered with bicarbonate and Hepes, 13 day old cultures are grown for 2 days (column 1) or 7 days (column 2) in the medium. DMEM 1 Glc or 4 Glc, DMEM with either 1 or 4 g/l glucose. SFM, serum free medium. *$P<0.001$ compared to cells grown in DMEM 1 g GLc/bicarbonate medium. (Student's t-test).

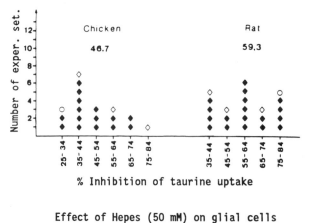

Effect of Hepes (50 mM) on glial cells

(DMEM - H/KRH)

o 7 day, ◆ 15 day, ◇ 30 day old cultures.

Figure 4. Variability of the effect of Hepes on taurine uptake. Cells were grown for 7, 15, or 30 days in DMEM-CO_2 and incubated in KR (control) or for 5, 13, and 28 days in DMEM-CO_2 followed by two days in DMEM-H and incubated in KRH. From Lleu and Rebel, 1989[11].

Figure 6 shows that the inhibition of taurine uptake linked to the presence of Hepes in the culture medium of rat cells increases with time and reaches a plateau after 7 days. This inhibition is slowly reversible when cells are returned to a DMEM-bicarbonate medium.

Supplementation of the Krebs-Ringer medium with increasing concentrations of Hepes induces an inhibition of taurine uptake which reaches a maximum around 10 mM in rat and chicken glial cells. Using a fixed concentration of the buffer, this inhibition increases linearly with time.

ß-Alanine and taurine are taken up by the same transport system. Hepes significantly inhibits ß-alanine uptake when it is present in the culture medium of rat glial cells. A weak inhibition of the ß-alanine uptake is observed when Hepes is added to the incubation medium of these cells. In the case of chicken cells, Hepes has no effect on ß-alanine uptake when present in the culture medium whereas it slightly decreases the uptake when present in the incubation medium. No matter what experimental conditions are utilized, Hepes does not affect the uptake of leucine (Figure 7).

EFFECT OF HEPES ON NEURONAL CELL CULTURES

Neuronal Cell Cultures

Neuronal cell cultures were established from 16 day-old rat embryo hemispheres by trypsin digestion[12]. The cells were plated on polylysine coated Petri dishes in DMEM supplemented with 20 percent heat inactivated selected fetal calf serum. The next day, cultures were changed to a synthetic medium (Bottenstein and Sato N2 medium[13]) supplemented with 5 g/l glucose and 15 mM KCl[12]. The culture media were buffered with 24 mM sodium bicarbonate or 20 mM Hepes.

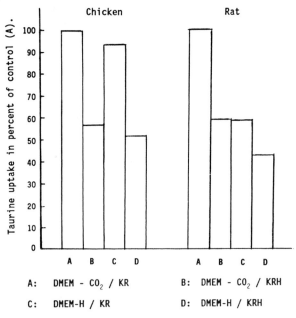

A: DMEM - CO₂ / KR
C: DMEM-H / KR
B: DMEM - CO₂ / KRH
D: DMEM-H / KRH

Each column is the mean of 16 values.

Figure 5. Effect of Hepes present in the incubation medium or culture medium. Cells are grown for 13 days in DMEM-CO$_2$ and the last two days as indicated in the figure.

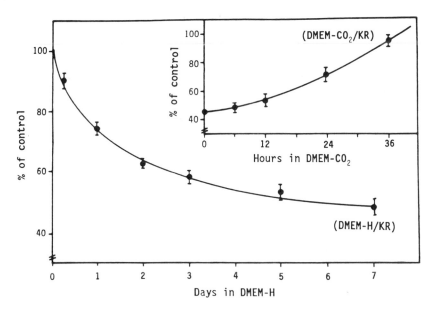

Effect of Hepes (20 mM) in the culture medium of rat glial cells

Figure 6. Effect of Hepes in the culture medium. Fifteen day old cultures are used. They are grown in DMEM-CO$_2$ (control) or DMEM-H for the last 7, 5, 3 days. Cells are incubated in KR. Insert shows the reversibility of the inhibitory effect of Hepes in cultures grown for two days in DMEM-H and returned to DMEM-CO$_2$ and incubated in KR. From Lleu and Rebel, 1989[11].

Modulation of the Uptake by Hepes

To avoid an overgrowth of glial cells, rat neuronal cells were usually cultured in the N2 medium (DMEM with insulin, transferrin, putrescine, selenium, and progesterone). Neuronal cells do not survive for a long time in N2 medium buffered with 20 mM Hepes, most of the cells degenerating quickly. Cultures are largely improved by supplementing N2 medium with glucose and KCl. However, even in these conditions, a slight inhibition of neuronal cell growth and maturation, as seen by quantitating the amount of DNA and protein was still observed in the presence of Hepes. However, since the quantity of protein per cell is not similar when the cells are grown in bicarbonate and Hepes media, protein cannot be used as reference. In these last culture conditions, Hepes added to the incubation or to the culture medium, has no effect on the taurine uptake expressed per quantity of DNA (Table 1).

Taurine Uptake in Chicken Neuronal Cells

We have looked for other culture conditions that could better support neuronal cells growth in a Hepes buffered medium. This would avoid the use of a KCl rich medium, since high concentrations of potassium many interfere with taurine transport.

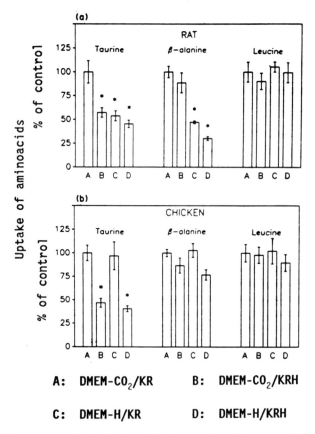

A: DMEM-CO_2/KR   B: DMEM-CO_2/KRH

C: DMEM-H/KR   D: DMEM-H/KRH

Figure 7. Effect of Hepes on taurine, ß-alanine and leucine uptake. Culture conditions are the same as in Figure 5.

Initially, we used chicken neuronal cultures in which the growth of glial cells remains very low. Cultures were established from 8 day old chicken embryo hemispheres and grown on polyethyleneimine (PEI) coated plates[14] in DMEM supplemented with 20 percent fetal calf serum. The next day, cultures were maintained in the same medium or changed to N2 medium. Using cell morphology, protein/DNA ratio, and ganglioside profiles as criteria, our results demonstrate that neuronal cells maturate faster on PEI coated plates than on polylysine coated plates[15]. Moreover, a better maturation is obtained when cells are maintained in a serum supplemented medium, compared to sister cultures in N2 medium. Table 2 shows that during the developmental phase the uptake of taurine is greater in cells grown on polyethyleneimine than on polylysine coated plates but is similar in the two mature cultures.

Table 1. Taurine uptake in rat neuronal cells grown for 7 days in modified N2-CO_2 medium and incubated in KR (control) and KRH (Hepes incubation media) or changed either on day 1 or 4 to modified N2-Hepes medium and incubated in KR (Hepes culture medium).

Taurine uptake expressed as	Control	Hepes incubation medium	Hepes culture medium for the last	
			6 days	3 days
pmole/min/mg protein	151 ± 19	136 ± 16		
	88 ± 6		57 ± 8	75 ± 2
nmole/min/mg DNA	1.25 ± 0.04		1.40 ± 0.09	1.16 ± 0.04

Table 2. Taurine uptake in chicken neuronal cells grown on polylysine or PEI coated plates.

Days in culture	Polylysine	PEI
2	22.9 [19.4 - 31.4]	47.3 [31.7 - 62.8]
4	38.3 [31.7 - 48.1]	88.0 [89.2 - 91.1]
7	215.7 [172.3 - 273.9]	193.3 [130.2 - 260.5]

Uptake is expressed as nmole/min/mg DNA. Numbers within the brackets are the extreme values.

GENERAL CONCLUSIONS

Hepes can be considered as a taurine derivative. Hepes inhibits taurine uptake in primary cultures of chicken and rat glial cells. This inhibition involves two different mechanisms. 1) A fast inhibition linked to the effect of Hepes on the taurine transport binding site which is specific for taurine since the uptake of ß-alanine is only weakly affected. These data demonstrate that Hepes is not a competitive inhibitor of the taurine uptake. The specificity of Hepes towards taurine could be explained by the existence of different domains in the ß amino acid transport binding site. Only one domain, specific for taurine uptake, is modified by Hepes (Figure 8). Another possibility is that Hepes interferes more with the sulfonate than

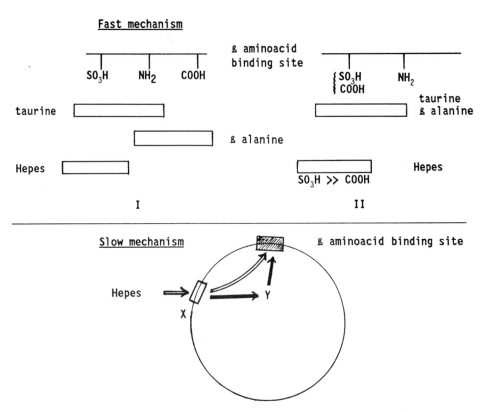

Figure 8. Possible mechanisms explaining the effects of Hepes on glial cells.

carboxylate moiety for binding to the transport site. 2) A slow inhibition, related to the presence of Hepes in the culture medium, which similarly affects both taurine and ß-alanine uptake but not α amino acid uptake. As an explanation, we hypothesize that Hepes modifies either directly or indirectly the concentration of an intracellular modulator of the taurine transporter (Figure 8). This slow mechanism is apparently absent from the chicken glial cells.

Hepes does not affect the uptake of taurine by cultured neuronal cells. However, in our culture conditions, Hepes slightly alters the growth and maturation of the neuronal cells. This could explain the insensitivity of these cells towards this buffer.

REFERENCES

1. N.M. van Gelder, A central mechanism of action for taurine: osmoregulation, bivalent cations, and excitation threshold, *Neurochem. Res.* 8:687-699 (1983).
2. C.E. Wright, H.H. Tallan, Y. Lin, and G.E. Gaull, Taurine: biological update, *Ann. Rev. Biochem.* 55:427-453 (1986).
3. D.L. Martin and W. Shain, High affinity transport of taurine and ß-alanine and low affinity transport of γ-aminobutyric acid by a single transport system in cultured glioma cells, *J. Biol. Chem.* 254:7076-7084 (1979).
4. O.M. Larsson, R. Griffiths, I.C. Allen, and A. Schousboe, Mutual inhibition kinetics analysis of γ-aminobutyric acid, taurine, and ß-alanine high-affinity transport into neurons and astrocytes: evidence for similarity between the taurine and ß-alanine carriers in both cell types, *J. Neurochem.* 47:426-432 (1986).
5. I. Holopainen and P. Kontro, Taurine and hypotaurine transport by a single system in cultured neuroblastoma cells, *A. Physiol. Scand.* 122:381-386 (1984).

6. R.J. Huxtable, H.E. Laird, and S.E. Lippincott, The transport of taurine in the heart and the rapid depletion of tissue taurine content by guanidinoethyl sulfonate, *J. Pharml. Exp. Therap.* 211:465-471 (1979).

7. O.H. Lowry, N.J. Rosebrough, A.L. Farr, and R.J. Randall, Protein measurement with the Folin phenol reagent, *J. Biol. Chem.* 193:265-275 (1951).

8. J. Booher and M. Sensenbrenner, Growth and cultivation of dissociated neurons and glial cells from embryonic chick, rat and human brain in flask cultures, *Neurobiology* 2:97-105 (1972).

9. S. Kim, J. Stern, M. Kim, and D. Pleasure, Culture of purified rat astrocytes in serum-free medium supplemented with mitogen, *Brain Res.* 274:79-86 (1983).

10. Lleu, P.L. and Rebel, G., 1990, Effect of HEPES on the Na^+, Cl^--dependent uptake of taurine, and ß-alanine by cultured glial cells. Modulation by composition and osmolarity of medium, *Neuropharmacol.* 29:719-725.

11. Lleu, P.L. and Rebel, G., 1989, Effect of HEPES on the taurine uptake by cultured glial cells, *J. Neurosci. Res.* 23:78-86.

12. D. Di Scala-Guenot, M.T. Strosser, M.J. Freund-Mercier and Ph. Richard, Characterization of oxytocin-binding sites in primary rat brain cell cultures, *Brain Res.* 524:10-16 (1990).

13. J.E. Bottenstein and G.H. Sato, Growth of a rat neuroblastoma cell line in serum-free supplemented medium, *Proc. Nat. Acad. Sci. US*, 76:514-517 (1979).

14. U.T. Rüegg and F. Hefti, Growth of dissociated neurons in culture dishes coated with synthetic polymeric amines, *Neurosci. Lett.* 49:319-324 (1984).

15. Lelong, I.H., Petegnief, V. and Rebel, G., 1991, Neuronal cells mature faster on polyethyleneimine coated plates than on polylysine plates, *J. Neurosci. Res.* (submitted).

THE TROPHIC ROLE OF TAURINE IN THE RETINA. A POSSIBLE MECHANISM OF ACTION

L. Lima, P. Matus and B. Drujan

Laboratorio de Neuroquimica
Instituto Venezolano de
Investigaciones Cientificas
Caracas, Venezuela

INTRODUCTION

Chronic taurine deficiency produces an irreversible degeneration in the retina of kittens (Hayes, et al., 1975; Hayes, 1985; Sturman et al., 1985a,b). It has also been reported that the lack of taurine results in loss of optic nerve axons in rats (Lake, 1988). An interesting finding in Rhesus monkeys is that animals fed a formula free of taurine develop poor visual acuity, suggesting a role for taurine in the central retina, which is involved in normal acuity (Neuringer and Sturman, 1987). Lowered levels of this amino acid lead to functional and structural changes in the central nervous system of the cat (Schmidt et al., 1975), the monkey (Sturman et al., 1984), and the rat (Pasantes-Morales et al., 1983; Quesada et al., 1984). Taurine depletion induced by blockers of its uptake, such as guanidinoethylsulfonate and ß-alanine, results in severe disruption of retinal structure (Pasantes-Morales et al., 1983) and modifications in the electroretinogram (Lake, 1986).

The visual system of goldfish have been used as a model to study central nervous system regeneration (Landreth and Agranoff, 1979), and has been useful in testing the possible trophic role of taurine. However, the majority of the information in the literature to date has been previously obtained by deficit and not by the addition of taurine.

The nerve growth index (NGI), which is the product of fiber length and density, was evaluated in post-crush goldfish retinal explants in the presence of taurine. The amino acid increases and accelerates the emission of neurites in a bell-shaped dose-dependent manner (Lima et al., 1988). The slope of outgrowth is higher between 0 to 5 days than from 5 to 10 days in culture, even if the NGI is significantly greater with respect to post-crush cultured retina in the absence of the amino acid. We also reported that the effect of taurine was substrate-dependent (Lima et al., 1989a).

Cell proliferation in teleosts and amphibians occurs throughout life but results in the formation of new neurons only in the peripheral zone of the retina (Easter, 1983; Johns, 1977; Johns and Fernald, 1981). In addition, catecholamine- and serotonin-containing neurons are lower in the central region than in the periphery of

Taurine, Edited by J.B. Lombardini *et al.*
Plenum Press, New York, 1992

various fish retina (Negishi et al., 1982). The concentration of taurine is elevated in the periphery of goldfish and *Eugerres plumieri* retina with respect to the center (Lima et al., 1989b). In goldfish the concentration of γ-aminobutyric acid is also higher in the periphery (Lima et al., 1989b); however, in rat and rabbit these variations were not observed. Differences in the uptake of taurine were also obtained between center and periphery in goldfish retina (Lima et al., 1991a).

The mechanisms of action of taurine as a trophic agent in the retina was investigated by the evaluation of *in vitro* neuritic outgrowth in the presence of extracellular and intracellular calcium chelators, and by the influence of taurine on calcium fluxes into cultured retinal cells.

Procedure

The optic nerve of goldfish (*Carassius auratus*) was crushed with fine forceps 10-14 days prior to explantation. The retina was dissected and sectioned into squares of $500 \mu m^2$. The nutrient medium was Leibowitz, 20 mM HEPES, 10% fetal calf serum and 0.1 mg/ml of gentamicin. Fluorodeoxyuridine (2 mM) was added 48 h after plating. The NGI was calculated as the product of fiber length and density (Lima et al., 1988).

The effect of calcium chelators was evaluated in explants which were growing since the moment of plating in the absence or in the presence of 4 mM taurine. The extracellular chelator, ethyleneglycol-bis-(ß-amino-ethyl ether)N,N'-tetraacetic acid (EGTA), was dissolved in the medium and added at day 0 (final concentrations 0.2, 0.5, 1 and 2 mM). The intracellular chelator, 1,2-bis(o-aminophenoxy)ethane-N,N,N,N-tetraacetic acid (BAPTA), was dissolved in dimethyl sulfoxide (0.025 mM), which was also present in control explants, and added on day 0 (final concentrations 0.01, 0.05, 0.1, 0.25, 0.5, and 1 μM).

Retinal cells were isolated by digestion with trypsin (0.25%, viability measured with Trypan blue > 96%) at 25°C for 30 min in Locke solution. The isolated cells were washed and resuspended (approximatively 200,000 cells/ml) in culture medium.

Cultured cells were detached by a mechanical procedure at 5 and 10 days in culture. Four groups of cells were analyzed for calcium influx: control (C), from intact retinas; control plus taurine (C+T), from intact retinas in the presence of taurine; lesioned (L), from post-crush retina; and lesioned plus taurine (L+T), from post-crush retina in the presence of taurine. The incubation medium for calcium influx was Ringer solution, $^{45}CaCl_2$ (21.2 Ci/mg, 0.1-0.2 μCi per tube), $CaCl_2$ in concentrations from 0.5-10 mM, and 450 μl of cell preparation. After a preincubation of 2 min calcium solution was added and the mixture further incubated for 3 min at 25°C. The reaction was terminated by the addition of cold Ringer solution and rapid filtration through glass fiber filters. The filters were dried and counted for radioactivity. Specific activity was calculated for each calcium preparation. Calcium flux was expressed in nmol/10^5 cells $\cdot$ min^{-1} (Lima et al., 1991b).

RESULTS

Effect of Calcium Chelators on Neuritic Outgrowth

Taurine stimulated the outgrowth from post-crush retinal explants (Fig. 1). EGTA 2 mM completely blocked the emission of neurites from explants in the absence and in the presence of taurine (Fig. 1 A,B). At 5 days in culture, up to 0.5 mM EGTA did not produce a significant decrease in the outgrowth from explants in the absence of taurine, but significantly reduced the trophic effect of the amino acid. By 10 days in culture EGTA blocked the outgrowth from explants with or without taurine in the medium. In addition, similar results were obtained with BAPTA, up

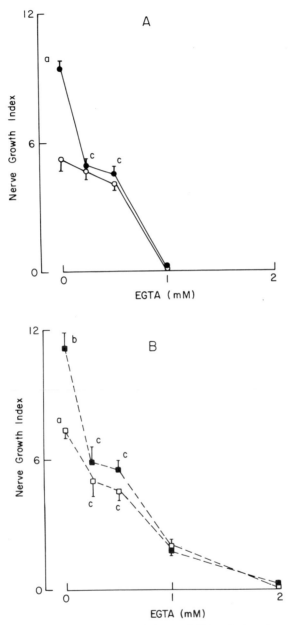

Figure 1. EGTA and BAPTA effect on outgrowth from goldfish retinal explants in the absence and in the presence of 4 mM taurine. Each value is mean ± S.E.M. (n = 11 - 27). (A,B) EGTA, (C,D) BAPTA, (o,□) (-)taurine, (●,■) (+)taurine, (——) 5 days in culture, (- - -) 10 days in culture. [a]P < 0.01 with respect to 5 days (-)taurine. [b]P < 0.025 with respect to 5 (-)taurine 10 days (-) or 10 days (+)taurine. [c]P < 0.001 with respect to corresponding condition at 0 chelator.

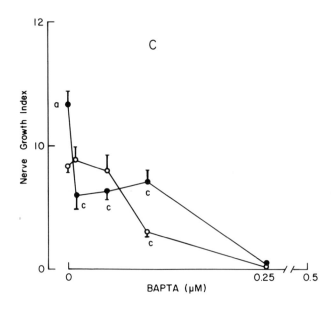

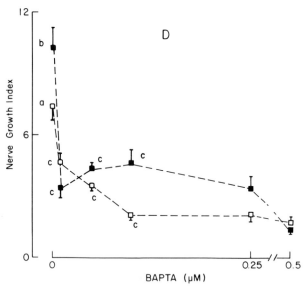

Figure 1. (continued).

to 0.05 μM, with significant effects on growth from explants in the absence of taurine at 5 days in culture and inhibition of the taurine effect (Fig. 1 C,D). The NGI in the presence of both chelators (Table 1) and taurine was reduced as compared with the explants in the presence of taurine and BAPTA (Fig. 1 C,D).

Table 1. Effect of EGTA and BAPTA on neuritic outgrowth from goldfish retinal explants in the presence of 4 mM taurine.

	Nerve Growth Index	
BAPTA (0.1 μM)	(-) EGTA	(+) EGTA (0.5 mM)
5 days in culture	4.4 ± 0.4	1.9 ± 0.1[a]
10 days in culture	4.6 ± 0.8	2.1 ± 0.2[b]

Each value is mean ± S.E.M (n = 11 - 20). [a]P < 0.001 with respect to (-) EGTA at 5 days. [b]P < 0.01 with respect to (-)EGTA at 10 days.

Calcium Flux into Cultured Retinal Cells

The influx of calcium was not significantly different in control, control in the presence of taurine, and lesioned retinas; however, the addition of the amino acid significantly increased the influx of calcium in post-crush retinal cells at 5 days in culture (Table 2). There was an increase in the influx of calcium between 5 and 10 days in culture with no significant differences among the four conditions analyzed for each day in culture (Table 3).

Table 2. Calcium influx into goldfish retinal cells cultured for 5 days.

Calcium Concentration (mM)	C	C + T	L	L + T
0.5	32.51 (14.96)	18.17 (5.21)	32.80 (5.22)	39.01 (2.88)
1.0	47.53 (11.99)	45.03 (13.57)	56.11 (9.58)	71.38 (17.73)
2.5	106.96 (32.64)	91.40 (22.92)	124.91 (24.89)	173.78[a] (17.84)
5.0	160.48 (36.24)	177.12 (45.18)	224.07 (15.49)	349.13[a] (21.27)
10.0	325.56 (111.50)	274.51 (56.17)	380.69 (31.33)	638.34[a] (21.71)

Each value is mean (S.E.M.), nmol·10^5 cells·min^{-1} (n = 4). C, intact retina and L, post-crush retina in the absence of taurine. C + T, intact retina and L + T, post-crush retina in the presence of 4 mM taurine. [a]P < 0.05 with respect to corresponding control.

Table 3. Calcium influx into goldfish retinal cells cultured for 10 days.

Calcium Concentration (mM)	C	C + T	L	L + T
0.5	23.55 (4.52)	31.89 (11.59)	29.42 (14.03)	24.64 (8.12)
1.0	38.23 (4.88)	37.58 (7.81)	93.63 (30.29)	41.09 (21.02)
2.5	150.63 (47.32)	115.56 (27.04)	174.31 (79.53)	130.73 (68.10)
5.0	247.25 (79.57)	230.99 (35.82)	300.25 (115.93)	221.82 (98.63)
10.0	586.17 (178.25)	338.09 (67.40)	559.48 (203.97)	539.92 (263.06)

Each value is mean (S.E.M.), $nmol \cdot 10^5$ cells$\cdot min^{-1}$ (n = 4). C, intact retina and L, post-crush retina in the absence of taurine. C + T, intact retina and L + T, post-crush retina in the presence of 4 mM taurine.

DISCUSSION

These results indicate that calcium is necessary for normal outgrowth of fibers, because certain concentrations of each chelator completely blocked the regeneration from post-crush retina. Low concentrations of EGTA decreased the outgrowth observed between 5 and 10 days in culture. The stimulatory effect of taurine on nervous outgrowth was blocked with 0.25 and 0.5 mM EGTA even if regeneration in control explants was still evident, indicating that the amino acid elevated the requirements of calcium from the extracellular space. BAPTA also inhibited the regeneration process at 5 days in culture, but at longer periods of incubation there was some stimulation of outgrowth. Lower concentrations of the chelator significantly impaired the increase of NGI produced by taurine, without reduction of outgrowth from control explants. The reduction of the intracellular calcium to a critical level produced by BAPTA might increase the influx of calcium and taurine could still elevate the outgrowth from explants when the culture is prolonged to 10 days. The reduction of the extracellular levels of calcium by the simultaneous addition of EGTA and BAPTA blocked the increase in NGI induced by taurine at 10 days in culture in the presence of the low concentration of BAPTA.

The growth cones of embryonic rat diencephalon (Connor, 1986) and molluscan preparations (Cohan et al., 1987; Connor et al., 1990) have a greater calcium concentration than the dormant growth cones. Moreover, it has been reported that the concentration of calcium in the growth cone of resting goldfish retinal ganglion cells is twofold higher than in the somata (Ishida et al., 1991). Also the increase in calcium fluxes has been reported to mediate the induction of expression of certain genes (Bartel et al., 1989; Sheng et al., 1990). In addition, a loss of calcium homeostasis may be responsible for neuronal degeneration in Alzheimer's disease (Ueda et al., 1990).

In the present work, it is also demonstrated that taurine increased calcium flux into explants from post-crush goldfish retina, but not from intact retina, suggesting that the amino acid elevates the influx of the ion only when regeneration is in process. It is well established that taurine stimulates ATP-dependent calcium flux at low concentration of the ion (Lombardini, 1983; 1985a,b; Pasantes-Morales and Ordóñez, 1982), and inhibits the ATP-dependent flux at high calcium concentration (Pasantes-Morales and Cruz, 1983). Taurine increases ATP-dependent calcium movement in retinal membrane homogenates and in a subcellular fraction enriched in synaptosomes (Lombardini, 1985a), and also stimulates the ATP-dependent influx of calcium into synaptosomal and mitochondrial preparations of rat retina (Lombardini, 1988).

The fact that the intracellular chelator has an effect on the regenerative action of taurine indicates that the levels of calcium in the cells are involved in the process. Moreover, the inhibitory effect of the extracellular chelator suggests that the influx of the cation is mainly responsible for the stimulation. Obviously, there are a great number of calcium channels and other routes for the maintenance of calcium intracellular levels (Kostyuk, 1989) that could be influenced by taurine.

There is a critical period of time in which the amino acid must be added to the medium to produce the increase in NGI, longer periods of incubation maintain the higher emission of neurites with no further stimulation (Lima et al., 1988). According to these observations, the effect of taurine on calcium influx is at 5 and not at 10 days in culture. Thus again, we demonstrate that there is a critical period for the regeneration of retinal cells *in vitro* in which taurine is effective.

ACKNOWLEDGEMENTS

This work was supported by the Grant S1-2018 from Consejo Nacional de Investigaciones Científicas y Tecnológicas, CONICIT.

REFERENCES

Bartel, D.P., Sheng, M., Lau, L.F., and Greenberg, M.E., 1989, Growth factors and membrane depolarization activate distinct programs of early response gene expression: dissociation of fos and jun induction, *Genes Dev.* 3:304-313.

Cohan, C.S., Connor, J.A., and Kater, S.B., 1987, Electrically and chemically mediated increases in intracellular calcium in neuronal growth cones, *J. Neurosci.* 7:3588-3599.

Connor, J.A., 1986, Digital imaging of free calcium changes and of spatial gradient in growing processes in single, mammalian central nervous system cells, *Proc. Natl. Acad. Sci.* USA 83:6179-6183.

Connor, J.A., Kater, S.B. Cohan, C., and Fink, L., 1990, Ca^{2+} dynamics in neuronal growth cones: regulation and changing patterns of Ca^{2+} entry, *Cell Calcium* 11:233-239.

Easter, S.S., 1983, Postnatal neurogenesis and changing connections, *Trends Neurosci.* 6:53-56.

Hayes, K.C., Carey, R., and Schmidt, S., 1975, Retinal degeneration associated with taurine deficiency in the cat, *Science* 188:949-951.

Hayes, K.C., 1985, Taurine requirement in primates, *Nutri. Rev.* 43:65-68.

Ishida, A.T., Binidokas, V.P., and Nuccitelli, R., 1991, Calcium ions levels in resting and depolarized goldfish retinal ganglion cell somata and growth cones, *J. Neurophysiol.* 65:968-979.

Johns, P.R., 1977, Growth of the adult goldfish. III. Source of the new retinal cells, *J. Comp. Neurol.* 176:343-358.

Johns, P.R. and Fernald, R.D., 1981, Genesis of rods in teleost fish retina, *Nature* 293:141-142.

Kostyuk, P.G., 1989, Diversity of calcium ion channels in cellular membranes, *Neuroscience* 28:253-261.

Lake, N., 1986, Electroretinographic deficits in rats treated with guanidinoethyl sulfonate, a depletor of taurine, *Exp. Eye Res.* 42:87-92.

Lake, N., 1988, Taurine depletion leads to loss of rat optic nerve axons, *Vision Res.* 28:1071-1076.

Landreth, G.E. and Agranoff, B.W., 1979, Explant culture of adult goldfish retina: a model for the study of CNS regeneration, *Brain Res.* 161:39-53.

Lima, L., Matus, P., Drujan, B., 1988, Taurine effect on neuritic growth from goldfish retinal explants, *Int. J. Dev. Neurosci.* 6(5):417-424.

Lima, L., Matus, P., Drujan, B., 1989a, The interaction of substrate and taurine modulates the outgrowth from regenerating goldfish retinal explants, *Int. J. Dev. Neurosci.* 7(4):295-300.

Lima, L., Matus, P., Drujan, B., 1989b, Spatial distribution of glutamate, taurine and gaba in teleosts and mammals retina: in vivo and in vitro study, *Int. J. Dev. Neurosci.* 7(3):295-300.

Lima, L., Matus, P., and Drujan, B., 1991a, Differential taurine uptake in central and peripheral regions of goldfish retina, *J. Neurosci. Res.* 28:422-427.

Lima, L., Matus, P., and Drujan, B., 1991b, Taurine-induced regeneration of goldfish retina in culture may involve a calcium-mediated mechanism, *J. Neurochem.* (in press).

Lombardini, J.B., 1983, Effects of ATP and taurine on calcium uptake by membrane preparations of the rat retina, *J. Neurochem.* 40:402-406.

Lombardini, J.B., 1985a, Opposite effects of 2-aminoethanesulfonic acid (taurine) and aminomethanesulfonic acid on calcium ion uptake in rat retinal preparations, *Eur. J. Pharmacol.* 110:385-387.

Lombardini, J.B., 1985b, Taurine effects on the transition temperature in Arrhenius plots of ATP-dependent calcium ion uptake in rat retinal membrane preparations, *Biochem. Pharmacol.* 34:3741-3745.

Lombardini, J.B., 1988, Effects of taurine and mitochondrial metabolic inhibitors on ATP-dependent Ca uptake in synaptosomal and mitochondrial subcellular fractions of rat retina, *J. Neurochem.* 51:200-205.

Negishi, K., Teranishi, T., and Kato, S., 1982, New dopaminergic and indolamine-accumulating cells in the growth ozone of goldfish retinas after neurotoxic destruction, *Science* 216:747-748.

Neuringer, M. and Sturman, J.A., 1987, Visual acuity loss in Rhesus monkey infants fed a taurine-free human infant formula, *J. Neurosci. Res.* 18:597-601.

Pasantes-Morales H. and Ordóñez, A., 1982, Taurine activation of a bicarbonate-dependent. ATP-supported calcium uptake in frog rod outer segments, *Neurochem. Res.* 7:317-328.

Pasantes-Morales, H., Quesada, O., Carabez, A., and Huxtable, R.J., 1983, Effect of the taurine transport antagonists, guanidinoethane sulfonate and beta-alanine, on the morphology of the rat retina, *J. Neurosci.* 9:135-143.

Pasantes-Morales, H. and Cruz, C., 1983, Possible mechanisms involved in the protective action of taurine on photoreceptor structure, in,"*Sulfur Amino Acids: Biochemical and Clinical Aspect*", K. Kuriyama, R.J. Huxtable, and H. Iwata, pp. 263-276, Alan R. Liss, New York.

Quesada, O., Huxtable, R.J., and Pasantes-Morales, H., 1984, Effect of guanidinoethane sulfonate taurine uptake by rat retina, *J. Neurosci. Res.* 11:179-186.

Schmidt, S., Berson, E.L., and Hayes, K.C., 1976, Retinal degeneration in cats fed casein. 1: taurine deficiency, *Invest. Opthalmol.* 15:47-52.

Sheng, M., McFadden, G., and Greenberg, M.E., 1990, Membrane depolarization and calcium induce c-fos transcription via phosphorylation of transcription factors CREB, *Neuron* 4, 571-582.

Sturman, J.A., Wen, G.Y., Wisniewski, H.M., and Neuringer, M.D., 1984, Retinal degeneration in primates raised on a synthetic human infant formual, *J. Dev. Neurosci* 2:121-129.

Sturman, J.A., Moretz, R.C., French, J.H., and Wisnewski, H.M., 1985a, Taurine defciency in the developing cat: persistence of the cerebellar external granule cell layer, *J. Neurosci. Res.* 13:405-416.

Sturman, J.A., Moretz, R.C., French, J.H., and Wisnewski, H.M., 1985b, Postnatal taurine deficiency in the kitten results in a persistence of the cerebellar external granule cell layer: correction by taurine feeding, *J. Neurosci. Res.* 13: 521-528.

Ueda, K., Masliah, E., Saitoh, T., Bakalis, S.L., Scoble, H., and Kosik, K.S., 1990, Alz-50 recognizes a phosphorylated epitope of tau protein, *J. Neurosci.* 10:3295-3304.

ENDOGENOUS REGULATION OF THE TAURINE RECEPTOR

X.W. Tang[1], Y.H. Lee[1], M. Yarom[1], B. Nathan[1], J. Bao[1],
A. Bhattacharyya[1], W.H. Tsai[2], and J.-Y. Wu[1,2]

[1]Department of Physiology and Cell Biology
University of Kansas
Lawrence, KS 66045

[2]Academia Sinica
Taipei, Taiwan, ROC

INTRODUCTION

Recently we have isolated and purified an endogenous brain modulator (EBM) which competes with ^{3}H-flunitrazepam (FNZP) for binding to the benzodiazepine (BZ) binding site[1]. In addition, we have also isolated and purified EBMs for the $GABA_A$ receptor as well as the glutamate receptor[2]. In light of numerous lines of evidence suggesting that taurine is involved in many important physiological functions such as in the development of the CNS[3,4] in maintaining the structural integrity of the membrane[5,6], in regulating calcium binding and transport[7,8] and serving as an osmoregulator[9,10] and inhibitory neurotransmitter[11,13], it is highly desirable to determine whether endogenous modulators are also involved in the regulation of the taurine receptor. In this communication we present evidence to show that the taurine receptor is indeed also regulated by EBMs.

METHODS AND MATERIALS

Preparation of Brain Extract, and P_2 Membranes (P_2M)

Pig brain tissue was blended in double distilled water for 3 x 15 sec (w/v = 1 gr/10 ml). The homogenate was centrifugated at 100,000 g for 1 hr. The supernatant was filtered through hollowfiber PM 10, with a cut off limit of 10,000 dalton (Amicon). The filtrate was concentrated by lyophilization (Virtis), usually 100 - 200 fold, and was referred to as FI. The pellet was re-extracted using the same procedure. The pellet thus obtained was resuspended in 50 mM Tris citrate pH 7.2 (v/v - 1/1), kept at -20°C and was referred to as P_2M.

Preparation of the Taurine Receptor

The membrane-bound taurine receptor was prepared from P_2M as previously described (see Wu et al. in this volume)[14].

Preparation of Endogenous Brain Modulators

P_2M was subjected to three freeze-thaw cycles, suspended in 5 mM Tris/HCl pH 7.2 (v/v = 1/20) and incubated 20 min at 22°C with gentle shaking. The suspension was centrifugated at 100,000 g for 30 min. The supernatant was filtered through hollowfiber PM 10. The filtrate was concentrated by lyophilization (10 - 100 fold) and was referred to as extract #1. Two more extractions of the pellet in the same manner yielded extracts #2-3. P_2M extracts and brain extract, FI, served as the starting material for EBM.

Assay for Activity of Endogenous Brain Modulators

The taurine receptor binding assay was conducted according to the previously described procedure (see Wu et al. in this volume)[14]. Briefly, for total binding (TB), the binding assay consisted of 800 μl crude taurine receptor (about 0.15 mg/ml), 100 μl of 5 mM Tris-Cl (pH 7.2), and 100 μl of ^{3}H-taurine (300 nM). For nonspecific binding (NB), the conditions were the same as TB except that 0.1 mM unlabeled taurine was included. To assay for EBM activity, unlabeled taurine was replaced by 100 μl of samples containing EBM and the activity of EBM was obtained from the following equation.

$$\text{EBM activity} = \frac{\text{Binding in the presence of EBM - NB}}{\text{TB - NB}} \times 100\%$$

Gel Filtration Chromatography

Concentrated FI or extract #1 from pig brain was applied to a Bio-Gel P-2 column, fractionation range 100 - 1800 dalton (Bio-Rad) which had been pre-equilibrated with 5 mM Tris/HCl pH 7.2. ^{3}H-Taurine was added to the sample as an external standard. The column was eluted with the same buffer. The details for each experiment are specified in the legends (see Results). The absorbance at 280 nm of the eluent was monitored; either simultaneously using a UV monitor (ISCO, Model UA-5 with type 10 optical unit and a 5 mm flow cell) or by measuring the individual fractions using a spectrophotometer (Beckman, Model 24) with a 10 mm cell. The fractions were assayed for activity of the endogenous modulators and radioactivity.

Cation Exchange Column Chromatography

Fractions from a Bio-Gel P-2 column were further purified by cation exchange chromatography (AG 50W X 8, Bio-Rad). The chromatography column was first eluted with 5mM Tris-Cl, followed by a gradient of potassium phosphate buffer from 0 - 200 mM, pH 7.2. Fractions were assayed for EBM activity and OD at 280 nm was also measured.

Chromatography on HPLC C18 column

A Beckman System Gold unit with a 166 detector was used. Active fractions from cation exchange (AG 50W X 8) columns were combined, concentrated and filtered (0.45 μm). One ml samples were applied to a preparative HPLC reverse phase C_{18} column (Zorbax ODS, 2.12 x 25 cm, Du Pont). The column was first pre-equilibrated with H_2O and then eluted with H_2O, followed by 10 mM trifluoroacetic acid (TFA) gradient at a flow rate of 4 ml/min. The absorbance of the eluent was monitored at 220 nm, and 4 ml fractions were collected. The fractions were concentrated to dryness by vacuum centrifugation (Savant), redissolved in 1 ml water and assayed for EBM activity.

RESULTS AND DISCUSSION

Effect of Brain Extracts on ^{3}H-Taurine Binding

As expected, ^{3}H-taurine receptor binding is inhibited by brain extracts. For instance, 100 ml of FI, extract #1, #2 and #3 were found to inhibit ^{3}H-taurine binding to an extent of 100, 81, 47 and 43% respectively. This is not surprising since it is known that taurine is one of the most abundant amino acids present in mammalian brain.

Separation of Endogenous Brain Modulators on the First Gel Filtration Column

In order to separate taurine from other EBMs we first employed a Bio-Gel P-2 gel filtration column to fractionate brain extracts according to molecular size. When FI was fractionated, two EBM peaks were obtained. The minor peak corresponds to the void volume suggesting a molecular weight greater than 1800-dalton and the major peak appears to have the same elution position as taurine (Figure 1). At least three EBM peaks were obtained when extract #1 was separated on a Bio-Gel-P-2 column (Figure 2). Whether these EBM peaks from extract #1 are related to those obtained from FI remains to be determined.

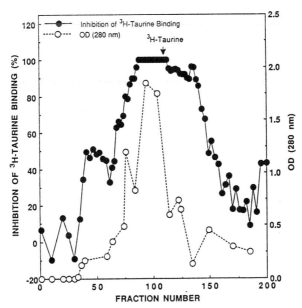

Figure 1. Separation of Endogenous Brain Modulators from FI on the first Bio-Gel P-2 Column. Fifty ml of 100 X concentrated FI was applied to the Bio-Gel P-2 column (100 X 5 cm). The column was eluted with 5 mM Tris-Cl (pH 7.2), and 12 ml per fraction were collected. The fractions were assayed for their effect on ^{3}H-taurine binding expressed as % inhibition and monitored for OD 280nm. ^{3}H-taurine (500,000 cpm) was applied to the same column as a standard, and the elution position is indicated by an arrow.

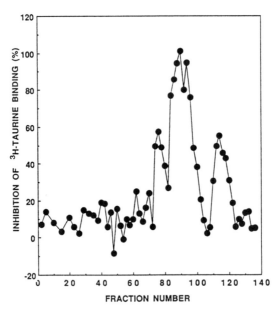

Figure 2. Separation of Endogenous Brain Modulators from extract #1 on the first Bio-Gel P-2 column. Thirty ml of 60 X concentrated extract #1 was applied to the Bio-Gel P-2 column (90 X 5 cm). The column was eluted with 5 mM Tris-Cl (pH 7.2) and 20 ml per fraction were collected. The fractions were assayed for their effect on ^{3}H-taurine binding, expressed as % inhibition.

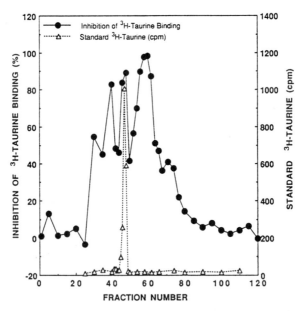

Figure 3. Separation of Endogenous Brain Modulators on the Second Gel Filtration Column. Fractions (#134 - #147) from the first Bio-Gel P-2 column (Figure 1) were concentrated to 3.5 ml. Two ml of the concentrated sample containing standard ^{3}H-taurine (350,000 cpm) were applied to the second Bio-Gel P-2 column (120 x 2.75 cm). The column was eluted with 5 mM Tris-Cl, pH 7.2, and 10 ml per fraction were collected. Fractions were assayed for their effect on ^{3}H-taurine binding, expressed as % inhibition. The elution profile of standard ^{3}H-taurine is shown.

Separation of Endogenous Brain Modulators on the Second Gel Filtration Column

When the major EBM peak fractions #134-147 from the first Bio-Gel P-2 column (Figure 1) were combined, concentrated and applied to a second Bio-Gel P-2 column, the EBM activity was further separated into three peaks (Figure 3). One of the three EBM peaks coincides with standard ^{3}H-taurine. However, the other two peaks of EBM are clearly separated from taurine and may represent a new set of endogenous brain substances that may regulate the binding of taurine to its receptor.

Separation of Endogenous Brain Modulators on a Cation Exchange Column

EBMs after elution from the second Bio-Gel P-2 column were further purified by cation exchange chromatography. Fifty ml of fractions #56-62 from the second Bio-Gel P-2 column (Figure 3) were applied to a cation exchange (AG W50 X 8) column. The column was first eluted with 5 mM Tris-Cl, followed by a gradient of phosphate buffer from 0 to 200 mM. The EBM activity was eluted before the start of the gradient (Figure 4). A substantial purification was achieved since most of the contaminants were removed as judged from the OD at 280 nm.

Purification of Endogenous Brain Modulators on a HPLC Column.

EBMs after cation exchange chromatography (Figure 4, fraction #25) were further purified on a C$_{18}$ HPLC column. The column was first eluted with H$_2$O, followed by a linear gradient, 0 - 10 mM, of trifluoroacetic acid (TFA) and finally continued with 10 mM TFA. EBM activity was recovered in the early fractions before the start of the TFA gradient.

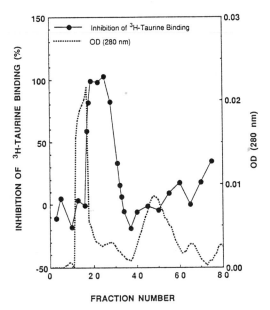

Figure 4. Separation of Endogenous Brain Modulators on a Cation Exchange Column. Fractions #56-62 from the second Bio-Gel P-2 column (Figure 3) were combined and applied to a AG W50 X 8 column (2.75 x 50 cm). The column was eluted with 5 mM Tris-Cl followed by a gradient of potassium phosphate buffer from 0 to 200 mM (pH 7.2). The gradient was started at fraction #60 and 10 ml per fraction were collected. The fractions were monitored for OD 280 nm and their effect on ^{3}H-taurine binding is expressed as % inhibition.

Characterization of Endogenous Brain Modulators

Stability -- EBMs appear to be quite stable since treatment with acid (pH 2.4, 1 hr), base (pH 11.5, 1 hr) or heat (95°C, 45 min) does not affect the EBM activity.

Specificity -- EBMs isolated and purified as described above appear to be quite specific for the taurine system. For instance, EBM which inhibits 80% of ^{3}H-taurine binding to the taurine receptor was found to have little effect on binding of ^{3}H-L-glutamate, ^{3}H-flunitrazepam (FNZP) and ^{3}H-muscimol to their respective receptors as shown in Table 1.

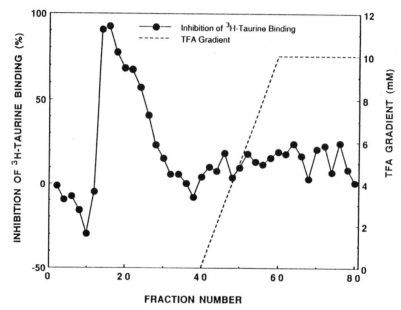

FRACTION NUMBER

Figure 5. Purification of Endogenous Brain Modulators ons C$_{18}$ HPLC column. Fraction #25 from the cation exchange column (Figure 4) was concentrated to 1 ml and applied to C$_{18}$ (2.12 X 25 cm) HPLC column. The column was eluted with H$_2$O for 40 min, followed by a 10 mM TFA gradient for 20 min, and finally was eluted with TFA for an additional 40 min. The flow rate were 4 ml/min and 4 ml per fraction was collected. The fractions were assayed for their effect on ^{3}H-taurine binding expressed as % inhibition.

Table 1. Specificity of Endogenous Brain Modulators

Receptor Type *	Ligands	Inhibition EBM (%)
Taurine	^{3}H-Taurine	80
GLU	^{3}H-L-Glutamate	12
BZ	^{3}H-Flunitrazepam	2
GABA$_A$	^{3}H-Muscimol	5

*Receptor binding assays were conducted as described elsewhere in this volume[14].

ACKNOWLEDGEMENT

This study was supported in part from grants NS20978 (NIH) and BNS-8820581 (NSF) USA and a grant from the National Science Council, Taiwan, ROC. The expert typing of the manuscript by Jan Elder and Judy Wiglesworth is gratefully acknowledged.

REFERENCES

1. C.C. Liao, H.S. Lin, J.-Y. Liu, L.S. Hibbard, and J.-Y. Wu, Purification and characterization of a benzodiazepine-like substance from mammalian brain, *Neurochem. Res.* 14:345 (1989).
2. J.-Y. Wu, M. Yarom, Y.H. Lee, J. Bao, and X.W. Tang, Endogenous modulators for taurine, GABA, and L-glutamate receptors. Abstract of the International Society for Developmental Neuroscience Satellite Symposium on "Neuroactive Amino Acids as Developmental Signals in the Nervous System", Cancun, Mexico, June 24-26, pp. 14 (1990).
3. J.A. Sturman, R.C. Moretz, J.H. French, and H.M. Wisniewski, Postnatal taurine deficiency in the kitten results in a persistence of the cerebellar external granule cell layer: correction by taurine feeding, *J. Neuro. Res.* 13:521 (1985).
4. J.A. Sturman, A.D. Gargano, J.M. Messing, and H. Imaki, Feline maternal taurine deficiency: effect on mother and offspring, *J. Nutr.* 116:655 (1986).
5. H. Pasantes-Morales and C. Cruz, Taurine and hypotaurine inhibit light-induced lipid peroxidation and protect rod outer segment structures, *Brain Res.* 330:154 (1985).
6. J. Moran, P. Salazar, and H. Pasantes-Morales, Effect of tocopherol and taurine on membrane fluidity of retinal rod outer segments, *Exp. Eye Res.* 45:769 (1988).
7. J.W. Lazarewicz, K. Noremberg, A. Lehmann, and A. Hamberger, Effects of taurine on calcium binding and accumulation in rabbit hippocampal and cortical synaptosomes. Neurochem. Int. 7:421 (1985).
8. J.B. Lombardini, Effects of taurine on calcium ion uptake and protein phosphorylation in rat retinal membrane preparations, *J. Neurochem.* 45:268 (1985).
9. J.M. Solis, A.S. Herranz, O. Herreras, J. Lerma, and R. Martin Del Río, Does taurine act as an osmoregulatory substance in the rat brain, *Neurosci. Lett.* 91:53 (1988).
10. J.V. Wade, J.P. Olson, F.E. Samson, S.R. Nelson, and T.L. Pazdernik, A possible role for taurine in osmoregulation within the brain, *J. Neurochem.* 51:740 (1988).
11. K. Okamoto, H. Kimura, and Y. Sakai, Evidence for taurine as an inhibitory neurotransmitter in cerebellar stellate interneurons: Selective antagonism by TAG (6-aminomethyl-3-methyl-4H,1,-2,4-benzothiadiazine-1,1-dioxide), *Brain Res.* 265(1):163 (1983).
12. C.-T. Lin, Y.Y.T. Su, G.-X Song, J.-Y. Wu, Is taurine a neurotransmitter in rabbit retina? *Brain Res.* 337:293 (1985).
13. T.C. Taber, C.-T. Lin, G.-X. Song, R.H. Thalman, J.-Y. Wu, Taurine in the rat hippocampus-localization and postsynaptic action, *Brain Res.* 386:113 (1986).
14. J.-Y. Wu, X. W. Tang, and W.H. Tsai, Taurine receptor: kinetic analysis and pharmacological studies, *in*: "Taurine: Nutritional Value and Mechanisms of Action," J.B. Lombardini, S.W. Schaffer, and J. Azuma, eds., Plenum Publishing Co., New York (in press).

LOCALIZATION OF TAURINE AND GLIAL FIBRILLARY ACIDIC PROTEIN IN HUMAN OPTIC NERVE USING IMMUNOCYTOCHEMICAL TECHNIQUES

Norma Lake

Departments of Physiology and Ophthalmology
McGill University, 3655 Drummond Street
Montreal, Quebec, Canada H3G 1Y6

INTRODUCTION

The well-known findings that taurine deficiency is associated with photoreceptor cell degeneration and loss of retinal function in cats, monkeys, rats, and humans, has focused attention on the role of taurine in photoreceptors (reviewed in Cocker and Lake, 1989). Although receptor cell degeneration is certainly a major effect, we have in addition observed shrinkage and cell loss in the inner retina (Lake and Malik, 1987; Lake et al., 1986) and fiber loss in the optic nerve of taurine-deleted adult rats (Lake et al., 1988). Morphometric studies of optic nerves from mother rats that had just raised a litter of pups have shown that in comparison to controls, in taurine-deleted dams there is a reduction of axon diameters and thinning of myelin sheaths, and a shift in the distributions of fiber size and myelin thickness (See Table 1). This thinning of myelin that we have observed, implicates optic nerve oligodendrocytes as targets of taurine deficiency; however, whether they are affected before, with, or secondary to axonal changes, remains to be established. It is also unknown whether the changes in the optic nerve axons occur secondary to the degeneration of photoreceptors, or are a primary effect. Hence, as a first step in resolving these questions, it is of obvious importance to determine the cytochemical location of the relatively high levels of taurine found in the nerve (e.g. Sturman et al., 1986; Ross et al., 1989). This we have done for the rat (Lake and Verdone-Smith, 1990a,b) and the micrographs which follow show the localization of taurine in human optic nerve tissue.

LOCALIZATION OF TAURINE

There are several technical difficulties involved in localizing taurine which arise from the fact that it is a constituent of the free amino acid pool, and has a low molecular weight. Hence it is soluble and not immunogenic. Antisera *can* be raised against taurine conjugated via glutaraldehyde (G) to a protein carrier such as bovine serum (BSA) (e.g. Madsen et al., 1985; Lake and Verdone-Smith, 1989). One can use

Taurine, Edited by J.B. Lombardini *et al.*
Plenum Press, New York, 1992

Table 1. Axon parameters in control and taurine-depleted rats

	Axon Diameter (% of 1500 fibres)			Myelin thickness (% of 1500 fibres)		
	.10-.69μ	.70-1.49μ	1.5-2.5μ	.02-.059μ	.06-.259μ	.26-.459μ
Control	20%	71%	9%	16%	80%	4%
Taurine-depleted	30%	60%	10%	24%	72%	4%

Axon characteristics of 1500 optic nerve axons from each of 4 control and 4 GES-treated (taurine-depleted) mother rats were measured from electron micrographs. Chi-square analysis indicated significant differences ($p<0.01$) in these distributions for depleted versus control.

the antisera (after suitable purification and characterization) for immunocytochemical localization in tissue which has been fixed with glutaraldehyde both to retain the amino acid, and to present it in the conjugated form which was the immunogen. We use high levels of glutaraldehyde (2.5% and above) since lower levels give much less taurine retention and allow its diffusion and redistribution. We also include magnesium in the fix, for amino acid and structural preservation. We have largely avoided post mortem changes resulting from delays before fixation (a drawback of much work with human tissues) since the material that we use is obtained at surgery and is immediately immersion fixed.

We have used ELISA (enzyme-linked immunosorbent assays) to determine relative titers and cross-reactivity of our sera to other compounds. We carry these out routinely on the sera from each bleed, and have been able to remove any cross-reactivity by adsorption with the appropriate G-BSA conjugate. Following pre-adsorption of our anti-sera and ELISA testing, further controls are carried out at the time of our immunocytochemistry. These include (i) substitution of pre-immune serum, or (ii) omission of the primary antiserum, or (iii) use of a primary anti-serum which has been pre-adsorbed with its target immunogen (e.g. taurine-G-BSA). In all of these control trials there has been no immunoreactive staining, which indicates the absence of non-specific binding and/or unreactivity of the tissues with the secondary antibodies, chromogens, etc., in our protocols.

The human optic nerve contains very few unmyelinated fibres and large numbers of myelinated axons separated into fascicles by glial cells and their processes, and less completely by septa derived from the pial meninges which encase the nerve (Hogan et al., 1971). There are at least two subtypes of macroglial cells: oligodendrocytes, which produce the myelin sheaths wrapping the axons; and astrocytes, whose processes line the septa, and expand at the periphery of the nerve just inside the *pia mater* to form a structure known as the *glia limitans* or glial mantle of Graefe (Hogan et al., 1971). Arterioles and capillaries of the blood supply of the nerve lie within the pial-derived septa.

Immunocytochemical processing using our taurine antibody and a fluorescently-labelled secondary antibody revealed prominent immunoreactivity (taurine-IR) within glial cells distributed throughout the nerve, especially in their nuclei and perikaryal cytoplasm and in their radiating, branching processes and extensions to the septa (Figure 1A). The glial nuclei tended to be in clusters, both within the nerve fascicles and at the periphery. The peripheral *glia limitans* was densely taurine-IR, whereas the nerve axons, meninges and pial septa were relatively unreactive. Control sections processed with substitution of the primary antiserum by BSA, preimmune sera, or serum preadsorbed with taurine-BSA conjugates showed no immunofluorescence.

Since this cellular distribution of taurine-IR included structures suspected to be of astrocytic origin, including the peripheral *glia limitans*, the patterns of immunofluorescence using a commercial antibody to glial fibrillary acidic protein

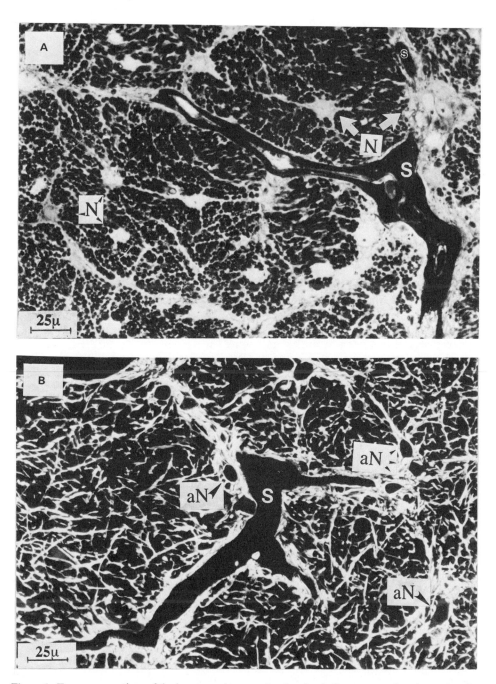

Figure 1. Transverse sections of the human optic nerve showing the similar pattern of taurine and GFAP immunofluorescence (white) in glial cells. For taurine (see A) clusters of glial cell nuclei (N) are diffusely stained, their processes surround the unstained axons and line the septa (S). GFAP immunoreactivity (see B) is more fibrillar and is excluded from the astrocyte nuclei (aN).

(GFAP), an astrocyte-specific marker (Bignami et al., 1972), was examined for comparison with that of taurine. Figure 1B shows a transverse section of the optic nerve in which the locations of astrocytic processes are indicted by a high intensity GFAP immunofluorescence. We found that the pattern of GFAP-IR was very similar to that for taurine-IR: in processes of astrocytes, and their expansions at the peripheral glial mantle and lining the septa. The differences are firstly that GFAP-IR appears more fibrillar, while taurine-IR appears more diffuse (compare Figure 1A,B), and secondly that glial nuclear regions are not GFAP-IR, but instead they appear as empty "ghosts" (Figure 1B), whereas taurine-IR was frequently localized to glial cell nuclei and the perinuclear region (Figure 1A).

COMMENTARY

The anti-serum that we used here was raised, preadsorbed and characterized as highly specific for taurine using ELISA and a panel of closely related targets, as we have described (Lake and Verdone-Smith, 1989). Pre-adsorption of our primary anti-taurine serum with taurine-BSA conjugates virtually eliminated all immunofluorescence, whereas pre-adsorption with for example, GABA-BSA conjugates did not. It thus seems likely that the immunoreactivity localizes sites of taurine concentration. Taurine-IR occurred predominantly in glial cells.

This immunocytochemical technique is not a quantitative one, so we do not exclude the presence of taurine from axons, but if present, it would appear to be at substantially lower levels than in the glia. Indeed, in the developing rat we have demonstrated prominent axonal staining with this anti-taurine serum during embryonic and early postnatal life (Lake and Verdone-Smith, 1990b), which transforms to the adult (glial) pattern by the time the animals are weaned. In contrast, all of our human material has been from donors from 27 to 76 years of age.

The similarity of the distributions of GFAP and taurine-IR suggests that at least a subset of taurine-containing glial cells are astrocytes, confirming in human tissue what we have previously described for the adult rat optic nerve (Lake and Verdone-Smith, 1990a). Correlative studies with an oligodendrocyte marker or a microglial marker have not been completed, so it is not known how widespread taurine immunoreactivity is amongst the various subclasses of glia.

There is little known of the role of taurine in glial cells. Some functions suggested for astrocytes are regulation of the external milieu of the nerve bundles through blood-nerve barrier and transport mechanisms, and perhaps even more direct participation in conduction velocity via processes at the nodes of Ranvier (Waxman and Black, 1984). Work in cultured glial cells and gliomas suggests that these cells possess high affinity uptake mechanisms for taurine, and that they release taurine in response to several types of stimuli: depolarization by high potassium (Philibert et al., 1988), stimulation of ß-adrenergic receptors (Shain et al., 1989), and exposure to hypotonic media (Pasantes-Morales and Schousboe, 1988). Whether these mechanisms operate *in vivo* and how they affect CNS function is unknown, although taurine does exert a potent depressant action (glycine-like) on neurons. It seems likely that the degeneration and loss of optic nerve axons and thinning of myelin sheaths during taurine deficiency that we have observed (Table 1 and Lake et al., 1988), may arise in part secondary to alterations in the functions of taurine-immunoreactive glial cells. Since the maintenance of cellular taurine levels depends in part on uptake via genetically-determined specific transport molecules, there may be a subset of patients e.g. with demyelinating disease for whom taurine deficiency is a contributing factor.

ACKNOWLEDGEMENTS

These studies were supported in part by funds from the Medical Research Council of Canada, RP Eye Research Foundation of Canada, and La Fondation OCULUS des Maladies de l'Oeil, Inc. I am grateful to Dr. S. Brownstein for providing access to human material, and to Carole Verdone-Smith for excellent assistance in immunocytochemistry, photography, and manuscript preparation.

REFERENCES

Bignami, A., Eng. L.F., Dahl, D., and Uyeda, C.T., 1972, Localization of the glial fibrillary acidic protein in astrocytes by immunofluorescence, *Brain Res.* 43:429-435.

Cocker, S.E. and Lake, N., 1989, Effects of dark maintenance on retinal biochemistry and function during taurine depletion in the adult rat, *Visual Neurosci.* 3:33-38.

Hogan, M.J., Alvarado, J.A., and Weddell, J.E., 1971, "Histology of the Human Eye: An Atlas and Textbook," W.B. Saunders Co., London.

Lake, N., Malik, N., and De Marte, L., 1986, Effects of taurine deficiency on the rat retina during development, *Soc. Neurosci. Abst.* 12:1560.

Lake, N. and Malik, N., 1987, Retinal morphology in rats treated with a taurine transport antagonist, *Exp. Eye Res.* 44:331-346.

Lake, N., Malik, N., and De Marte, L., 1988, Taurine depletion leads to loss of optic nerve axons, *Vision Res.* 28:1071-1076.

Lake, N. and Verdone-Smith, C., 1989, Immunocytochemical localization of taurine in the mammalian retina, *Curr. Eye Res.* 8:163-173.

Lake, N. and Verdone-Smith, C., 1990a, Immunocytochemical localization of taurine within glial cells in the optic nerve of adult albino rats, *Curr. Eye Res.* 9:1115-1120.

Lake, N. and Verdone-Smith, C., 1990b, Immunocytochemical localization of taurine and GFAP in the developing rat optic nerve, *Invest. Ophthalmol. Vis. Sci.* 31:159.

Madsen, S., Ottersen, O.P., and Storm-Mathisen, J., 1985, Immunocytochemical visualization of taurine: neuronal localization in the rat cerebellum, *Neurosci. Lett.* 60:255-260.

Pasantes-Morales, H. and Schousboe, A., 1988, Volume regulation in astrocytes: a role for taurine as an osmoeffector, *J. of Neurosci. Res.* 20:505-509.

Philibert, R.A., Rogers, K.L., Allen, A.J., and Dutton, G.R., 1988, Dose-dependent, K^+-stimulated efflux of endogenous taurine from primary astrocyte cultures is Ca^{2+}-dependent, *J. Neurochem.* 51:122-126.

Ross, C.D., Parli, J.A., and Godfrey, D.A., 1989, Quantitative distribution of six amino acids in rat retina layers, *Vision Res.* 29:1079-1984.

Shain, W., Connor, J.N., Madelian, V., and Martin, D.L., 1989, Spontaneous and beta-adrenergic receptor-stimulated taurine release from astroglial cells are independent of manipulations of intracellular calcium, *J. Neurosci.* 9:2306-2312.

Sturman, J.A., Gargano, A.D. Messing, J.M., and Imaki, H., 1986, Feline maternal taurine deficiency: effect on mother and offspring, *J. Nutr.* 166:655-667.

Waxman, S.G. and Black, J.A., 1984, Freeze-fracture ultrastructure of the perinodal astrocyte and associated glial junctions, *Brain Res.* 308:77-87.

EFFECTS OF TAURINE ON PROTEIN PHOSPHORYLATION
IN MAMMALIAN TISSUES

John B. Lombardini

Departments of Pharmacology and Ophthalmology
  & Visual Sciences
Texas Tech University Health Sciences Center
Lubbock, Texas 79430

INTRODUCTION

It has been proposed that taurine has various biological functions such as influencing neurotransmitter release, stabilizing membranes, modulating Ca^{2+} fluxes, inhibiting protein phosphorylation, and regulating osmolarity (for reviews, see ref. 1 and 2). Up until recently, there has been no well-defined role for taurine at the molecular level, and thus the function of taurine in excitable tissues remained unclear. However, results from our laboratory indicate that taurine has an inhibitory effect on the phosphorylation of specific proteins in the retina[3-5], brain[6-8], and heart[9].

We have recently demonstrated in a P_2 fraction of the cortex of the rat brain that taurine inhibits the Ca^{2+}-dependent, protein kinase C-catalyzed phosphorylation of a ~20 K apparent molecular weight (M_r) protein[7]. In the intrasynaptosomal cytosolic fraction of the P_2 preparation, taurine did not have an inhibitory effect on protein phosphorylation unless mitochondria were reconstituted with the cytosol[8]. In addition, taurine inhibited the accumulation of ^{32}P-labeled phosphatidic acid in synaptosomes. These results suggest that taurine may inhibit specific protein phosphorylation by reducing the optimal cytosolic Ca^{2+} levels through stimulation of a mitochondrial Ca^{2+} uptake system and by inhibiting the turnover of phosphoinositides[8].

In this chapter, we present additional studies on the phosphorylation of specific proteins in the rat retina, rat heart, and rabbit retina.

METHODS

The retinal tissue homogenate, mitochondrial fraction, and rod outer segment fraction were prepared as previously described[3]. The osmotically shocked P_2 fraction of the rat brain was prepared as described by Li and Lombardini[6]. The rat heart mitochondrial fraction was prepared by homogenizing the heart tissue in 10 vol of buffer (220 mM mannitol, 70 mM sucrose, 5 mM MOPS, pH 7.4) and then centrifuging at 500 g for 10 min to remove cell debris. The supernatant was centrifuged at

3000 g for 10 min to sediment the mitochondria. The cardiac mitochondrial pellet was then washed once in a Krebs-bicarbonate buffer (NaCl, 118 mM; KCl, 4.7 mM; KH_2PO_4, 1.2 mM; $MgSO_4$, 1.17 mM; $NaHCO_3$, 25 mM, pH 7.4) and finally suspended in the Krebs-bicarbonate buffer.

The incubation system for the phosphorylation assays contained Krebs-bicarbonate buffer and 20 μCi of $[\gamma\text{-}^{32}P]ATP$ (10 μM). EGTA and $CaCl_2$ were added as indicated. Final volume was 0.25 ml. The incubation tubes were preincubated for 2 min at 37°C and the reaction was initiated by the addition of the $[\gamma\text{-}^{32}P]ATP$. After an incubation period of 1 to 10 min (depending upon the experiment) the reaction was stopped by the addition of 0.5 ml of sodium dodecyl sulfate (SDS) solubilization buffer [0.05 M 2-(N-cyclohexylamino)ethanesulfonic acid, 2% SDS, 10% glycerol, 2% 2-mercaptoethanol, 0.00125% bromophenol blue, pH 9.5]. The incubation mixtures were agitated thoroughly on a Vortex mixer and then immediately boiled for 5 minutes. Aliquots of the boiled incubation mixtures were subjected to 1- or 2-dimensional polyacrylamide gel electrophoresis (PAGE) according to the method of Laemmli[10] or O'Farrell[11]. The gels were dried and exposed to X-ray film to visualize the incorporation of radioactive phosphate into the various proteins. Quantitation of the radioactivity in specific proteins was determined by densitometry.

RESULTS AND DISCUSSION

Retina

When 20 mM taurine was added to rat retinal homogenates, the inhibition of the phosphorylation of specific proteins was observed in 1-dimensional (PAGE) gels (Fig. 1). The greatest effect of taurine appeared to be on the protein(s) located in a band (designated a) with a molecular weight of ~20 K. Gels obtained after 2-dimensional PAGE of the retinal homogenate revealed a series of 3 proteins (designated a, b, and c) which were affected by taurine (Fig. 2A and 2B).

Phosphorylation of the ~20 K M_r protein was also observed in each of three subcellular fractions of the rat retinal homogenate[3,12]. These fractions were: a P_1 fraction (Fig. 3) which contains synaptosomes from the photoreceptors. Mitochondria and rod outer segments are also present in this fraction. The greatest inhibitory effect (73%) of taurine (20 mM) on the phosphorylation of the ~20 K M_r protein was observed in the P_1 fraction (Fig. 3) and the mitochondrial fraction (autoradiographs not shown) (43%). The taurine effect in the rod outer segment fraction (autoradiographs not shown) was only 21%.

The observation that the ~20 K phosphoprotein was present in the P_1 fraction, the rod outer segments, and the mitochondrial fraction suggests some general importance for this particular protein. Alternatively, contamination of the P_1 fraction with mitochondria is certainly present[13] and is also possible for the rod outer segment fraction.

Protein kinase activators such as cAMP, cGMP, calmodulin[12], and phorbol ester[12] had no effect on the phosphorylation of the ~20 K M_r protein.

A rabbit retinal homogenate was also tested for the effects of taurine (20 mM) on the phosphorylation of specific proteins (Fig. 4). As in the rat retina taurine inhibited the phosphorylation of a ~20 K M_r protein (designated b). However, the phosphorylation of proteins with approximate molecular weights of 17 K (designated a) and 38 K (designated c) were stimulated. Proteins with higher molecular weights (not designated) also appeared to be affected.

Brain

The taurine (10 mM) effect on phosphorylation of a ~20 K M_r protein in a P_2

synaptosomal fraction of the rat cortex was slight when observed in 1-dimensional (PAGE) gels (Fig. 5). However, when the P_2 synaptosomal fraction was chromatographed by 2-dimensional PAGE (Fig. 6A and 6B), taurine inhibited the phosphorylation of the ~20 K M_r protein by $72.4 \pm 5.4\%$[6]. In addition there is a second protein with an apparent molecular weight of ~140 K which is also inhibited. This higher molecular weight protein is difficult to resolve by PAGE and thus has not been studied further.

The phosphorylation of the ~20K M_r protein in the cortical P_2 fraction is dependent upon Ca^{2+} and is stimulated (30-fold) by 30 mM potassium chloride[7]. Phosphorylation of the ~20 K M_r protein is also stimulated by phorbol ester together with phosphatidylserine, but cAMP, cGMP, and calmodulin have no effect[7]. The addition of taurine (10 mM) to the phorbol ester stimulated system blocks phosphorylation by approximately 30%. Moreover, we have demonstrated that the inhibitory effects of taurine (10 mM) on the phosphorylation of the ~20 K M_r protein in the cortex P_2 synaptosomal preparation appear to require the presence of mitochondria. This conclusion was obtained by performing reconstitution experiments that demonstrated taurine has no effect on protein phosphorylation in a subcellular fraction of mitochondrial-free intrasynaptosomal cytosol unless the mitochondria are added back[8].

It is also known that taurine stimulates Ca^{2+} uptake in brain subcellular fractions[8,14] and inhibits the accumulation of phosphatidic acid in synaptosomes[8]. Pasantes-Morales[14] observed that the taurine effect is mainly present in ATP-dependent Ca^{2+} accumulation although we[8] have found that taurine also stimulates ATP-independent Ca^{2+} uptake. Furthermore, Kuriyama and colleagues[15] demonstrated that taurine inhibits both inositol 1-phosphate and inositol 1,4-bisphosphate accumulation in cardiac slices, and they suggested that taurine may directly inhibit phospholipase C activity. These results indicate that taurine may inhibit the phosphorylation of specific proteins by reducing cytosolic Ca^{2+} levels and by inhibiting the turnover of phosphoinositides.

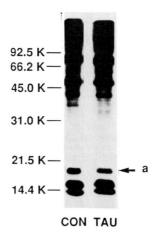

CON TAU

Figure 1. Autoradiographs from a 1-dimensional (PAGE) gel (12% slab gel) of the effects of 20 mM taurine on the phosphorylation of rat retinal proteins (homogenate preparation). Incubation time was 10 min. Marker proteins with molecular weights ranging from 14.4 to 92.5 K are indicated (reprinted with permission from ref. 3).

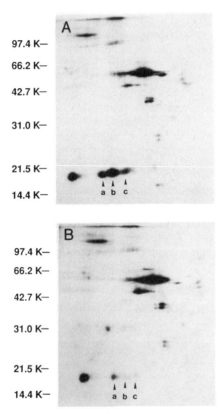

Figure 2. Autoradiographs from 2-dimensional (PAGE) gels of the effects of 20 mM taurine on the phosphorylation of rat retinal proteins (homogenate preparation). Incubation time was 6 min. Marker proteins with molecular weights ranging from 14.4 to 97.4 K are indicated. Isoelectric focusing on tube gels was used in the first electrophoresis dimension and 12% slab gels in the second electrophoresis dimension. A = control. B = 20 mM taurine.

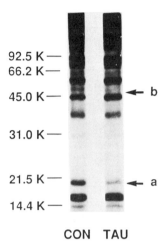

92.5 K —
66.2 K —

45.0 K —   ← b

31.0 K —

21.5 K —   ← a

14.4 K —

CON TAU

Figure 3. Autoradiographs from a 1-dimensional (PAGE) gel of the effects of 20 mM taurine on the phosphorylation of rat retinal proteins (P_1 preparation). Incubation time was 10 min. Marker proteins with molecular weights ranging from 14.4 to 97.4 K are indicated (reprinted with permission from ref. 3).

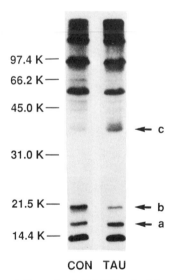

97.4 K —
66.2 K —

45.0 K —

                                    ← c

31.0 K —

21.5 K —   ← b
                                    ← a
14.4 K —

CON TAU

Figure 4. Autoradiographs from a 1-dimensional (PAGE) gel of the effects of 20 mM taurine on the phosphorylation of rabbit retinal proteins (homogenate preparation). Incubation time was 6 min. Marker proteins with molecular weights ranging from 14.4 to 97.4 K are indicated.

It was observed when comparing the ~20 K phosphoprotein in the brain with the phosphoprotein in the retina that they do not have identical molecular weights although we, unfortunately, refer to both of them as ~20 K M_r phosphoproteins. The retinal ~20 K M_r phosphoprotein migrates slightly faster on PAGE than the brain ~20 K M_r phosphoprotein and thus has a lower molecular weight.

Comparisons between the two ~20 K M_r phosphoproteins in the retina and the brain demonstrate a number of differences indicating that they are not the same protein. First, the molecular weights as described above are not identical. Second, the isoelectric point of the phosphoprotein in the brain cortex is 5.6[6] while the isoelectric point of the retinal phosphoprotein is higher. Third, phosphorylation of the brain ~20 K M_r protein is stimulated by phorbol ester[8] while the phosphorylation of the retinal ~20 K M_r protein is unaffected by phorbol ester. Fourth, the retinal ~20 K M_r phosphoprotein appears to be a family of three proteins (Fig. 2) while in the brain (Fig. 6) there is only one phosphoprotein.

Heart

In a rat heart mitochondrial preparation, taurine (20 mM) inhibited the phosphorylation of a ~44 K M_r protein (Fig. 7). cAMP, cGMP, and phorbol ester had no effect on the phosphorylation of the ~44 K M_r protein. The phosphorylation of the ~44 K M_r protein in our system appears to be stimulated by calmodulin (Fig. 7). However, this observation may be artifactual due to the observation that exogenous Ca^{2+} inhibited the phosphorylation of the ~44 K M_r protein. This latter result thus suggests that the observed stimulatory effect of calmodulin (which was tested in the presence of Ca^{2+}) may be due to the removal of Ca^{2+} from the incubation system by binding the Ca^{2+} to the calmodulin.

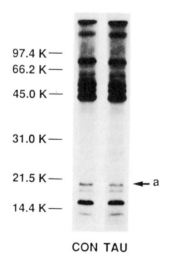

CON TAU

Figure 5. Autoradiographs from a 1-dimensional (PAGE) gel of the effects of 20 mM taurine on the phosphorylation of rat brain retinal proteins (P_2 fraction). Incubation time was 1 min. Marker proteins with molecular weights ranging from 14.4 to 97.4 K are indicated.

RAT BRAIN: P$_2$ FRACTION

pH 7.0 6.8 6.6 6.4 6.2 6.0 5.8 5.6 5.2 4.6

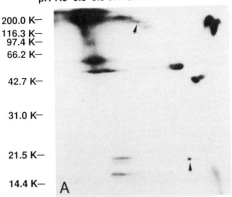

200.0 K—
116.3 K—
97.4 K—
66.2 K—
42.7 K—
31.0 K—
21.5 K—
14.4 K— A

pH 7.0 6.8 6.6 6.4 6.2 6.0 5.8 5.6 5.2 4.6

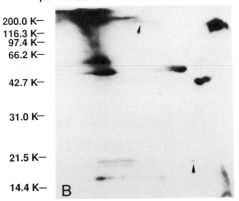

200.0 K—
116.3 K—
97.4 K—
66.2 K—
42.7 K—
31.0 K—
21.5 K—
14.4 K— B

Figure 6. Autoradiographs from 2-dimensional (PAGE) gels of the effects of 10 mM taurine on the phosphorylation of rat brain proteins (cortex preparation). Incubation time was 1 min. Marker proteins with molecular weights ranging from 14.4 to 97.4 K and the pH gradient of the isoelectric focusing on the gels are indicated. Isoelectric focusing on tube gels was used in the first electrophoresis dimension and 12% slab gels in the second electrophoresis dimension. A = control. B = 10 mM taurine. (Reprinted with permission from ref. 6.)

It has also been previously reported by Schaffer and colleagues[16] that rat heart sarcolemma contains a calmodulin-dependent protein kinase that phosphorylates three membrane proteins with molecular weights of ~44 K, ~54 K, and ~190 K. The phosphorylation of all three of these proteins was inhibited by 20 mM taurine. These results have led Schaffer and colleagues[16] to further speculate that the effects of taurine on myocardial metabolism may be mediated through the inhibition of calmodulin-dependent processes. Their earlier studies demonstrated that taurine altered glycogen metabolism by inhibiting calmodulin action[17]. Sperelakis et al.[18] have shown that taurine blocks the slow component and activates a fast (transient) component of the Ca^{2+} current in embryonic chick heart cells. This slow Ca^{2+} channel is activated by calmodulin[19]. Finally, Segawa et al.[20,21] have reported that taurine interferes with the Ca^{2+}-calmodulin system in cerebral cortex adrenergic neurons, and thus decreases norepinephrine release from nerve terminals. However, the mode of action of taurine is not necessarily by affecting the influx of Ca^{2+} into the nerve terminal, and thus the mechanism is not clear. Therefore, the observations that taurine affects calmodulin-dependent processes in cardiac tissue and perhaps in other excitable tissues are viable starting points for studying the effects of taurine at the molecular level and should be pursued.

In all of the above studies involving retina, brain, and heart, the effects of taurine on protein phosphorylation appear to be specific for the taurine structure as analogues of taurine such as isethionic acid and guanidinoethanesulfonic acid do not have any inhibitory effect[3,6,9]. Thus, the possibility that taurine is inhibiting phosphorylation through an effect on ionic strength has been ruled out.

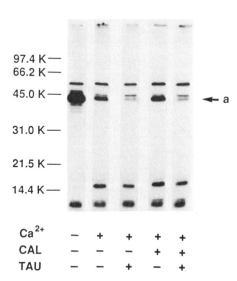

Figure 7. Autoradiographs from a 1-dimensional (PAGE) gel of the effects of 20 mM taurine on the phosphorylation of rat heart proteins (mitochondrial preparation). Incubation time was 6 min. Lane 1, control; lane 2, 0.7 mM Ca^{2+} plus 0.4 mM EGTA; lane 3, 0.7 mM Ca^{2+}, 0.4 mM EGTA, plus 20 mM taurine; lane 4, 0.7 mM Ca^{2+}, 0.4 mM EGTA, plus 50 units calmodulin; lane 5, 0.7 mM Ca^{2+}, 0.4 mM EGTA, 50 units calmodulin plus 20 mM taurine. Marker proteins with molecular weights ranging from 14.4 to 97.4 K are indicated.

It is well established that phosphorylation of key substrate proteins involving protein kinases is an important regulatory process of cellular function. In the nervous system, extracellular signals (neurotransmitters or hormones) activate specific receptors in the plasma membrane and result in augmentation of second messengers, which in turn regulate specific protein kinases. Numerous studies have indicated that cAMP, cGMP, calmodulin, and phorbol esters are activators of various classes of protein kinases and hence play a regulatory role in signal transduction in the brain[22]. In our laboratory over the last few years, we have actively pursued the role of taurine as a modulator of the phosphorylation of proteins in the retina, brain, and heart. In these studies we observed that taurine is an inhibitor of the phosphorylation of specific proteins although in some instances taurine stimulates protein phosphorylation. In addition, it has now been demonstrated that taurine affects both Ca^{2+} binding and uptake in excitable tissues[3,9,14,23,24] and inhibits the release of various neurotransmitters[25-28]. Thus, taurine may have a function in the complex and interconnected series of reactions involved in protein phosphorylation, Ca^{2+} regulation, and neurotransmitter release. The present results further support a regulatory role for taurine in mammalian tissues.

DEDICATION

I would like to dedicate this paper to Professor Doriano Cavallini for his 75th birthday and in recognition of his numerous contributions in advancing the knowledge of sulfur metabolism.

ACKNOWLEDGEMENTS

This work was supported in part by NEI Grant EYO4780 and the Tarbox Institute of Texas Tech University Health Sciences Center. Mr. J. Landon Farley and Mr. Ernest Redding are thanked for the photography of the autoradiograms.

REFERENCES

1. R.J. Huxtable, Taurine in the central nervous system and the mammalian actions of taurine, *Prog. Neurobiol.* 32:471-533 (1989).
2. J.B. Lombardini, Taurine: retinal function, *Brain Research Reviews* 16:157-169 (1991).
3. J.B. Lombardini, Effects of taurine on calcium ion uptake and protein phosphorylation in rat retinal membrane preparations, *J. Neurochem.* 45:268-275 (1985).
4. J.B. Lombardini, Inhibition by taurine of the phosphorylation of rat retinal membranes, *in*: "Taurine: Biological Actions and Clinical Perspectives," S.S. Oja, L. Ahtee, P. Kontro, and M.K. Paasonen, eds., Alan R. Liss, New York, pp. 384-393 (1985).
5. S.M. Liebowitz, J.B. Lombardini, and C.I. Allen, Effects of aminocycloalkanesulfonic acid analogs of taurine on ATP-dependent calcium ion uptake and protein phosphorylation, *Biochem. Pharmacol.* 37:1303-1309 (1988).
6. Y.P. Li and J.B. Lombardini, Taurine inhibits the phosphorylation of two endogenous proteins (M_r ~140 and ~20 K) in subcellular preparations of rat cortex, *Neurochem. Int.* 17:389-399.
7. Y.-P. Li and J.B. Lombardini, Taurine inhibits protein kinase C-catalyzed phosphorylation of specific proteins in a rat cortical P_2 fraction, *J. Neurochem.* 56:1747-1753 (1991).
8. Y.-P. Li and J.B. Lombardini, Inhibition by taurine of the phosphorylation of specific synaptosomal proteins in the rat cortex: Effects of taurine on the stimulation of calcium uptake in mitochondria and inhibition of phosphoinositide turnover, *Brain Res.* 553:89-96 (1991).
9. J.B. Lombardini and S.J. Liebowitz, Taurine modifies calcium ion uptake and protein phosphorylation in rat heart, *in*: "Taurine and the Heart", H. Iwata, J.B. Lombardini, and T. Segawa, eds., Kluwer Academic Publishers, Boston, pp. 117-137 (1989).
10. U.K. Laemmli, Cleavage of structural proteins during the assembly of the head of bacteriophage T_4, *Nature* 227:680-685 (1970).
11. P.H. O'Farrell, High resolution two-dimensional electrophoresis of proteins, *J. Biol. Chem.* 250:4007-

4021 (1975).

12. J.B. Lombardini, Effects of taurine on the phosphorylation of specific proteins in subcellular fractions of the rat retina, *Neurochem. Res.* (In Press).

13. J.B. Lombardini, Effects of taurine and mitochondrial metabolic inhibitors on ATP-dependent Ca^{2+} uptake in synaptosomal and mitochondrial subcellular fractions of rat retina, *J. Neurochem.* 51: 200-205 (1988).

14. H. Pasantes-Morales, N.E. Arzate, and C. Cruz, The role of taurine in nervous tissue: its effects on ionic fluxes, *in*: "Taurine in Nutrition and Neurology", R.J. Huxtable and H. Pasantes-Morales, eds., Plenum, New York, pp. 273-292 (1982).

15. K. Kuriyama, T. Hashimoto, M. Kimori, Y. Nakamura, and S.-I. Yamamoto, Taurine and receptor mechanisms in the heart: possible correlates with the occurrence of ischemic myocardial damages, *in*: "Taurine and the Heart, H. Iwata, J.B. Lombardini, and T. Segawa, eds., Kluwer Academic, Boston, pp. 139-158 (1989).

16. S.W. Schaffer, S. Allo, H. Harada, and J. Azuma, Regulation of calcium homeostasis by taurine: role of calmodulin, *in*: "Taurine: Functional Neurochemistry, Physiology, and Cardiology," H. Pasantes-Morales, D.L. Martin, W. Shain, and R. Martin del Rio, eds., Wiley-Liss, New York, pp. 217-225 (1990).

17. S.W. Schaffer, J.H. Kramer, W.G. Lampson, E. Kulakowski, and Y. Sakane, Effect of taurine on myocardial metabolism: role of calmodulin, *in*: "Sulfur Amino Acids: Biochemical and Clinical Aspects," K. Kuriyama, R.J. Huxtable, H. Iwata, eds., Alan R. Liss, New York, pp. 39-50 (1982).

18. N. Sperelakis, T. Yamamoto, G. Bkaily, H. Sada, A. Sawamura, and J. Azuma, Taurine effects on action potentials and ionic currents in chick myocadial cells, *in*: "Taurine and the Heart," H. Iwata, J.B. Lombardini, T. Segawa, Kluwer Academic, Boston, pp 1-20 (1989).

19. G. Bkaily, and N. Sperelakis, Calmodulin is required for a full activation of the calcium slow channels in heart cells, *J. Cyclic Nucleotide Protein Phosphorylation Res.* 11:25-34 (1986).

20. T. Segawa, Y. Nomura, and I. Shimazaki, Possible involvement of calmodulin in modulatory role of taurine in rat cerebral ß-adrenergic neurons, *in*: "Taurine: Biological Actions and Clinical Perspectives," Alan R. Liss, New York, pp. 321-330 (1985).

21. T. Segawa, Y. Momura, and I. Shimazaki, Further observations on the interaction of taurine and calmodulin on the central adrenergic neuron, *in*: "The Biology of Taurine", R.J. Huxtable, F. Franconi, A. Giotti, eds., Plenum Press, New York, pp. 341-346 (1987).

22. H.C. Hemmings, Jr., A.C. Nairn, A.C., T.L. McGuinness, R.L. Huganir, and P. Greengard, Role of protein phosphorylation in neuronal signal transduction, *FASEB J.*3:1538-1592 (1989).

23. J.B. Lombardini and S.D. Prien, Taurine binding by rat retinal membranes, *Exp. Eye Res.* 37:239-250 (1983).

24. J.B. Lombardini, S.M. Liebowitz, and T.-C. Chou, Analogues of taurine as stimulators and inhibitors of ATP-dependent calcium ion uptake in rat retina: Combination kinetics, *Mol. Pharmacol.* 36:256-264 (1989).

25. Arzate, M.E., Morán, J., and Pasantes-Morales, H., 1986, Inhibitory effect of taurine on 4-aminopyridine-stimulated release of labelled dopamine from striatal synaptosomes, *Neuropharmacology* 25:689-694 (1986).

26. J.R. Cunningham and M.J. Neal, Effect of γ-aminobutyric acid agonists, glycine, taurine, and neuropeptides on acetylcholine release from the rabbit retina, *J. Physiol.* 33:563-577 (1983).

27. K. Kuriyama, M. Muramatsu, K. Nakagawa, and K. Kakita, Modulating role of taurine on release of neurotransmitters and calcium transport in excitable tissues, *in*: "Taurine and Neurological Disorders", A. Barbeau and R.J. Huxtable, eds., Raven, New York, pp. 201-206 (1978).

28. P.S. Whitton, R.A. Nicholson, and H.C. Strang, Effect of taurine on neurotransmitter release from insect synaptosomes, *J. Neurochem.* 51:1356-1360 (1988).

TAURINE PROTECTION OF LUNGS IN HAMSTER MODELS
OF OXIDANT INJURY: A MORPHOLOGIC TIME STUDY
OF PARAQUAT AND BLEOMYCIN TREATMENT

Ronald E. Gordon, Richard F. Heller
and Rachel F. Heller

Mount Sinai School of Medicine
Department of Pathology - Box 1194
New York, New York 10029

INTRODUCTION

The intracellular reduction of oxygen is a natural part of nutrient conversion into energy and results in the production of small amounts of short-lived intermediates of oxygen; superoxide [O_2-], singlet oxygen [$^1O_2$], hydrogen peroxide [H_2O_2] and hydroxide radical [OH][1]. Naturally occurring enzymatic "scavengers" serve as the primary protection against oxidant injury[2]. When the production of oxygen intermediates exceeds the scavenging and detoxifying capacity of these enzymes, then lung parenchymal cell injury results[3].

The anticancer drug, bleomycin, is known to cause pulmonary fibrosis in humans[4,5]. A single intratracheal instillation of bleomycin in the hamster is a well established model for changes that are histologically and biochemically similar to the histopathologic changes in human lung exposed to bleomycin[5-7]. Although not fully understood, it is hypothesized that bleomycin, by its molecular structure, initially acts as a free radical. In the presence of iron and oxygen, bleomycin can also generate superoxide, hydroxyl radicals and other free radicals[8-10]. The action produces a sequence of events including an acute phase with epithelial injury, recruitment of inflammatory cells and edema, followed by proliferation of fibrocytes and accumulation of collagen, ultimately resulting in pulmonary fibrosis[5-7,11,12].

The broad spectrum herbicide paraquat (1,1-dimethyl-4,4'-bipyridulium dichloride) was shown to be injurious to the lung when inhaled and associated with marijuana smoke. Paraquat-induced lung injury resembles that caused by molecular oxygen, and bleomycin and is therefore useful as a model for the study of mechanisms of oxidant injury to the lung[13]. It can accumulate in the lung after I.P. and I.V. injection[3,14]. An energy dependent process may be responsible for lung tissue levels of paraquat that are higher than plasma levels of paraquat[15]. Paraquat in the lung appears to increase the turnover of cytoplasmic NADPH and increase mitochondrial activity changing the cytoplasmic redox state[16].

Taurine is a naturally occurring free-amino acid that exhibits antioxidant properties[17,18], appears to suppress membrane lipid peroxidation[19], and may have the properties of a general detoxifier[20]. In this study we examined the relationship between concentrations of taurine in the lungs of hamsters and the degree of acute injury and the regenerative response to a single dose of bleomycin or paraquat.

Guanidinoethyl sulfonate (GES) which acts as a competitive inhibitor of taurine binding and transport into cells acts as an effective agent to deplete cellular taurine levels. GES in the cell does not appear to have any metabolic function or induce any other side effects[21,22].

To evaluate the possible role of taurine in bleomycin and paraquat-induced injury to the lung, we studied the effects of bleomycin and paraquat on the lungs of hamsters on normal diets, hamsters given GES in drinking water and hamsters maintained on taurine supplemented diets.

MATERIALS AND METHODS

Ninety male Golden Syrian hamsters (70-100 g, Charles River Laboratories, Inc., Boston, Mass.) were randomized and prior to treatment, acclimated for four days in our laboratory facilities. Throughout the study, water and food (Purina Hamster Chow) were available *ad libitum*. For 28 days prior to injection, 30 hamsters were given a 1% solution of taurine (Sigma Chemical Co., St. Louis, Mo.) in their drinking water, 30 were given a 1% solution of guanidinoethyl sulfonate (GES, Pharmatech, Inc., West Orange, N.J.) in their drinking water and the remaining 30 were given tap water for drinking.

Paraquat (methyl viologen - Sigma Chemical Co., St. Louis, mo.) was dissolved in 0.9% sodium chloride solution and was given as a single 0.2 ml I.P. injection, total dosage 25 mg/Kg body weight, to each of 15 animals from each group. The remainder of the animals received a 0.2 ml intraperitoneal injection of a 0.9% sodium chloride solution. One day, 7 days and 28 days after injection 5 animals from each subgroup were anesthetized with sodium pentobarbital and exsanguinated. Serum, liver, and 1 lung were frozen and sent to Dr. John Sturman, Institute for Basic Research, Staten Island, NY, for biochemical assay of taurine content. The remaining lung was inflation fixed with a solution consisting of 3% glutaraldehyde buffered with a 0.2 M cacodylate solution.

Coronal lung sections were processed for embedment in paraffin, sectioned at 4 μm, stained with H & E and examined by light microscope. For each animal, ten blocks of lung tissue, each containing a bronchiole measuring approximately 2 mm x 2 mm x 1 mm were dissected so bronchioles could be oriented for exact cross-sectioning. The tissue blocks were processed, embedded in Epon, 1 μm section cut, stained with methylene blue and examined in a light microscope.

Statistical analysis of taurine content in lung tissue, serum and liver was based on a multiple analyses of variance (MANOVA). Data are reported as means $\pm$ SE. Differences were considered statistically significant at $p < 0.05$.

RESULTS

ANIMALS GIVEN GES IN DRINKING WATER

One day after a single I.P. injection of paraquat or tracheal instillation of bleomycin, light microscopy of the lung revealed a generalized, severe edema and aggregation of large numbers of inflammatory cells (Figure 1). Electron microscopy confirmed that the inflammatory cells were macrophages and neutrophils. The paraquat lungs had a greater degree of edema when compared with the bleomycin

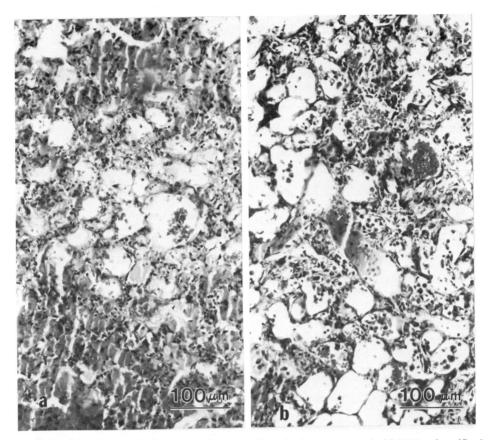

Figure 1. Light micrographs of representative areas of lungs for hamsters treated with GES and sacrificed 1 day after paraquat treatment (a) or bleomycin treatment (b). It is possible to discriminate a greater amount of edema in the paraquat treated lung and a greater cellular infiltrate in the bleomycin treated lung. Sections were stained with hematoxylin and eosin. X200

treated lungs (Figure 1). After seven days, it was evident that the edema was reduced, sites of alveolar bronchiolarization were apparent throughout the lung, and there was substantial interstitial fibrosis and alveolar bronchiolarization throughout the lung (Figure 2). Twenty-eight days after paraquat treatment, there was no edema present, but there were persistent alveolar and interstitial inflammatory cells and a severe interstitial fibrosis and alveolar bronchiolarization (Figure 3).

ANIMALS GIVEN CLEAN DRINKING WATER

One day after a single I.P. injection of paraquat and intratracheal instillation of bleomycin, light microscopy of the lung revealed a localized, mild edema, and aggregation of large numbers of inflammatory cells. The response exhibited macrophages and neutrophils, but was less severe than that seen in animals treated with GES. After seven days, there was no evidence of edema, and only small focal areas of alveolar bronchiolarization and interstitial fibrosis of the lung, mainly at the lung periphery (Figure 4). Twenty-eight days after treatment, there was no edema present, but there was a persistence of alveolar and interstitial inflammatory cells and localized interstitial fibrosis qualitatively equivalent to what was observed in the lungs of animals treated with GES and sacrificed 7 days after oxidant treatment (Figure 5).

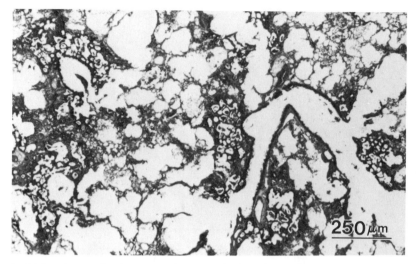

Figure 2. Light micrograph of representative area of a lung from a hamster treated with GES and sacrificed 7 days after treatment with bleomycin. Large areas of alveolar bronchiolarization and fibrosis are established. Sections were stained with hematoxylin and eosin. X80

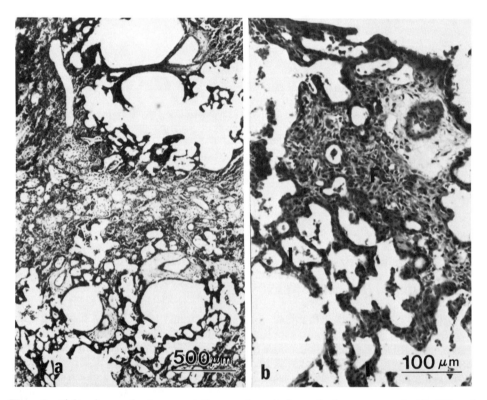

Figure 3. Light micrograph of representative area from the lung of a hamster treated with GES and sacrificed 28 days after treatment with paraquat. At low magnification (a), it is apparent that the entire lung field is effected. At higher magnification (b), it is possible to see the well-established peribronchiolar fibrosis (F), interstitial fibrosis (I) and the alveolar bronchiolarization. Sections were stained with hematoxylin and eosin. X40 and X200

322

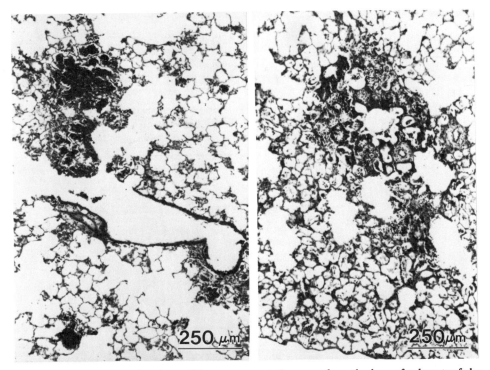

Figure 4. Light micrography of two different representative areas from the lung of a hamster fed a normal diet and sacrificed 7 days after treatment with bleomycin. In both areas it is possible to identify focal peribronchial and bronchiolar inflammatory infiltrate without evidence of edema. There is no evidence of interstitial fibrosis or alveolar bronchiolarization Sections were stained with hematoxylin and eosin. X80

Figure 5. Light micrograph of representative area from the lung of a hamster fed a normal diet and sacrificed 28 days after treatment with paraquat. Peripheral and more centrally located areas (arrows) of interstitial fibrosis and alveolar bronchiolarization are exhibited. There are substantial areas of normal lung. Sections were stained with hematoxylin and eosin. X40

323

ANIMALS GIVEN TAURINE IN DRINKING WATER

One day after a single I.P. injection of paraquat or intratracheal instillation of bleomycin, light microscopy of the lung revealed a few small, local foci of peribronchiolar inflammation, mild edema, and a slight accumulation of inflammatory cells. After seven days, foci of inflammatory cells were seen with no evidence of edema, alveolar bronchiolarization, or fibrosis (Figure 6). Twenty-eight days after paraquat treatment, there was a clear resolution of the inflammatory response and only a few small foci of fibrosis and alveolar bronchiolarization at the lung periphery (Figure 7).

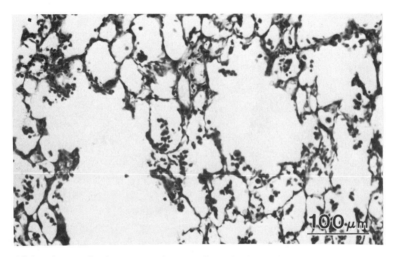

Figure 6. Light micrograph of representative area from the lung of a hamster fed taurine in drinking water and sacrificed 7 days after treatment with bleomycin. There was a minor inflammatory infiltrate without edema. There was no evidence of any other lesions. Sections were stained with hematoxylin and eosin. X200

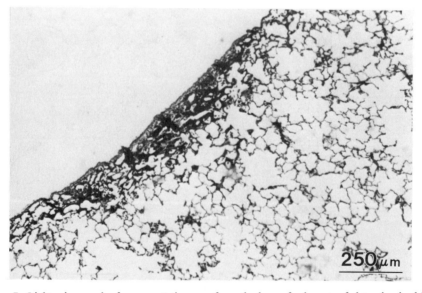

Figure 7. Light micrograph of representative area from the lung of a hamster fed taurine in drinking water and sacrificed 28 days after treatment with paraquat. Only small, peripheral sites of interstitial fibrosis were identified. There was no evidence of any other lesions. Sections were stained with hematoxylin and eosin. X80

Animals treated with taurine alone were morphologically no different from saline controls. However, animals treated with GES alone exhibit a qualitatively mild generalized increase in inflammatory cells and on occasion there was a peripheral focus of peribronchiolar and interstitial fibrosis.

TISSUE AND PLASMA TAURINE LEVELS

Lung means taurine levels measured in the animals treated with bleomycin, paraquat, or saline are plotted in Figure 8a, 8b, and 8c. The GES was effective only

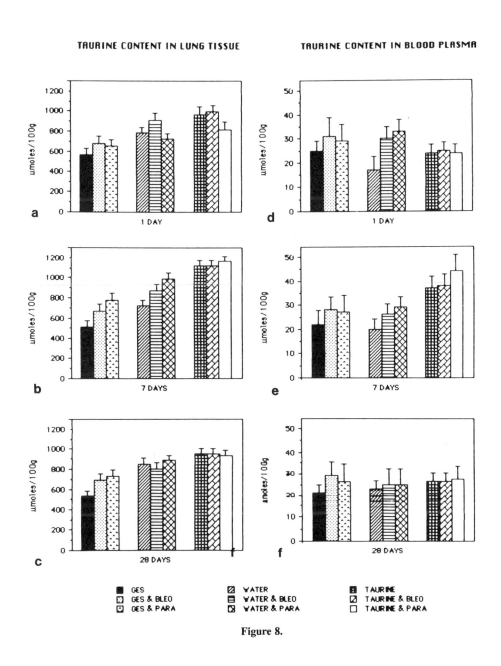

Figure 8.

in reducing taurine levels. Dietary taurine supplementation increased taurine levels in the animals. The most interesting finding was the increase in lung taurine levels in the animals treated with oxidants. This is not significant in taurine supplemented animals. Taurine supplementation and deprivation itself had little effect on plasma levels (Figure 8d, 8e, and 8f). The most significant effects were observed in the animals receiving oxidant treatments and particularly those animals sacrificed 7 days after paraquat or bleomycin treatment.

DISCUSSION

Bleomycin and paraquat lung injury in hamsters have been used as a model in the study of possible mechanisms in the development of pulmonary fibrosis and to test agents that may prevent the development of these lesions[9,14,15,23-32]. Both bleomycin and paraquat act to form superoxides and free radicals. However, the mechanisms by which the superoxides and free radicals are formed may be different, but the results are clearly similar if not the same.

In the present study, we have used these models to test the effectiveness of taurine in the prevention of inflammation and fibrosis caused by these agents. We found that taurine was significant in the prevention of the initial acute inflammatory response and the later development of fibrosis. We also conclude based on the effects of GES-treated animals compared with animals on normal diets, that endogenous taurine levels are important under normal circumstances in preventing oxidant injury. Maintaining tissue or cellular taurine levels was critical to the prevention of oxidant injury to the lung.

Based on qualitative evidence it was obvious that the taurine was preventing the early edema and inflammatory response. This would indicate that the action of the taurine was to deactivate or detoxify the paraquat or bleomycin before they had a chance to interact with the tissue and form oxidants and free radicals. The taurine could be interacting with the oxidants or free radicals directly or it could be also interacting with the cells in the tissue to prevent the bleomycin or paraquat interaction with the cellular components resulting in the production of the oxidants and the free radicals.

Paraquat's major effect appears to be lipid peroxidation of plasma membranes[33]. Based on this mechanism it could be assumed that at least one site of taurine action is at the plasma membrane. Its function at that site is speculative, but it could intercalate into the membrane thereby stabilizing it and not allowing lipid peroxidation to occur by blocking the appropriate interaction or it is acting as a detoxifier to bind the paraquat or bleomycin before it acts on the membrane. Less likely but also possible, is that there is lipid peroxidation, but the taurine is scavenging the oxidants produced. The best test for such a mechanism would be to look for products of lipid peroxidation of plasma membrane other than the oxidants themselves, such products include chemotactic factors and cytokines. It is these factors which ultimately result in recruitment of inflammatory cells and, migration and proliferation of fibrocytes responsible for the development of fibrosis.

Based on the data presented in this manuscript, designating any singular molecular mechanism for the protection imposed by taurine would be purely speculative and beyond the scope of the data contained in this study. The one interesting and yet unexplainable phenomenon was the increase in taurine content in the lungs of hamsters treated with bleomycin and paraquat as compared to their untreated controls. This observation is similar to those reported by Wang, et al.[34]. They attributed the increase to release of cytosolic content including taurine from injury to epithelial and connective tissue cells. However, such a mechanism would be significant and would, therefore, indicate that taurine is not protecting the cells. In our study the taurine supplemented animals did not show a significant increase in lung taurine after oxidant treatments. This would be further evidence that the tissue taurine had protected the lung cells. An alternate scenario would be that plasma taurine may be mobilized to the lungs. Minor injury to endothelium making it leaky would lead to some degree of edema as a source of the increased taurine levels. The edema would not necessarily lead to any additional lung injury because of increased taurine levels. It would also account for the increase in BALF levels seen by Wang, et al.[34].

Our observations (unpublished) were that oxidant injury results in cell enlargement or swelling. It is also known that when cells become hypotonic taurine is released which may account for the increased taurine levels observed by Wang, et al[34] in the BALF. Such release during injury may make cells more vulnerable if insult should continue. This could be the mechanism for permanent damage to the lung resulting from long-term intermittent or continuous exposure. It is well documented that the liver is a major site of taurine synthesis and storage and it may be the source of the mobilized taurine similar to alpha-1-antiprotease in other cases of oxidant injury to the lung and most important in the prevention of emphysema. The plasma taurine levels correlate well with increased lung tissue levels and increased levels associated with oxidant treatment. Since the bleomycin and paraquat become systemic, a less likely but possible process may just represent liver injury as well as lung injury and liver stores of taurine are redistributed throughout the body.

The results of this study indicate that taurine significantly abated the inflammation and fibrosis caused by bleomycin and paraquat. Future studies are being designed around the determination of specific sites of interaction of taurine with cellular components and whether taurine's interaction will prevent the injury and release of molecules that could influence the development of inflammation and fibrosis.

ACKNOWLEDGEMENTS

The authors would like to thank Mr. Norman Katz, Mr. Ron Uson, Ms. Ana Bazan, Ms. Sharon Smith, and Ms. Audrey Stedford for their technical assistance and Mr. Norman Katz and Mr. Glenn Maffei with the photographic portions of this manuscript. Special thanks to Dr. John Sturman for his work in the determination of taurine in the blood, liver, and lung samples. This grant was supported by National Institute of General Medical Sciences grant GM 08174.

REFERENCES

1. I. Fridovich, The biology of oxygen radicals, *Sci.* 201:875-880 (1978).
2. J.M. McCord, The biology and pathology of oxygen radicals, *Ann. Intern. Med.* 898:12-127 (1978).
3. M.S. Rose, L.L. Smith, and L. Wyatt, Evidence for energy-dependent accumulation of paraquat in rat lung, *Nature* (London) 252:314-315 (1974).
4. M. Luna, C. Bedrossian, L. Lichtiger, and P.A. Salem, Interstitial pneumonitis associated with bleomycin therapy, *Am. J. Clin. Pathol.* 58:501-510 (1974).
5. G.L. Snider, B.R. Celli, R.H. Goldstein, J.J. O'Brien, E.C. Lucy, Chronic interstitial pulmonary fibrosis produced in hamsters by endotracheal bleomycin, *Am. Rev. Respir. Dis.* 1170-289-297.
6. G.L. Snider, J.A. Hayes, a.L. Kortly, Chronic interstitial pulmonary fibrosis produced in hamsters by endotracheal bleomycin: Pathology and stereology, *Am. Rev. Respir. Dis.* 117:1099-1108.
7. J.E. Zuckerman, J.E., M.A. Hollinger, S.N. Giri, Evaluation of antibiotic drugs in bleomycin-induced pulmonary fibrosis in hamsters, *J. Pharmacol. Exp. Ther.* 213:425-431 (1980).
8. E.A. Sausville, K. peisach, S.B. Horowitz, Effect of chelating agents and metal ions on the degradation of DNA by bleomycin, *Biochem.* 17:2740-2746 (1978).
9. H. Ekimoto, H., H. Kuramochi, K. Takahashi, A. Matsuda, H. Umezawa, Kinetics of the ration of bleomycin-Fe(II)-O_2 complex with DNA, *J. Antibiot.* (Tokyo); 33:426-434 (1980).
10. J.M.C. Gutteridge, T. Xiai-Chang, Protection of iron-catalyzed free radical damage to DNA and lipids by copper (II)-bleomycin, *Biochem. Biophys. Res. Commun.* 99:1354-1360 (1981).

11. D.B. Chandler, D.M. Hyde, S.N. Giri, Morphometric estimates of infiltrative cellular changes during the development of bleomycin-induced pulmonary fibrosis in hamsters, *Am. J. pathol.* 112-170-177 (1983).

12. R.S. Thrall, R.W. Barton, a comarison of lymphocyte populations in lung tissue and in bronchiolar lavage fluid of rats at various times during the development of bleomycin-induced pulmonary fibrosis, *Am. Rev. Respir. Dis.* 129: 279-283.

13. J.E. Bus and J.E. Gison, Paraquat: Model of oxidant-initiated toxicity, *Environ. Health Perspectives* 55:37-46 (1984).

14. M.S. Rose, E.L. Lock, L.L. Smith, and L. Wyatt, Paraquat accumulation: tissue and species specificity, *Biochem. Pharmacol.* 25:419-423 (1976).

15. L.L. Smith, M.S. Rose, I. Wyatt, The pathology and biochemistry of paraquat, *CIBA Found Symp.* 65:321-341 (1978).

16. D.J.P. Bassett and A.B. Fisher, Alterations of glucose metabolism during perfusion of rat lung with paraquat, *Am. J. Physiol.* 234(6):E653-E659 (1978).

17. J.G. Alvarez and B.T. Storey, Taurine, hypotaurine and albumin inhibit lipid peroxidation in rabbit spermatozoa and protect against loss of motility, *Biol. Reprod.* 291:548-555 (1983).

18. M.J. McBroom and J.D. Welty, Comparision of taurine-verapamil interaction in hamsters and rats, *Comp. Biochem. Physiol.* 80C(2):217-219 (1985).

19. T. Nakashima, T. Takino, and K. Kuriyama, Therapeutic and prophylactic effects of taurine administration on experimental liver injury, *in:* "Sulfur Amino Acids: Biochemical and Clinical Aspects," K. Kuriyama, R.J. Huxtable, and H. Iwata, eds., pp. 449-459, Alan R. Liss Inc. NY (1985).

20. C.E. Wright, T.T. Lin, Y.Y. Lin, J.A. Sturman, and G.E. Gaull, Taurine scavengers oxidized chlorine in biological systems, *in:* "Taurine: Biologial Actions and Chemical Perspectives," S.S. Oja, L. Ahtee, P. Kontrol, and M.K. Paasonen, eds., pp. 137-147, Alan R. Liss Inc., NY (1985).

21. R.J. Huxtable, H.E. Laird, and S.E. Lippincott, The transport of taurine in the heart and the rapid depletion of tissue taurine by guanidinoethyl sulfonate, *J. Pharmac. Exp. Ther.* 211:465-471 (1979).

22. J.D. Welty and M.J. McBroom, Effects of taurine and verapamil on heart calcium in normal (F_1B) and cardiomyopathic (BIO 14.6) hamsters, *Fedn. Proc. Fedn. Am. Soc. Exp. Biol.* 41:1524 (1982).

23. C.W.N. Bedrossian, M.A. Luna, B. MacKay, and B. Lichtiger, Ultrastructure of pulmonary bleomycintoxicity, *Cancer* 32:44-51.

24. H.F. Krous and W.B. Hanlin, Pulmonary toxicity due to bleomycin, *Arch. Pathol.* 95:407-410 (1973).

25. S.N. Giri, D.M. Hyde, and B.J. Marafio, Ameliorating effect of murine interferon gamma on bleomycin-induced lung collagen fibrosis in mice, *Biochem. Med. Metab. Biol.* 36:194-197.

26. D.B. Chandler, T.W. Butler, D.D. Brigs III, W.E. Grizzle, J.C. Barton, and J.D. Fuller, Modulation of the development of bleomycin-induced fibrosis by deferoxamine, *Toxicol. Appl. Pharmacol.* 92:358-367 (1988).

27. S.N. Giri, Z. Chen, W.R. Younker, and M.J. Schmidt, Effects of intratracheal administration of bleomycin on GSH-shuttle enzymes, catalase, lipid peroxidation and collagen content in the lungs of hamsters, *Toxicol. Appl. Pharmacol.* 71:132-141 (1983).

28. S.N. Giri, D.M. Hyde, and J.M. Nakashima, Analysis of bronchiolar lavage fluid from bleomycin-induced pulmonary fibrosis in hamsters, *Toxicol. Pathol.* 14:149-157 (1986).

29. S.N. Giri, J.M. Nakashima, and D.L. Curry, Effects of intratracheal administration of bleomycin or saline in pair-fed and control-fed hamsters on daily food intake and on plasma levels of glucose, *Toxicol. Pathol.* 15:163-172.

30. D.J. Schreier and S.H. Phan, Modulation of bleomycin-induced pulmonary fibrosis in the BALB/c mouse by cyclophosphamide-sensitive T cells, *Am. J. Pathol.* 116:270-278 (1984).

31. J.M. Nakashima, D.M. Hyde, and S.N. Giri, Effects of a calmodulin inhibitor on bleomycin-induced lung inflammation in hamsters, Biochemical morphometric and bronchio-alveolar lavage data, *Am. J. Pathol.* 124:528-536 (1986).

32. P. Smith, D. Heath, and J.M. Kay, The patholgenesis and development of paraquat-induced pulmonary fibrosis in rats, *J. Path.* 114:57-67 (1974).

33. T.K. Aldrich, A.B. Fisher, E. Cadenas, and B. Chance, Evidence for lipid peroxidation by paraquat in the perfused rat lung, *J. Lab. Clin. Med.* 101:66-73 (1983).

34. Q. Wang, S.N. Giri, D.M. Hyde, and J.M. Nakashima, Effects of taurine on bleomycin-induced lung fibrosis in hamsters, *Proc. Soc. Exper. Biol. Med.* 190:330-338 (1989).

TAURINE AND NIACIN OFFER A NOVEL THERAPEUTIC
MODALITY IN PREVENTION OF CHEMICALLY-INDUCED
PULMONARY FIBROSIS IN HAMSTERS

S.N. Giri and Q. Wang

Department of Veterinary Pharmacology and Toxicology
University of California
Davis, CA 95616

INTRODUCTION

Taurine, a sulfur containing ß-amino acid, is present in high concentrations in most mammalian tissues. Although taurine is not a constituent of mammalian proteins, it plays an important role in the pathophysiology of cardiovascular and nervous systems[1,2] and conjugation of bile acids[3]. The beneficial effects of taurine against oxidant-induced tissue injury have been attributed to its ability to stabilize biomembranes[4,5] and scavenge reactive oxygen species (ROS) including hypochlorous acid[6-8]. Several effects of taurine on the cardiovascular and central nervous systems may be related to calcium redistribution since taurine alters calcium transport[9,10]. It has been demonstrated that exogenously administered taurine would offer protection against oxidant-induced lung damage[11,12]. In addition, Nakashima et al.[13] reported therapeutic and prophylactic effects of taurine on experimentally-induced liver injury.

Although the underlying mechanisms for the protective effect of taurine against tissue damage are not clearly understood, its antioxidant, membrane-stabilizing and ROS and HOCl scavenging properties seem to be somehow involved[4-8]. It has been suggested that taurine acts as a superoxide scavenger and reduces the formation of ROS in rabbit spermatozoa[14] and this was attributed to the ability of hypotaurine to scavenge hydroxy radicals by its sulfinate moiety[15]. Although taurine per se is a poor scavenger of free radicals[15], its ability to scavenge HOCl in biological systems is well established[6-8].

Niacin, a B vitamin, is a precursor of NAD and NADP[16] and shown to prevent cytotoxicity and DNA damage by maintaining intracellular level of NAD[17]. It has been demonstrated in the cell culture study that non-toxic concentrations of H_2O_2 caused an early and reversible drop in NAD and ATP levels while toxic concentrations caused an irreversible depletion[18,19]. Hypochlorous acid released from activated PMN caused depletion of NAD and ATP and this led to cell injury[20,21].

One of the postulated mechanisms for bleomycin(BL)-induced pneumotoxicity which eventuates in fibrosis, is DNA damage[22]. We have demonstrated that intratracheal (IT) instillations of BL in hamsters increased the lung poly(ADP-ribose) polymerase activity, secondary to DNA strand breaks and caused a marked depletion

Taurine, Edited by J.B. Lombardini *et al.*
Plenum Press, New York, 1992

of lung NAD during the first week of the study [23]. It has been shown that H_2O_2-induced DNA strand scission was also associated with the depletion of NAD and ATP and it was suggested that the overactivation of poly(ADP-ribose) polymerase in response to DNA strand breaks was responsible for the depletion[24]. Poly(ADP-ribose) polymerase catalyzes the transfer of ADP-ribose residues from NAD to acceptor proteins to form ADP-ribose polymer which facilitates DNA repair by activating DNA ligase and maintaining proper chromosomal conformation essential to the DNA repair process[25]. Thus, an overactivation of poly(ADP-ribose) polymerase leads to intracellular depletion of NAD which serves as a substrate for the synthesis of poly(ADP-ribose)[26]. Agents that cause DNA strand breaks stimulate the synthesis of this polymer which helps repair the damaged DNA. However, a sustained synthesis of this polymer would lead to intracellular depletion of NAD and decrease in poly(ADP-ribose) synthesis[27]. Intracellular depletion of NAD would also inhibit the glycolytic pathway and ATP formation[28]. It has been demonstrated that cytotoxicity in isolated hepatocytes results from a marked depletion of intracellular NAD and its maintenance offered protection against the cytotoxicity[24].

The mechanisms for BL-induced lung injury and the subsequent development of fibrosis are not clearly understood. However, it is known that BL binds to DNA and iron and generates ROS under aerobic conditions[29]. These ROS being in close proximity, cause DNA damage and lipid peroxidation followed by the light and ultrastructural microscopic histopathologic changes of the lung[30]. We have reported in our earlier papers that administration of either taurine[12] or niacin[31] partially reduced lung inflammation and fibrosis in the BL-hamster model. Since taurine and niacin produce their antifibrotic effects presumably by different mechanisms, it was hypothesized that the combined treatment with these two compounds could greatly ameliorate the lung fibrosis in the BL-hamster model. An attempt will be made not only to report this aspect of our study in this symposium but also to discuss the beneficial effect of taurine and/or niacin against the lung fibrosis caused by another therapeutic agent, amiodarone. Amiodarone is used to treat cardiac arrhythmia[32] but, unfortunately, produces undesirable side effects such as pulmonary fibrosis[33] and phospholipidosis[34].

METHODS

The combined effect of taurine and niacin against BL-induced lung fibrosis was evaluated in two independent studies. In Study #1, taurine (T) was given in drinking water (1% W/V) and niacin (N) by IP route (250 mg/kg) for 2 days prior to the first IT instillation of BL or saline (SA) and thereafter daily. In Study #2, T (2.5% W/W) and N (2.5% W/W) were thoroughly mixed with the pulverized chow and given 2 days prior to the first IT instillation of BL. Bleomycin (2.5, 2, 1.5 units/5ml/kg) or an equivalent volume of SA was instilled IT in three consecutive doses at weekly intervals. Twenty days after the last IT instillation, the hamsters were anesthetized and blood and tissue samples were collected for all biochemical measurements as previously described[35]. Bronchoalveolar lavage (BAL) followed by fixation of lung using cacodylate-buffered glutaraldehyde paraformaldehyde fixative (400 mOsm) via trachea were carried out in some hamsters. Total cell counts in BALF (BAL fluid) was estimated by Coulter counter and slides for differential cell counts were prepared by the method of Willcox et al.[36]. The remaining BALF was centrifuged at 1500 g for 20 min at 4°C. The resulting supernatant was used for the assay of acid phosphatase activity[37] and protein content[38]. Differential cell counts were based on a minimum of 250 cells counted per slide. Histopathologic and morphometric analysis including the volume of parenchymal lesions of the lung sections were done as described[39].

In the amiodarone (AD) study, hamsters were given pulverized chow supplemented with taurine (2.5% W/W) and/or niacin (2.5% W/W) for 6 days before the

IT instillation of AD (15 μmol/5 ml/kg) or saline and thereafter daily through the entire period of the study; hamsters were killed at 21 days after the IT instillation. The lungs were processed for all biochemical measurements as described above.

All data are expressed on a per total lung basis in order to avoid the artificial lowering of the values in BL or AD treated animals[42]. The values are reported as the mean $\pm$ SE and analyzed by a one-way analysis of variance and Duncan's multiple range test[40]. A value of P $\leq$ 0.05 was considered significant.

RESULTS

Effects of combined treatment with taurine (in drinking water) and niacin (IP) in Study #1 and their dietary supplementation in Study #2 on BL-induced changes in various biochemical and morphometric measurements are summarized in Tables 1 and 2, respectively. Bleomycin significantly increased the lung collagen content, measured as hydroxyproline, to 152 and 194% of the SA control in Study #1 and Study #2, respectively. Combined treatment with taurine and niacin ameliorated the BL-induced increases in the lung collagen content in both studies. Similarly, this combined treatment also blocked the BL-induced increases in the lung prolyl hydroxylase activity from 166% in Study #1 and 181% in Study #2 to 121 and 108% of the SA control, respectively. The lung malondialdehyde equivalent (MDAE), an index of lipid peroxidation, and superoxide dismutase (SOD) activity were significantly increased in BL-treated hamsters to 260% and 122% in Study #1 and 194 and 145% of the SA control in Study #2, respectively. The combined treatment with taurine and niacin caused significant reductions in the BL-induced increases in the lung MDAE and SOD activity in both studies. The combined treatment also markedly decreased the BL-induced increases in the total lung Ca^{2+} and poly(ADP-ribose) polymerase activity from 171 and 194% to 122 and 115% in Study #1 and from 187 and 183% to 113 and 105% of the SA control, in Study #2, respectively. The BL-induced increases in the pulmonary vascular permeability (measured by the protein content in BALF supernatant), acid phosphatase activity and volume of lesions were markedly suppressed in both studies by the combined treatment with taurine and niacin.

The effects of the combined treatment with taurine and niacin on BL-induced changes in total and differential cell counts of BALF are summarized in Table 3 for Study #1 and Table 4 for Study #2. In both studies, increases in the total and differential cell counts in the BALF of BL-treated hamsters were significantly suppressed by the combined treatment, especially as reflected in total cell, neutrophil, monocyte and lymphocyte counts.

Effects of dietary supplementation with taurine and/or niacin on AD-induced lung fibrosis and phospholipidosis are summarized in Table 5. Intratracheal instillation of AD increased the lung hydroxyproline and phospholipid content to 154 and 133% of the SA control, respectively. Taurine, niacin and taurine plus niacin decreased the AD-induced increase in hydroxyproline content from 154 to 125, 117 and 116% of the SA control, respectively. Similarly, phospholipid content in the corresponding groups were significantly decreased from 133 to 123, 113 and 114% of the SA control. The taurine and/or niacin treatment also suppressed the AD-induced increases in the lung MDAE and SOD activity.

DISCUSSION

The two therapeutic agents bleomycin, an anticancer drug, and amiodarone, an antiarrhythmic drug, share one thing in common; they cause lung toxicity which

Table 1. Effects of Combined Treatment With Taurine (in Drinking Water) and Niacin (IP) on Bleomycin-Induced Increases in Various Biochemical and Morphometric Parameters in Hamster Lungs.

Parameters	SA	TNSA	BL	TNBL
Hydroxyproline (µg/Lung)	989.7 ± 51.20(7)	1038.0 ± 39.0(8)	1508.0 ± 63.0[a](14)	1018.0 ± 39.00 (14)
Prolyl Hydroxylase (dpm x 10^{-3}/30min/lung)	162.0 ± 22.00(7)	165.0 ± 8.0(8)	268.0 ± 18.0[a](14)	197.0 ± 10.00 (14)
Malondialdehyde Equivalent (nmol/lung)	103.4 ± 3.27(7)	116.9 ± 14.2(8)	268.0 ± 26.1[a](14)	165.4 ± 10.70[b](14)
Superoxide Dismutase (units/Lung)	408.1 ± 23.90(7)	417.8 ± 15.0(8)	498.2 ± 19.5[a](14)	416.4 ± 18.30 (14)
Lung Total Calcium (µg/lung)	29.1 ± 2.10(6)	32.4 ± 1.7(8)	49.6 ± 2.2[a](14)	35.6 ± 2.20 (14)
Poly(ADP-Ribose) Polymerase (pmol/Lung)	827.2 ± 202.40(7)	869.4 ± 115.0(7)	1604.4 ± 293.1[a](14)	954.9 ± 104.60 (14)
Plasma Taurine (nmol/ml)	115.3 ± 3.76(7)	236.9 ± 19.9[c](8)	109.6 ± 5.97(14)	225.1 ± 16.90[c](14)
BALF Supernatant Protein (µg/lung)	1789.0 ± 212.00(4)	1551.0 ± 141.0(5)	4738.0 ± 302.0[a](6)	3125.0 ± 269.00[d](6)
BALF Acid Phosphatase (nmol/h/lung)	253.6 ± 9.10(4)	216.3 ± 6.6(5)	610.8 ± 112.4[a](6)	349.1 ± 86.30 (6)
Lung Volume (cm^3)	6.1 ± 0.25(4)	5.6 ± 0.1(5)	5.0 ± 0.28[e](6)	5.8 ± 0.03 (6)
Volume of Lesions (cm^3)	0 (4)	0 (5)	0.8 ± 0.17[a](6)	0.04 ± 0.01 (6)

Hamsters were treated with taurine in drinking water (1%) and niacin (250mg/kg, IP daily) and given bleomycin IT (2.5, 2.0 and 1.5 U/5ml/kg bw) in three consecutive doses at weekly intervals. Twenty days after the last IT instillation, hamsters were killed and lung and bronchoalveolar lavage fluid (BALF) processed for various biochemical and morphometric analysis as described in Methods. SA: Saline control; TNSA: Taurine + Niacin + Saline; BL: Bleomycin alone; TNBL: Taurine + Niacin + Bleomycin. The number of animals in each group is shown in parentheses. Values are expressed as mean ± SEM. [a] Significantly higher ($P < 0.05$) than all other corresponding groups. [b] Significantly lower ($P < 0.05$) than the corresponding BL group, but higher ($P < 0.05$) than corresponding SA group. [c] Significantly higher ($P < 0.05$) than corresponding SA and BL groups. [d] Significantly lower ($P < 0.05$) than the corresponding BL group, but higher ($P < 0.05$) than corresponding SA and TNSA groups. [e] Significantly lower ($P < 0.05$) than all other corresponding groups.

Table 2. Effects of Dietary Treatment With Taurine and Niacin on Bleomycin-Induced Increases in Various Biochemical and Morphometric Parameters in Hamster Lungs.

Parameters	SA	TNSA	BL	TNBL
Hydroxyproline (µg/Lung)	908.6 ± 56.5(7)	848.4 ± 36.4(7)	1763.0 ± 122.0[a](9)	896.3 ± 25.2 (13)
Prolyl Hydroxylase (dpmx10^{-3}/30min/Lung)	112.0 ± 5.2(7)	107.0 ± 9.0(7)	203.0 ± 10.0[a](9)	121.0 ± 6.7 (13)
Malondialdehyde Equivalent (nmol/Lung)	109.4 ± 6.98(7)	109.5 ± 8.9(7)	211.1 ± 10.5[a](9)	146.4 ± 8.1[b](13)
Superoxide Dismutase (units/Lung)	400.5 ± 13.8(7)	360.3 ± 12.2(7)	582.2 ± 22.2[a](9)	385.6 ± 7.1(13)
Lung Total Calcium (µg/Lung)	33.9 ± 1.3(6)	31.7 ± 0.8(6)	63.7 ± 4.1[a](9)	38.2 ± 1.6(13)
Poly(ADP-Ribose) Polymerase (pmol/Lung)	872.4 ± 223.8(7)	773.2 ± 110.7(7)	1595.5 ± 86.0[a](9)	915.8 ± 151.1(13)
Plasma Taurine (nmol/mL)	119.9 ± 7.7(7)	344.6 ± 47.2[c](7)	159.2 ± 15.5(7)	373.5 ± 68.8[c](7)
BALF Supernatant Protein (µg/Lung)	1520 ± 95(5)	1602.0 ± 113.0(5)	5043.0 ± 619.0[a](6)	3249.0 ± 298.0[b](6)
BALF Acid Phosphatase (nmol/h/Lung)	235.3 ± 36.1(5)	296.8 ± 66.70(5)	710.7 ± 141.2[a](6)	380.4 ± 115.9(6)
Lung Volume (cm^3)	6.72 ± 0.27[a](5)	5.74 ± 0.25(5)	4.7 ± 0.19[d](6)	5.61 ± 0.13(6)
Volume of Lesions (cm^3)	0 (5)	0 (5)	0.92 ± 0.14[a](6)	0.127 ± 0.02(6)

Hamsters were fed taurine (2.5%) and niacin (2.5%) in the pulverized diet and given bleomycin IT (2.5, 2.0 and 1.5 U/5ml/kg bw) in three consecutive doses at weekly intervals. Twenty days after the last IT instillation, hamsters were killed and lung and bronchoalveolar lavage fluid (BALF) processed for various biochemical and morphometric analysis as described in Methods. SA: Saline control; TNSA: Taurine + Niacin + Saline; BL: Bleomycin alone; TNBL: Taurine + Niacin + Bleomycin. The number of animals in each group is shown in parentheses. Values are expressed as mean ± SEM. [a] Significantly higher (P < 0.05) than all other corresponding groups. [b] Significantly lower (P < 0.05) than the corresponding BL group, but higher (P < 0.05) than corresponding SA and TNSA groups. [c] Significantly higher (P < 0.05) than corresponding SA and BL groups. [d] significantly lower (P < 0.05) than all other corresponding groups.

eventuates in lung fibrosis. In this symposium, we have presented data which suggest that administration of taurine and/or niacin either completely or partially ameliorated the lung collagen accumulation, a hallmark of fibrosis. The mechanisms for their antifibrotic effects are not clearly understood. However, it is possible that the antioxidant, membrane stabilizing and ROS and HOCl scavenging properties of taurine[4-8], and the ability of niacin to increase tissue levels of NAD[17] and inhibit overstimulation of poly(ADP-ribose) polymerase[31] in response to BL treatment might be somehow involved. There is some evidence that BL-induced pulmonary toxicity partly depends upon the generation of ROS[41] as reflected by an increased lung SOD activity, presumably, as a defense mechanism to protect tissue against the injurious effect of superoxide radicals[42]. The combined treatment with taurine and niacin in both studies had remarkable inhibitory effects on the BL-induced increases in lipid peroxidation and SOD activity. This suggests that taurine and/or niacin may prevent the initial BL-induced generation of ROS or block their deleterious effects by stabilizing the biomembranes.

Although there is no direct evidence that taurine scavenges free radicals, its ability to trap hypochlorous acid is well known[6-8]. BL-induced acute inflammatory phase is always associated with an excessive accumulation of neutrophils in vascular, interstitial and intraalveolar compartments of the lung[43]. The activated neurophils by virtue of the enzyme myeloperoxidase are known to oxidize halides Cl-, Br-, or I- to their corresponding hypohalous acid in a reaction involving H_2O_2[44]. The biocidal

Table 3. Effects of Combined Treatment With Taurine (in Drinking Water) and Niacin (IP) on Bleomycin-Induced Increases in Total and Differential Cell Counts in the Hamster Bronchoalveolar Lavage Fluid

Treatment Groups ()	Total cells X 10^{-6}	Neutrophils X 10^{-4}	Monocytes X 10^{-4}	Macrophage X 10^{-6}	Lymphocytes X 10^{-5}
SA (4)	2.54 ± 0.75	4.80 ± 2.80	0.00 ± 0.00	2.44 ± 0.71	0.47 ± 0.10
TNSA (5)	2.09 ± 0.25	3.80 ± 0.67	0.21 ± 0.21	1.99 ± 0.25	0.41 ± 0.13
BL (6)	5.46 ± 0.64[a]	39.7 ± 8.18[a]	5.75 ± 0.86[a]	4.19 ± 0.53[b]	5.97 ± 0.68[a]
TNBL (6)	3.71 ± 0.21[d]	11.8 ± 2.66	1.71 ± 0.60	3.32 ± 0.19	2.30 ± 0.37[c]

Hamsters were treated with taurine in drinking water (1%) and niacin (250mg/kg, IP daily) and given bleomycin IT (2.5, 2.0 and 1.5 U/5ml/kg bw) in three consecutive doses at weekly intervals. Twenty days after the last IT instillation, bronchoalveolar lavage was carried out and total and differential cell counts analyzed as described in Methods. SA: Saline control; TNSA: Taurine + Niacin + Saline; BL: Bleomycin alone; TNBL: Taurine + Niacin + Bleomycin. The number of animals in each group is shown in parenthesis. Values are expressed as mean ± SEM.

[a] Significantly higher (P < 0.05) than all other corresponding groups.

[b] Significantly higher (P < 0.05) than corresponding SA and TNSA groups.

[c] Significantly lower (P < 0.05) than the corresponding BL group, but higher (P < 0.05) than corresponding SA and TNSA groups.

[d] Significantly lower (P < 0.05) than the corresponding BL group, but higher (P < 0.05) than corresponding TNSA group.

property of HOCl in oxidizing a variety of biologically significant substances including carbohydrates, nucleic acids, peptide linkages and amino acids is well known. The beneficial effect of the combined treatment with taurine and niacin against BL-induced lung damage may also be attributed to a decreased influx of neutrophils in the various compartments of the lung as well as to the ability of taurine to act as a trap for HOCl. Recently it has been demonstrated that of all the amino acids, taurine was the most effective inhibitor of HOCl-induced lysis of erythrocytes[7]. Thus it is possible that taurine could be effective in preventing the biomembrane damage caused by BL. This would then also explain why the BL-induced increases in pulmonary vascular permeability and acid phosphatase activity of the BALF were markedly suppressed by the combined treatment with taurine and niacin. The ability of taurine to prevent an excess influx of Ca^{2+} by its membrane stabilizing effect might have also played a key role in ameliorating BL-induced lung fibrosis. This hypothesis draws support from two lines of evidence: 1) BL-induced lung fibrosis is always associated with an intracellular overload of Ca^{2+}[45]; and 2) biochemical[5,46] and morphological findings[47] provided by other investigators suggest that taurine does indeed prevent an excess influx of Ca^{2+}. In the presence of an excessive amount of intracellular Ca^{2+}, the ability of mitochondria to synthesize ATP is impaired and Ca^{2+} accumulates at the inner surface of the cell membrane and activates the membrane bound enzymes including protease and phospholipases which in turn damage the cell membrane leading to cell death.

Calcium also stimulates phospholipase-A_2, a rate limiting enzyme, in the metabolism of arachidonic acid. Metabolites of arachidonic acid appear to be involved in the BL-induced lung fibrosis[48]. Activation of PLA_2 stimulates the peroxidation of membrane lipids[49] which is measured by the amount of MDAE. The generation of MDAE, in addition to being an index of lipid peroxidation also reflects DNA damage[30]. A significant suppression in BL-induced increases in the lung MDAE

Table 4. Effects of Dietary Treatment with Taurine and Niacin on Bleomycin-Induced Increases in Total and Differential Cell Counts in the Hamster Bronchoalveolar Lavage Fluid

Treatment Groups ()	Total cells X 10^{-6}	Neutrophils X 10^{-4}	Monocytes X 10^{-4}	Macrophage X 10^{-6}	Lymphocytes X 10^{-5}
SA (5)	2.76 ± 0.22	0.37 ± 0.37	0.00 ± 0.00	2.66 ± 0.23	0.77 ± 0.12
TNSA (5)	2.63 ± 0.24	3.99 ± 1.39	0.22 ± 0.22	2.55 ± 0.24	0.39 ± 0.10
BL (6)	8.63 ± 1.41[b]	90.50 ± 40.8[a]	5.05 ± 1.86[b]	6.75 ± 1.17[b]	6.19 ± 3.18[b]
TNBL (6)	5.76 ± 1.12	11.20 ± 3.29	3.29 ± 1.70	5.24 ± 0.96[b]	3.18 ± 0.94

Hamsters were fed taurine (2.5%) and niacin (2.5%) in the pulverized diet and given bleomycin IT (2.5, 2.0 and 1.5 U/5ml/kg bw) in three consecutive doses at weekly intervals. Twenty days after the last IT instillation, bronchoalveolar lavage was carried out and total and differential cell counts analyzed as described in Methods. SA: Saline control; TNSA: Taurine + Niacin + Saline; BL: Bleomycin alone; TNBL: Taurine + Niacin + Bleomycin. The number of animals in each group is shown in parenthesis. Values are expressed as mean ± SEM.

[a] Significantly higher (P < 0.05) than all other corresponding groups.

[b] Significantly higher (P < 0.05) than corresponding SA and TNSA groups.

by combined treatment with taurine and niacin as reported in this symposium indicates that this treatment might have inhibited BL-induced lipid peroxidation and DNA damage. This is not surprising since taurine and niacin are known to inhibit lipid peroxidation[13,46] and DNA damage[17], respectively.

The beneficial effect of niacin against paraquat-induced lung toxicity has been previously reported[50]. It is generally believed that niacin offers protection against tissue damage caused by a variety of oxidants by virtue of its ability to increase intracellular level of NAD and to inhibit overactivation of poly(ADP-ribose) polymerase[17,31]. An overstimulation of this chromosomal enzyme leads to intracellular depletion of NAD[24]. This hypothesis is consistent with our earlier findings that IT instillation of BL increased the lung poly(ADP-ribose) polymerase activity and caused a marked depletion of NAD[23]. This will explain why the combined treatment with taurine and niacin markedly suppressed the BL-induced overstimulation of poly(ADP-ribose) polymerase, and allowed an adequate availability of NAD, as reported in the present study. The availability of NAD would maintain the critical vital cell function including the DNA repair of injured lung epithelium. It is believed that the proliferation of interstitial cells in the lung is normally hampered by the intact epithelial cells; under normal conditions, the lung injury is repaired[51]. However, if the injured epithelial cells remain unrepaired, it would trigger the proliferation of interstitial cells leading to excessive synthesis and deposition of collagen in the lung[52,53]. Our data suggest that the combined treatment with taurine and niacin helps repair the BL-induced injury of epithelium and thus minimizes the interstitial cell proliferation and subsequent build-up of collagen.

The mechanisms for the beneficial effects of taurine and/or niacin against AD-induced lung fibrosis are not presently known. However, it is possible that the membrane stabilizing, ROS and HOCl scavenging properties of taurine and increased tissue level of NAD in response to niacin treatment might contribute to their beneficial effects. Regardless of the mechanism, it is tempting to speculate that taurine and/or niacin offer a novel therapeutic approach in the prevention of chemically-induced pulmonary fibrosis.

SUMMARY

The bleomycin(BL)-hamster model of interstitial pulmonary fibrosis (IPF) is generally associated with increased lung lipid peroxidation, measured as malondialdehyde equivalent (MDAE), calcium and collagen content; and superoxide dismutase (SOD), prolyl hydroxylase (PH) and poly(ADP-ribose) polymerase activities. We found that combined treatment with taurine in drinking water (1%) and niacin IP (250 mg/kg) daily, significantly decreased the BL-induced increases in lung MDAE and calcium content, and SOD, PH and poly(ADP-ribose) polymerase activities. This treatment almost completely ameliorated the BL-induced increases in the lung collagen accumulation as well. Findings of a similar nature were also demonstrated when taurine (2.5%) and niacin (2.5%) were supplemented in the diet of hamsters used in the same BL model of IPF. The diet supplemented with taurine (2.5%), niacin (2.5%), or taurine (2.5%) + niacin (2.5%) also reduced AD-induced increases in lung collagen accumulation, phospholipids, MDAE and SOD activity. It was concluded that diet supplemented with taurine and/or niacin would completely or partially ameliorate chemically-induced pulmonary fibrosis.

ACKNOWLEDGEMENT

The data reported in this symposium and the papers cited in the symposium article resulted from the research grants awarded to Shri N. Giri by the National Institute of Health, NHLBI Grant #2R01 HL27354.

Table 5. Effects of Dietary Treatment With Taurine and/or Niacin on Amiodarone-Induced Increases in Various Biochemical Parameters in Hamster Lungs.

Parameters	SA	T + SA	N + SA	TN + SA	AD	T + AD	N + AD	TN + AD
Hydroxyproline (µg/Lung)	765.0 ± 22.0	734.0 ± 15.0	707.0 ± 10.0	685.0 ± 6.0	1177.0 ± 73.0[a]	954.0 ± 38.0[b]	891.00 ± 40[b]	884.00 ± 33.0[b]
Prolyl Hydroxylase (dpmx10⁻³/30min/Lung)	164.0 ± 10.0	158.0 ± 7.0	156.0 ± 11.0	147.0 ± 8.0	247.0 ± 9.0[a]	187.0 ± 9.0[c]	186.00 ± 10[c]	189.00 ± 7.0[c]
Malondialdehyde Equivalent (nmol/Lung)	111.3 ± 3.2	100.3 ± 2.9	102.4 ± 4.2	99.8 ± 7.8	165.5 ± 11.6[d]	155.9 ± 9.1[d]	134.90 ± 8.0[e]	145.10 ± 5.0[e]
Superoxide Dismutase (units/Lung)	323.7 ± 26.9	285.9 ± 13.6	283.0 ± 8.6	291.4 ± 7.2	405.5 ± 24.1[a]	348.4 ± 19.1[c]	352.30 ± 16.1[c]	355.70 ± 10.0[c]
Phospholipid (µmol/Lung)	9.7 ± 0.4	9.7 ± 0.3	9.5 ± 0.4	9.5 ± 0.4	12.9 ± 0.6[f]	11.9 ± 0.46[e]	10.95 ± 0.51[c]	11.08 ± 0.3[c]

Hamsters were fed taurine (2.5%) and/or niacin (2.5%) in the diet. Six days after taurine and/or niacin treatment, hamsters were given a single IT instillation of amiodarone (15 µmol/5ml/kg). Twenty one days after the IT instillation, hamsters were killed and the lungs processed for various biochemical assay as described in METHODS. SA: Saline control (n=5); T+SA: Taurine + Saline (n=6); N+SA: Niacin + Saline (n=6); TN+SA: Taurine + Niacin + Saline (n=5); AD: Amiodarone alone (n=6); T+AD: Taurine + Amiodarone (n=8); N+AD: Niacin + Amiodarone (n=7); TN+AD: Taurine + Niacin + Amiodarone (n=6). Values are expressed as mean ± SEM. [a] Significantly higher (P < 0.05) than all other corresponding groups. [b] Significantly lower (P < 0.05) than the corresponding AD group, but higher (P < 0.05) than corresponding SA, T+SA, N+SA and TN+SA groups. [c] Significantly lower (P < 0.05) than the corresponding AD group, but higher (P < 0.05) than corresponding T+SA, N+SA and TN+SA groups. [d] Significantly higher (P < 0.05) than corresponding SA, T+SA, N+SA, TN+SA and N+AD groups. [e] Significantly higher (P < 0.05) than corresponding SA, T+SA, N+SA and TN+SA groups. [f] Significantly higher (P < 0.05) than all other corresponding groups except T+AD group.

REFERENCES

1. P. Pion, M.D. Kittleson, Q.R. Rogers, and J.G. Morris, Myocardial failure in cats asociated with low plasma taurine: a reversible cardiomyopathy, *Science* 237:764-768 (1987).
2. A. Barbeau and R.J. Huxtable, Taurine and Neurological Disorders, Raven Press, New York (1978).
3. R.W. Chesney, Taurine: its biological role and clinical implications, *in*: "Advances in Pediatrics," L.A. Barness, ed., Year Book Medical Publishers, Chicago, Vol. 32, pp. 1-42 (1985).
4. C.E. Wright, H.H. Tallan, Y.Y. Lin, and G.E. Gaull, Taurine: biological update, *Annu. Rev. Biochem.* 55:427-453 (1986).
5. R.J. Huxtable, From heart to hypothesis: a mechanism for the calcium modulatory actions of taurine, *Adv. Exp. Med. Biol.* 217:371-387 (1987).
6. C.E. Wright, T. Lin, Y.Y. Lin, J.A. Sturman, and G.E. Gaull, Taurine scavenges oxidized chlorine in biological systems, *in*: "Taurine: Biological Actions and Clinical Perspectives," S.S. Oja, L. Ahtee, P. Kontro, and M.K. Paasonen, eds., Alan R. Liss, New York, Vol. 179, pp. 112-123 (1985).
7. K. Nakamori, I. Koyama, T. Nakamura, T. Yoshida, M. Umeda, and K. Inoue, Effectiveness of taurine in protecting biomembrane against oxidant, *Chem. Pharm. Bull.* (Tokyo) 38:3116-3119 (1990).
8. D.M. McLoughlin, P.P. Stapleton, and F.J. Bloomfield, Influence of taurine and a substituted taurine on the respiratory burst pathway in the inflammatory response, *Biochem. Soc. Trans.* 19:73-78 (1991).
9. P. Dolara, A. Agresti, A. Giotti, and E. Sorace, The effect of taurine on calcium exchange of sarcoplasmic reticulum of guinea pig heart studied by means of dialysis kinetics, *Can. J. Physiol. Pharmacol.* 54:529-533 (1975).
10. H. Pasantes-Morales and A. Gamboa, Effect of taurine on $^{45}Ca^{2+}$-accumulation in rat brain synaptosomes, *J. Neurochem.* 34:244-246 (1980).
11. R.E. Gordon, A.A. Shaker, and D.F. Solano, Taurine protects hamster bronchioles from acute NO_2-induced alterations. A histologic, ultrastructural, and freeze-fracture study, *Am. J. Pathol.* 125:585-600 (1986).
12. Q. Wang, S.N. Giri, D.M. Hyde, and J.M. Nakashima, Effects of taurine on bleomycin-induced lung fibrosis in hamsters, *Proc. Soc. Exp. Biol. Med.* 190:330-338 (1989).
13. T. Nakashima, T. Takino, and K. Kuriyama, Therapeutic and prophylactic effects of taurine administration on experimental liver injury, *in*: "Sulfur Amino Acids, Biochemical and Clinical Aspects," K. Kuriyama, R. Huxtable, and H. Iwata, eds., Alan R. Liss, New York, pp. 449-459 (1983).
14. J.G. Alvarez and B.T. Storey, Taurine, hypotaurine, epinephrine and albumin inhibit liquid peroxidation in rabbit spermatozoa and protect against loss of motility, *Biol. Reprod.* 29:548-555 (1983).
15. J.H. Fellman and E.S. Roth, The biological oxidation of hypotaurine to taurine: Hypotaurine as an antioxidant, *Prog. Clin. Biol. Res.* 179:71-82 (1985).
16. D.A. Bender, B.I. Magboul, and D. Wynick, Probable mechanisms of regulation of utilization of dietary tryptophan, nicotinamide and nicotinic acid as precursors of nicotinamide nucleotides in the rat, *Br. J. Nutr.* 48:119-127 (1982).
17. A.B. Weitberg, Effect of nicotinic acid supplementation in vivo on oxygen radical-induced genetic damage in human lymphocytes, *Mutat. Res.* 216:197-201 (1989).
18. I.U. Schraufstatter, D.B. Hinshaw, P.A. Hyslop, R.G. Spragg, and C.G. Cochrane, Oxidant injury of cells. DNA strand-breaks activate polyadenosine diphosphate-ribose polymerase and lead to depletion of nicotinamide adenine dinucleotide, *J. Clin. Invest.* 77:1312-1320 (1986).
19. I.U. Schraufstatter, P.A. Hyslop, D.B. Hinshaw, R.G. Spragg, L.A. Sklar, and C.G. Cochrane, Hydrogen peroxide-induced injury of cells and its prevention by inhibitors of poly(ADP-ribose) polymerase, *Proc. Natl. Acad. Sci.* USA 83:4908-4912 (1986).
20. F. Dallegri, R. Goretti, A. Ballestrero, L. Ottonello, and F. Patrone, Neutrophil-induced depletion of adenosine triphosphate in target cells: evidence for a hypochlorous acid-mediated process, *J. Lab. Clin. Med.* 112:765-772 (1988).
21. F. Dallegri, L. Ottonello, A. Ballestrero, F. Bogliolo, F. Ferrando, and F. Patrone, Cytoprotection against neutrophil derived hypochlorous acid: a potential mechanism for the therapeutic action of 5-aminosalicylic acid in ulcerative colitis, *Gut* 31:184-186 (1990).
22. P.L. Moseley, Augmentation of bleomycin-induced DNA damage in intact cells, *Am. J. Physiol.* 257:C882-C887 (1989).
23. M.Z. Hussain, S.N. Giri, and R.S. Bhatnager, Poly(ADP-ribose) synthetase activity during bleomycin-induced lung fibrosis in hamsters, *Exp. Mol. Pathol.* 43:162-176 (1985).

24. C.R. Stubberfield and G.M. Cohen, NAD$^+$ depletion and cytotoxicity in isolated hepatocytes, *Biochem. Pharmacol.* 37:3967-3974 (1988).

25. Y. Ohashi, K. Ueda, M. Kawaichi, and O. Hayaishi, Activation of DNA ligase by poly(ADP-ribose) in chromatin, *Proc. Natl. Acad. Sci.* USA 80:3604-3607 (1984).

26. O. Hayaishi and K. Ueda, Poly- and mono(ADP-ribosyl)ation reactions: Their significance in molecular biology, *in:* "ADP-Ribosylation Reactions," O. Hayaishi and K. Ueda, eds., Academic Press, New York, pp. 3-14 (1982).

27. N.A. Berger, G.W. Sikorski, S.J. Petzold, and K.K. Kurohara, Association of poly(ADP-ribose) synthesis with DNA damage and repair in normal human lymph, *J. Clin. Invest.,* 63:1164-1171 (1979).

28. I.M. Roitt, The inhibition of carbohydrate metabolism in ascites-tumour cells by ethylenimines, *Biochem. J.,* 63:300-307 (1956).

29. Y. Sugiura and T. Kikuchi, Formation of superoxide and hydroxy radicals in iron (II)-bleomycin-oxygen system: Electron spin resonance detection by spin trapping, *J. Antibiot.* (Tokyo) 31:1310-1312 (1978).

30. M.A. Trush, E.G. Mimnaugh, E. Ginsburg, and T.E. Gram, Studies on the interaction of bleomycin A$_2$ with rat lung microsomes. II. Involvement of adventitious iron and reactive oxygen in bleomycin-mediated DNA chain breakage, *J. Pharmacol. Exp. Ther.* 221:159-165 (1982).

31. Q. Wang, S.N. Giri, D.M. Hyde, J.M. Nakashima, and I. Javadi, Niacin attenuates bleomycin-induced lung fibrosis in the hamster, *J. Biochem. Toxicol.* 5:13-22 (1990).

32. J.J. Heger, E.N. Prystowsky, and D.P. Zipes, Clinical efficacy of amiodarone in treatment of recurrent ventricular tachycardia and ventricular fibrillation, *Am. Heart J.* 106:887-894 (1983).

33. S.M. Sobol and L.L. Rakita, Pneumonitis and pulmonary fibrosis associated with amiodarone treatment: a possible complication of a new antiarrhythmic drug, *Circulation* 65:819-824 (1982).

34. M.J. Reasor, C.L. Ogle, E.R. Walker, and S. Kacew, Amiodarone-induced phospholipidosis in rat alveolar macrophages, *Am. Rev. Respir. Dis.* 137:510-518 (1988).

35. Q. Wang, S.N. Giri, D.M. Hyde and C. Li, Amelioration of bleomycin-induced pulmonary fibrosis in hamsters by combined treatment with taurine and niacin, *Biochem. Pharmacol.* 42:1115-1122 (1991).

36. M. Willcox, A. Kervitsky, L.C. Watters, and T.E. King, Jr., Quantification of cells recovered by bronchoalveolar lavage. Comparison of cytocentrifuge prepartaions with the filter method, *Am. Rev. Respir. Dis.* 138:74-80 (1988).

37. D.W. Moss, Acid phosphatases, *in:* "Methods of Enzymatic Analysis," H.U. Bergmeyer, J. Bergmeyer and M. Grassl, eds., Chemie Verlag, Deerfield Beach, Fl., pp. 92-106 (1983).

38. O.H. Lowry, N.J. Rosebrough, A.L. Farr, and R.J. Randall, Protein measurement with the Folin phenol reagent, *J. Biol. Chem.* 193:265-275 (1951).

39. H. Elias and D.M. Hyde, A Guide to Practical Stereology, Karger, New York, pp. 25-34 (1983).

40. SAS Institute Inc. SAS/STAT$^{\underline{TM}}$ Guide for Personal Computers, Version 6th Edition. SAS Institute, Cary, N.C., pp. 183-260 (1985).

41. S.N. Giri and Q. Wang, Mechanisms of bleomycin-induced lung injury, *Comments Toxicol.* 3:145-176 (1989).

42. S.N. Giri, H.P. Misra, D.B. Chandler, and Z. Chen, Increases in lung prolyl hydroxylase and superoxide dismutase activities during bleomycin-induced lung fibrosis in hamsters, *Exp. Mol. Pathol.* 39:317-326 (1983).

43. S.N. Giri, D.M. Hyde, and J.M. Nakashima, Analysis of bronchoalveolar lavage fluid from bleomycin-induced pulmonary fibrosis in hamsters, *Toxicol. Path.* 14:149-157 (1986).

44. S.J. Klebanoff, Phagocytic cells: Products of oxygen metabolism, *in:* "Inflammation: Basic principles and clinical correlates," J.I. Gallin, I.M. Goldstein and R. Snyderman, eds., Raven Press, New York, pp. 391-444 (1988).

45. S.N. Giri, J.M. Nakashima, and D.L. Curry, Effects of intratracheal administration of bleomycin or saline in pair-fed and control-fed hamsters on daily food intake and on plasma levels of glucose, cortisol and insulin and lung levels of calmodulin, calcium, and collagen, *Exp. Mol. Path.* 42:206-219 (1985).

46. J. Azuma, T. Hamaguchi, H. Ohta, K. Takihara, N. Awata, A. Sawamura, H. Harada, Y. Tanaka, and S. Kishimoto, Calcium overload-induced myocardial damage caused by isoproterenol and by adriamycin: possible role of taurine in its prevention, *Adv. Exp. Med. Biol.* 217:167-179 (1987).

47. R.E. Gordon, R.F. Heller, and J.R. Del Valle, Membrane perturbations and mediation of gap junction formation in response to taurine treatment in normal and injured alveolar epithelia, *Exp. Lung Res.* 15:895-908 (1989).

48. Q. Wang, S.N. Giri, and D.M. Hyde, Characterization of a phospholipase A$_2$ in hamster lung and in vitro and in vivo effects of bleomycin on this enzyme, *Prostaglandins Leukot. Essential Fatty*

Acids 36:85-92 (1989).

49. K.R. Chien, J. Abrams, A. Serroni, J.T. Martin, and J.L. Farber, Accelerated phospholipid degradation and associated membrane dysfunction in irreversible, ischemic liver cell injury, *J. Biol. Chem.* 253:4809-4819 (1978).

50. O.R. Brown, M. Heitkamp, and C.S. Song, Niacin reduces paraquat toxicity in rats, *Science* 212:1510-1512 (1981).

51. H.P. Witschi, W.M. Haschek, A.J.P. Klein-Szanto, and P.J. Hakkinen, Potentiation of diffuse lung damage by oxygen: Determining variables, *Am. Rev. Respir. Dis.* 123:98-103 (1981).

52. W.M. Haschek and H.P. Witschi, Pulmonary fibrosis: A possible mechanism, *Toxicol. Appl. Pharmacol.* 51:475-487 (1979).

53. H. Witschi, G. Godfrey, E. Frome, and R.C. Lindenschmidt, Pulmonary toxicity of cytostatic drugs: Cell kinetics, *Fundam. Appl. Toxicol.* 8:253-262 (1987).

TAURINE PROTECTS AGAINST OXIDANT INJURY
TO RAT ALVEOLAR PNEUMOCYTES

Melanie A. Banks[1], Dale W. Porter[2], William G. Martin[2],
and Vincent Castranova[3]

[1]Division of Food Chemistry
American Bacteriological and Chemical Research Corp.
Gainesville, FL 32608

[2]Division of Animal and Veterinary Science
College of Agriculture and Forestry
West Virginia University
Morgantown, WV 26506

[3]Division of Respiratory Disease Studies
National Institute for Occupational Safety and Health
944 Chestnut Ridge Road
Morgantown, WV 26505

INTRODUCTION

Taurine (2-aminoethanesulfonic acid) is an unusual amino acid found in a wide variety of animal species. Its precise role in human and animal nutrition has remained unclear despite intensive investigation. Recently, however, evidence has accumulated supporting the hypothesis that taurine protects cellular membranes against toxic compounds, including bile acids, xenobiotics, and oxidants. In different experimental systems, taurine has been demonstrated to act as a direct (primary) antioxidant that scavenges oxygen-free radicals and as an indirect (secondary) antioxidant that prevents changes in ion transport and membrane permeability which result from oxidant injury[1-3].

Exposure to oxidant gases, including hyperbaric oxygen, nitrogen dioxide, and ozone, results in pulmonary injury[4-7]. The morphological changes which occur in lung tissue include destruction of capillary endothelial cells, edema, hypertrophy and hyperplasia of the bronchiolar epithelium, bronchiolization of the alveolar duct epithelium, and an influx of macrophages and polymorphonuclear leukocytes into the alveolar air spaces. Metabolic changes which result from oxidant injury to the lung include lipid peroxidation and mobilization of cellular antioxidants, such as glutathione, ascorbic acid, and vitamin E.

Although hyperoxia and NO_2 or ozone exposures result in similar morphological

Taurine, Edited by J.B. Lombardini *et al.*
Plenum Press, New York, 1992

and metabolic changes in the lung, the distribution of the injury varies; NO_2 and ozone toxicities primarily involve the epithelium of the conducting airways and alveoli near the terminal bronchioles (proximal alveoli), whereas exposure to high oxygen tensions induces damage at the levels of the trachea through the distal alveoli. Dietary taurine supplementation has previously been shown to protect against bronchiolar damage induced by the oxidant gas NO_2[8].

Pneumocytes found on the alveolar surface exhibit a wide range of sensitivities to oxidant injury. Alveolar type II epithelial cells are relatively resistant to oxidant injury while type I cells are the most susceptible to it. In fact, type II cells are thought to repair oxidant induced damage to the alveolar epithelium by replacing injured type I cells. Alveolar macrophages are relatively insensitive to damage by oxidant gases *in vivo*[9]; however, *in vitro*, they appear to be more sensitive than type II cells to oxidant injury[10].

The studies described here were conducted to investigate the possible role of taurine as an antioxidant in alveolar macrophages.

METHODS

Isolation of Lung Cells and Exposure to Ozone

Rats (male Sprague Dawley) weighing 200-300 g were anesthetized with pentobarbital sodium (65 mg/kg body wt). Alveolar macrophages were obtained by pulmonary lavage, as previously described[11-13]. Cell yield for fast preparation was determined with a Coulter electronic cell counter. Mean cell volumes and the purity of each preparation were determined using an electronic cell sizing attachment. The cellular preparations averaged $91 \pm 1\%$ alveolar macrophages.

Isolated alveolar macrophages were allowed to adhere to the bottom surface of a 75-cm^2 tissue culture flask. The flask was mounted on an orbital mixer, and the cells were exposed to 0.45 ± 0.05 ppm ozone for 60 min at room temperature while the flask was slowly rocked from side-to-side (22 cycles/min) to allow direct contact between the cells and the oxidant gas. After ozone exposure, the medium was recovered and the non-adherent and adherent cell fractions were combined prior to conducting assays on the medium and the cells[12,13].

Measurement of Lung, Plasma, Extracellular and Intracellular Taurine Levels

Taurine was isolated from tissue, cells and media in the experiments using dual-bed ion-exchange chromatography[14], and taurine contents were measured by an HPLC technique which we developed[15]. This method involves deproteinized samples and therefore measures free (i.e. not protein-bound) taurine, rather than total taurine.

Cell and Medium Assays

To measure cell viability, the exclusion of trypan blue dye was determined microscopically as previously described[12,13]. Chemiluminescence (both resting and symosan-stimulated) was measured as previously described[12,13], using luminol and a Packard Tri-Carb scintillation counter operated in the out-of-coincidence mode. Membrane ATPase activities were measured as inorganic phosphate liberated from ATP, as previously described[12,13]. To estimate the activity of Na^+/K^+ ATPase by difference, ouabain was added to half of the samples.

Lipid peroxides released into the medium were measured spectrophotometrically as thiobarbituric acid-reactive substances, by the method previously described. Protein leakage into the medium was estimated with the Bio-Rad technique as previously described. Leakage of reduced and oxidized glutathione into the medium was determined fluorometrically, by the method previously described. Potassium ion leakage was measured by atomic absorption spectroscopy, as previously described[12,13].

Statistical Analyses

Data reported are expressed as means $\pm$ standard errors of at least three separate experiments using cells or tissues pooled from groups of rats euthanized on at least 3 separate days. Statistical differences between data points were estimated using either the Student's t test or one-way ANOVA. The significance level was set at $p < 0.05$.

RESULTS

Effects of In Vitro Ozone Exposure on Peroxidative Damage, Membrane Leakage, and Taurine Content of Rat Alveolar Macrophages

Recovery of macrophages from the culture flask averaged 57.7% with cell populations that were not exposed to ozone. Recovery was not affected after a 15-min exposure to ozone but decreased significantly after 30 min of ozone exposure. Initially, adherent cells represented 60% of the recoverable macrophages. The ratio of adherent to non-adherent macrophages remained relatively constant for the first 30 min of ozone exposure. However, after a 60-min treatment with ozone, only 44% of recoverable cells were adherent. Similarly, cell viability, as estimated by trypan blue exclusion, was maintained for the first 30 min of ozone exposure and declined thereafter (Figure 1). The decreasing yield of macrophages from the culture flask

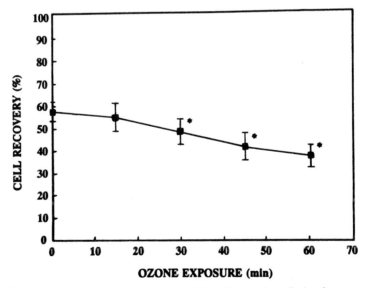

Figure 1. Viability of alveolar macrophages recovered from tissue culture flasks after exposure to ozone for 0-60 min (mean $\pm$ SEM for N = 5-7 observations). *A significant decrease from the value at 0 min at the $p < 0.05$ level (t test).

343

with increasing ozone exposure may reflect decreasing cell viability and/or changes in the adherent properties of the cells.

Cellular chemiluminescence, which is a reflection of free radical generation (including reactive oxygen and lipid metabolites), is given in Table 1. Initially, ozone exposure caused increased chemiluminescence in both resting and zymosan-stimulated macrophages. This response peaked at 15-30 min of exposure and declined to the basal level by 60 min of exposure to ozone. At 15 through 45 min of exposure, zymosan-stimulated chemiluminescence greatly exceeded resting values, indicating that the ability of particles to enhance the production of reactive oxygen metabolites is maintained in ozone-exposed macrophages.

Total membrane ATPase activity decreased with increasing length of ozone exposure (Table 2) and reached a maximum at 30 min of exposure. The Na^+/K^+ ATPase activity (which is ouabain-sensitive) also decreased with time and was undetectable by 30 min of exposure.

Lipid peroxides were detectable in the extracellular medium. Lipid peroxidation products were significantly elevated after 15 min of exposure to ozone and their production increased linearly with increasing exposure time (Figure 2).

Protein leakage from cultured alveolar macrophages also linearly increased with increasing length of ozone exposure (Figure 3). However, protein leakage was not significantly elevated until 30 min of exposure.

Leakage of reduced glutathione into the medium (Figure 4) increased significantly after 30 min of exposure to ozone and significantly declined thereafter. Leakage of oxidized glutathione into the medium (Figure 4) increased by 15 min and remained constant from 15-60 min of ozone exposure.

Table 1. Effect of ozone exposure on chemiluminescence in rat alveolar macrophages[a].

| Ozone exposure time (min) | Chemiluminescence[b] | | Difference[e] | Ratio[f] |
	Resting[c]	Zymosan-stimulated[d]		
0	0.036 ± 0.002	0.121 ± 0.007	0.084 ± 0.006	3.3 ± 0.1
15	0.266 ± 0.017 (7.4)[g]	1.639 ± 0.032 (13.5)[g]	1.373 ± 0.024[g]	6.2 ± 0.3[g]
30	0.293 ± 0.006 (8.1)[g]	1.404 ± 0.176 (11.6)[g]	1.226 ± 0.137[g]	4.8 ± 0.5[g]
45	0.178 ± 0.027 (4.9)[g]	0.882 ± 0.186 (7.3)[g]	0.705 ± 0.179[g]	4.9 ± 0.9[g]
60	0.093 ± 0.005 (2.6)	0.147 ± 0.007 (1.2)	0.054 ± 0.007	1.6 ± 0.1

[a] Values represent means ± SEM for three observations.
[b] Relative chemiluminescence determined gravimetrically as the area under the curve of cpm vs time.
[c] Activity in the absence of 2 mg/ml zymosan.
[d] Activity in the presence of 2 mg/ml zymosan.
[e] Zymosan-stimulated minus resting values.
[f] Zymosan-stimulated values divided by resting values.
[g] Increase relative to value at 0 min.

The intracellular taurine concentration increased with increasing length of exposure to ozone, peaked at 30 min, and then declined (Figure 5). Extracellular taurine levels increased with increasing length of ozone exposure (Figure 6).

Potassium ion leakage increased with increasing length of exposure to ozone (Figure 7). The increase in extracellular K^+ concentration can be correlated with the decrease in activity of the Na^+/K^+ ATPase from 0 to 30 min of ozone exposure (Table 2).

Table 2. Effect of ozone exposure on membrane ATPase activity of rat alveolar macrophages[a]

Ozone exposure time (min)	Total ATPase[b]	Na$^+$/K$^+$ ATPase[c]
	(nmol PO$_4$/hr/10^6 cells)	
0	50.7 ± 2.1	45.2 ± 2.3 (89%)[d]
15	31.8 ± 2.0*	16.4 ± 1.8 (52%)*
30	10.0 ± 3.0*	−1.5 ± 3.7 (~0)*
45	9.9 ± 0.7*	−3.1 ± 2.8 (~0)*
60	9.9 ± 0.8*	−2.6 ± 1.6 (~0)*

[a] Values represent means ± SEM for three observations.
[b] Activity in the absence of 1 mM ouabain.
[c] Difference between total activity and activity in the presence of 1 mM ouabain.
[d] Percentage of total ATPase activity.
* A significant decrease from the value at 0 min at the $p < 0.05$ level (t test).

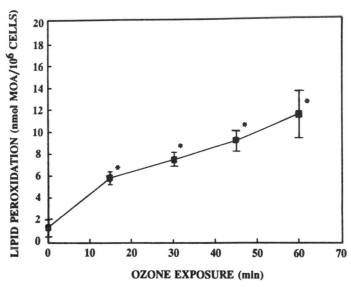

Figure 2. Appearance of lipid peroxidation products in the medium after exposure of alveolar macrophages to ozone for 0-60 min (mean ± SEM for N = 5-7 observations). *A significant increase from the value at 0 min in the $p < 0.05$ level (t test).

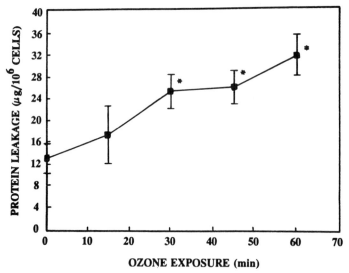

Figure 3. Protein leakage into the medium from alveolar macrophages exposed to ozone for 0-60 min (mean $\pm$ SEM for N = 4-6 observations). *A significant increase from the value at 0 min at the $p < 0.05$ level (t test).

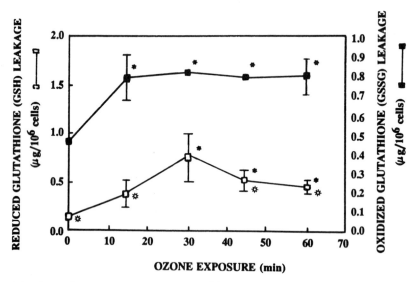

Figure 4. Leakage of reduced glutathione, GSH (☐), and oxidized glutathione, GSSG (☐), from alveolar macrophages exposed to ozone for 0-60 min (mean $\pm$ SEM for N = 3-7 observations). *A significant increase from the value at 0 min at the $p < 0.05$ levels (t test). ✳A significant decrease from the value at 30 min at the $p < 0.05$ level (t test).

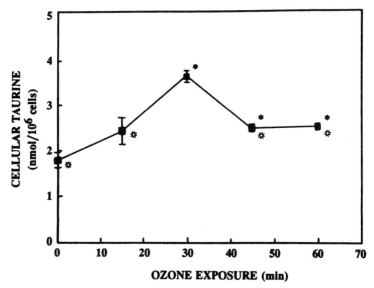

Figure 5. Intracellular taurine concentrations in alveolar macrophages exposed to ozone for 0-60 min (mean ± SEM for N = 3 observations). *A significant increase from the value at 0 min at the p < 0.05 level (*t* test). ✻A significant decrease from the value at 30 min at the p < 0.05 level (*t* test).

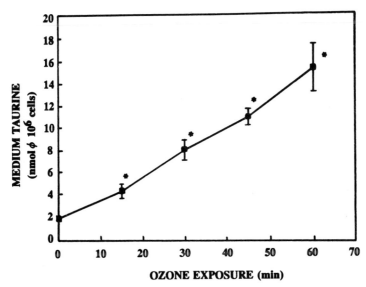

Figure 6. Taurine leakage into the medium from alveolar macrophages exposed to ozone for 0-60 min (mean ± SEM for N = 3 observations). *A significant increase from the value at 0 min at the p < 0.05 level (*t* test).

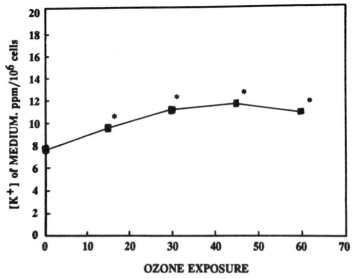

Figure 7. Potassium ion leakage into the medium from alveolar macrophages exposed to ozone for 0-60 min (mean $\pm$ SEM for N = 3 observations). *A significant increase from the value at 0 min at the $p < 0.05$ level (t test).

Protective Effects of Taurine on Ozone-Induced Lipid Peroxidation and Membrane Leakage in Isolated Rat Alveolar Macrophages

The actual intracellular taurine concentrations of alveolar macrophages preincubated in 0-500 μM taurine prior to ozone exposure are given in Figure 8. The taurine content of the cells increased with increasing extracellular taurine. This result was expected since rat alveolar macrophages have been shown to actively transport and accumulate this nutrient via a specialized sodium-taurine co-transport mechanism[11].

The free intracellular taurine concentrations of alveolar macrophages after exposure to ozone are given in Figure 9. Comparing data from Figures 8 and 9, cytoplasmic taurine rose by approximately 1.6, 1.3, 5, and 7-fold in ozone-exposed cells incubated at 0, 100, 250, and 500 μM extracellular taurine, respectively. This ozone-induced taurine mobilization was significant in macrophages supplemented with 250 and 500 μM extracellular taurine.

Ozone exposure of alveolar macrophages has been shown to decrease cellular viability, induce lipid peroxidation, decrease total ATPase and Na^+/K^+ ATPase levels, and cause leakage of glutathione and protein[12]. Data in Table 3 indicate that taurine enrichment decreased ozone-induced damage as measured by these cellular parameters. At 100 μM extracellular taurine (i.e., the approximated plasma level of this nutrient)[11], viability was increased by 38%, lipid peroxidation decreased by 70%, total ATPase increased by 113%, Na^+/K^+ ATPase increased by 62%, GSH leak decreased by 93%, and protein leak decreased by 67% compared to levels measured for ozone-exposed cells incubated in the absence of extracellular taurine. In the case of viability, lipid peroxidation, total ATPase, and GSH leak, this protection from oxidant injury was also significant at 250 μM and 500 μM; high extracellular taurine levels did not significantly elevate Na^+/K^+ ATPase or prevent protein leak.

As shown previously, ozone exposure of alveolar macrophages increased chemiluminescence, decreased cell recovery, and increased leakage of GSSG and potassium ions[12]. In the present study, taurine supplementation at any level failed to significantly alter these ozone-induced changes (data not shown).

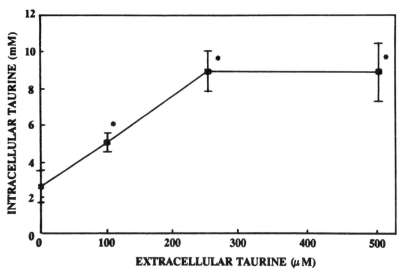

Figure 8. Effect of preincubation of rat alveolar macrophages with variable extracellular taurine concentrations on intracellular taurine content prior to ozone exposure. Accumulation of taurine is both time and concentration dependent. The asterisk (*) indicates a significant increase above the value for cells incubated in $0 \mu M$ taurine at the $p < 0.05$ level.

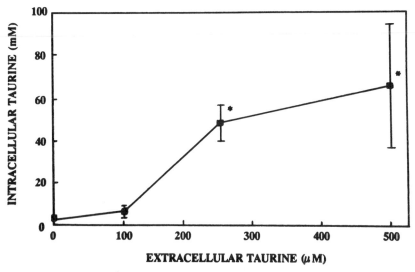

Figure 9. Intracellular taurine content of alveolar macrophages exposed to 0.45 ± 0.05 ppm ozone for 30 min. Cells were preincubated in media of different taurine concentrations at $37°C$ for 30 min prior to being cultured at various taurine levels and exposed to ozone. At all extracellular taurine levels, free intracellular taurine rises in response to ozone exposure. The asterisk (*) indicates a significant increase from the value for unexposed cells supplemented at the respective extracellular taurine level ($p < 0.05$).

Table 3. Effect of taurine supplementation on ozone-induced injury to alveolar macrophages[a].

Parameter	Extracellular Taurine Concentration (μM)[b]			
	0	100	250	500
Cell Viability (%)	60.6 ± 3.0	84.1 ± 1.7[*]	86.0 ± 1.3[*]	78.3 ± 1.3[*]
Lipid Peroxidation (nmol MDA/10⁶ cells)	4.6 ± 0.3	1.4 ± 0.4[*]	2.1 ± 0.9[*]	2.3 ± 0.7[*]
Total ATPase (nmol Pi/hr/10⁶ cells)	30.8 ± 7.0	65.9 ± 4.0[*]	64.0 ± 2.0[*]	91.3 ± 4.2[*]
GSH Leak (μg/10⁶ cells)	0.42 ± 0.14	0.03 ± 0.01[*]	0.04 ± 0.02[*]	0.05 ± 0.02[*]
Protein Leak (μg/10⁶ cells)	15.9 ± 4.2	5.2 ± 1.4[*]	10.0 ± 1.0	11.4 ± 2.6
Na⁺/K⁺ ATPase (nmol/Pi/hr/10⁶ cells)	4.4 ± 6.4	29.5 ± 3.4[*]	2.3 ± 3.0	4.3 ± 8.4

[a] Alveolar macrophages were preincubated in media of different taurine concentrations at 37°C for 30 min prior to being cultured at various taurine levels and exposed to ozone (0.45 ppm for 30 min at 25°C).

[b] Values are means ± standard errors from 3-8 separate experiments.

[*] Significantly different from the value at 0 μM taurine (p < 0.05).

DISCUSSION

Effects of In Vitro Ozone Exposure on Peroxidative Damage, Membrane Leakage, and Taurine Content of Rat Alveolar Macrophages

The results which we observed following exposure of isolated rat alveolar macrophages to ozone included decreased cell viability, increased resting and particle-stimulated chemiluminescence, inactivation of the Na⁺/K⁺ ATPase, appearance of lipid peroxides in the medium, increased leakage of protein, reduced and oxidized glutathione, taurine, and potassium ions into the medium, and increased intracellular taurine levels.

Our results are consistent with those of Van Der Zee et al.[16-18] who have described the toxic effects of ozone exposure on murine fibroblasts and human erythrocytes. When the fibroblasts were exposed to ozone (10 μumol/min for 0-80 min) in a system similar to the one we used, leakage of protein and potassium ions into the medium was shown to increase with length of exposure. The cellular level of reduced glutathione decreased with time, while that of oxidized glutathione initially increased (presumably due to the oxidation of GSH) and then decreased[16]. In erythrocytes, ozone exposure (4 μmol/min for 0-120 min conducted over a stirred cell suspension) resulted in lipid peroxidation, K⁺ leakage, and a reduction in the cellular level of reduced glutathione[18].

Chronic exposure (3 hr/day, 5 days/week for 3 months) to nitrogen dioxide (0.5 ppm) or ozone (0.1 ppm) resulted in decreased viability of alveolar macrophages that were lavaged from the lungs of the exposed mice[19]. However, there was no effect of a short (2 hr) in vitro ozone (0.29-0.61 ppm) exposure on the viability of alveolar macrophages in monolayer culture in a chamber[20], or of a short (2-3 hr) in vitro NO₂ (1.3-17.0 ppm) or ozone (0.9-3.5 ppm) exposure on rat alveolar macrophage viability[21]. The latter in vivo exposure may have been too short to affect cell viability. In the former study, the petri dishes containing the cells were rotated such that the medium covered half of the cells for 30 sec and the remaining half of the cells were exposed to ozone for 30 sec. Thus, the exposures were more intermittent than in our study, where half of the cells were exposed approximately every 3 sec.

A decrease in cellular protein content and an increase in cellular GSH were observed in isolated alveolar macrophages from guinea pigs exposed to hyperoxia (85% for 90 hr)[22]. Exposure of isolated, perfused whole rat lung to hyperoxic conditions (0.4×10^6 Pa/min, 90 min) resulted in an increase in glutathione release from the organ[23]. Conversely, glutathione depletion of isolated rat alveolar macrophages has been shown to increase the susceptibility of the cells to oxidant injury[24]. Our results show a time-dependent leak of both GSH and GSSG upon exposure of alveolar macrophages to ozone. It should be noted that GSSG levels exceeded GSH levels. Much of the oxidized glutathione which was detected in the extracellular medium in our study may have been produced as a result of the reaction of ozone with reduced glutathione. This would explain the artifactually high GSSG/GSH ratios reported in this investigation.

Dowell et al.[25] were unable to detect lipid peroxidation in alveolar macrophages isolated from ozone-exposed rabbits (2-10 ppm for 3 hr or 0.5-2 ppm 8 hr/day for 8 days). However, lipid peroxidation has been strongly implicated to be the mechanism of toxic action of O_3 or NO_2 exposures in studies where isolated alveolar macrophages were preincubated with polyunsaturated fatty acids, arachidonic acid, vitamin C, or vitamin E, or in glutathione-depleted cells[26,27]. Pigmented alveolar macrophages have been observed in the lungs of animals exposed to NO_2,[11] and presumably this pigmentation is similar to the accumulation of lipofuscin or ceroid pigment which results from lipid peroxidation in various organs of vitamin E-deficient animals.

Chemiluminescence is a function of the production of reactive oxygen metabolites in alveolar macrophages during the respiratory burst, the activation of which results in phagocytosis and bacterial killing[17]. Zymosan-stimulated chemiluminescence was increased by ozone to a greater extent than at rest, i.e., 13- and 8-fold, respectively, therefore, ozone treatment must activate the cells in some manner so that they are more reactive in the presence of particles. It is possible that this sensitization is mediated via ozone-induced changes in membrane structure, since substantial lipid peroxidation results from ozone exposure (Figure 2). The ability of particle exposure to enhance the production of reactive oxygen metabolites was maintained in exposed alveolar macrophages even after a 45-min treatment with ozone. These data suggest that exposure of alveolar macrophages to ozone may not have affected the bactericidal capacity of the surviving phagocytes. However, since the number of surviving macrophages had declined, the antibacterial defenses of the lung may be compromised by oxidant exposure. Indeed, the enhanced generation of reactive species by ozone-exposed alveolar macrophages (Table 1) would be expected to potentiate the oxidant damage in the lung.

In the present study, the cytoplasmic taurine level of alveolar macrophages increased in response to ozone exposure (Figure 9). Whether the alveolar macrophage has the ability to synthesize taurine has not, to our knowledge, been investigated. However, it is unlikely that induction of enzymes which synthesize taurine would occur as early as 15 min after exposure to ozone, when the increase in intracellular taurine was initially seen. Furthermore, no extracellular taurine was included in the medium in our experiments, so uptake of taurine cannot account for this result. Even so, the ozone-induced inactivation of the Na^+/K^+ ATPase which we observed would prevent uptake of taurine uptake since it is a sodium and energy-dependent process. Finally, two pools of intracellular taurine, a rapidly exchangeable free pool and a slowly exchangeable bound pool, have been identified[28]. Shifts of cytosolic to membrane-bound pools of taurine have been demonstrated to occur in response to magnesium deficiency[29,30]. Therefore, our results suggest the ozone induced mobilization of taurine from bound stores. This mobilization and the resulting increase in cytoplasmic taurine may be significant steps in the process by which this nutrient acts as a possible antioxidant.

An increase in extracellular taurine levels was noted in this study (Figure 6). We have shown that the membrane of the alveolar macrophage is relatively leaky with respect to taurine[11]. Therefore, initially some of the taurine detected in the extracellular medium may have leaked out of viable cells in response to an increase in cytoplasmic taurine. Later, the continued rise of taurine in the medium may have been a consequence of increasing membrane permeability and eventual lysis of macrophages as a result of prolonged ozone exposure.

Protective Effects of Taurine on Ozone-Induced Lipid Peroxidation and Membrane Leakage in Isolated Rat Alveolar Macrophages

The results of this study indicate that taurine may function as an antioxidant in rat alveolar macrophages at its physiological concentration of $100 \mu M$ (i.e., the plasma concentration of taurine in the rat). At this level of supplementation, taurine significantly protected alveolar macrophages from ozone-induced damage. That is, recovered cells demonstrated an increase in viability as judged by trypan blue exclusion, a decrease in lipid peroxidation, a decrease in the ozone-induced decline in total and Na^+-K^+ ATPase activity, and a lessening of the leak of reduced glutathione and protein. Protection against ozone-induced cell damage was less obvious at pharmacological concentrations of external taurine, since the decline in Na^+/K^+ ATPase and the leak of protein were not significantly prevented at 250 and 500 μM extracellular taurine. The suggestion that taurine may protect alveolar macrophages from oxidant injury agrees with its ability to protect bronchioles from oxidant injury due to NO_2 exposure[8].

Taurine has been proposed as both a direct and indirect antioxidant[1-3]. As a direct antioxidant, taurine would act to quench radicals derived from the interaction of ozone with membrane lipids. As an indirect antioxidant, taurine would act to stabilize the plasma membrane and thus prevent oxidant-induced increases in membrane permeability. In support of taurine's role as a membrane stabilizer, taurine has been shown to prevent Ca^{2+} influx in cat cerebral cortex resulting from treatment with ouabain[31] and to prevent K^+ leakage in dog heart[32]. In contrast, support for taurine as a direct antioxidant was given by Nakashima et al.[33] who reported that taurine was able to mitigate CCl_4-induced lipid peroxidation in rat liver. Data from the present study are consistent with both views, since taurine significantly reduced lipid peroxidation (direct effect) and significantly increased membrane integrity (indirect effect).

The data on intracellular taurine concentrations after exposure to ozone deserve attention. In our previous study of the time course of metabolic changes occurring in alveolar macrophages during ozone exposure, free cytoplasmic taurine increased with ozone exposure. Therefore, we concluded that taurine was mobilized from cellular bound stores to the free state in response to oxidant injury[12]. In the present study, extremely high intracellular taurine concentrations resulted in alveolar macrophages incubated with 250 μM or 500 μM taurine prior to ozone exposure. There are two possible explanations for this effect: either (1) cells incubated in high taurine levels were stimulated to increase their rate of taurine uptake from the medium in response to ozone exposure; or (2) the cells were mobilizing taurine from bound stores to the free state in response to ozone exposure. Taurine uptake is Na^+ and energy dependent[11]. Since Na^+-K^+ ATPase activity in ozone-exposed cells at 250 or 500 μM taurine was very low, the size of the inwardly directed concentration gradient for Na^+ should have decreased. Thus, it is unlikely that taurine uptake had increased. Therefore, the second explanation seems more likely.

ACKNOWLEDGEMENTS

The authors thank Victor Robinson and Dr. David Frazer of the Physiology Section, NIOSH, for technical advice and construction of the modified tissue culture flasks. We are grateful to Dr. Knox Van Dyke of the Pharmacology and Toxicology

Department of West Virginia University for the gift of luminol. We thank Gunnar Shogren and Ghazi Hussein of the Divisions of Animal and Veterinary Sciences and Plant and Soil Sciences (respectively) of West Virginia University for performing the atomic absorption spectroscopy. Finally, the preparation of this manuscript by Linnea Danielsen, of the Technical Services Department, Health Science Center Library, University of Florida is appreciated.

REFERENCES

1. G.G. Gaull, H. Pasantes-Morales, and C.E. Wright, Taurine in human nutrition: overview, *in*: "Taurine: Biological Actions and Clinical Perspectives," S.S. Ota, L. Ahtel, P. Kontro, and M.K. Paasonen, eds., Liss, New York (1985).
2. K.C. Hayes and J.A. Sturman, Taurine in metabolism, *Annu. Rev. Nutr.* 1:401-425 (1981).
3. C.E. Wright, H.H. Tallan, and Y.Y. Lin, Taurine: biological update, *Annu. Rev. Biochem.* 55:427-453 (1986).
4. C.K. Chow, M.Z. Hussian, C.E. Cross, D.L. Dungworth, and M.G. Mustafa, Effect of low levels of ozone on rat lungs. I. Biochemical responses during recovery and reexposure, *Exp. Mol. Pathol.* 25:182-188 (1976).
5. J.D. Crapo, B.E. Barry, H.A. Foscue, and J. Shelburne, Structural and biochemical changes in rat lungs occurring during exposures to lethal and adaptive doses of oxygen, *Am. Rev. Respir. Dis.* 122:123-143 (1980).
6. M.G. Mustafa, and D.F. Tirney, Biochemical and metabolic changes in the lung with oxygen, ozone and nitrogen dioxide toxicity, *Am. Rev. Respir. Dis.* 118:1061-1090 (1978).
7. C.C. Plopper, C.K. Chow, D.L. Dungworth, M. Brummer, and T.J. Nemeth, Effect of low level of ozone on rat lungs II. Morphological responses during recovery and reexposure, *Exp. Mol. Pathol.* 29:400-411 (1978).
8. R.E. Gordon, A.A. Shaked, and D.F. Solano, Taurine protects hamster bronchioles from acute NO_2-induced alterations. A histological, ultrastructural and freeze-fracture study, *Amer. J. Pathol.* 125:585-600 (1986).
9. J. Klienerman, M.P.C. Ip, and J. Sorensen, Nitrogen dioxide exposure and alveolar macrophage elastase in hamsters, *Amer. Rev. Respir. Dis.* 125:203-207 (1982).
10. J.E. Sturrock, J.R. Nunn, and A.J. Jones, Effects of oxygen on pulmonary macrophages and alveolar epithelial type II cells in culture, *Respir. Physiol.* 41:381-390 (1980).
11. M.A. Banks, W.G. Martin, W.H. Pailes, and V. Castranova, Taurine uptake by isolated alveolar macrophages and type II cells, *J. Appl. Physiol.* 66:1079-1086 (1989).
12. M.A. Banks, D.W. Porter, W.G. Martin, and V. Castranova, Effects of *in vitro* ozone exposure on peroxidative damage, membrane leakage and taurine content of rat alveolar macrophages, *J. Toxicol. Appl. Pharmacol.* 105:55-65 (1990).
13. M.A. Banks, D.W. Porter, W.G. Martin, and V. Castranova, Ozone-induced lipid peroxidation and membrane leakage in isolated rat alveolar macrophages: protective effects of taurine, *J. Nutr. Biochem.* 2:308-313 (1991).
14. Z. K. Shihabi and J. P. White, *Clin. Chem.* 25:1368 (1979).
15. D.W. Porter, M.A. Banks, V. Castranova, and W.G. Martin, Reversed-phase high-performance liquid chromatography technique for taurine quantitation, *J. Chromatogr.* 454:311-316.
16. J. van der Zee, T.M.A.R. Dubbleman, T.K. Raap, and J. van Steveninck, Toxic effects of ozone on murine L929 fibroblasts. Enzyme inactivation and glutathione depletion, *Biochem. J.* 242:707-712 (1987).
17. J. van der Zee, T.M.A.R. Dubbleman, and J. van Steveninck, Toxic effects of ozone on murine L929 fibroblasts. Damaging action on transmembrane transport systems, *Biochem J.* 245:301-304 (1987).
18. J. van der Zee, K. Tijssen-Christainse, T.M.A.R. Dubbleman, and J. van Steveninck, The influence of ozone on human red blood cells. Comparison with other mechanisms of oxidants stress, *Biochim. Biophys. Acta.* 924:111-118 (1987).
19. R. Erlich, J.C. Findley, and D.E. Gardner, Effects of repeated exposures to peak concentrations of nitrogen dioxide and ozone on resistance to streptococcal pneumonia, *J. Toxicol. Environ. Health* 5:631-642 (1979).
20. R. Valentine, An *in vitro* system for exposure of lung cells to gases: Effects of ozone on rat macrophages, *J. Toxicol. Environ. Health* 16:115-126 (1985).
21. M.A. Amoruso, G. Witx, and B.D. Goldstein, Decreased superoxide anion radical production by rat alveolar macrophages following inhalation of ozone or nitrogen dioxide, *Life Sci.* 28:2215-2221 (1981).
22. M. Rister and C. Wustrow, Effect of hyperoxia on reduced glutathione in alveolar macrophages and polymorphonuclear leukocytes, *Res. Exp. Med.* 185:445-450 (1985).
23. K. Nishiki, D. Jamieson, N. Oshino, and B. Chance, Oxygen toxicity in the perfused rat liver and lung under hyperbaric conditions, *Biochem. J.* 160:343-355 (1976).

24. C. Voisin, C. Aerts, and B. Wallaert, Prevention in *in vitro* oxidant-mediated alveolar macrophage injury by cellular glutathione and precursors, *Bull. Eur. Physiopathol. Respir.* 23:309-313 (1987).

25. A.R. Dowell, L.A. Lohrbauer, D. Hurst, and S.C. Lee, Rabbit alveolar macrophage damage caused by *in vivo* ozone inhalation, *Arch. Environ. Health.* 21:121-127 (1970).

26. I.M. Rietjens, H.H. Lemmink, G.M. Aunk, and P.J. van Bladeren, The role of glutathione and glutathione S-transferases in fatty acid ozonide detoxification, *Chem. Biol. Interact.* 62:2-14 (1987).

27. I.M. Rietjens, C.A. van Tilburg, T.M. Coenen, G.M. Alink, and A.W. Konings, Influence of polyunsaturated fatty acid supplementation and membrane fluidity on ozone and nitrogen dioxide sensitivity of rat alveolar macrophages, *J. Toxicol. Environ. Health* 21:45-46 (1987).

28. J.A. Sturman, G.W. Hepner, A.L. Hufmann, and P.J. Thomas, Metabolism of ^{35}S taurine in man, *J. Nutr.* 105:1206-1214 (1975).

29. B.L. Robeson, W.G. Martin, and M.H. Friedman, A biochemical and ultrastructural study of skeletal muscle from rats fed a magnesium-deficient diet, *J. Nutr.* 110:2078-2084.

30. J. Durlach, J.R. Rapin, M. Leponcin-Lafitte, Y. Rayssiguier, M. Bata, and A. Guiet-Bara, ^{3}H-taurine distribution in various organs of magnesium deficient adult rats, *in*: "Magnesium Deficiency. Physiopahtology and Treatment Implications, 1st European Congress on Magnesium," M.J. Halpern and J. Durlach, eds., Karger, Paris (1985).

31. D.B. Tower, Ouabain and the distribution of calcium and magnesium in cerebral tissues *in vitro*, *Exp. Brain. Res.* 6:275-283 (1968).

32. W.O. Read and J.D. Welty, Studies on some cardiac effects of taurine, *J. Pharmacol. Exp. Thera.* 139:283 (1963).

33. T. Nakashima, T. Takino, and K. Kuriyama, Therapeutic and prophylactic effects of taurine administration on experimental liver injury, *in*: "Sulfure Amino Acids: Biochemical and Clinical Aspects," K. Kuriyama, R.J. Huxtable, and H. Iwata, eds., Liss, New York (1983).

THE PROTECTIVE EFFECT OF TAURINE ON THE
BIOMEMBRANE AGAINST DAMAGE PRODUCED BY THE
OXYGEN RADICAL

Ikuo Koyama, Tomomi Nakamura, Mizuho Ogasawara,
Masami Nemoto and Tsuguchika Yoshida

Research Center
Taisho Pharmaceutical Co., Ltd.
Tokyo, Japan

INTRODUCTION

Recently, the oxygen radical has attracted attention as a factor that is capable of inducing a variety of diseases. Taurine, on the other hand, is an amino acid present in blood in the highest amount among the free amino acids. However, while the physiological functions of taurine are of special interest, the mechanisms of its action are for the most part unknown. We have studied the possible protective effect of taurine on the biomembrane against damage caused by the action of the oxygen radical.

Dog erythrocytes were utilized as the test biomembranes while liposomes were used as a model membrane. The oxygen radical was generated by a water soluble azo-compound, 2,2'-azobis(2-amidinopropane) dihydrochloride [AAPH][1] which is known to oxidize biomembranes from outside the cells. The extent by which taurine suppressed hemolysis and leakage of glucose from liposomes after treatment with AAPH was studied. These results with taurine were compared with α-alanine.

METHODS

Erythrocyte Suspension

Fresh dog erythrocytes were washed with physiological saline and suspended in physiological saline or PBS.

Erythrocyte Hemolysis with AAPH

Two ml of 40 mM taurine and 1 ml of 25 mM AAPH were added to 1 ml of an erythrocyte suspension (4×10^7 cells/ml). The mixture was incubated at 37°C to initiate the radical reaction.

Taurine, Edited by J.B. Lombardini *et al*.
Plenum Press, New York, 1992

Estimation of the Hemolytic Rate

After a defined period of treatment, the erythrocyte suspension was centrifuged (2000 x g, 5 min) and the supernatant was used for the estimation of hemoglobin-Fe by atomic absorption spectrometry. The hemolytic rate was calculated.

Preparation of Glucose-Containing Liposomes

A thin film was prepared from a chloroform solution containing 20 μmol of egg yolk lecithin, 25 μmole of dicetyl-phosphate and 5 μmol of cholesterol. After hydration with 5 ml of 30 mM glucose solution, the artificial membranes were submitted to an extruder for sizing at 200 nm.

Estimation of Glucose Leakage from the Liposomes

The liposomes prepared as above were dialyzed and submitted to leakage experiments. Fifty μl of 40 mM taurine and 25 μl of 25 mM AAPH were added to 25 μl of the liposome suspension. After incubating the mixture for a defined period at 37°C, 20 μl of the suspension was removed and measured for glucose leakage. Determination of glucose was performed by the method of Inoue et al.[2]

Oxidation of Lecithin by AAPH

A thin film prepared from a chloroform solution of 43.1 μmol of egg yolk lecithin and 5 μmol of dicetyl-phosphate was hydrated with 5 ml of either PBS or 40 mM taurine solution to obtain a liposome suspension. Two hundred fifty μl of 50 mM AAPH was added to 0.5 ml of the above suspension and after allowing the mixture to stand for a defined period at 37°C it was used for estimation of residual lecithin.

Estimation of Residual Lecithin

Residual lecithin was estimated by HPLC as described by Niki et al.[1]. The HPLC column employed was Chromosorb Si60. The conditions of chromatography were as follows: temperature 25°, eluting solution MeOH: 40 mM NaH_2PO_4/96:4, wavelength 210 nm, and volume of injection 25 μl.

Hemolysis Produced by Osmotic Shock

An erythrocyte suspension (4 x 10^8 cells/ml) was added to 10 times as much volume of physiological saline or 20 mM aqueous amino acid solution and the mixture was allowed to stand for 30 min at 4°C. The mixture was then added to 3 volumes of solutions with different osmotic pressures (120-190 mOsm) and after standing for 10 min at 37°C the hemolytic rates were estimated.

RESULTS

Effect of Taurine on the Suppression of Erythrocyte Hemolysis Due to Oxidative Damage

Taurine demonstrated a more potent ability to suppress erythrocyte hemolysis due to oxidative damage than α-alanine which was used as a reference compound (Figure 1).

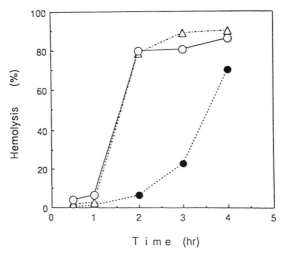

Figure 1. Effect of taurine on the suppression of hemolysis induced by AAPH. O, No addition; ●, 20 mM taurine (final concentration); Δ, 20 mM α-alanine (final concentration).

Effect of Taurine on the Suppression of Glucose Leakage from Liposomes Due to Oxidative Damage

Taurine demonstrated a tendency to suppress the leakage of glucose from liposomes due to oxidative damage (Figure 2).

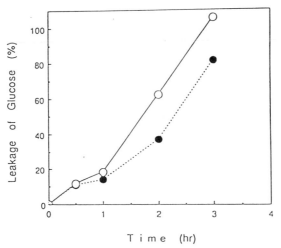

Figure 2. Effect of taurine on the suppression of glucose leakage from liposomes which was induced by AAPH. O, No addition; ●, 20 mM taurine (final concentration).

Effect of Taurine on the Protection of Liposomal Lecithin Against Peroxidation

Taurine did not demonstrate any protective effect on liposomal lecithin against peroxidation (Figure 3).

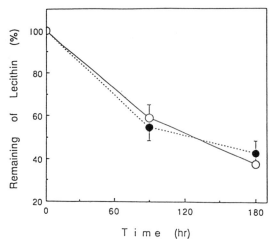

Figure 3. Effect of taurine on the protection of liposomal lecithin against peroxidation. O, No addition; ●, 20 mM taurine (final concentration).

Effect of Taurine on the Suppression of Erythrocyte Hemolysis Induced by Osmotic Shock

Taurine demonstrated a more potent action on the suppression of erythrocyte hemolysis caused by osmotic shock than α-alanine which was used as a reference compound (Figure 4).

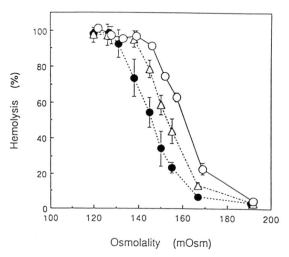

Figure 4. Effect of pretreatment with taurine on the suppression of erythrocyte hemolysis due to osmotic shock. O, No addition; ●, 4.5 mM taurine (final concentration); Δ, 4.5 mM α-alanine (final concentration).

CONCLUSIONS

1.   Taurine demonstrated a more potent action on the suppression of erythrocyte hemolysis caused by oxidative damage than α-alanine which was used as a reference compound.
2.   Taurine demonstrated a tendency to suppress glucose leakage from liposomes after oxidative damage.
3.   Taurine failed to demonstrate any effect on the peroxidation of liposomal lecithin.
4.   Taurine demonstrated a more potent action on the suppression of erythrocyte hemolysis due to osmotic shock than α-alanine which was used as a reference compound.
5.   These results suggest that taurine has a protective effect on the biomembrane by directly acting on the membrane but not by acting as an antioxidant.

REFERENCES

1.   M. Niki, H. Tamai, M. Mino, Y. Yamamoto and E. Niki, Free-radical chain oxidation of rat red blood cells by molecular oxygen and its inhibition by α-tocopherol, *Arch. Biochem. Biophys.* 258:373-380 (1987).
2.   K. Inoue, Permeability properties of liposomes prepared from dipalmitoyllecithin, dimyristoyllecithin, egg lecithin, rat liver lecithin and beef brain sphingomyelin, *Biochim. Biophys. Acta.* 339:390-402 (1974).

VOLUME REGULATORY FLUXES IN GLIAL AND RENAL CELLS

Herminia Pasantes-Morales, Julio Morán, and Roberto Sánchez-Olea

Institute of Cell Physiology
National A. University of Mexico
Mexico City, Mexico

INTRODUCTION

A variety of vertebrate cells have the ability to adjust to changes in cell volume by activating transmembrane fluxes of osmotically active solutes in the necessary direction to correct the deviations in cell volume (Macnight, 1988; Hoffmann and Simonsen, 1989). The osmotically active compounds in animal cells fall into two main groups: inorganic ions, Na^+, K^+ and Cl^- and a large variety organic molecules, including polyols, methylamines and amino acids. Among these later, taurine appears to be predominantly involved in volume regulatory processes (Pasantes-Morales and Martín del Río, 1990).

The relative contribution of these two groups of osmolytes to the process of cell volume regulation may be different in the various cells and tissues. In nervous tissue, the concentration of inorganic ions in the intracellular and extracellular compartments has to be carefully controlled to maintain the proper excitability of neurons, and therefore, it is unlikely that these ions play an important role as osmolytes. In contrast, organic molecules may significantly contribute to the process of cell volume adjustment in this tissue. To examine this possibility, in the present work we compared the properties of the volume activated fluxes of taurine and of ^{86}Rb, used as a tracer for K^+, in two types of cells in culture: cerebellar astrocytes and MDCK cells, a transformed line of renal distal cells. These two types of cells exhibit volume regulatory processes (Kimelberg and Ransom, 1986; Roy and Sauvé, 1987) and the mechanisms by which the cells accomplish these regulatory response are now actively investigated.

Concentration of Potassium, Chloride and Free Amino Acids in Astrocytes and MDCK Cells

The concentrations of the main intracellular inorganic ions K^+ and Cl^- in astrocytes and MDCK cells are shown in Table 1. A somewhat higher content of these two ions is found in the renal cells as compared to astrocytes. The opposite is observed for free amino acids, which are more concentrated in astrocytes.

Taurine is the most abundant free amino acid in these cells, accounting for almost 40% of the total pool whereas in MDCK cells, the four most abundant free

Taurine, Edited by J.B. Lombardini *et al.*
Plenum Press, New York, 1992

Table 1. Potassium, chloride and free amino acid content of astrocytes and MDCK cells.

	Astrocytes	MDCK
Potassium	130 mM	160 mM
Chloride	30 mM	60 mM
Amino Acids	102 mM	70 mM

Data for potassium and chloride in MDCK cells are from Roy and Sauvé (1987) and those in astrocytes from Kimelberg (1990). The concentration of free amino acids was measured by HPLC as described in Sanchez Olea et al. (1990), considering a cell volume of 4.5 μl/mg protein for astrocytes and 6 μl/mg protein for MDCK cells.

amino acids, taurine, glycine, alanine and glutamate contribute with about the same proportion (20%) to the free amino acid pool.

Renal cells contain relatively high amounts of polyols, particularly sorbitol, and myo-inositol as well as betaine and glycero-phosphorylcholine (Nakanishi et al., 1988).

Volume Activated Fluxes of ^{86}Rb and ^{3}H-Taurine from Astrocytes and MDCK Cells

Exposure of cultured astrocytes to solutions of reduced osmolarity (50%) results in a large increase of ^{3}H-taurine efflux of more than 10 times the release in isosmotic solutions. This hyposmolarity-evoked efflux of taurine is very fast reaching the maximal peak release as early as 1-2 min after the stimulus. Then, the efflux rate decreases despite the persistence of the hyposmotic condition (Figure 1). A rather

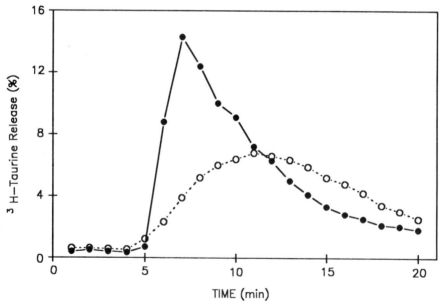

Figure 1. Time course of the hyposmolarity evoked release of taurine from astrocytes (●) and MDCK cells (O). After loading with ^{3}H-taurine, cells were superfused as described by Sánchez Olea et al. (1991). After a washing period to attain the baseline, the isosmotic superfusion medium was replaced by a medium with 50% reduced osmolarity. Results are expressed as fractional release, i.e., the radioactivity present in the samples as % of total radioactivity accumulated by cells during the loading period.

different release pattern of ^{3}H-taurine is observed in MDCK cells. The time course of taurine efflux in these cells is slow in onset and offset, attaining the maximal release only after 10 min of exposure to hyposmolarity (Figure 1). In both types of cells, the amount of taurine released from the volume-sensitive pool is related to the extent of reduction in osmolarity (See Figure 3 for astrocytes and Roy and Sauvé, 1987, for MDCK cells).

In astrocytes, detectable increases in taurine efflux are observed with reductions in osmolarity as low as 10%. In the two types of cells, a decrease in osmolarity of 50% essentially depletes the intracellular pool of taurine.

The volume associated fluxes of K^+ from astrocytes and MDCK cells were followed using ^{86}Rb as a tracer for K^+. Also marked differences were observed between astrocytes and MDCK cells in the release of Rb evoked by hyposmolarity but in the opposite direction as for taurine. In MDCK cells, a decrease in osmolarity of 30% evokes an increase in Rb efflux, rapid in onset and offset, of about 3-times over the efflux in isosmotic conditions. In contrast, in astrocytes the same reduction in osmolarity only slightly increases Rb efflux and the response is delayed (Figure 2). In these cells, substantial enhancement in Rb efflux can be evoked only by solutions with low osmolarity, of 50% or less (Figure 3).

These observations indicate that the handling of inorganic ions and organic molecules in the process of cell volume adjustment is clearly different in astrocytes and MDCK cells. In the neural cells, the mechanism of taurine release is more sensitive to hyposmolarity than that of Rb whereas the opposite is observed in MDCK cells. This suggests that the contribution of organic osmolytes such as taurine, may be predominant in excitable tissues, whereas that of inorganic ions is likely to be more restricted. This interpretation is in accordance with the crucial role played by K^+, Cl^- and Na^+ in the control of nervous excitability. An alteration in the extracellular concentration of these ions as a consequence of volume activated fluxes would result in changes in the membrane potential incompatible with normal neuronal function. In contrast, the release of taurine does not lead to any gross disturbance in

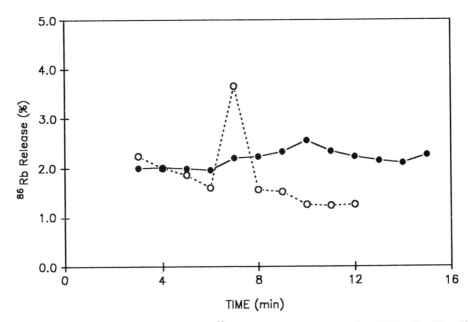

Figure 2. Hyposmolarity evoked release of ^{86}Rb from astrocytes (O) and MDCK cells (O). The experimental procedure is as described in Figure 1 except that osmolarity was reduced only 30%.

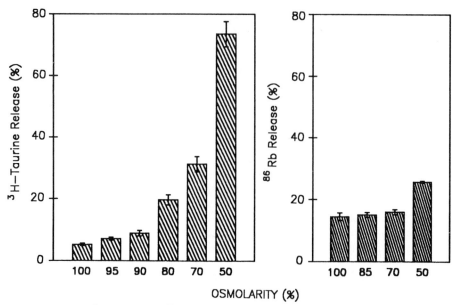

OSMOLARITY (%)

Figure 3. Release of ^{3}H-taurine and ^{86}Rb from astrocytes elicited by reductions in osmolarity. The experimental procedure is as described in Figure 1. Results represent radioactivity released in a superfusion period of 10 min with solutions of the indicated osmolarity.

nervous excitability. On the contrary, since many conditions resulting in astrocyte swelling are associated with neuronal hyperexcitability (Kimelberg and Ransom, 1986),the release into the extracellular space of a compound such as taurine, with a mild, generalized depressant action on neuronal activity, may contribute to restore the normal levels of excitability. In addition, the well known effects of taurine on protecting cell membranes from deleterious conditions such as increased Ca^{++} permeability and lipid peroxidation (Nakagawa, 1990), which also often occur associated with hyperexcitability and cell swelling, may represent an additional benefit to that of contributing to counteract the altered volume in cells.

Many questions are still unresolved regarding the volume-sensitive fluxes of osmolytes in vertebrate cells. It is unknown whether swelling elicits changes in intracellular messengers, which in turn may influence the activation of the volume regulatory fluxes of osmolytes. For many cells it is still unclear whether these fluxes occur through common pathways or by separated but interconnected pathways. Obviously, the properties of each one of the various volume sensitive fluxes have to be well defined before a complete picture of the volume regulatory process is depicted. To contribute to clarify these questions, in the present work we examined how blockers of K^+ and Cl^- carriers or channels, some of which also affect the volume regulatory process, influence the release of ^{86}Rb and ^{3}H-taurine. We also intend to investigate the mechanism of this release and we report here results on this respect for the volume sensitive efflux of taurine. With comparative purposes, similarities and dissimilarities between astrocytes and MDCK cells in these different aspects of cell volume regulation are discussed.

Interactions Between the Fluxes of Osmolytes

The question of whether the volume activated fluxes of osmolytes are interconnected has not been addressed so far. This aspect may be of interest particularly in the MDCK cells, in which the release of taurine is clearly delayed with respect to that of Rb. It is then possible that a signal for activating the efflux of taurine is related to a change in K^+ and/or Cl^- intracellular concentrations. To clarify this point it is necessary to characterize the mechanisms responsible for the activation of the volume sensitive ionic fluxes and identify drugs that can be used as pharmacological tools for this purpose. Some information in this respect exists for MDCK cells, in which the effects of inhibitors of K^+ and Cl^- channels and transporters on the volume activated transmembrane fluxes have been examined (Roy and Sauvé, 1987; Knoblauch et al., 1989; Völk et al., 1988). This is not the case for astrocytes for which such information is not available.

Results in Table 2 indicate that, from the inhibitors of K^+ channels, only quinidine showed a significance effect on the hyposmolarity activated release of Rb. Barium, TEA, and 4AP had no noticeable effect. These results are similar to those observed in MDCK cells where the K^+ efflux evoked by hyposmolarity was insensitive to these three inhibitors, but reduced by quinidine. In many cells, the K^+ fluxes associated to cell volume regulation are interconnected with Cl^- fluxes, also elicited by swelling. In some cases the two ions are cotransported by a carrier but in many others, the activated conductances occur through separate channels which nonetheless seem to be closely interrelated. This notion is supported by the effect of a number of compounds known to inhibit Cl^- channels on the K^+ activated efflux. The effect of these compounds on the volume-sensitive fluxes of Rb in MDCK cells and astrocytes is shown in Table 3. All the drugs examined similarly affected the efflux of Rb from the two types of cells. DIDS and DPC were effective in reducing Rb release whereas no inhibitory action was shown by 9-AC.

The involvement of an electroneutral carrier in the volume activated fluxes of K^+ and Cl^- is not supported by the observation that the specific inhibitors of this carrier, furosemide and bumetanide, failed to modify the efflux of K^+ and Cl^- associated to cell swelling (results not shown).

Table 2. Effect of potassium channel blockers on the efflux of ^{86}Rb and ^3H-taurine evoked by hyposmolarity.

Drug	mM	^{86}Rb Release %		^3H-Taurine Release %	
		Astrocytes	MDCK	Astrocytes	MDCK
None		100	100	100	100
Barium	15	99	98	94	91
TEA	15	96	95	104	94
4-AP	1	95	103	101	
Quinidine	0.2	76*	72*	63*	74*

Astrocytes were obtained from rat cerebellum and cultured in 24-well multitest dishes following the procedure of Drejer et al. (1987). MDCK cells were cultured as described by Sanchez Olea et al. (1991). Cells were loaded with ^{86}Rb or ^3H-taurine, washed and incubated in isosmotic Krebs-HEPES (100% osmolarity) for 15 min. Then this medium was replaced by a hyposmotic medium (50% osmolarity). The efflux is calculated as ^{86}Rb or ^3H-taurine released in 15 min of exposure to the hyposmolar condition, expressed as a percentage of total radioactivity accumulated by cells. Drugs were present in both isosmotic and hyposmotic media. The efflux evoked by hyposmolarity without addition of drugs is taken as 100%. TEA: tetraethylammonium, 4-AP: 4-aminopyridine. * Significantly different from the control P<0.05.

Table 3. Effect of chloride channel blockers on the efflux of ^{86}Rb and ^{3}H-taurine from astrocytes and MDCK cells evoked by hyposmolarity

Drug	mM	^{86}Rb Release		^{3}H-Taurine Release	
		Astrocytes	MDCK	Astrocyes	MDCK
None		100	100	100	100
DIDS	0.1	48**	70	49**	61**
DPC	1.0		102		104
9-AC	1.0	79*	94	32*	102
DIP	0.05	81*		24*	

Cells were preincubated with DIDS during the last 10 min of the loading period. Otherwise, the experimental procedure is as described in Table 3. Abbreviations: DIDS, 4-4-diisothiocyanostilbene-2,2-disulfonate; DPC, diphenylamino-2-carboxylate; 9-AC, 9-anthracene carboxylate; DIP, dipyridamol. Significantly different from the control ** $P<0.001$; * $P<0.05$.

Interestingly, the release of taurine from both MDCK cells and astrocytes was also affected by inhibitors of K$^+$ as well as of Cl$^-$ fluxes. As shown in Tables 2 and 3, taurine release in the two types of cells was inhibited by quinidine but not by Ba, TEA, or 4-AP. It was also reduced by DIDS and dipyridamol and unaffected by DPC or 9-AC. These observations suggest that the fluxes of the three osmolytes are closely interconnected. Another possibility is that the efflux of both inorganic ions and taurine occurs by the same pathway. It has been suggested that the swelling evoked taurine release from erythrocytes may take place through the anion exchanger band 3 (Goldstein and Brill, 1991). This may also be a possibility for other cells but further studies are necessary to clarify this question.

Mechanism of Taurine Release from the Volume-Sensitive Pool

Taurine is accumulated in animal cells by a widely distributed transporter which uses the energy driven by the Na$^+$ gradient to translocate taurine (Schousboe et al., 1976). Accordingly, taurine transport by this carrier is saturable and Na$^+$-dependent. A diffusional component, Na$^+$-independent and non-saturable, also contributes to accumulate taurine into the cells. These two systems of taurine transport may be affected by changes due to cell swelling and could be involved in the release of taurine from the volume sensitive pool. To investigate the contribution of the two components of taurine transport to the efflux of taurine evoked by hyposmolarity, the kinetics of taurine uptake was examined in isosmotic and hyposmotic conditions (Table 4).

Experiments were carried out measuring ^{3}H-taurine uptake in conditions of identical Na$^+$ concentrations with osmolarity adjusted with choline-chloride. In astrocytes, hyposmolarity did not affect the saturable component of taurine transport but markedly enhanced the diffusional component of the process (Table 4). In MDCK cells, the hyposmotic condition induced the same increase in diffusion but in addition, the saturable component was essentially abolished. These results seem to exclude the involvement of the active carrier in the transmembrane fluxes of taurine activated by hyposmolarity in the two types of cells. The Na$^+$ and temperature independency of taurine release from the volume sensitive pool previously observed (Pasantes-Morales et al., 1990) supports this interpretation. The large effects on the

Table 4. Effect of hyposmolarity on the kinetic constants of taurine uptake by astrocytes and MDCK cells

	Km μM	Vmax nmol/min/g	k ml/min/g
ISOSMOLAR			
Astrocytes	38	0.189	2.1×10^{-4}
MDCK	110	0.112	5.7×10^{-5}
HYPOSMOLAR			
Astrocytes	71	0.162	5.5×10^{-4}
MDCK			4.8×10^{-4}

Cells were incubated for 5 min in isosmotic or hyposmotic (50%) media containing taurine concentrations from 5-1000 μM and subsequently for 2 min in analogous solutions containing ^{3}H-taurine (4 μCi/ml). To calculate the kinetic constants the curves were fitted to the experimental points as described by Schousboe et al. (1976).

diffusion component suggest another mechanism, possibly involving the opening of a membrane pore. The similarities in the effects of volume increase of taurine fluxes in astrocytes and MDCK, indicate that the mechanism of taurine release is the same in the two types of cells.

CONCLUSIONS

The present comparative study lead to the following conclusions:

1. Astrocytes and MDCK cells release K$^+$ and taurine to adjust volume during exposure to hyposmotic conditions, but the differences in time course and sensitivity to osmolarity suggest that K$^+$ is preferentially used over taurine as an osmolyte in MDCK cells whereas the opposite occurs in astrocytes.

2. Rb and taurine fluxes associated with volume adjustment in MDCK cells and astrocytes were similarly affected by some antagonists of K$^+$ and Cl$^-$ channels, suggesting either that the release of the two osmolytes is coordinately regulated or that the mechanism of release has some common sites sensitive to the drugs.

3. The mechanism of taurine release from the volume sensitive pool is similar in MDCK cells and astrocytes and seems to involve diffusion rather than an active carrier.

ACKNOWLEDGEMENTS

The skillful work of Claudia Rodríguez, Irma López Martínez and Amparo Lázaro is greatly appreciated. The work was supported by grants from DGAPA-UNAM (IN-024589) and from the Stiftung Volkswagenverk. We are indebted to Prof. Bernd Hamprecht who acted as the German partner in this program.

REFERENCES

Drejer, J., Honoré, T., and Schousboe, A., 1987, Excitatory amino acid induced release of ^{3}H-GABA from cultured mouse cerebral cortex interneurons, *J. Neurosci.* 7:2910-2916.

Goldstein, L. and Brill, S.R., 1991, Volume-activated taurine efflux from skate erythrocytes: possible band 3 involvement, *Amer. Physiol. Soc.* 1014-1020.

Hoffmann, E.K., Simonson, L.O. and Lambert, I.H., 1984, Volume induced increase of K^+ and Cl^- permeabilities in Ehrlich ascites tumor cells. Role of internal Ca^{2+}, *J. Membr. Biol.* 78:211-222.

Kimelberg, H.K., 1990, Chloride transport across glial membranes, *in*: "Chloride Channels and Carriers in Nerve, Muscle, and Glial Cells," F.J. Alvarez-Leefmans and J.M. Russell, eds., Plenum Press, New York, 159-191.

Kimelberg, H.K. and Ransom, B.R., 1986 Physiological and pathological aspects of astrocytic swelling, *in*: "Astrocytes," S. Fedoroff and A. Vernadakis, eds., Academic Press, New York, 3:129-165.

Knoblauch, C., Marschall, Montrose, M.H., and Murer, H., 1989, Regulatory volume decrease by cultured renal cells, *Amer. Physiol. Soc.* 252-259.

Macknight, A.D.C., 1988, Principles of cell volume regulation, *Renal Physiol. Biochem.* 5:114-141.

Nakagawa, M.D., 1990, Homeostatic and protective effects of taurine, *in*: "Taurine: Functional Neurochemistry, Physiology, and Cardiology," H. Pasantes-Morales, W. Shain, D.L. Martin and R. Martín del Río, eds., Wiley-Liss, New York, 447-449.

Pasantes-Morales, H. and Martín del Río, R. 1990, Taurine and mechanisms of cell volume regulation, *in*: "Taurine: Functional Neurochemistry, Physiology and Cardiology," H. Pasantes-Morales, W. Shain, D.L. Martin and R. Martín del Río, eds., Wiley-Liss, New York, 317-320.

Pasantes-Morales, H., Morán, J., and Schousboe, A., 1990, Volume-sensitive release of taurine from cultured astrocytes: Properties and mechanism, *GLIA* 3:427-432.

Roy, G. and Sauvé, R., 1983, Stable membrane potentials and mechanical K^+ responses activated by internal Ca^{2+} in HeLa cells, *Can. J. Physiol. Pharmacol.* 61:144-148.

Sánchez Olea, R., Pasantes-Morales, H., Lázaro, A., and Cereijido, M., 1991, Osmolarity-sensitive release of free amino acids from cultured renal cells (MDCK), *J. Membrane Biol.* 121:1-9.

Schousboe, A., Fosmark, H., and Svenneby, G., 1976, Taurine uptake in astrocytes cultured from dissociated mouse brain hemispheres, *Brain Res.* 116:158-164.

Volkl, H., Paulmichl, M., and Lang, F., 1988, Cell volume regulation in renal cortical cells, *Renal Physiol. Biochem.* 11:158-173.

CELL VOLUME CHANGES AND TAURINE RELEASE
IN CEREBRAL CORTICAL SLICES

S.S. Oja and Pirjo Saransaari

Tampere Brain Research Center
Department of Biomedical Sciences
University of Tampere, Box 607
SF-33101 Tampere, Finland

INTRODUCTION

A number of years ago taurine was recognized as an osmoregulator in marine animals[1] and attention has been drawn to its possible participation in cell volume regulation also in terrestial animals.[2] It has been suggested that taurine participates in the regulation of extracellular ionic homeostasis in the central nervous system, and in particular, in the active uptake of K^+ by astroglial cells.[3,4] Recently, some investigators have thought that the role of taurine in the mammalian brain is to act as an osmolyte[5,6] which is released as a response to cell swelling.[7] Since depolarizing stimuli are known to release taurine from various brain preparations,[8,9] this response has been interpreted to ensue from depolarization-induced cell swelling.[10]

The studies which have coupled the release of taurine with cellular swelling have been carried out on astrocytes and neurons in culture. We have thus now reinvestigated the release of taurine from mouse cerebral cortex slices in which the structural integrity is to some extent preserved and which may thus better reflect relationships prevailing *in vivo*. To obtain an insight into the mechanisms involved in the release of taurine, the slices were exposed to media of widely varying ionic composition and osmotic strength and to several excitatory agents. We also assessed the alterations of cell volumes in the slices and related these changes to the magnitudes of simultaneous release of taurine.

MATERIAL AND METHODS

Slices 0.4 mm thick were manually cut from the cerebral cortices of young adult 3-month-old white NMRI mice of both sexes. The slices were preloaded for 30 min with 10 μM [³H]taurine and subsequently superfused for 50 min as detailed in Oja and Kontro.[11] The standard medium used was Hepes-buffered Krebs-Ringer solution, containing (in mM) NaCl 127, KCl 5, $CaCl_2$ 0.8, $MgSO_4$ 1.3, Hepes 15, NaOH 11, and D-glucose 10 (pH 7.4), modified in most experiments as indicated elsewhere. The efflux rate constants for taurine were computed from the amounts of label released in the course of superfusion as described in Kontro and Oja.[12] The superfusion period of 20 to 30 min represented the baseline prestimulation release of labeled taurine. The effectors were added and the ionic composition of medium was altered for the last 20 min of superfusion. In order to minimize random experimental variation the calculated efflux rate constants k_2 for the superfusion period of 34 to 50 min were calculated and given in per cent of the constant k_1 (20-30 min) of each individual slice.

The extracellular spaces in the slices were estimated as in Laakso and Oja[13] by adding 2 g/l [^{3}H]inulin to medium for the last 20 min of experiments. The total volume (weight) minus the inulin space was assumed to represent the intracellular space in the slices. Total swelling or shrinking of the slices during experiments was determined by measuring the fresh and dry weights of the slices as in Vahvelainen and Oja[14] in all different experimental conditions. Intracellular swelling or shrinking of the slices during the experiments could then be calculated with the aid of estimated alterations in the extracellular spaces and volumes (weights). All these results are given in per cent per reference volume as defined later.

For comparison, glial cells (cerebral astrocytes) and neurons (cerebellar granule cells) were also preloaded in culture with [^{3}H]taurine and the release of taurine studied in the above medium precisely as in Holopainen et al.[15] The loading time with [^{3}H]taurine was 30 min with astrocytes and 4 days with granule cells. Here with both cells, the efflux period up to 10 min represented the baseline prestimulation release and the period from 10 to 30 min the stimulation phase. The constants k_1 and k_2 were computed for the periods of 4 to 10 min and 14 to 30 min, respectively.

RESULTS

In astrocytes the K^+ stimulation of taurine release was totally abolished in hyperosmotic and Cl$^-$-free media and in media in which the ionic product Cl$^-$ x K^+ was kept constant (Table 1). The results were the same when Cl$^-$ was replaced by either the highly permeant anion acetate or the relatively impermeant gluconate. The release of taurine was markedly enhanced by hypoosmotic medium. Also in granule cells hypoosmotic medium was equally effective. The replacement of 50 mM Na$^+$ by equimolar K$^+$ caused a release of [^{3}H]taurine which was relatively small in magnitude. It was also totally abolished in hyperosmotic high-K$^+$ medium, but partially preserved in media in which the ionic product Cl$^-$ x K$^+$ was kept constant. The reason for this preservation seems to be the concominant reduction in Cl$^-$, since the release of taurine was also enhanced in Cl$^-$-free-, low-K$^+$ (5 mM) media.

The replacement of 50 mM Na$^+$ by equimolar K$^+$ evoked an about threefold increase in the release of [^{3}H]taurine from cerebral cortical slices from young adult mice (Table 2). The magnitude of the evoked release was not diminished in hyperosmotic media and in media in which the ionic product Cl$^-$ x K$^+$ was kept constant. A reduction in the Cl$^-$ content in medium also evoked release of taurine, as did exposure to hypoosmotic medium. A total or partial omission of Cl$^-$ from medium also enhanced the release of taurine. It did not matter whether the Cl$^-$ deficit was compensated by the permeant anion acetate or by the impermeant anion gluconate. The release of taurine could also be markedly enhanced by glutamate agonists. In all the above cases there was a statistically significant (P<0.01) enhancement. On the other hand, hyperosmotic media slightly but significantly (P<0.01) diminished the release. If the NaCl omitted was equimolarly compensated by sucrose, no statistically significant change was discernible.

Table 1. Release of [³H]taurine from cultured astrocytes and granule cells

Treatment of cells	Efflux rate constants k_2 (% of k_1)	
	Astrocytes	Granule cells
50 mM K	137.8 ± 7.3	94.6 ± 5.6
+ 50 mM KCl	44.8 ± 5.4	46.4 ± 2.5
0 mM Cl (acetate)	53.8 ± 5.4	116.1 ± 4.4
0 mM Cl (gluconate)	53.3 ± 5.7	80.4 ± 5.7
- 25 mM NaCl	134.0 ± 14.9	137.3 ± 1.8
13 mM Cl, 50 mM K, 88 mM Na (acetate)	72.6 ± 12.4	101.3 ± 2.7
13 mM Cl, 50 mM K, 88 mM Na (gluconate)	68.2 ± 9.1	83.4 ± 3.1
13 mM Cl, 50 mM K, 88 mM Na, 50 mM sucrose (acetate)	54.4 ± 3.3	82.5 ± 0.9
13 mM Cl, 50 mM K, 88 mM Na, 50 mM sucrose (gluconate)	36.4 ± 2.2	94.4 ± 2.1
13 mM Cl, 50 mM K, 138 mM Na (acetate)	40.1 ± 6.4	83.3 ± 2.2
13 mM Cl, 50 mM K, 138 mM Na (gluconate)	49.8 ± 5.9	79.2 ± 3.1

The cells were preincubated for 10 min with the standard medium and then exposed to the medium indicated for 20 min. Cl⁻ ions were replaced by acetate or gluconate as shown. Mean values ± SEM of 4 experiments. The average efflux rate constants k_1 for baseline efflux were $(5.70 ± 0.13) \times 10^{-3}$ min⁻¹ (n=44) with astrocytes and $(4.00 ± 0.15) \times 10^{-3}$ min⁻¹ (n=44) with granule cells.

Table 2. Release of [³H]taurine from cerebral cortical slices under different experimental conditions

Treatment of slices	Efflux rate constants k_2 (% of k_1)	
None (control)	85.9 ± 4.7	(9)
50 mM K	173.2 ± 5.9	(12)
+ 50 mM KCl	169.3 ± 13.2	(5)
50 mM K, 50 mM sucrose	216.0 ± 17.9	(9)
13 mM Cl, 50 mM K (acetate)	163.4 ± 14.4	(3)
13 mM Cl, 50 mM K (gluconate)	166.2 ± 10.7	(4)
0 mM Cl, 50 mM K (acetate)	221.5 ± 26.8	(7)
0 mM Cl, 50 mM K (gluconate)	198.6 ± 16.7	(4)
0 mM Cl (acetate)	224.8 ± 18.9	(6)
0 mM Cl (gluconate)	177.9 ± 15.3	(4)
50 mM Cl (acetate)	208.0 ± 16.7	(4)
50 mM Cl (gluconate)	138.6 ± 10.2	(4)
- 25 mM NaCl	192.0 ± 8.9	(9)
+ 50 mM sucrose	69.5 ± 5.7	(8)
- 25 mM NaCl, + 50 mM sucrose	78.7 ± 3.3	(8)
+ 100 mM sucrose	58.2 ± 4.6	(4)
0.1 mM quisqualate	106.3 ± 4.1	(12)
1.0 mM kainate	162.0 ± 13.1	(12)
1.0 mM NMDA	167.2 ± 11.1	(4)
1.0 mM NMDA, 50 mM K	286.1 ± 32.8	(5)

The slices were exposed to the treatments indicated during the last 20 min of superfusion. Mean values ± SEM are shown. Number of experiments in parentheses. The average baseline efflux rate constant k_1 was $(1.58 ± 0.14) \times 10^{-3}$ min⁻¹ (n=68). NMDA = N-methyl-D-aspartate.

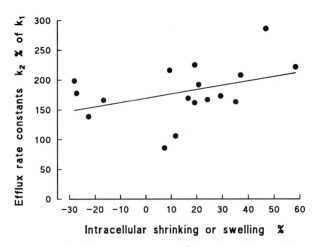

Figure 1. Correlation between the efflux rate constants of [³H]taurine and intracellular swelling in cerebral cortical slices during the last 20 min of superfusion. The efflux rate constants are from Table 2. Intracellular swelling was calculated as described in the text and given in per cent of the size of intracellular space of superfused slices prior to the stimulation phase.

The efflux rate constants of [³H]taurine in different superfusion conditions were related to the magnitude of intracellular swelling of slices during the stimulation period under the same experimental conditions (Fig. 1). The calculated correlation coefficient was 0.39. It did not significantly (P>0.1) differ from zero.

DISCUSSION

The present data on cultured cells are in keeping with the previous findings of others on the same preparations. Our results confirm that in astrocytes the release of taurine can be evoked only in hypoosmotic media[16] and in media in which K⁺ stimulation causes cell swelling.[10] However, no significance whatsoever seems to obtain, whether a fairly permeant anion, acetate, or an impermeant anion, gluconate, is used as replacer of Cl⁻. The earlier experiments on cultured cells in gluconate media, which have directly related the release of taurine to changes in cell volumes, should thus be repeated in the presence of acetate, since now in slices the intracellular swelling was converted to shrinking in the presence of gluconate and in spite of this the apparent release of taurine was practically the same. Our results also suggest that some K⁺-stimulated release of taurine may also occur from granule cells under experimental conditions in which astrocytes do not respond. Schousboe et al.[17] have failed to detect any K⁺-stimulated release of [³H]taurine from granule cells which had been preloaded for 60 min. The K⁺ stimulation in granule cells is discernible only after a prolonged preloading for 4 days with labeled taurine but not after a short-term preloading.[15] This may signify that exogenous [³H]taurine does not label identical cellular pools after different preloading periods.

The present poor or non-existent correlation between the release of taurine and intracellular swelling does not exclude the possibility that under certain isolated conditions such a correlation might exist. For instance, we obtained a fairly good positive correlation between these two parameters when only the osmolarity of medium was gradually reduced (data not shown). A very strong positive correlation was also obtained within experiments on glutamate agonists,[18] but no correlation prevailed when Cl⁻ ions were increasingly replaced by acetate, and the correlation tended to be negative when gluconate was used as replacer. Since a great variety of experimental conditions have been tested by us we can be fairly confident in claiming that in brain slices the apparent release of taurine does not reflect cell swelling alone.

Another possible explanation for the enhancement of taurine release in different experimental conditions could be a concomitant reduction in the reuptake of labeled molecules released from slices. Most of the release occurs into extracellular spaces and not directly to superfusion medium. The released molecules must traverse the extracellular space and escape reuptake in order to reach their final destination, the superfusion medium.[19] The measured release thus represents some kind of an 'overflow'. Many of the experimental conditions tested are known to inhibit the uptake of taurine into brain slices, e.g., high-K^+ medium, hypoosmolarity, and reduction in Na^+ and Cl^-.[20] We have measured molar amounts of endogenous taurine released under the present experimental conditions.[21] For example, when the relatively low capacity of both high- and low-affinity uptake systems of taurine[20,22] is taken into account, the extracellular concentrations of taurine should be within the millimolar range to constitute the sole explanation for the apparent stimulation of release in high-K^+ media. This is not likely. Moreover, our superfusion system with freely floating and vigorously agitated slices is specifically designed to minimize the reuptake.[23]

In conclusion, we may take our data to indicate that taurine is not released from cerebral cortical slices only as a response to cell swelling but also owing to K^+ depolarization. This assumption does not necessarily imply that exocytosis is the mechanism responsible for the release, since cell membrane carriers operating in an outward direction are also likely to participate, at least in media with modified ionic composition. It has already been shown that Na^+ gradients govern the release under certain experimental conditions, possibly through redistributed orientation of taurine carriers in cell membranes.[24]

ACKNOWLEDGEMENTS

The skilful technical assistance of Ms Paula Kosonen, Ms Oili Pääkkönen and Ms Irma Rantamaa and the financial support of the Emil Aaltonen Foundation, Finland, are gratefully acknowledged.

REFERENCES

1. J.W. Simpson, K. Allen, and J. Awapara, Free amino acids in some aquatic invertebrates, *Biol. Bull.* 117:371 (1959).
2. E.K. Hoffmann and K.B. Hendil, The role of amino acids and taurine in isosmotic intracellular regulation in Ehrlich ascites mouse tumor cells, *J. Comp. Physiol.* 108:279 (1976).
3. W. Walz and E.C. Hinks, Carrier-mediated KCl accumulation accompanied by water movements is involved in the control of physiological K^+ levels by astrocytes, *Brain Res.* 343:44 (1985).
4. W. Walz, Swelling and potassium uptake in cultured astrocytes, *Can. J. Physiol. Pharmacol.* 65:1051 (1987).
5. N.M. Van Gelder, Brain taurine content as a function of cerebral metabolic rate: osmotic regulation of glucose derived water production, *Neurochem. Res.* 14:495 (1989).
6. J.E. Olson and M.D. Goldfinger, Amino acid content of rat cerebral astrocytes adapted to hyperosmotic medium *in vitro, Neurosci. Res.* 27:241 (1990).
7. H. Pasantes-Morales, J. Moran, and A. Schousboe, Volume-sensitive release of taurine from cultured astrocytes: properties and mechanism. *Glia* 3:427 (1990).
8. S.S. Oja and P. Kontro, Taurine, *in*: "Handbook of Neurochemistry", 2nd edn, vol. 3, p. 501, A. Lajtha, ed., Plenum, New York (1983).
9. P. Saransaari and S.S. Oja, Release of GABA and taurine from brain slices, *Prog. Neurobiol.,* in press (1992).

10. H. Pasantes-Morales and A. Schousboe, Release of taurine from astrocytes during potassium-evoked swelling, *Glia* 2:45 (1989).

11. S.S. Oja and P. Kontro, Cation effects on taurine release from brain slices: comparison to GABA, *J. Neurosci. Res.* 17:302 (1987).

12. P. Kontro and S.S. Oja, Taurine and GABA release from mouse cerebral cortex slices: potassium stimulation releases more taurine than GABA from developing brain, *Devl Brain Res.* 37:277-291 (1987).

13. M.-L. Laakso and S.S. Oja, Factors influencing the inulin space in cerebral cortex slices from adult and 7-day-old rats, *Acta Physiol. Scand.* 97:486 (1976).

14. M.-L. Vahvelainen and S.S. Oja, Kinetics of influx of phenylalanine, tyrosine, tryptophan, histidine and leucine into slices of brain cortex from adult and 7-day-old rats, *Brain Res.* 40:477 (1972).

15. I. Holopainen, P. Kontro, and S.S. Oja, Release of taurine from cultured cerebellar granule cells and astrocytes: co-release with glutamate, *Neuroscience* 29:425 (1989).

16. H. Pasantes Morales and A. Schousboe, Volume regulation in astrocytes: a role for taurine as an osmoeffector, *J. Neurosci. Res.* 20:505 (1988).

17. A. Schousboe, J. Morán, and H. Pasantes-Morales, Potassium-stimulated release of taurine from cultured cerebellar granule neurons is associated with cell swelling. *J. Neurosci. Res.* 27:71 (1990).

18. P. Saransaari and S.S. Oja, Excitatory amino acids evoke taurine release from cerebral cortex slices from adult and developing mice, *Neuroscience*, in press (1991).

19. S.S. Oja and M.-L. Vahvelainen, Transport of amino acids in brain slices, *in*: "Research Methods in Neurochemistry", vol. 3, p. 67, N. Marks and R. Rodnight, eds, Plenum, New York (1975).

20. P. Kontro, Effects of cations on taurine, hypotaurine, and GABA uptake in mouse brain slices, *Neurochem. Res.* 7:1391 (1982).

21. S.S. Oja and P. Kontro, Release of endogenous taurine and γ-aminobutyric acid from brain slices from the adult and developing mouse, *J. Neurochem.* 52:1018 (1989).

22. P. Kontro and S.S. Oja, S.S. Hypotaurine transport in brain slices: comparison with taurine and GABA, *Neurochem. Res.* 6:1179 (1981).

23. E.R. Korpi and S.S. Oja, Comparison of two superfusion systems for study of neurotransmitter release from rat cerebral cortex slices, *J. Neurochem.* 43:236 (1984).

24. E.R. Korpi and S.S. Oja, Characteristics of taurine release from cerebral cortex slices induced by sodium-deficient media, *Brain Res.* 289:197-204 (1983).

L-GLUTAMATE-INDUCED SWELLING OF CULTURED ASTROCYTES

Yutaka Koyama, Tadashi Ishibashi and Akemichi Baba

Department of Pharmacology, Faculty of Pharmaceutical
Sciences, Osaka University, 1-6 Yamada-Oka, Suita
Osaka, Japan

INTRODUCTION

Neurotransmission is influenced by various astrocytic functions. One of the more important astrocytic functions is control of the ion composition in extracellular fluid. Changes in astrocytic volume are responsible for the maintenance of extracellular K^+ and H^+ concentrations in the CNS (Dietzel et al., 1982; Walz and Hinks, 1985). In fact, disturbance of astrocytic volume regulation is suggested to be involved in brain damage (Hertz, 1981). Taurine (Tau) is reported to be an osmoregulatory substance in the brain since Tau is released from astrocytes into the extracellular space by both high-K^+ and hypotonic-induced volume increase (Pasantes-Morales and Schousboe, 1988; 1989; Martin et al., 1990). Numerous possible roles of the released Tau in neuronal pathological states have been suggested.

It is accepted that disruption of L-glutamate (L-Glu) neurotransmission underlies the neuronal injury observed in brain damage such as ischemia and head trauma (Choi, 1988). *In vivo* experiments using a microdialysis method demonstrated that excitatory amino acids (EAAs) stimulate Tau release in brain, suggesting multiple mechanisms for the Tau release (Lehmann et al., 1985; Menéndez et al., 1990). Astrocytes have several types of EAA receptors. Activation of the EAA receptors induces astrocytic depolarization and an increase in cytosolic Ca^{2+} (Bowman and Kimelberg, 1984; Cornell-Bell et al., 1990). At neurotoxic doses, L-Glu causes swelling of cultured astrocytes (Chan and Chu, 1989; Koyama et al., 1991a; 1991b). In the studies reported in this communication, we examined possible mechanisms of Tau release by EAAs from rat cerebral cultured astrocytes in relation to cell swelling.

EXPERIMENTAL PROCEDURE

Cell Culture

Cultured astrocytes were prepared by the previously described method (Koyama et al., 1991a). Briefly, cells from cerebrum of 1-2 day old Sprague-Dawley rats were enzymatically dissociated, and cultured for 2 weeks in Eagle's minimum-essential medium containing 10% fetal bovine serum. Dibutyryl cAMP (0.25 mM) was then added to the medium and the cells were further cultured for 10-14 days. At this stage, there were approximately 95% glial fibrillary acidic protein positive cells most of which were protoplasmic type astrocytes (Type 1).

Release Experiments

After the culture medium was removed, the astrocytic monolayer was rinsed with HEPES-buffered balanced solution (HBBS: NaCl, 128; KCl, 5; $CaCl_2$, 1; $MgSO_4$, 0.5; KH_2PO_4, 1; D-glucose, 10; HEPES/NaCl pH 7.4, 20; in mM) and incubated with 50 μM ^{3}H-Tau (17 Ci/mmole, Amersham) for 60 min. The astrocytes were then washed and pre-incubated at 37°C for 30 min to stabilize the basal ^{3}H-Tau efflux. EAAs were included in HBBS (1.0 ml/well) and the EAA-containing HBBS was replaced with fresh solution every 10 min. In some experiments, astrocytes were incubated with EAAs for 60 min without exchange of the HBBS. Radioactivity release in each fraction and that remaining in the cells were determined by scintillation counting. Results were expressed as percentages of released ^{3}H-Tau to the total radioactivity taken up by the astrocytes. Release of endogenous Tau was determined by the o-phthalaldehyde derivative method utilizing reverse-phase HPLC. Cl$^-$-free HBBS was prepared by substituting $Ca(NO_3)_2$ and sodium gluconate for $CaCl_2$ and NaCl, respectively. Ca^{2+}-free HBBS was prepared by replacing $MnCl_2$ for $CaCl_2$.

Cell Volume Measurement

The volume of the astrocytes were determined according to the previously described method of Koyama et al. (1991a) using ^{3}H-O-methyl-D-glucose (OMG, 85 Ci/mmole, ARC).

RESULTS

^{3}H-Tau release from pre-loaded astrocytes was stimulated by incubation with 0.5-1 mM L-Glu, kainate (KA), DL-homocysteate (HCA), L-cysteate (CA), L-aspartate (Asp) and quisqualate (QA), but not with NMDA. Release of endogenous Tau from cultured astrocytes, but not that of pre-loaded ^{3}H-hydroxy-tryptamine (5-HT), was also stimulated by L-Glu (Table 1). At the same concentrations that stimulated Tau release, L-Glu, L-Asp, HCA, L-CA and QA also increased by approximately 2 fold the astrocytic intracellular water space (^{3}H-OMG space) after 60 min of treatment. These EAAs induced morphological changes in astrocytes observed under phase-contrast microscopy, i.e., enlargement of the astrocytic processes and cell bodies with swollen nuclei. NMDA and KA had no effect on the astrocytic ^{3}H-OMG space.

Figure 1A shows the time course of ^{3}H-Tau release stimulated by L-Glu and KA. Stimulated ^{3}H-Tau release was first observed 10 min after L-Glu addition, and the

Table 1. Effect of L-Glu on release of Tau and 5-HT from cultured astrocytes.

	Release (%)	
	Basal	L-Glu
^{3}H-Tau	11.3 ± 0.6 (10)	26.8 ± 4.1 (6)*
^{3}H-5HT	11.2 ± 1.8 (4)	7.5 ± 0.9 (4)
Endogenous Tau	10.6 ± 1.0 (3)	33.2 ± 1.2 (3)*

Astrocytes were incubated with 0.5 mM L-Glu for 60 min without exchange of the medium. Results are means ± SEM. The number of experiments is given in the parentheses. * P<0.001 for L-Glu treatment vs basal release.

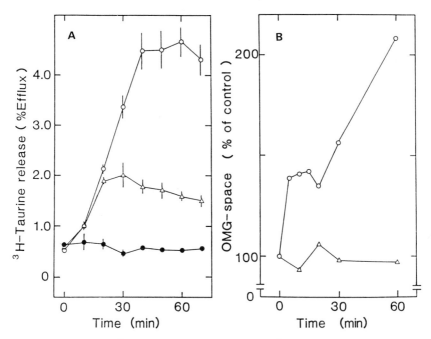

Figure 1. The time-courses of ^{3}H-Tau release (A) and increase in ^{3}H-OMG space (B) by L-Glu and KA. (A): Astrocytes were incubated with 0.5 mM L-Glu (O) or 0.5 mM KA (△) for 60 min. The EAA-containing HBBSs were exchanged every 10 min and radioactivity in each fraction was measured. Basal ^{3}H-Tau release is presented by closed circles. Results are means ± SEM of 4 experiments. **(B):** Astrocytes were incubated with 0.5 mM L-Glu (O) and 1 mM KA (△). Results are means of 3-5 experiments.

rate of ^{3}H-Tau release gradually increased reaching a maximum at 40 min. Stimulated rate of ^{3}H-Tau release by KA was also first observed at 10 min, reached a maximum at 20-30 min, and then decreased with time. L-Glu-induced increase in astrocytic volume had two onsets: ^{3}H-OMG space increased 5 min after L-Glu addition, plateaued for 15 min, and then further increased for 40 min (Figure 1B). KA did not increase astrocytic ^{3}H-OMG space irrespective of the treatment length. Characteristics of each component of L-Glu-induced swelling of cultured astrocytes are summarized in Table 2. The increase in astrocytic ^{3}H-OMG space induced by 10 min of L-Glu treatment (rapid increase) was prevented by furosemide, an anion transport inhibitor, and by removal of extracellular Cl$^-$, but not by verapamil, a Ca^{2+} channel blocker, and removal of Ca^{2+}. The effect of 60 min of L-Glu treatment was attenuated by removal of Ca^{2+} and verapamil as well as by the Cl$^-$-free condition and furosemide.

The L-Glu stimulated ^{3}H-Tau release from cultured astrocytes was reduced by removal of either Cl$^-$ or Ca^{2+} from the HBBS (Figure 2). Addition of 50 mM sucrose to HBBS attenuated the effect of L-Glu. The KA-stimulated ^{3}H-Tau release was not attenuated by removal of Ca^{2+} and addition of sucrose, however, the Cl$^-$-free condition diminished the KA-stimulated release.

DISCUSSION

EAA-stimulated Tau release, determined by using the *in vivo* microdialysis technique, has been reported for the hippocampus (Lehmann et al., 1985; Menéndez

Table 2. Characteristics of the rapid and delayed swelling of cultured astrocytes induced by L-Glu

	L-Glu Treatment	
	10 min (rapid)	60 min (delayed)
Dependence for		
Na$^+$	N.D.	+
Cl$^-$	+	+
Ca^{2+}	-	partial
Furosemide	inhibition	inhibition
Verapamil	no effect	partial inhibition

Swelling of cultured astrocytes was evaluated by determination of ^{3}H-OMG space. +, dependent; -, independent; N.D., not determined.

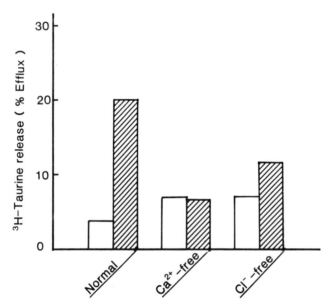

Figure 2. Dependence of L-Glu-stimulated ^{3}H-Tau release on extracellular Ca^{2+} and Cl$^-$. Astrocytes were incubated with 0.5 mM L-Glu for 60 min without exchanging the HBBS. Results are means of 3 experiments. Open columns: basal release; hatched columns: 0.5 mM L-Glu.

et al., 1990). However, mechanisms of the Tau release by EAAs are not known. Astrocytes have a high-affinity uptake system for Tau, which enables astrocytes to maintain a high level of cytosolic Tau. Tau is released from astrocytes during high-K$^+$- or hypoosmolarity-induced swelling (Pasantes-Morales and Schousboe, 1988,1989; Martin et al., 1990), suggesting an osmoregulatory role for this amino acid in the brain (Wade et al., 1988). Several types of L-Glu receptors are present in astrocytes and we previously demonstrated that overactivation of the L-Glu receptors induces the swelling of rat cultured astrocytes (Koyama et al., 1991a,1991b). In the present study, L-Glu stimulated the Tau release from cultured astrocytes. Lack of the effect of L-Glu on ^{3}H-5HT release (Table 1) indicates that the stimulated Tau release is not due to astrocytic damage. We consider that the L-Glu-stimulated Tau release is caused by the swelling of the astrocytes. The reasons are: 1) the slow onset of increase in ^{3}H-Tau release rate by L-Glu corresponds to that of the astrocytic volume increase, 2) removal of extracellular Cl$^-$ and Ca^{2+}, which prevents L-Glu-induced swelling (Koyama et al., 1991b), attenuated the stimulated Tau release (Figure 2), and 3) hyperosmolarity induced by the addition of sucrose diminished the effect of L-Glu on Tau release. Treatment of cultured astrocytes with L-Asp, DL-HCA, L-CA and QA increased both ^{3}H-Tau release and ^{3}H-OMG space.

KA increased the rate of ^{3}H-Tau release from the astrocytes but did not increase astrocytic ^{3}H-OMG space (Figure 1B). The increase in ^{3}H-Tau release rate by KA showed a different time-course from that of L-Glu (Figure 1A). These results indicate that, unlike L-Glu, KA stimulates Tau release independently of the cell swelling. Thus, the present study clearly shows two different mechanisms of astrocytic Tau release by EAAs; one mechanism is cell volume dependent and the other is cell volume independent.

Astrocytic swelling is observed in brain ischemia and head trauma (Chan and Fishman, 1985; Kimelberg and Ransom, 1986). In such neuropathological states, excess L-Glu is released from the nerve terminals and the resulting overactivation of L-Glu receptors causes neuronal death. L-Glu release may also induce cell swelling and Tau release. Thus, Tau may also share a role in osmoregulation in some pathological states of the brain.

ACKNOWLEDGEMENTS

This work was supported by grants from the Ministry of Education, Science and Culture of Japan, and The Cell Science Research Foundation.

REFERENCES

Bowman, C.L. and Kimelberg, H.K., 1984, Excitatory amino acids directly depolarize rat brain astrocytes in primary culture, *Nature* (Lond.), 311:656-659.

Chan, P.H. and Chu, L., 1989, Ketamine protects cultured astrocytes from glutamate-induced swelling, *Brain Res.* 487:380-383.

Chan, P.H. and Fishman, R.A., 1985, Brain Edema, *in*: "Handbook of Neurochemistry", A. Lajtha, ed., pp. 153-174, Plenum Press, New York.

Choi, D.W., 1988, Glutamate neurotoxicity and diseases of the nervous system, *Neuron* 1:623-634.

Cornell-Bell, A.H., Finkbeiner, S.M., Cooper, M.S., and Smith, S.J., 1990, Glutamate induces calcium waves in cultured astrocytes: long-range glial signaling, *Science* 247:470-473.

Dietzel, I., Heinemann, U., Hofmeier, G., and Lux, H.D., 1982, Stimulus-induced changes in extracellular Na$^+$ and Cl$^-$ concentration in relation to changes in the size of the extracellular space, *Exp. Brain Res.* 46:73-84.

Hertz, L., 1981, Features of astrocytic function apparently involved in the response of central nervous tissue to ischemia-hypoxa, *J. Cereb. Blood Flow Metab.* 1:143-153.

Kimelberg, H.K. and Ransom, A.R., 1986, Physiological and pathological aspect of astrocytic swelling, *in* "Astrocyte", Vol. 3, S. Fedroff and A. Vernadakis, ed., pp. 129-166, Academic Press, Orlando.

Koyama, Y., Baba, A., and Iwata, H., 1991a, L-Glutamate-induced swelling of cultured astrocytes is dependent on extracellular Ca^{2+}, *Neurosci. Lett.* 122:210-212.

Koyama, Y., Sugimoto, T., Shigenaga, Y., Baba, A., and Iwata, H., 1991b, A morphological study of glutamate-induced swelling of cultured astrocytes: involvement of chloride and calcium mechanisms, *Neurosci. Lett.* 124:235-238.

Lehmann, A., Lazarewicz, J.W., and Zeise, M., 1985, N-Methylaspartate-evoked liberation of taurine and phosphoethanolamine *in vivo*: site of release, *J. Neurochem.* 45:1172-1177.

Martin, D.L., Madelian, V., Seligmann, B., and Shain, W., 1990, The role of osmotic pressure and membrane potential in K$^+$-stimulated taurine release from cultured astrocytes and LRM55 cells, *J. Neurosci.* 10:571-577.

Menéndez, N., Solis, J.M., Herreras, O., Herranz, A.S., and Martin del Rio, R., 1990, Role of endogenous taurine on the glutamate analogue-induced neurotoxicity in the rat hippocampus *in vivo*, *J. Neurochem.* 55:714-717.

Pasantes-Morales, H. and Schousboe, A., 1988, Volume regulation in astrocytes: a role for taurine as osmoeffector, *J. Neurosci. Res.* 20:505-509.

Pasantes-Morales, H. and Schousboe, A., 1989, Release of taurine from astrocytes during potassium-evoked swelling, *Glia* 2:45-50.

Wade, J.V., Olson, J.P. Samson, F.E., Nelson, S.R., and Pazdernik, T.L., 1988, A possible role for taurine in osmoregulation within the brain, *J. Neurochem.* 51:740-745.

Walz, W. and Hinks, E.C., 1985, Carrier-mediated KCl accumulation accompanied by water movements is involved in the control of physiological K^+ levels by astrocytes, *Brain Res.* 343:44-51.

TAURINE AND VOLUME REGULATION IN
ISOLATED NERVE ENDINGS

Roberto Sánchez-Olea and Herminia Pasantes-Morales

Institute of Cell Physiology
National A. University of Mexico
Mexico City, Mexico

INTRODUCTION

Taurine is released from nerve endings isolated from different brain regions by a number of depolarizing agents and conditions, including high concentrations of potassium, ouabain, veratridine and electrical stimulation (Rev. in Huxtable, 1989). This response has been considered as evidence supporting a role for taurine as a neurotransmitter. However, the features of this release do not fit well with those currently exhibited by synaptic transmitters. It has been consistently observed that taurine release from synaptosomes is essentially calcium independent (Huxtable, 1989) in opposition to the calcium requirement for exocytotic release of neurotransmitters. In contrast, a predominant chloride dependent component of the potassium stimulated efflux of taurine is observed and evidence has been provided suggesting that this release is associated with synaptosomal swelling (Sánchez-Olea and Pasantes-Morales, 1990). This observation together with the above mentioned properties of taurine release raised the question of whether a component of the taurine efflux elicited by depolarizing conditions is a response to swelling and reflects an involvement of taurine in mechanisms of volume regulation in the nerve endings.

Isolated Nerve Endings Exhibit Volume Regulatory Processes

A variety of cells are able to preserve constant cell size in face of variations in osmolarity of the medium (Gilles, 1988). It was initially considered that this was a property of cells in species which are naturally exposed to such changes in osmolarity. However, there is increasing evidence that the ability to adjust cell volume is a general feature of cells even when they are always exposed to conditions of controlled osmolarity. In these conditions, though, cells tend to passively gain volume due to the occurrence of impermeant, polyanionic macromolecules in the cytosol which generate oncotic pressure, drawing water and permeant solutes into the cell. It is assumed that this tendency of cells to swell is counteracted by active extrusion of ions by the Na/K ATPase but there is evidence of deviation of this theory since inhibition of the pump does not result in marked cell swelling (Negendank, 1982).

Taurine, Edited by J.B. Lombardini *et al.*
Plenum Press, New York, 1992

It is then possible that transmembrane fluxes of other osmotically active solutes, particularly organic molecules, contribute, at least in part, to counteract this passive tendency to cell swelling. Among them, taurine may be particularly suitable due to its high intracellular concentration and its relative metabolic inertness.

It is now well established that the intracellular taurine pool is mobilized by hyposmolarity-induced swelling in cells such as lymphocytes, renal cells and astrocytes, which actively adjust cell volume under these conditions (García et al., Sánchez-Olea et al., 1991, Pasantes-Morales and Schousboe, 1989). A recent study by Babila et al. (1990) has shown that synaptosomes also possess this ability and regulate their volume following exposure to anisosmotic solutions. Synaptosomes incubated in a hyposmolar solution, rapidly swell increasing their initial volume by about 25%. Within 2 min after the swelling, synaptosomal volume is recovered to almost the initial value.

Taurine Efflux of Taurine is Associated to Volume Regulation in Synaptosomes

When synaptosomes preloaded with ^{3}H-taurine were exposed to a solution of reduced osmolarity (200 mOsm), a marked increase in the efflux of labeled taurine was observed. The release of taurine increased more than 13-fold over the baseline in isosmotic conditions (Figure 1). This efflux of taurine evoked by hyposmolarity is rapid, reaching a maximum within 1 min after exposure to the hyposmotic solution. After this peak release, taurine efflux decreased despite the persistence of the

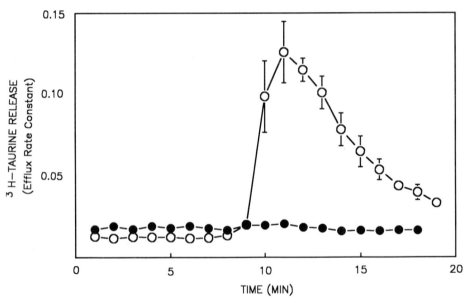

Figure 1. Hyposmolarity evoked release of taurine from synaptosomes. A purified synaptosomal fraction was obtained from rat brain cerebral cortex and incubated with ^{3}H-taurine. After loading, synaptosomes were separated by filtration and superfused with isosmotic Krebs-HEPES medium, 300 mOsm, for 8 min and with medium with reduced osmolarity, 150 mOsm for 12 min (O). The curve with filled circles (●) corresponds to taurine release from synaptosomes superfused with medium containing reduced NaCl concentrations at the same extent as in the hyposmotic medium but made isosmotic with sucrose. Results represent fractional release (% ± SE) of the radioactivity accumulated by synaptosomes during the loading period.

hyposmotic conditions to again attain the baseline in about 3 min. The reduction in osmolarity is obtained by decreasing the concentration of NaCl but this reduction per se is not responsible for the release of taurine as solutions with decreased NaCl but made isosmotic with sucrose did not elicit taurine release (Figure 1). The time course of taurine release follows a pattern similar to that of the volume regulatory decrease (Babila et al., 1990).

Taurine is the most abundant free amino acid in rat brain nerve endings (Simler et al., 1990). The concentration of taurine in synaptosomes incubated in isosmotic medium (300 mOsm) is about 100 nmoles/mg protein. Following a decrease in osmolarity to 200 mOsm, 40 % of this pool is released. Reductions in osmolarity to only 240 and 270 mOsm still induce the release of 32% and 19% of the taurine content in synaptosomes. Other amino acids such as alanine, serine, glutamine and alanine were much less affected by hyposmolarity.

Concluding Remarks

The present results which show a significant release of taurine subsequent to synaptosomal swelling suggest that this amino acid is involved in the volume regulatory decrease which has been recently described in isolated nerve endings. It has been known for a long time that astrocytes adjust cell volume following a hyposmotic shock (Kimelberg and Frangakis, 1986). It is still unclear whether neurons have the same ability or if the volume regulatory processes are restricted to nerve endings.

The question is raised whether functions in nerve endings may lead to volume changes requiring adjustment. Besides the regulation of the oncotic pressure in the cytosol mentioned above, similar processes might occur during the normal operation of synaptic transmission. Membrane potential changes associated with nerve stimulus and synaptic transmission may alter the steady-state levels of osmolytes. At the various steps of neurotransmitter secretion, the redistribution of osmotically active or inactive species within the cell may generate internal anisotonicities. The storage of transmitters within the vesicles, the process of exocytosis, the transport of metabolites and enzymes from the pericaryon to the nerve ending, the insertion of receptors and carriers are all examples of processes generating microscopic transient osmotic gradients leading to local volume changes requiring compensatory mechanisms.

Taurine at very high concentrations is present in nerve endings of the Torpedo electric plate (Osborn et al., 1975), which contains a pure population of cholinergic terminals. It is also released from cerebellar granule neurons which are glutamatergic and from cortical neurons which are predominantly GABAergic (Schousboe and Pasantes-Morales, 1989). These observations may be interpreted as if taurine release from synapses handling different neurotransmitters is related to a basic function of the nerve ending which may well be associated with volume regulatory processes.

ACKNOWLEDGMENTS

The technical assistance of Irma López Martinez is greatly acknowledged. The work was supported in part by grants from DGAPA-UNAM (IN-024589) and from the Stiftung Volkswagensverk. We are indebted to Prof. Bernd Hamprecht who acted as the German partner in this program.

REFERENCES

Babila, T., Atlan, H., Fromer, I., Schwalb, H., Uretzky, G., and Litchstein, G., 1990, Volume regulation of nerve terminals, *J. Neurochem.* 55:2058-2062.

García, J.J., Sánchez-Olea, R., and Pasantes-Morales, H., 1991, Taurine release associated to volume regulation in rabbit lymphocytes, *J. Cell Biochem*. 45:207-212.

Gilles, R., 1988, Comparative aspects of cell osmoregulation and volume control, *Renal Physiol. Biochem*. 5: 277-288.

Huxtable, R.J., 1989, Taurine in the central nervous system and the mammalian actions of taurine, *Progr. Neurobiol*. 32:472-533.

Kimelberg, H.K. and Frangakis, M.V., 1986, Volume regulation in primary astrocyte cultures, *Adv. Biosci*. 61:177-186.

Negendank, W., 1982, Studies of ions and water in human lymphocytes, *Biochim. Biophys. Acta*, 694:123-161.

Osborn, N.N., Zimmermann, H., Dowdall, M.J., and Seiler, N., 1975, GABA and amino acids in the electric organ of Torpedo, *Brain Res*. 88:115-119.

Pasantes-Morales, H. and Schousboe, A., 1988, Volume regulation in astrocytes: a role for taurine as osmoeffector, *J. Neurosci. Res*. 20:505-509.

Pasantes- Morales, H. and Martín del Río, R. 1990. Taurine and mechanisms of cell volume regulation. in:"Taurine: Functional Neurochemistry, Physiology and Cardiology", H. Pasantes-Morales, W. Shain, D. Martin, and R. Martín del Río, eds., Wiley-Liss, New York, 317-328.

Sánchez-Olea, R. and Pasantes-Morales, H, 1990, Chloride dependence of the K^+-stimulated release of taurine from synaptosomes, *Neurochem. Res*. 15:535-540.

Sánchez-Olea, R., Pasantes-Morales, H., Lázaro A., and Cereijido, M., 1991, Osmolarity sensitive release of free amino acids from cultured kidney cells (MDCK), *J. Membrane Biol*. 121:1-9.

Schousboe, A. and Pasantes-Morales, H., 1989, Potassium-stimulated release of ^{3}H-taurine from cultured GABAergic and glutamatergic neurons.

Simler, S. Ciesielski, L., Gobaille, S., and Mandel, P., 1990, Alterations in synaptosomal neurotransmitter amino acids in petit-mal rats at daytime and nighttime, *Neurochem. Res*. 15:1079-1084.

HYPEROSMOLARITY AND TAURINE CONTENT, UPTAKE AND RELEASE IN ASTROCYTES

Julio Morán, Roberto Sánchez-Olea and
Herminia Pasantes-Morales

Institute of Cell Physiology
National University of Mexico
Mexico City, Mexico

INTRODUCTION

There is increasing evidence suggesting an association of free amino acids with the mechanisms of cell volume regulation in the nervous system[1]. A response of tissue cells to hyposmolarity by massive release of taurine and other free amino acids has been documented in a wide variety of prepartions that includes cultured cells and intact brain[2,3,4]. There is also an indication that as part of an adaptative response of the brain to hyperosmolarity there is an increase in the intracellular concentration of amino acids, including taurine. Thurston et al.[5] showed that in chronic hypernatremic dehydrated mice a decrease in the water content of the brain is followed by a marked increment in the concentration of free amino acids, particularly taurine. Similarly, in chronically dehydrated Brattelboro rats, taurine concentration in the brain is markedly enhanced[6]. Recently a study by Olson and Goldfinger[7] demonstrated an increase of taurine content in cultured astrocytes grown in hyperosmotic solutions. The mechanism by which these increases in taurine content occur is still unknown. In this work we explored the possible mechanisms responsible for the increase of taurine levels induced by hyperosmolarity in cerebellar astrocytes.

Effect of Hyperosmolarity on Free Amino Acids Content

When cells are incubated in a medium made 50% hyperosmotic with NaCl or sucrose during 24 h there is a remarkable enhancement in the concentration of some amino acids (Table I). In hypernatremic conditions, taurine content is increased by more than 70% and glutamine by about 100%. Under the same conditions, the concentrations of glutamate, alanine + tyrosine, phenylalanine + glycine + threonine also increase, but to a lesser extent. Some amino acids are not affected by hyperosmolarity including serine, histidine, leucine, isoleucine, valine and lysine (not shown). The same results are obtained when hyperosmolar media are prepared with sucrose, but changes are smaller than with NaCl. Similar results were observed by Olson and Goldfinger[7]. Increasing NaCl concentration from 135 mM to 185 mM caused only a

Table 1. Effect of hyperosmolar conditions on free amino acid content of cultured cerebral cortex astrocytes.

Amino acid	Isosmolar NaCl 135 mM	Hyperosmolar NaCl 210 mM	Hyperosmolar Sucrose 150 mM
Glu	100.5 ± 10.3	135.4 ± 8.4	-
Gln	228.5 ± 11.9	457.8 ± 45.3	396.3 ± 27.9
Gly + Thr	32.5 ± 2.2	54.5 ± 4.2	46.5 ± 3.9
Tau	290.0 ± 19.2	501.7 ± 36.2	382.0 ± 19.5
Phe	5.9 ± 1.5	10.0 ± 1.4	11.5 ± 1.6

Cells were grown in isosmolar medium for two weeks. After this time, 75 mM NaCl or 150 mM sucrose were added to the culture medium. After 24 h, the free amino acid content of the cells was measured in alcoholic extracts and assayed by HPLC. Results are expressed as nmol/mg protein and are the means ± SEM of 6-12 cultures. (From Sánchez-Olea et al., J. Neurosci. Res., 1992, with permission).

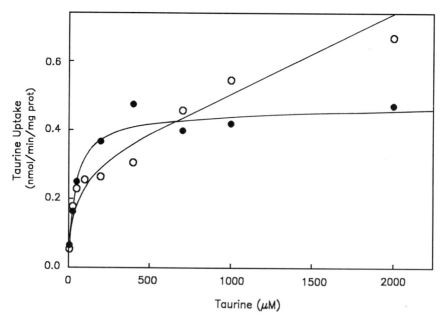

Figure 1. Effect of hyperosmolarity on taurine uptake by astrocytes. Cells were grown for two weeks *in vitro* and then exposed to medium made hyperosmotic with NaCl (450 mOsm) for 3 days. After this time, taurine uptake was measured in hyperosmotic media containing ^{3}H-taurine (3-6 μCi/ml) and unlabelled taurine in concentrations ranging from 5-2000 μM. Control cells were grown in isosmotic medium (300 mOsm), but taurine uptake was measured in media made hyperosmotic with NaCl (450 mOsm). Results are means of 12 cultures. (From Sánchez-Olea et al., J. Neurosci. Res., 1992, with permission).

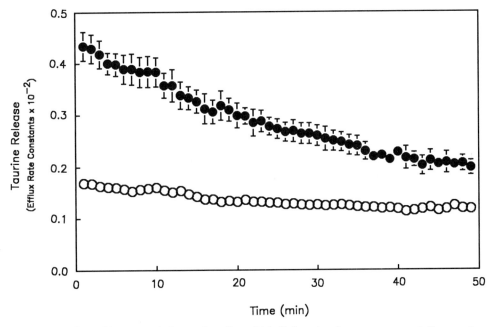

Figure 2. Effect of hyperosmolarity on the efflux of labelled taurine from astrocytes. Cells treated as described in Figure 1 were loaded with labelled taurine and then superfused with isosmotic (open circles) or hyperosmotic (filled circles) media for 50 min. Results represent efflux rate constants and are means $\pm$ SEM of 12 cultures. (From Sánchez-Olea et al., J. Neurosci. Res., 1992, with permission).

small rise in taurine content. The maximal increase in taurine is achieved at a NaCl concentration of 210 mM (450 mOsm). The changes induced by hyperosmolarity are significant only after 6 h of treatment and the maximal effect is reached after about 12 h of exposure to hyperosmotic medium. Two possible mechanisms could be responsible for the observed effects. One is the stimulation of the taurine uptake system and the other is an inhibition of taurine efflux. These possibilities were studied here.

Hyperosmolarity Affects Taurine Uptake

Taurine uptake in astrocytes occurs through an active and a diffusional component[8,9]. In our preparation taurine uptake shows a K_m of 38.6 μM, a V_{max} of 0.3 nmol/min/mg protein and a diffusional component of 2.26 x 10^{-4} ml/min/mg protein. The kinetic parameters of astrocytes exposed to hyperosmolar conditions for 3 days are strikingly different from those of control cells (Figure 1). The most pronounced difference found was a marked reduction of about 97% in the diffusional component of treated cells. Hyperosmotic conditions also lead to an increase of the saturable component, that is the result of a rise in the V_{max} value (50% over control), without affecting significantly K_m (Figure 1). The same has been found for the transport of other osmolytes in renal cells cultured in hyperosmotic conditions[10].

The enhancement of the V_{max} values suggested that the number of taurine transporters in the plasmatic membrane could be increasing. This might be occurring through an activation of either the synthesis or the insertion in the membrane of preexisting transporters as it has been proposed by Chesney et al.[11] for renal cells. However, no variation was seen when cells are incubated with hyperosmotic solutions in the presence of cycloheximide or colchicine, conditions that block protein synthesis and intracellular transport of macromolecules, respectively.

Hyperosmolarity Reduces Taurine Efflux

It is clear from the kinetics of taurine transport that the main alteration induced by hyperosmolarity is the reduction in the diffusional component (Figure 1). As transport through this mechanism is bidirectional, one could expect a reduction not only in influx but also in efflux of taurine in cells grown in a hyperosmotic medium. Figure 2 shows that spontaneous ^{3}H-taurine release is remarkably lower in cells previously treated with a hyperosmotic medium. This suggests that the observed hyperosmolarity-induced taurine increase is the result of a strong blockade of a leak pathway.

Interestingly, the changes induced by hyperosmolarity are opposite to those found under hyposmolar conditions. It has been shown that taurine levels are markedly reduced by hyposmolarity[1,2] subsequent to strong stimulation of taurine efflux. Hyposmolarity causes a slight inhibition of the active transport and a marked increase in the diffusional pathway, suggesting that the hyposmolarity-sensitive efflux of taurine occurs via the diffusional component[9]. It is still unclear whether this component represents a transbilayer bidirectional diffusive translocation or a passive, protein-mediated transport.

CONCLUSIONS

From the above results it can be concluded that:
1) Hyperosmotic conditions induce an increase of some free amino acids, glutamine and taurine being the most affected.
2) The mechanisms subserving the hyperosmolar induced taurine increase involve an enhancement of Na^+-dependent and carrier-mediated uptake and a reduction in taurine efflux occurring through a diffusional pathway.
3) Changes observed are opposite to changes that occur under hyposmotic conditions.
4) These results show that, as in hyposmotic conditions, free amino acids and particularly taurine play a primary role in the processes of cell adaptation to osmotic variations.

ACKNOWLEDGMENTS

This work was supported by grants from DGAPA-UNAM (IN-0245589) and from Stiftung Volkswagenverk. We are indebted to Prof. Bernd Hamprecht who acted as the German partner in the Program of the Volkswagenverk. We acknowledge the technical assistance of Ms. Claudia Rodríguez.

REFERENCES

1. H. Pasantes-Morales, M. Martín del Río, Taurine and mechanisms of cell volumen regulation, in:"Taurine: Functional Neurochemistry, Physiology, and Cardiology", H. Pasantes-Morales, D. Martin, W. Shain, and R. Martín del Río, eds., Wiley-Liss, New York, pp 317 (1990).
2. H. Pasantes-Morales, A. Schousboe, Volume regulation in astrocytes: A role of taurine as an osmoeffector, J. Neurosci. Res. 20:505 (1988).
3. J.V. Wade, J.P. Olson, S.R. Nelson, T.L. Pazdernik, A possible role for taurine in osmoregulation within the brain, J. Neurochem. 51:740 (1988).
4. J. Morán, S. Hurtado, H. Pasantes-Morales, Similar properties of taurine release induced by potassium and hyposmolarity in the rat retina, Exp. Eye Res. 53:347 (1991).
5. J.H. Thurston, R.E. Hauhart, J.A. Dirco, Taurine a role in osmotic regulation of mammalian brain

and possible clinical significance, *Life Sci.* 26:1561 (1980).

6. M.L. Nieminen, L. Tuomisto, E. Solatunturi, L. Eriksson L., M.K. Paasonen, Taurine in the osmoregulation of the Brattleboro rat, *Life Sci.* 42:2137 (1988).

7. J.E. Olson, M. D. Goldfinger, Amino acid content of rat cerebral astrocytes adapted to hyperosmotic medium in vitro, *J. Neurosci. Res.* 27:241 (1990).

8. A. Schousboe, Fosmark H, G. Svenneby, Taurine uptake in astrocytes cultured from dissociated mouse brain hemisferes, *Brain Res.* 116:158 (1976).

9. R. Sánchez-Olea, Morán, A. Schousboe, H. Pasantes-Morales, Hyposmolarity-activated fluxes of taurine in astrocytes are mediated by diffusion, *Neurosci. Lett.* In press.

10. T.R. Nakanishi, R.J. Turner, M.B. Burger, Osmoregulatory changes in myoinositol transport by renal cells, *Proc. Natl. Acad. Sci.* USA 86:6002 (1989).

11. R.W. Chesney, K. Jolly, I. Zelikovic, C. Iwahashi, P. Lohstroh, Increased Na$^+$-taurine symporter in rat renal brush border membranes: performed or newly synthesized? *FASEB J.* 3:2081 (1989).

GABA AND TAURINE SERVE AS RESPECTIVELY A NEUROTRANSMITTER AND AN OSMOLYTE IN CULTURED CEREBRAL CORTICAL NEURONS

Arne Schousboe[1], Clara Lopez Apreza[2] and
Herminia Pasantes-Morales[2]

[1]PharmaBiotec Research Center, The Neurobiology
Unit, Department of Biological Sciences, The Royal
Danish School of Pharmacy, DK-2100 Copenhagen,
Denmark

[2]Institute of Cellular Physiology, The National
University of Mexico, 04510 Mexico City, D.F., Mexico

INTRODUCTION

It is well established that cultures of cerebral cortical neurons consist for the major part of GABAergic neurons which during the culture period of 1-2 weeks undergo a pronounced differentiation to attain properties of mature neurons utilizing GABA as their neurotransmitter[1-10].

It has recently been demonstrated by Schousboe & Pasantes-Morales[11] that these cultured neurons in addition to a stimulus-coupled GABA release also exhibit a pronounced K^+-stimulated taurine release. The finding that the time courses of the K^+-stimulated release of the two amino acids were different (Figure 1), combined with the observation of differences in ionic requirements for the two release processes, led to the suggestion that the mechanisms by which GABA and taurine are released during exposure to 55 mM K^+ could be different. It has recently been convincingly demonstrated in preparations of astrocytes and cerebellar granule neurons[12,13] that K^+-stimulated taurine release originates from an osmotically sensitive taurine pool which is also mobilized during hypotonicity induced swelling[13,14-17]. To extend this research to brain cortex, it was further investigated whether K^+-stimulated taurine release from cerebral cortical neurons in culture as well as cerebral cortical slices has properties linking it to a neurotransmitter related taurine pool or to an osmotically active taurine pool involved in cell volume control.

HYPOSMOLARITY INDUCED RELEASE OF TAURINE AND GABA

Cerebral cortical slices

Table 1 shows that when slices from cerebral cortex of 3-day-old mice are incubated in isosmotic medium (300 mOsm), endogenous free amino acids are released, in a proportion of about 10% for GABA, glutamate and taurine and 20-30% for glycine and alanine. A reduction in the osmolarity of the incubation media of 25% (210 mOsm) and 50% (150 mOsm) leads to a pronounced release of the first group of amino acids, taurine, GABA and glutamate whereas the efflux of glycine and alanine is much less affected by hyposmolarity.

The pattern of hyposmolarity induced release of these amino acids changes as a function of development. In cortical slices from adult animals the hyposmolarity-induced release was much more pronounced for taurine than for the other amino acids, particularly GABA, the release of which was not sensitive to changes in osmolarity. The efflux pattern of glycine and alanine is similar to that in slices from immature animals, i.e. it is characterized by a large efflux in isosmotic conditions and no further release in hyposmotic conditions. These observations are consistent with the general notion that a substantial proportion of the taurine pool in brain cortex is associated to volume regulatory processes whereas GABA, at least in brain tissue at mature stages behaves like a neurotransmitter amino acid[18]. It should, however, be noted that in immature nervous tissue, in addition to taurine, GABA and other amino acids may also play a role as osmolytes.

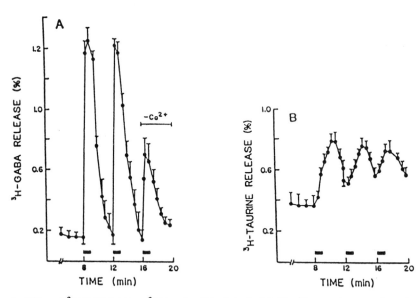

Figure 1. Efflux of ^{3}H-GABA (A) or ^{3}H-taurine (B) elicited with 56 mM potassium in cultured cerebral cortex neurons. After an initial incubation for 30 min with ^{3}H-GABA or ^{3}H-taurine ($5\,\mu$M, $1\,\mu$Ci/dish) cells were superfused with Krebs-HEPES medium at a rate of 2 ml/min. The perfusate was collected at 1-min intervals. After an initial washing period (6 min), perfusate from the basal efflux was collected (4 min) and cultures were then stimulated at the time indicated by the black bar with a solution containing 56 mM KCl, replacing an equimolar amount of NaCl in the presence or absence of calcium. The points represent amount released at each interval as a percentage of the total ^{3}H-GABA or ^{3}H-taurine accumulated during loading (released plus tissue content). Results are the mean $\pm$ S.E.M. of 7-8 experiments and S.E.M. are shown by vertical bars. From Schousboe and Pasantes-Morales[11].

Table 1. Hyposmolarity evoked release of endogenous amino acids from cerebral cortical slices.

Amino acid	Osmolarity (mOsm)					
	300		210		150	
	Release (% of total content)					
	3-day-old	adult	3-day-old	adult	3-day-old	adult
Glu	11.4	10.7	27.6	14.6	43.7	17.1
Gly	23.0	20.4	33.8	23.4	37.3	25.6
Tau	11.2	11.2	28.1	21.2	46.8	30.7
Ala	31.2	25.6	35.2	26.0	35.8	28.1
GABA	7.0	7.4	18.5	10.9	26.6	11.4

Cortical brain slices from 3-day-old or adult mice were incubated for 15 min in media of the indicated osmolarity and at the end of the incubation period, free amino acids released into the medium and those remaining in tissue were extracted with 70% ethanol, derivatized with O-pthaldialdehyde and measured by HPLC. Results are means of 6-10 experiments.

Cultured Cortical Neurons

As can be seen from Table 2, GABA and taurine pools in immature cultures of cerebral cortical neurons are also sensitive to a reduction in the osmolarity of the incubation medium since exposure to 50% hyposmolarity leads to a considerable increase in the release of preloaded ^{3}H-GABA and ^{3}H-taurine.

In agreement with the observation *in vivo*, in the more differentiated neurons cultured for 8 days, the taurine pool was found to be more sensitive to a reduction in the osmolarity of the incubation medium than the GABA pool. This is consistent with the notion that taurine and GABA reside in intracellular pools serving different purposes and that taurine plays a role as an osmolyte in these neurons as well as in astrocytes and other types of neurons[12,18]. This may, in turn, be an indication that at least part of the K^+-stimulated taurine release observed in these neurons (Figure 1) could be linked to an osmotically sensitive taurine pool. This question will be addressed below.

K⁺-STIMULATED GABA AND TAURINE RELEASE

As shown in Figure 1 exposure of mature cerebral cortical neurons to 56 mM K^+ leads to a pronounced increase in GABA and taurine efflux. The K^+-stimulated GABA efflux is for the major part Ca^{++} dependent[5,9,10,19] and represents release from a vesicular pool[9] (B. Belhage and A. Schousboe, unpublished results). As can be seen from Table 3, this K^+-stimulated GABA release is independent of an increase in the osmolarity of the incubation medium brought about by addition of 100 mM sucrose, confirming that the K^+-releasable GABA pool is not osmotically sensitive. On the contrary, the K^+-stimulated taurine release could be totally abolished by increasing the osmolarity of the medium by addition of sucrose (Table 3). This is compatible with observations on K^+-stimulated taurine release from astrocytes[12] and cerebellar granule neurons[13] where this release of taurine has been unequivocally linked to an osmotically sensitive taurine pool. It may, therefore, be concluded that K^+-stimulated taurine release in GABAergic cerebral cortical neurons represents release from an osmotically sensitive taurine pool whereas the corresponding GABA release represents mostly a release of a neurotransmitter.

Table 2. Effect of hyposmolar conditions on release of ^{3}H-GABA and ^{3}H-taurine from cerebral cortical neurons cultured for 3 or 8 days.

Condition	Release (% of total)			
	^{3}H-GABA		^{3}H-taurine	
	3-day-old	8-day old	3-day-old	8-day-old
Isotonic	8.9 ± 0.1	5.0 ± 0.3	7.2 ± 0.7	6.2 ± 0.3
Hypotonic (25%)	12.3 ± 1.4	15.7 ± 1.1	18.1 ± 0.5	28.8 ± 2.0
Hypotonic (50%)	54.8 ± 1.7	45.3 ± 1.3	69.4 ± 1.3	65.4 ± 1.1

Cerebral cortical neurons were cultured for 3 or 8 days and preloaded with ^{3}H-GABA or ^{3}H-taurine (1 μCi per culture) for 30 minutes. Subsequently, release of radioactivity was followed for 10 min in isotonic media and for 10 min under the conditions indicated. Release is expressed as fractional release, i.e. percent of the total amount of radioactivity present at the start of the release experiment (cf. Schousboe & Pasantes-Morales[11]). Results are averages ± SEM of 4 experiments.

Table 3. Effect of 55 mM KCl on ^{3}H-GABA and ^{3}H-taurine release from cerebral cortical neurons in the presence or absence of 100 mM sucrose.

Condition	Release (% of total)	
	^{3}H-GABA	^{3}H-taurine
Basal	5.5 ± 0.4	8.8 ± 1.5
55 mM KCl	19.9 ± 1.0	15.3 ± 2.2
55 mM KCl, 100 mM sucrose	18.6 ± 1.2	6.8 ± 0.4

Cerebral cortical neurons were cultured for 8 days and preloaded with GABA or taurine as described in Table 1. The release experiment was performed as described in Table 1. Results are averages ± SEM of 4 experiments.

MECHANISMS FOR OSMOLARITY INDUCED TAURINE RELEASE

The previously reported insensitivity of the swelling induced taurine release in astrocytes to reduced temperature and to removal of sodium ions[15] has led to the assumption that this taurine release is somehow governed by diffusional, concentration gradient dependent forces rather than the high-affinity taurine carrier. That this is indeed the case has recently been directly demonstrated both in astrocytes and in cerebellar granule neurons[20,21]. It was shown that after exposure of the cells to hyposmotic media high-affinity taurine uptake was largely unaffected while the diffusion constant was increased by 2 to 3 fold. Moreover, in the astrocytes it was demonstrated that in sodium free media, taurine fluxes under hyposmolar conditions were solely determined by the transmembrane taurine gradient, i.e. at external taurine concentrations lower than the normal internal concentration (30 mM) swelling was associated with taurine efflux whereas at higher external taurine concentrations the flux was reversed[20].

Analogous results have been obtained for taurine transport in cerebral cortical neurons (Table 4). In mature cells where taurine was released by K$^+$ from a volume sensitive pool whereas GABA was only moderately released from such a pool (cf. Table 2) exposure of the neurons to hypotonic media had no effect on the saturable uptake of taurine and GABA (results not shown) whereas the diffusion constant for taurine uptake was increased.

Table 4. Effect of hyposmotic conditions on the diffusion constant (K_{diff}) for GABA and taurine uptake in cerebral cortical neurons.

Culture period (Days)	K_{diff} ($\mu l \times min^{-1} \times mg^{-1}$)			
	GABA uptake		Taurine uptake	
	Isotonic	Hypotonic	Isotonic	Hypotonic
3	1.4	1.7	-	-
8	2.6	1.7	0.8	1.7

Cerebral cortical neurons were cultured for 3 or 8 days and the kinetics of ^{3}H-GABA or ^{3}H-taurine uptake under isosmotic and hyposmotic conditions (50% osmolarity) were studied as described by Sánchez-Olea et al.[21] and Schousboe et al.[22]. Constants for the saturable, high-affinity component (K_m, V_{max}) were either only marginally (K_m) affected or somewhat (V_{max}) reduced by exposure of the cells to hyposmotic conditions.

In the case of GABA the diffusion constant was less affected by the hyposmotic media. This is in agreement with the previous conclusion[20,21] that the volume sensitive release of taurine is brought about by diffusion.

That release of taurine and GABA from volume sensitive pools is not mediated by a reversal of the carrier is further substantiated by the demonstration (Table 5) that the hyposmolarity induced release of these amino acids was independent of Na$^+$ and inhibited by DIDS (4,4'-diisothiocyanatostilbene-2,2'-disulfonate) which has been shown to affect the diffusion of taurine in cerebellar granule cells and astrocytes[20,21]. It should particularly be noted that in the case of GABA, the potent, non-transportable blocker of high-affinity GABA transport SKF 89976A[22] had no effect on the volume-sensitive GABA release. This GABA transport blocker has recently been used to selectively inhibit the depolarization-mediated GABA release in cerebral cortical neurons mediated by a reversal of the carrier[23].

Table 5. Effect of Na$^+$, SKF 89976A and DIDS on hypotonicity-induced [^{3}H]GABA and [^{3}H]taurine release in 8-day-old cerebral cortical neurons.

Condition	(Release (% of total))	
	^{3}H-GABA	^{3}H-taurine
Isotonic	5.0 ± 0.3	6.2 ± 0.3
Hypotonic (25%)	15.7 ± 1.1	28.8 ± 2.0
Na$^+$-free, hypotonic	14.9 ± 0.8	24.7 ± 2.1
100 μM DIDS, hypotonic	8.0 ± 2.3	19.1 ± 2.0
10 μM SK89976A, hypotonic	18.2 ± 0.9	-

Cerebral cortical neurons were cultured for 8 days before loading with ^{3}H-GABA or ^{3}H-taurine (Table 1) and subsequent determination of fractional release (Table 1). Results are averages ± SEM of 4 experiments.

CONCLUDING REMARKS

In agreement with previous studies it has been shown that taurine release from cerebral cortical neurons induced by high concentrations of potassium can be mimicked by hyposmotic conditions and is likely to represent taurine release from a volume-sensitive pool. It is unlikely that in these neurons, as well as in other types of neurons, taurine is released from a vesicular, neurotransmitter related pool. This is in contrast to GABA which clearly is preferentially released from such a pool in mature GABAergic neurons both *in vivo* and *in vitro*. It is, however, interesting that in immature cerebral cortex *in vivo* as well as in immature cerebral cortical neurons in culture, GABA is mobilized from an osmotically sensitive pool. This latter phenomenon may also apply to other neurotransmitter amino acids such as glutamate and glycine. The functional implications of this may not be obvious at the present time, but it should be kept in mind that all of these amino acids may serve as important signalling molecules for early neuronal development and differentiation[24].

ACKNOWLEDGMENTS

The secretarial and technical assistance of Ms. Hanne Danø, Charlotte F. Andersen and Inge Damgaard is gratefully acknowledged. The work has been supported financially by the Danish State Biotechnology Program (1991-95) and the Lundbeck Foundation.

REFERENCES

1. M.A. Dichter, Physiological identification of GABA as the inhibitory transmitter for mammalian cortical neurons in cell culture, *Brain Res.* 190:111 (1980).
2. S.R. Snodgrass, W.F. White, B. Biales and M.A. Dichter, Biochemical correlates of GABA function in rat cortical neurons in culture, *Brain Res.* 190:123 (1980).
3. K. Hauser, V.J. Balcar and R. Bernasconi, Development of GABA neurons in dissociated cell culture of rat cerebral cortex, *in:* "Current Developments in Physiology and Neurochemistry", Brain Res. Bull. 5 Suppl. 2," H. Lal, S.Fielding, J.Malick, E. Roberts, N. Shah and E. Usdin, eds., p. 35, Ankho International, Fayetteville, New York (1980).
4. A.C.H. Yu, E. Hertz, and L. Hertz, Alterations in uptake and release rates for GABA, glutamate and glutamine during biochemical maturation of highly purified cultures of cerebral cortical neurons, a GABAergic preparation, *J. Neurochem.* 42:951 (1984).
5. O.M. Larsson, L. Hertz, and a. Schousboe, Uptake of GABA and nipecotic acid in astrocytes and neurons in primary cultures: Changes in the sodium coupling ratio during differentiation, *J. Neurosci. Res.* 16:699-708 (1986).
6. J. Drejer, T. Honoré, and A. Schousboe, Excitatory amino acid-induced release of ³H-GABA from cultured mouse cerebral cortex interneurons. *J. Neurosci.* 7:2910 (1987).
7. A. Schousboe and L. Hertz, Primary cultures of GABAergic and glutamatergic neurons as model systems to study neurotransmitter functions. II. Developmental aspects, *in* "Model Systems of Development and Aging of the Nervous System," A. Vernadakis, A. Privat, J.M. Lauder, P. Timiras and E. Giacobini, eds., p. 33, Martinus Nijhoff Publ., Amsterdam (1987).
8. L. Hertz and A. Schousboe, Primary cultures of GABAergic and glutamatergic neurons as model systems to study neurotransmitter functions. I. Differentiated cells, *in:* "Model Systems of Development and Aging of the Nervous System," A. Vernadakis, A. Privat, J.M. Lauder, P. Timiras and E. Giacobini, eds., p. 19, Martinus Nijhoff Publ., Amsterdam (1987).
9. K. Kuriyama and S. Ohkuma, Development of cerebral cortical GABAergic neurons in vitro, *in:* "Model Systems of Development and Aging of the Nervous System," A. Vernadakis, A. Privat, J.M. Lauder, P. Timiras and E. Giacobini, eds., p. 43, Martinus Nijhoff Publ., Boston (1987).
10. M. Ehrhart-Bornstein, M. Treiman, G.H. Hansen, A. Schousboe, N.A. Thorn and A. Frandsen, Parallel expression of synaptophysin and evoked neurotransmitter release during development of cultured neurons, *Int. J. Dev. Neuroscience* 9: 430 (1991).
11. N. Westergaard, O.M. Larsson, B. Jensen, and A. Schousboe, Synthesis and release of GABA in cerebral cortical neurons co-cultured with astrocytes from cerebral cortex or cerebellum, *Neurochem. Int.* (1992), in press.
12. A. Schousboe, and H. Pasantes-Morales, Potassium-stimulated release of [³H]-taurine from cultured GABAergic and glutamatergic neurons, *J. Neurochem.* 53:1309 (1989).
13. H. Pasantes-Morales and A. Schousboe, Release of taurine from astrocytes during potassium-evoked swelling, *Glia* 2:45 (1989).
14. A. Schousboe, R. Sánchez-Olea and H. Pasantes-Morales, Depolarization induced neuronal release of taurne in relation to synaptic transmission: Comparison with GABA and glutamate, *in:* "Functional Neurochemistry of Taurine," H Pasantes-Morales, W. Shain, D.L. Martin, and R.M. Del Rio, eds., p. 289, Alan R. Liss, New York (1990).

15. H. Pasantes-Morales, and A. Schousboe, Volume regulation in astrocytes: A role for taurine as an osmoeffector, *J. Neurosci. Res.* 20:505 (1988).

16. H. Pasantes-Morales, J. Morán, and A. Schousboe, Volume sensitive release of taurine from cultured astrocytes: Properties and mechanisms, *Glia* 3:427 (1990).

17. H.K. Kimelberg, S.K. Goderie, S. Higman, S. Pang and R.A. Waniewski, Swelling-induced release of glutamate, aspartate, and taurine from astrocyte cultures, *J. Neurosci.* 10:1583 (1990).

18. S. Shain, V. Madelian, R.A. Wainewski, and D.L. Martin, Characteristics of taurine release from astroglial cells, *in:* Functional Neurochemistry of Taurine," H. Pasantes-Morales, W. Shain, D. Martin, and R. Martin del Rio, eds., p. 299, Alan R. Liss, Inc., New York (1990).

19. A. Schousboe, J. Morán, and H. Pasantes-Morales, Potassium-stimulated release of taurine from cultured cerebellar granule neurons is associated with cell swelling, *J. Neurosci. Res.* 27:71 (1990).

20. L. Gram, O.M. Larsson, A.H. Johnsen, and A. Schousboe, Effects of valproate, vigabatrin and aminooxyacetic acid on release of endogenous and exogenous GABA from cultured neurons, *Epilepsy Res.* 2:87 (1988).

21. R. Sánchez-Olea, J. Morán, A. Schousboe and H. Pasantes-Morales, Hyposmolarity-activated fluxes of taurine in astrocytes are mediated by diffusion, *Neurosci. Lett.* 130:233 (1991).

22. A. Schousboe, R. Sánchez-Olea, J. Morán, and H. Pasantes-Morales, Hyposmolarity induced taurine release in cerebellar granule cells is associated with diffusion and not with high affinity transport, *J. Neurosci. Res.* 30:661 (1991).

23. O.M. Larsson, E. Falch, P. Krogsgaard-Larsen, and A. Schousboe, Kinetic characterization of inhibition of γ-aminobutyric acid uptake into cultured neurons and astrocytes by 4,4-diphenyl-3-butenyl derivatives of nipecotic acid and guvacine, *J. Neurochem.* 50:818 (1988).

24. J. Dunlop, A. Grieve, A. Schousboe and Griffiths, R., Stimulation of [^{3}H]GABA release from cultured mouse cerebral cortex neurons by sulphur-containing excitatory amino acid transmitter candidates: receptor activation mediates two distinct mechanisms of release, *J. Neurochem.* 57:1388 (1991).

25. E. Meier, L. Hertz, and A. Schousboe, Neurotransmitters as developmental signals, *Neurochem. Int.* 19:1 (1991).

POTASSIUM-STIMULATED RELEASE OF TAURINE IN
A CRUDE RETINAL PREPARATION OBTAINED FROM
THE RAT IS CALCIUM INDEPENDENT

John B. Lombardini

Departments of Pharmacology and Ophthalmology
  & Visual Sciences
Texas Tech University Health Sciences Center
Lubbock, Texas 79430

INTRODUCTION

The role of taurine as a potential neurotransmitter in the retina has been questioned in numerous literature reports. Thus, crude retinal homogenates prepared from rat retina were preloaded with [^{3}H]taurine under either high- or low-affinity uptake conditions and were then subjected to superfusion techniques to determine whether KCl (56 mM), veratridine (100 μM), 4-aminopyridine (1 mM), or LiCl (56 mM) could evoke the release of [^{3}H]taurine. Dependence of the K^+-evoked release of [^{3}H]taurine on Ca^{2+} was also measured since one of the essential criteria for a neurotransmitter is Ca^{2+}-dependent release of the substance from presynaptic nerve terminals.

METHODS

Materials

1,2[^{3}H]Taurine (20.9 Ci/mmole) was purchased from New England Nuclear. Veratridine and 4-aminopyridine were obtained from Sigma Chemical Co.

Preparation of Retinal Homogenates

Young adult Wistar rats (weighing 120-150 g) were used in all experiments. After anesthetizing the animals with ether followed by decapitation, the eyes were removed and placed in ice-cold 0.32 M sucrose. In all subsequent procedures, the retinal tissue was maintained at 2°C, except for the final incubation. The retinal tissue was removed from the eye cup, gently homogenized in 0.32 M sucrose and centrifuged for 10 min at 16,000 g. The resulting pellet was washed in 20 mM sodium bicarbonate and recentrifuged as above. The retinal membranes located in the pellet were then washed in Krebs-Ringer medium, pH 7.4 (containing NaCl, 128 mM; KCl, 5 mM; CaCl$_2$, 2.7 mM; MgCl$_2$, 1.2 mM; Tris-HCl, 15, and glucose, 10), recentrifuged, and finally homogenized in a glass-glass homogenizer in the Krebs-Ringer medium.

Taurine Uptake System

The incubation system contained the above Krebs-Ringer medium and [^{3}H]taurine (10 μCi). The final taurine concentration was either 1 μM or 2 mM. The reaction mixture was preincubated for 2 min at 37°C in glass test tubes treated with Prosil-28, a surface-treating agent for preparation of a water repellent surface. The reaction was initiated with the retinal preparation ($\sim$ 3 mg) and incubated for 30 min. The reaction was terminated by adding 3 ml of ice-cold Krebs-Ringer medium to the incubation system and filtering on a Millipore vacuum manifold using Whatman GF/B glass fiber filters. The filter was washed 3 times with 3 ml of the above buffer.

Taurine Release Assay

The moist filters containing taurine-loaded retinal homogenate (prepared from the taurine uptake system) were transferred to incubation chambers[1] and 1.4 ml of Krebs-Ringer medium were added to the chambers. The chambers (kept at 37°C in a water bath) were capped and the filters superfused with Krebs-Ringer medium (0.5 ml/min). Two-min fractions of the superfusates were collected directly into liquid scintillation vials.

At indicated times, the original Krebs-Ringer medium was rapidly changed to a medium containing either 56 mM KCl, 100 μM veratridine, or 1 mM 4-aminopyridine. In the experiments involving 56 mM KCl the Krebs-Ringer medium was reduced by 56 mM NaCl to maintain the original ionic strength of the buffer. The retinal preparation contained on the filters was superfused with the new medium for 5 min and then superfusion was continued with the original Krebs-Ringer medium.

Data Analysis

Release was expressed as a fractional rate defined as the radioactivity released in each 2-min fraction divided by the radioactivity remaining in the retinal preparation in the preceding 2-min fraction[2].

RESULTS

Stimulated Release of Taurine From the Retinal Preparation Preloaded with [^{3}H]Taurine in the High-Affinity (1 μM) Uptake Range.

The effects of 56 mM K$^+$ on release of [^{3}H]taurine are shown in Fig. 1A. When the retinal preparation preloaded with 1 μM [^{3}H] taurine was superfused with Krebs-Ringer medium containing Ca$^+$ (2.7 mM) and then exposed to 56 mM K$^+$ between 25-30 and 50-55 minutes, a release of [^{3}H]taurine was observed. Exposure of the preloaded retinal preparations to 56 mM Li$^+$ (Fig. 1A) produced no increase in the release of [^{3}H]taurine.

Potassium stimulated release of [^{3}H]taurine still occurred even when Ca^{2+} was omitted from the superfusion medium, and 1 mM EGTA was present (Fig. 1B).

Veratridine (100 μM) and 4-aminopyridine (1 mM) had no effect on the spontaneous release of taurine from the retinal preparation (Fig. 1C).

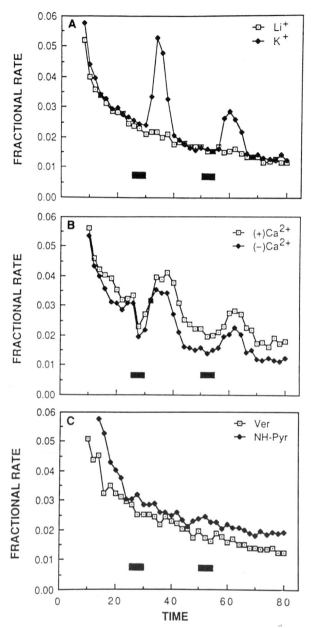

Figure 1. Release of [³H]taurine in the high-affinity uptake range from rat retina. The retinal homogenate was preloaded with [³H]taurine (1.0 μM) and superfused as described in Experimental Procedures. A. Calcium (2.7 mM) was present throughout the superfusion period. At the indicated times (bars), the medium was switched to one containing either potassium chloride (closed diamonds: K⁺, 56 mM) or lithium chloride (open squares: Li⁺, 56 mM) for 5 min, and then switched back to the original medium. During the 5 min period the sodium chloride concentration of the medium was reduced by 56 mM. Fractional rates were calculated from the radioactivity in the superfusates as described in Experimental Procedures. B. Calcium (2.7 mM) was present throughout the superfusion period for the experiment designated by the open squares. Calcium was omitted and 1 mM EGTA included throughout the entire superfusion period for the experiment designated by the closed diamonds. At the indicated times (bars) the medium was switched to one containing potassium chloride (K⁺, 56 mM). C. At the indicated times (bars) the medium was switched to one containing veratridine (open squares: Ver, 100 μM) or 4-aminopyridine (closed diamonds: NH-Pyr, 1 mM).

Stimulated Release of Taurine From the Retinal Preparation Preloaded with [³H]Taurine in the Low Affinity (2 mM) Uptake Range.

High K^+ concentration (56 mM) stimulated [³H]taurine release in the low-affinity uptake range from retinal homogenates (Fig. 2A). Li^+ (56 mM) had no effect (Fig. 2A). Veratridine (100 μM) and 4-aminopyridine (1 mM) were also without effect on [³H]taurine release (Fig. 2B).

DISCUSSION

The possibility that taurine may be a neurotransmitter in the central nervous system has been considered by many investigators (reviewed in ref. 3 and 4). Accordingly, one of the prime criteria for a compound to be a neurotransmitter is the Ca^{2+}-dependent, K^+-evoked release of the substance from presynaptic terminals.

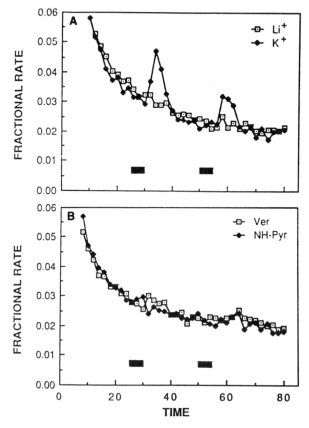

Figure 2. Release of [³H]taurine in the low-affinity uptake range from the rat retina. The retinal homogenate was preloaded with [³H]taurine (2 mM) and superfused as described. Typical experiments are shown in each graph. At the indicated times (bars) the medium was switched to one containing potassium chloride (closed diamonds: K^+, 56 mM) or lithium chloride (open squares: Li^+, 56 mM). B. At the indicated times (bars) the medium was switched to one containing veratridine (open squares: Ver, 100 μM) or 4-aminopyridine (closed diamonds: NH-Pyr, 1 mM).

Our aim was thus to test whether three depolarizing agents (veratridine, 4-amino-pyridine, and LiCl) with different mechanisms of action, would augment the release of taurine by a Ca^{2+}-dependent process. If so, it would be possible to satisfy the criteria that taurine was being released from nerve terminals in the retina. It was also of interest to determine whether high K^+ might stimulate taurine release from the retinal synaptosomes loaded with [^{3}H]taurine in the low-affinity range. Such a result might indicate that taurine was being released from glial rather than neuronal cells.

In this report we have observed that the release of [^{3}H]taurine from a crude homogenate of rat retina that was preloaded with [^{3}H]taurine in the high-affinity (1 μM) uptake range is stimulated by potassium but the release is Ca^{2+} independent. This observation agrees with literature reports that K^+-evoked release from superfused brains, brain slices, and glial culture systems is generally Ca^{2+} dependent while release of taurine from various brain homogenates is most often reported to be Ca^{2+}-independent (reviewed in ref. 3 and 5). Thus we conclude from our results that taurine release from the high-affinity uptake compartment is not vesicular in origin.

Veratridine, an in vitro depolarizing agent which has been suggested to be more physiological than elevated potassium levels[6], and aminopyridine had no effect on the release of taurine from the crude retinal preparation (either under high- or low-affinity preloading conditions with taurine). On the contrary, veratridine released taurine in a hypothalamic preparation which was preloaded with taurine in the high-affinity uptake range but did not have an effect under the low-affinity conditions[1,3].

These results are consistent with the observations of Pasantes-Morales and colleagues[7-9] who reported that K^+-stimulated taurine release from chick retinas was volume-dependent and chloride sensitive. These investigators concluded that the major component of [^{3}H]taurine release was due to changes in cell volume while a minor component of release may be due to the depolarizing effects of K^+. Thus, KCl causes both depolarization and synaptosomal swelling which subsequently allows the release of taurine. Veratridine and 4-aminopyridine cause depolarization but no synaptosomal swelling and do not stimulate taurine release. LiCl causes depolariza-tion by affecting the K^+-equilibrium in excitable tissues[10], but does not cause swelling and consequently no [^{3}H]taurine release was observed.

ACKNOWLEDGEMENTS

This work was supported in part by NEI Grant EYO4780. Mrs. S. Paulette Decker is thanked for her skillful technical assistance and Mr. J. Landon Farley for photographing the figures.

REFERENCES

1. A.T. Hanretta and J.B. Lombardini, Properties of spontaneous and evoked release of taurine from hypothalamic crude P_2 synaptosomal preparations, *Brain Res.* 378:205-215 (1986).
2. G. Levi, V. Gallo, and M. Raiteri, A re-evaluation of veratridine as a tool for studying the depolarization-induced release of neurotransmitters from nerve endings, *Neurochem. Res.* 5:281-295 (1980).
3. A.T. Hanretta and J.B. Lombardini, Is taurine a hypothalamic neurotransmitter?: a model of the differential uptake and compartmentalization of taurine by neuronal and glial cell particles from the rat hypothalamus, *Brain Res. Rev.* 12:167-201 (1987).
4. R.J. Huxtable, Taurine in the central nervous system and the mammalian actions of taurine, *Prog. Neurobiol.* 32:471-533 (1989).
5. R.A. Philibert, K.L. Rogers, A.J. Allen, and G.R. Dutton, Dose-dependent, K^+-stimulated efflux of endogenous taurine from primary astrocyte cultures is Ca^{2+}-dependent, *J. Neurochem.* 51:122-126 (1988).
6. M.P. Blaustein, Effects of potassium, veratridine, and scorpion venom on calcium accumulation and transmitter release by nerve terminals in vitro, *J. Physiol.* (London) 47:617-655 (1975).
7. L. Domínguez, J. Montenegro, and H. Pasantes-Morales, A volume-dependent, chloride-sensitive

component of taurine release stimulated by potassium from retina, *J. Neurosci. Res.* 22:356-361 (1989).

8. R.S. Olea and H. Pasantes-Morales, Chloride dependence of the K^+-stimulated release of taurine from synaptosomes, *Neurochem. Res.* 15:535-540 (1990).

9. H. Pasantes-Morales, J. Morán, and A. Schousboe, Taurine release associated to cell swelling in the nervous system, *in*: "Taurine: Functional Neurochemistry, Physiology, and Cardiology," H. Pasantes-Morales, D.L. Martin, W. Shain, and R. Martín del Río, eds., Wiley-Liss, New York, pp. 369-376 (1990).

10. V. Adam-Vizi, M. Banay-Schwart, I. Wajda, and A. Lajtha, Depolarization of brain cortex slices and synaptosomes by lithium. Determination of K^+-equilibrium potential in cortex slices, *Brain Res.* 410:257-263 (1987).

CHARACTERISTICS OF TAURINE TRANSPORT IN CULTURED RENAL EPITHELIAL CELL LINES: ASYMMETRIC POLARITY OF PROXIMAL AND DISTAL CELL LINES

Deborah P. Jones, Leslie A. Miller,
Andrea Budreau, and Russell W. Chesney

Department of Pediatrics, The University of Tennessee,
Memphis College of Medicine, Center for Pediatric
Pharmacokinetics and Therapeutics, and Le Bonheur
Children's Medical Center, Memphis, Tennessee USA

INTRODUCTION

Taurine is present in millimolar quantities in the intracellular space of many tissues (Chesney 1985) and is postulated to serve many functions including osmotic cell volume regulation, cell detoxification, and enhancing the development of normal neural and retinal tissues (Chesney 1985). In mammals, taurine total body pool is regulated by the renal proximal tubule where taurine is transported by the neutral ß-amino acid transport system (Chesney et al., 1983; Chesney 1985; Rozen and Scriver 1982). When dietary intake of taurine or its precursor amino acids is reduced, urinary taurine levels decline due to enhanced renal tubular reabsorption of taurine (Chesney et al., 1983; Chesney 1985; Rozen and Scriver 1982).

Taurine transport by two continuous renal epithelial cell lines: LLC-PK1, derived from the proximal tubule of the pig, and the Madin-Darby canine kidney cell (MDCK) from the distal tubule of the dog was characterized (Jones and Miller 1990a; Jones and Miller 1990b). The ß-amino acid transporter in these cell lines has similar characteristics to that described for the mouse and rat brush border membranes (Rozen and Scriver 1982; Zelikovic et al., 1989). Transport is dependent on both sodium and chloride in the external medium, and is stereospecific for ß-amino acids (Jones and Miller 1990ab). Renal cells in culture exhibit an adaptive response to the medium taurine concentration comparable to the renal adaptation seen in animals after manipulation of dietary taurine and its precursor amino acids. When LLC-PK1 cells for MDCK cells are deprived of taurine for 8-24 hours, an increase in the apparent transport maximum for taurine (Jmax) is evident. Conversely, cell incubation in high taurine reduces Jmax. Changes in extracellular taurine concentration do not alter the Km for taurine transport.

Taurine, Edited by J.B. Lombardini *et al.*
Plenum Press, New York, 1992

The nature of apical and basolateral taurine transport in these two continuous renal epithelial cell lines was studied to determine if transporter activity was dependent upon the cell surface or cell origin studied. We speculate that the function of the ß-amino acid transporter will differ depending on the origin of the cell: in the proximal cell, net transepithelial reabsorption will permit addition of filtered taurine to its body pool. In contrast, taurine accumulation by distal cells would regulate cell volume in response to osmotic stress.

MATERIALS AND METHODS

LLC-PKl and MDCK cell lines were seeded onto 0.4 μm polycarbonate filter supports (Costar, Transwell) (Boerner et al., 1986). For experiments, medium was replaced with a hormonally-defined, serum-free formulation described by Taub (1979).

Uptake studies were performed on confluent monolayers 10-14 days after seeding (Boerner 1986). Uptake was terminated by the removal of uptake solution followed by three rapid washes with cold Earl's balanced salt solution (EBSS). The uptake solution contained ^{14}C-inulin (0.1 μCi/ml) in addition to ^{3}H-taurine to measure leak. Luminal uptake was measured by addition of uptake solution to the upper surface of the monolayer with an equal volume of EBSS on the lower chamber. For basolateral uptake, the uptake solution was added to the lower chamber and the EBSS with or without sodium to the upper chamber. ^{14}C-Inulin was used as a measure of monolayer integrity. Monolayers were intact when inulin leakage was under 1% (Boerner et al., 1986). Uptake was expressed as nanomoles taurine per mg cell protein. The Eadie-Hofstee method was employed for kinetic analysis.

The concentration of sodium in the uptake solution varied between 0 and 200 mM while holding the concentration of chloride constant at 124 mM. Hill plots were drawn and the slope of the line was taken to represent the relative amount of the ion cotransported with each taurine molecule. Confluent cell monolayers were exposed for 24 hours to hormonally-defined, serum-free medium with 0, 50, or 500 μM taurine. Cell monolayers were exposed for 24 hours to 500 mOsm by the addition of mannitol or raffinose.

RESULTS

Sodium-dependent taurine uptake into the LLC-PK1 cells (Figure 1) was greater from the apical than the basolateral surface. In the MDCK cells (Figure 1), taurine

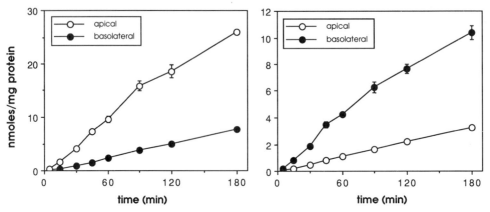

Figure 1. Polarity of taurine uptake by LLC-PK1 (left) and MDCK cells (right). Apical (closed) and basolateral (open) sodium-dependent uptake into filter-grown cells was measured over time (n-4).

uptake from the basolateral was greater than uptake from the apical surface. In the uptake from the basolateral surface was greater than uptake from the apical surface. In the LLC-PK1 cells, LLC-PK1 cells, taurine uptake increased in a linear fashion up to a sodium concentration of 100 mM on the apical and to a lesser extent on the basolateral surface. Both apical and basolateral uptake in the MDCK cells was highly sodium-dependent reaching maximum uptake values at 100 mM.

Hill plot analyses revealed that the ratio of Na: taurine for the apical surface of the LLC-PK1 cells is 2:1 and for the basolateral surface it is 1:1. In the MDCK cells, the ratio for Na: taurine cotransport is 2:1 for both the apical and the basolateral surfaces. Substitution of uptake medium NaCl by either sodium gluconate (less permeant anion compared to chloride) or sodium nitrate (more permeant anion compared to chloride) resulted in a significant reduction in taurine uptake by both the apical and the basolateral surfaces of the LLC-PK1 and MDCK cells.

In the LLC-PK1 cells, taurine uptake by the apical surface was greater than that of the basolateral surface at all substrate concentrations and exhibited a pattern suggestive of a saturable carrier. Initial (20 min) sodium-dependent taurine uptake by the MDCK cells was consistently greater on the basolateral surface.

The initial rate, sodium dependent transport maximum for the LLC-PK1 cells was 1.06 nmoles/mg protein on the apical and 0.15 nmoles/mg protein on the basolateral surface (P=0.001). The Km values for the apical and basolateral surfaces were 14.8 μM and 14.5 μM respectively, not significantly different. In the MDCK cells the Jmax for the apical surface was 0.38 nmoles/mg protein and for the basolateral surface 1.33 nmoles/mg protein (P=0.0028). Km values were 65.9 μM and 37.2 μM for the apical and the basolateral surfaces, respectively (P=0.068).

Either ß- or L-α-alanine at 500 μM was added to the uptake solution in addition to 50 μM taurine. Taurine transport by both the apical and the basolateral surfaces of both cells lines was preferentially reduced by the ß-analog only.

Sodium-dependent taurine uptake by both surfaces of the LLC-PK1 cells was significantly increased following incubation in taurine-free medium for 24 hours (Figure 2). Conversely, taurine uptake was decreased on both surfaces of the LLC-PK1 monolayers following 24 hours in high taurine medium (500 μM). A similar pattern was seen in the MDCK cells although the actual magnitude of the increase was greater on the basolateral surface than the apical surface (Figure 2).

Confluent filter-grown cell monolayers were incubated in either raffinose or mannitol to achieve a final osmolality of 500 mOsm. In the LLC-PK1 cells (Figure 3), apical sodium-dependent uptake was reduced when cells were placed in the hypertonic medium regardless of the osmotic agent. In contrast, sodium-dependent,

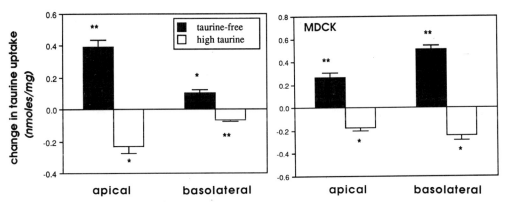

Figure 2. Adaptive response to changes in external taurine concentration: LLC-PK1 (left) and MDCK cells (right) were incubated for 24 hours in the presence of medium without taurine (solid bars) or in medium containing 50 μm (baseline) or 500 μm (open bars) taurine. Comparisons are made between the taurine-free and 50 μm taurine groups and between the 50 μm and 500 μm groups (n=6-8; * p $\leq$ 0.01, ** p = 0.001).

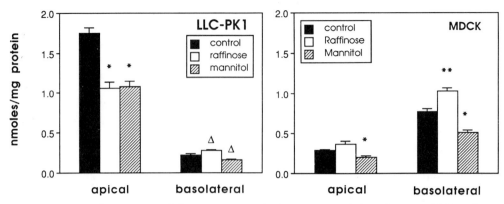

Figure 3. Effect of medium osmolality on taurine uptake into LLC-PK1 (left) and MDCK (right) cells. Apical and basolateral taurine uptake was measured following incubation of filter-grown cells for 24 hours in 350 mOsm (solid) and 500 mOsm medium made hypertonic with raffinose (open) or mannitol (hatched) (n=4; *p < 0.05, **p ≤ 0.01).

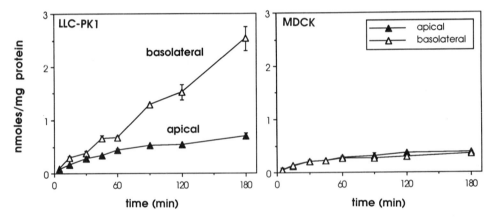

Figure 4. Contralateral efflux of taurine: the appearance of taurine in the contralateral bathing solution was monitored over time in the LLC-PK1 (left) and the MDCK (right). Either the apical or basolateral surface of cell monolayers was incubated in standard uptake medium ± sodium. The appearance of taurine in the contralateral apical (closed triangles) or the basolateral (open triangles) bathing solution was measured over time (n=4).

basolateral taurine accumulation was increased in the presence of raffinose yet decreased in the presence of mannitol. When MDCK cells were placed in medium made hyperosomolar by raffinose (Figure 3), apical uptake was unchanged and basolateral uptake was significantly increased. If mannitol was used as the osmotic agent, both apical and basolateral taurine uptake in the MDCK cells were diminished compared to controls (Figure 3).

In the LLC-PK1 cells, more net taurine transport occurred in the direction of apical to basolateral as more taurine appeared in the basolateral bathing medium than the apical side over time (Figure 4). In contrast, in the MDCK cells, efflux of taurine from the apical and the basolateral cell surfaces was equal (Figure 4) indicating lack of net transepithelial movement.

DISCUSSION

In LLC-PK1 cells, ß-amino acid transport is concentrated on the apical surface of the cell with net transepithelial movement of taurine from the apical to the basolateral surface. Taurine is cotransported with two sodium ions on the apical surface and with one on the basolateral surface. This asymmetry of sodium stoichiometry favors the transepithelial movement of the amino acid from apical to basolateral. A ratio of 1 Na^+:1 taurine favors the efflux of taurine from the cell to the contraluminal side while a ratio of 2 Na^+:1 taurine promotes the influx at the luminal membrane, thus providing the driving force for net reabsorption (King et al., 1985).

In contrast, taurine uptake by the distally-derived, MDCK cells is concentrated on the basolateral surface with little net transepithelial movement of taurine. The stoichiometry of Na:taurine cotransport is 2:1 for both the apical and the basolateral surfaces. Both surfaces of the MDCK cells adapt to changes in extracellular taurine while only the basolateral surface increases taurine accumulation when cells are exposed to high external osmolality.

Studies using brush border membrane vesicles from kidney have demonstrated sodium dependent taurine transport primarily on the apical surface (Wolff and Kinne 1988; Zelikovic et al., 1989), however, difficulties with obtaining a pure preparation of basolateral membrane vesicles has impeded the study of the basolateral systems. A ß-specific, sodium-dependent taurine transporter has been detected in basolateral brush border membrane vesicles from mice (Mandla et al., 1988). This transporter has a rather high Km (0.36 mM) compared to the Km values reported for the taurine transporter from apical renal brush border membrane vesicles. In isolated microperfused tubules, taurine uptake was localized to the proximal tubule (Silbernagl 1988). These studies support the concept of net transepithelial transport of taurine and other ß-amino acids from the lumen of the proximal tubule to the basolateral surface. Such a process takes place by active concentrative cotransport of taurine with Na^+ and Cl- to achieve high intracellular concentrations of the amino acids with passive movement out of the cell from the basolateral surface in a manner similar to other neutral amino acids. This process is important to the reclamation of filtered taurine as well as the addition of amino acids to the total body pool in times of taurine deficiency.

In marine animals, taurine is secreted by the renal tubule as a mechanism of osmotic cell volume regulation (King 1985). Upon exposure to hypotonic medium, taurine efflux increases along with a rise in plasma taurine levels (King 1982). In the dogfish, flounder and little skate, taurine is actively transported against a concentration gradient by a basolateral ß-specific amino acid transporter with 2 Na^+. Net secretion is achieved as taurine moves down its concentration gradient with movement from the cell to the tubular lumen. Taurine transport by the flounder tubule is in the opposite direction compared to that observed for the mammalian proximal renal tubule (King 1985). It has been proposed that secretion of taurine by marine animals is related to the ability of the marine animals to accumulate taurine from a basolateral location as well as the relatively high tissue to medium ratio of taurine (15:1) in marine animals (King et al., 1985). There is no net secretion of taurine by the mammalian proximal tubule because conditions favor net reabsorption rather than secretion: taurine uptake by the mammalian proximal renal tubule is concentrated on the luminal surface of the cell, and the tissue to medium ratio of taurine is relatively low (7:1) in mammals compared to marine animals.

Our studies with the MDCK cell line indicate that the mammalian kidney can concentrate taurine by means of a basolaterally located transport system, however net secretion did not occur in our experience. Lack of net secretion may be related to the characteristics of the apical membrane with respect to taurine permeability.

The basolateral transport system in MDCK cells is clearly regulated by medium osmolality, and in this manner, they are similar to the marine animal's renal tubular epithelial cells. MDCK cells responded to hypertonicity induced by addition of raffinose to the external medium with enhanced accumulation of taurine by the basolateral surface. This response has been previously reported by other investigators (Uchida et al., 1990). In contrast, the LLC-PK1 cells responded to hypertonic medium with reduced apical taurine accumulation. This differential response of renal epithelial cells to hypertonicity may imply a difference in the function of the ß-amino acid transport system in cells depending on their site of origin. In the distal tubule, which is commonly exposed to external hypertonicity, the ß-amino acid transporter may respond to osmotic stress. In contrast, this adaptive mechanism may be unimportant in the setting of a proximal tubule cell. The osmolyte used appears to have an effect on the response of taurine transporter activity. Raffinose is an impermeant molecule whereas mannitol is able to diffuse through cell tight junctions. The accumulation of other osmolytes has been shown to be dependent upon the type of compound used to manipulate medium osmolality (Heilig et al., 1990).

In our study, apical and basolateral surfaces of both cell types responded to changes in medium taurine concentration with reciprocal changes in taurine transport. However, only the basolateral surface of the MDCK cell responded to hyperosmolality with increased taurine accumulation. This may indicate differential control of the ß-amino acid transport system by substrate and external tonicity.

The asymmetry of taurine transporter activity in these cells may be related to the tubular site of cell origin and the particular function of taurine in the proximal versus the distal tubular cell. In proximal tubule (Figure 5), taurine is reabsorbed against a concentration gradient along with two sodium ions, and passively exits the cell. This process is controlled by substrate availability. In contrast, taurine transport in the distal tubule (Figure 5) occurs primarily from the basolateral side of the cell with net cellular accumulation and minimal transepithelial movement. Distal tubular taurine transport is regulated by taurine availability as well as external osmolality.

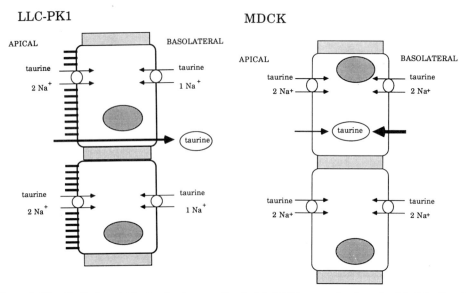

Figure 5. Hypothetical model for taurine transport by the LLC-PK1, or proximal renal epithelial cell and by the MDCK, or distant renal epithelial cell.

SUMMARY

Taurine transport was determined in two continuous, renal epithelial cell lines: LLC-PK1 derived from the proximal tubule of the pig, and the Madin-Darby canine kidney cell (MDCK) from the distal tubule of the dog. In LLC-PK1, taurine transport is maximal at the apical surface, whereas in MDCK cells, transport is greatest at the basolateral surface. Transport is highly dependent on both sodium and chloride in the external medium, and is specific for ß-amino acids. The apical and basolateral surfaces of both cell lines show an adaptive response to extracellular taurine concentration, but only the basolateral surface of the MDCK cell responds to hyperosomolality by increased taurine accumulation. Thus, differential control of the ß-amino acid transport system by substrate and external tonicity exists.

The role of the ß-amino acid transport system may differ according to the origin of the cell: in the proximal renal tubular cell, net transepithelial reabsorption of filtered taurine increases the body pool. By contrast, taurine accumulation by distal tubular cells may form a mechanism of cell volume regulation in response to osmotic stress.

REFERENCES

Boerner, P., Evans-Layng, M., Snag, U-H., 1986, Polarity of neutral amino acid transport and characterization of a broad specificity transport activity in a kidney epithelial cell line, MDCK, *J. Biol. Chem.* 261:13957-13962.

Chesney, R.W., 1985, Taurine: Its biological role and clinical implications, *Adv. Pediatr.* 32:1-42.

Chesney, R.W., Gusowski, N., Friedman, A.L., 1983, Renal adaptation to altered dietary sulfur amino acid intake occurs at luminal brushborder membrane,*Kidney Int.* 24:588-594.

Chesney, R.W., Gusowski, N., Dabbagh, S., 1985, Renal cortex taurine content regulates renal adaptive response to altered dietary intake of sulfur amino acids, *J. Clin. Invest.* 76:2213-2221.

Heilig, C.W., Brenner, B.M., Yu, A.S.L., Kine, B.C., Gullans, S.R., 1990, Modulation of osmolytes in MDCK cells by solutes, inhibitors, and vasopressin, *Am. J. Physiol.* 259:F653-F659.

Jones, D.P., Miller, L.A., Chesney, R.W., 1990a, Adaptive regulation of taurine transport in two continuous renal epithelial cell lines, *Kidney Int.* 38:219-226.

Jones, D.P., Miller, L.A., Chesney, R.W., 1990b, Asymmetric polarity of taurine transport in cultured renal epithelium, *J. Am. Soc. Nephrology* 1:701.

King, P.A., Kinne, R., Goldstein, L., 1985, Taurine transport by brush border membrane vesicles isolated from the flounder kidney, *J. Comp. Physiol. B.* 155:185-193.

King, P.A., Beyenbach, K.W., Goldstein, L., 1982, Taurine transport by isolated flounder renal tubules, *J. Exp. Zool.* 223:103-114.

Mandla, S., Scriver, C.R., Tenenhouse, H.S., 1988, Decreased transport in renal basolateral membrane vesicles from hypertaurinuric mice, *Am. J. Physiol.* 255:F88-F95.

Rozen, R., Scriver, C.R., 1982, Renal transport of taurine adapts to perturbed taurine homeostasis, *Proc. Natl. Acad. Sci. USA* 79:2101-2105.

Silbernagl, S., 1988, The renal handling of amino acids and oligopeptides, *Phys. Rev.* 68:911-1007.

Uchida, S., Nakanishi, T., Kwon, H.M., Preston, A.S., Handler, J.S., 1990, Taurine behaves as an osmolyte in Madin-Darby canine cells. Protection by polarized, regulated transport of taurine, *J. Clin. Invest.* 88:656-662, 1991.

Wolff, N.A., Kinne, R., 1988, Taurine transport by rabbit kidney brush-border membrane: coupling to sodium, chloride, and the membrane potential, *J. Membr. Biol.* 102:131-139.

Zelikovic, I., Stejskal-Lorenz, E., Lohstroh, P., Budreau, A., Chesney, R.w., 1989, Anion dependence of taurine transport by rat renal brush-border membrane vesicles,*Am. J. Physiol.* 256:F646-F655.

METABOLISM OF CYSTEINE TO TAURINE BY
RAT HEPATOCYTES

Martha H. Stipanuk, Pamela J. Bagley,
Relicardo M. Coloso, and Mark F. Banks

Division of Nutritional Sciences
Cornell University
Ithaca, New York 14853

INTRODUCTION

During the past two decades, many investigators have assumed that the major locus of regulation of cysteine catabolism is the partitioning of cysteinesulfinate between its decarboxylation and transamination pathways. Hepatic cysteinesulfinate decarboxylase activity correlates well with the capacity of animals to synthesize taurine[1-4], and low cysteinesulfinate decarboxylase activity in the cat has been associated with its nutritional requirement for dietary taurine[5]. More recent studies in our laboratory have indicated that cysteinesulfinate-independent pathways also play a major role in cysteine metabolism[6,7]. In contrast to cysteinesulfinate-dependent metabolism of cysteine, which leads to both taurine and sulfate production, the cysteinesulfinate-independent pathways all result in release of reduced inorganic sulfur and its subsequent oxidation to sulfate. This evidence revealing a contribution of cysteinesulfinate-independent pathways to cysteine catabolism suggested that partitioning of cysteine between cysteinesulfinate formation and metabolism by cysteinesulfinate-independent pathways may also be important in the regulation of cysteine metabolism to taurine.

In this paper we describe a series of studies of cysteine metabolism in isolated rat hepatocytes. The results indicate that both cysteine availability and cysteine dioxygenase activity play key roles in determining the rate of conversion of cysteine to taurine.

METHODS

Male Sprague-Dawley rats (Blue Spruce Farms, Altamont, NY) were fed ad libitum either a nonpurified diet (Prolab RMH 1000, Agway, Syracuse, NY) or a semi-purified diet based on the AIN-76A formulation (Dyets, Inc., Bethlehem, PA) for three weeks. Rats were used for preparation of hepatocytes when they weighed approximately 250 g.

Taurine, Edited by J.B. Lombardini *et al.*
Plenum Press, New York, 1992

Hepatocytes were isolated by the method of Berry and Friend[8] as modified by Krebs et al.[9]. Hepatocytes were incubated at 37°C with either L-[35S]cysteine or L-[35S]cysteinesulfinate in Krebs-Henseleit buffer, pH 7.4, and in an atmosphere of 95% O_2/5% CO_2 as described previously[7]. Maintenance of a cellular ATP concentration of $>2.0 \mu$mol/g throughout the 20-min incubation period was used as the criterion for viability. Incubations were stopped by addition of sulfosalicylic acid. Acid supernatants were assayed for [35S]sulfate and other sulfur anions by the method of Coloso and Stipanuk[10] and for [35S]taurine (including hypotaurine oxidized to taurine) by the method of Stipanuk et al.[11]. Bathocuproine disulfonate (BCS) was added to some incubations with cysteine as this copper chelator minimized the oxidation of cysteine to cystine[7]. In one study, hepatocytes were preincubated for 20 min with 2 mmol/L DL-propargylglycine, an irreversible inhibitor of cystathionine γ-lyase[6].

Cysteine dioxygenase (EC 1.13.11.20) activity was assayed as described by Bagley and Stipanuk[12]. Cysteinesulfinate decarboxylase (EC 4.1.1.29) activity and cystathionine γ-lyase (EC 4.4.1.1) activity were assayed as described by De La Rosa and Stipanuk[13] and Stipanuk[14].

Data were expressed on the basis of wet weight using a dry to wet weight ratio of 3.7 as calculated by Krebs et al.[9] or on the basis of protein as determined by the method of Lowry et al.[15]. The reported values for sulfate include the sulfur equivalent of thiosulfate that accumulated in the incubation system during incubations with the higher concentrations ($>$1mmol/L) of cysteine; values for taurine include hypotaurine that accumulated in the system and was oxidized to taurine with H_2O_2 prior to analysis of taurine. Data were analyzed by analysis of variance and Tukey's ω-procedure or by the Student's t-test (Minitab 81.1, Pennsylvania State University, State College, PA).

RESULTS AND DISCUSSION

Production of [35S]sulfate and [35S]taurine increased 10-fold and 45-fold, respectively, as the cysteine concentration was increased from 0.05 to 1.0 mmol/L in the absence of BCS and 12.5-fold and 26-fold, respectively, in the presence of BCS (Table 1). Values for production of sulfate or taurine by hepatocytes incubated with the same total exogenous cyst(e)ine concentration were always higher for those incubations that contained BCS. The designated concentration of cysteine was added to the incubation mixtures as substrate, but variable amounts of cysteine were oxidized to cystine and mixed disulfides such that the cysteine concentration was not constant. This oxidation was minimized by the addition of BCS. The approximate concentration of cysteine (thiol) at the mid-point of the incubation period was assessed by measuring DTNB-reactive thiol concentrations, and these values for cysteine concentrations are given in parentheses. The higher rates of cysteine catabolism observed in incubations with BCS were consistent with higher cysteine (thiol) concentrations in these incubations. Hence, the percentage partitioning of cysteine between taurine and sulfate production responded to cysteine concentration; low cysteine concentrations favored sulfate production, whereas higher cysteine concentrations favored taurine production. This effect of cysteine concentration was not due to activation of cysteine dioxygenase activity, as the V_{max} activity of this enzyme did not change when hepatocytes were incubated with the various concentrations of cysteine. This increased partitioning of cysteine to taurine as cysteine availability was increased probably was due to increased saturation of cysteine dioxygenase with substrate, because the K_m of rat liver cysteine dioxygenase for cysteine (0.45 mmol/L) is within the range of exogenous substrate concentrations tested (0.05 to 1.0 mmol/L)[16].

A separate series of studies allowed us to estimate the percentages of cysteine catabolism attributable to cysteinesulfinate-dependent versus cysteinesulfinate-independent pathways. As shown in Table 2, cysteinesulfinate-dependent catabolism of cysteine was greater for hepatocytes incubated with 1 mmol/L cysteine than for those incubated with 0.1 mmol/L cysteine and for incubations with BCS than for those without BCS; cysteinesulfinate-dependent metabolism ranged from 21 to 55% of the total catabolism. The majority of the cysteinesulfinate-independent catabolism of cysteine was inhibited by propargylglycine (an irreversible and specific inhibitor of cystathionine γ-lyase[6,17]), and was thus attributed to the cystathionine γ-lyase-catalyzed β-cleavage pathway of cyst(e)ine catabolism. At an extremely high cysteine concentration, 25 mmol/L, cysteinesulfinate-independent catabolism by other pathways (possibly transamination) appeared to increase, and this caused a decrease in the percentage flux (although not the absolute flux) through the cysteinesulfinate-dependent pathways.

To further assess the roles of cysteine dioxygenase and cysteinesulfinate decarboxylase in the regulation of cysteine catabolic flux, rats were fed semi-purified diets that contained either 10% or 30% casein (supplemented with 0.15% or 0.45% DL-methionine, respectively) for 3 weeks prior to hepatocyte isolation. The hepatocytes

Table 1. Effect of cysteine concentration and bathocuproine disulfonate on the catabolism of L-[^{34}S]cysteine by rat hepatocytes.

[^{35}S]Cyst(e)ine concentration	BCS	[^{35}S]Sulfate	[^{35}S]Taurine
mmol/L (~ cysteine concn.)	+/-	% of total catabolic flux (pmol·min^{-1}·mg cells^{-1})	
0.05 (0.01)	-	84 ± 4[a] (4 ± 1)	16 ± 4[a] (0.4 ± 0.1)
0.2 (0.03)	-	71 ± 4[ab] (11 ± 1)	29 ± 5[ab] (3 ± 1)
1.0 (0.10)	-	61 ± 4[abc] (44 ± 7)	39 ± 4[abc] (18 ± 2)
0.05 (0.04)	+	66 ± 5[ab] (6 ± 0.4)	34 ± 5[ab] (2 ± 0.7)
0.2 (0.15)	+	58 ± 4[bc] (23 ± 1)	42 ± 4[bc] (11 ± 2)
1.0 (0.70)	+	48 ± 5[c] (75 ± 11)	52 ± 5[c] (52 ± 3)

Values are means ± SEM for five experiments. BCS = 0.05 mM bathocuproine disulfonate. Approximate cysteine concentrations (in parentheses) were based on DTNB-reactive thiol concentrations in the incubation medium at the mid-point of the incubation period, DTNB = 5,5'-dithiolbis-(2-nitrobenzoic acid). Percentage values within a column not followed by the same superscript letter are significantly different (P<0.05) by Tukey's ω-procedure. Values were recalculated from data of Stipanuk et al. (11).

Table 2. Estimations of catabolism of cysteine by various cysteinesulfinate-dependent and cysteinesulfinate-independent pathways in freshly isolated rat hepatocytes.

Cysteine	BCS	CSA-Dependent	CSA-Independent β-Cleavage	Other	Total catabolic flux
mmol/L	mmol/L	% of total catabolic flux			$(pmol \cdot min^{-1} \cdot mg\ cells^{-1})$
0.1	--	21 ± 9[a]	60 ± 5[c]	19 ± 5[a]	6
0.1	0.05	41 ± 13[ab]	41 ± 3[b]	18 ± 3[a]	9
1.0	--	37 ± 8[ab]	32 ± 5[ab]	31 ± 5[ab]	33
1.0	0.05	55 ± 6[a]	21 ± 2[a]	24 ± 1[a]	99
25	--	15 ± 4[a]	40 ± 3[b]	45 ± 3[b]	526
25	0.05	38 ± 12[ab]	28 ± 1[ab]	34 ± 1[ab]	665

Values are means ± SEM for three hepatocyte preparations. Values within the same column that are not followed by the same superscript are significantly different (P<0.05). CSA = Cysteinesulfinate. Data are from Coloso et al., Am. J. Physiol. 259: E443 (1990).

were assayed for activities of cysteine-degrading enzymes (Table 3), and the production of sulfate and taurine from L-[^{35}S]cysteine and L-[^{35}S]cysteinesulfinate was determined (Tables 4 and 5). Cysteine dioxygenase activity in hepatocytes isolated from rats fed the 30% casein diet was 21-times that for hepatocytes from rats fed the 10% casein diet, whereas cysteinesulfinate decarboxylase activity in hepatocytes from rats fed the 30% casein diet was only one-fourth that found in hepatocytes from rats fed the lower protein diet. Activity of cystathionine γ-lyase, the major enzyme that catalyzed cysteinesulfinate-independent catabolism of cysteine in rat liver, was the same for hepatocytes from rats in the two dietary groups.

Table 3. Activities of cysteine-degrading enzymes in rat hepatocytes isolated from rats fed diets containing 10% or 30% casein.

Enzyme	10% Casein	30% Casein
	$nmol \cdot min^{-1} \cdot mg\ protein^{-1}$	
Cysteine dioxygenase	0.09 ± 0.02**	1.9 ± 0.3**
Cysteinesulfinate decarboxylase	30 ± 5**	7.9 ± 1.4**
Cystathionine γ-lyase	18 ± 4	19 ± 3

Values are means ± SEM for six or seven rats. Values followed by ** are significantly different (P<0.05) for the 10% casein vs. the 30% casein groups.

Both taurine and sulfate production from cysteine were more rapid in hepatocytes isolated from rats fed the higher protein diet; this was observed in incubations of hepatocytes with 0.2 mmol/L and in incubations with 1.0 mmol/L cysteine. Taurine production in incubations with 0.2 mmol/L and 1 mmol/L cysteine was 37- and 13-times, respectively, the rates observed for hepatocytes isolated from rats fed 10% casein. Sulfate production was only 3.7- and 2.6-times the rates observed for hepatocytes from rats fed 10% casein. Thus, the percentage of total catabolism accounted for by taurine formation increased from 3.8 or 6.6% for the 10% casein group to 32 or 28% for the 30% casein group. This increase in taurine formation occurred in conjunction with the 20-fold increase in cysteine dioxygenase activity and in spite of the 74% decrease in cysteinesulfinate decarboxylase activity, suggesting that the change in cysteine dioxygenase activity had a greater effect than the change in cysteinesulfinate decarboxylase activity in rats fed different levels of protein.

Table 4. Metabolism of cysteine by hepatocytes isolated from rats fed diets containing 10% or 30% casein.

Diet	Cysteine concen.	Taurine production	Sulfate production	Taurine / Sulfate+Taurine
% casein	mmol/L	pmol·min^{-1}·mg protein^{-1}		
10	0.2	0.96 ± 0.68*	28 ± 5**	0.038 ± 0.019*
30	0.2	37 ± 18*	103 ± 7**	0.32 ± 0.10*
10	1.0	10 ± 6*	162 ± 24**	0.066 ± 0.042
30	1.0	131 ± 52*	425 ± 57**	0.28 ± 0.11

Values are means ± SEM for five rats. Values followed by ** (P<0.05) or * (P<0.10) are significantly different for the 10 % casein vs. the 30% casein groups.

Table 5. Metabolism of cysteinesulfinate by hepatocytes isolated from rats fed diets containing 10% or 30% casein.

Diet	Cysteinesulfinate concentration	Taurine production	Sulfate production	Taurine / Sulfate+Taurine
% casein	mmol/L	pmol·min^{-1}·mg protein^{-1}		
10	0.2	360 ± 58	223 ± 41	0.60 ± 0.04
30	0.2	208 ± 41	267 ± 24	0.44 ± 0.04
10	1.0	840 ± 64**	224 ± 27	0.74 ± 0.06
30	1.0	290 ± 52**	241 ± 20	0.51 ± 0.06

Values are means ± SEM for seven rats. Values followed by ** are significantly different (P<0.05) for the 10% casein vs. the 30% casein group.

When hepatocytes from rats fed the same two diets were incubated with [^{35}S]cysteinesulfinate, the first intermediate in the cysteinesulfinate-dependent metabolism of cysteine, cysteinesulfinate was rapidly metabolized to taurine and sulfate. Taurine formation tended to be lower for hepatocytes isolated from rats fed the 30% casein diet although this was statistically significant only in incubations with the higher (1 mmol/L) concentration of exogenous cysteinesulfinate. The percentage of total cysteinesulfinate catabolism resulting in taurine production appeared to be slightly, but not significantly, less for incubations of cysteinesulfinate with hepatocytes from rats fed the higher protein diet. This trend of decreased taurine formation from cysteinesulfinate in hepatocytes isolated from rats fed the 30% casein diet was associated with the 74% decrease in cysteinesulfinate decarboxylase activity suggesting that cysteinesulfinate decarboxylase activity or regulation at the level of cysteinesulfinate may superimpose some control over that brought about at the cysteine locus.

The results of the incubations of hepatocytes from rats fed 10% or 30% casein diets with 0.2 mmol/L cysteine or cysteinesulfinate have been summarized in Figures 1 and 2. Taurine and sulfate production from cysteine were measured directly. We assumed that all taurine was derived from cysteinesulfinate and used the partitioning observed in incubations with [^{35}S]cysteinesulfinate to calculate the portion of the total sulfate production that occurred by the cysteinesulfinate-dependent pathway. The remainder of the sulfate was then assumed to have resulted from cysteinesulfinate-independent metabolism of cysteine.

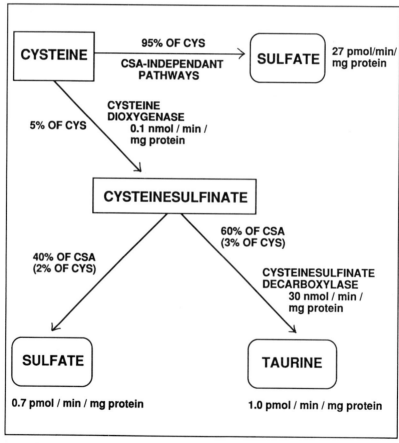

Figure 1. Metabolism of cysteine by rat hepatocytes isolated from rats fed a semi-purified diet with 10% casein. Hepatocytes were incubated with 0.2 mmol/L cysteine. Calculations are based on the data shown in Tables 3-5. CYS = cysteine; CSA = cysteinesulfinate

The percentage partitioning of cysteine among various pathways in hepatocytes of rats fed the 10% casein diet is shown in Figure 1. In the presence of low cysteine dioxygenase activity, only 5% of cysteine was catabolized by cysteinesulfinate-dependent pathways; most of the cysteine was catabolized to sulfate by cysteinesulfinate-independent pathways. Of the 5% of cysteine that was converted to cysteinesulfinate, 60% of the cysteinesulfinate (3% of the cysteine) was further metabolized to taurine, resulting in formation of 1 pmol taurine·min^{-1}·mg protein^{-1}. A small amount of sulfate was also produced from cysteinesulfinate, but the major product of cysteine catabolism was clearly sulfate derived from cysteinesulfinate-independent pathways.

The percentage partitioning of cysteine catabolism in hepatocytes from rats fed 30% casein is illustrated in Figure 2. In contrast to the observations for rats fed 10% casein, and in the presence of high cysteine dioxygenase activity, 60% of cysteine was oxidized to cysteinesulfinate; 40% of cysteine was catabolized by cysteinesulfinate-independent routes. Due to the lower activity of cysteinesulfinate decarboxylase, only 44% (versus 60%) of the cysteinesulfinate was decarboxylated leading to taurine formation. However, the much higher cysteine dioxygenase activity or increased formation of cysteinesulfinate in hepatocytes from rats fed 30% casein resulted in production of 37 pmol taurine·min^{-1}·mg protein^{-1}. Hence, taurine production was 36-

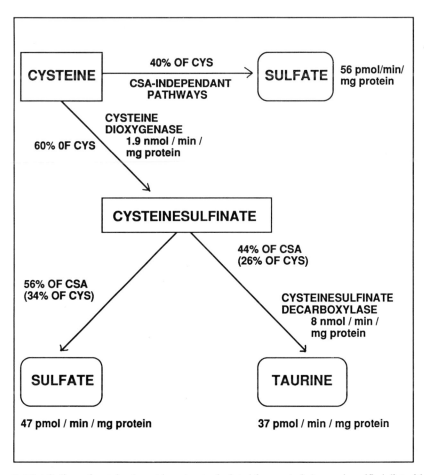

Figure 2. Metabolism of cysteine by rat hepatocytes isolated from rats fed a semi-purified diet with 30% casein. Hepatocytes were incubated with 0.2 mmol/L cysteine. Calculations are based on the data shown in Tables 3-5. CYS = cysteine; CSA = cysteinesulfinate.

Table 6. Comparison of hepatic cysteine dioxygenase and cysteine-sulfinate decarboxylase activities.

	Cysteine dioxygenase	Cysteinesulfinate decarboxylase	Reference
	nmol·min^{-1}·mg protein^{-1}		
Male Rat	11 ± 2	4.4 ± 2.2	# 19
Female Rat	4.1 ± 0.2	1.3 ± 0.7	# 13
Fox	1.8 ± 0.4	2.1 ± 0.1	# 20
Kitten	0.53 ± 0.14	0.62 ± 0.21	# 13

Values are means ± SD for five to fourteen animals.

fold greater in hepatocytes isolated from rats fed 30% casein than in hepatocytes from rats fed 10% casein despite the somewhat lower cysteinesulfinate decarboxylase activity. The large increase in cysteine dioxygenase activity effected a large increase in cysteinesulfinate formation which much more than compensated for the small decrease in further metabolism of cysteinesulfinate to taurine.

Our observations for cysteine metabolism by freshly isolated rat hepatocytes are in agreement with the observations of Hosokawa et al.[18] for apparent cysteine metabolism in intact rats. They reported that hepatic cysteine dioxygenase activity was 23-times as great in rats fed 30% casein as in rats fed 10% casein, and that this was associated with excretion of 29-times as much taurine in the urine and with hepatic taurine concentrations that were 2.4-times those of rats fed 10% casein.

The relative roles of cysteine dioxygenase and cysteinesulfinate decarboxylase in regulation of cysteine metabolism to taurine have not been extensively studied in animals other than rodents. However, for the limited number of species studied in our laboratory, we have found that animals with low hepatic cysteinesulfinate decarboxylase activity also have low hepatic cysteine dioxygenase activity[13,19,20] as summarized in Table 6. Thus, it is possible that the relative ability of various animals to synthesize taurine may be more related to their lack of cysteine dioxygenase activity than to their lack of cysteinesulfinate decarboxylase activity.

Clearly, cysteine dioxygenase activity plays a major role in the regulation of cysteine catabolism in rat liver. Both the large capacity of cysteine dioxygenase (presumably amount of enzyme protein) to respond to changes in dietary protein or sulfur amino acid load and the fact that the K_m of cysteine dioxygenase for cysteine is near the upper end of cysteine concentrations observed in rat liver[18] allow the flux of cysteine to cysteinesulfinate, and hence to taurine, to increase with an increase in cysteine load. Our studies of cysteine and cysteinesulfinate metabolism suggest that cysteinesulfinate decarboxylase plays a lesser role than cysteine dioxygenase in the regulation of catabolic flux of cysteine to taurine in rat liver. Taurine production appears to be favored when cysteine availability is high, suggesting that taurine formation and excretion may be an important route of disposal of excess sulfur. These observations also suggest that the synthesis of taurine may not be conserved metabolically in the face of a limited supply of cysteine and that the diet may be the major source of taurine under such conditions.

ACKNOWLEDGMENTS

This research was supported by National Institutes of Health Research Grant No. DK-26959, by the Cooperative State Research Service, United States Department of Agriculture under Agreement No. GAM900895, and by New York State Hatch Project No. 399-492.

DEDICATION

The authors are pleased to dedicate this paper to Dr. Doriano Cavallini on the occasion of his 75th birthday in recognition of his long and distinguished career in the area of metabolism of sulfur-containing compounds.

REFERENCES

1. J.G. Jacobsen and L.H. Smith, Jr., Comparison of decarboxylation of cysteine sulphinic acid-1-14C and cysteic acid-1-14C by human, dog, and rat liver and brain, *Nature*, Lond. 200:575 (1963).
2. J.G. Jacobsen, L.L. Thomas, and L.H. Smith, Jr., Properties and distribution of mammalian L-cysteine sulfinate carboxy-lyases, *Biochim. Biophys. Acta* 85:103 (1974).
3. G.E.Gaull, D.K. Rassin, N.C.R. Raiha, and K. Heinonen, Milk protein quantity and quality in low-birth-weight infants. III. Effects on sulfur amino acids in plasma and urine, *J. Pediatr.* 90:348 (1977).
4. W.G.M. Hardison, C.A. Wood, and J.H. Proffitt, Quantification of taurine synthesis in the intact rat and cat liver, *Proc. Soc. Exp. Biol. Med.* 155:55 (1977).
5. K. Knopf, J.A. Sturman, M. Armstrong, and K.C. Hayes, Taurine: an essential nutrient for the cat, *J. Nutr.* 108:773 (1978).
6. M.R. Drake, J. De La Rosa, and M.H. Stipanuk, Metabolism of cysteine in hepatocytes. Evidence for cysteinesulphinate-independent pathways, *Biochem. J.* 244:279 (1987).
7. R. M. Coloso, M.R. Drake, and M.H. Stipanuk, Effect of bathocuproine disulfonate, a copper chelator, on cyst(e)ine metabolism by freshly isolated rat hepatocytes, *Am. J. Physiol.* 259:E443 (1990).
8. M.N. Berry, and D.S. Friend, High-yield preparation of isolated rat liver parenchymal cells. A biochemical and fine structural study, *J. Cell. Biol.* 43:506 (1969).
9. H.A. Krebs, N.W. Cornell, P. Lund, and R. Hems, Isolated liver cells as experimental material, in: "Regulation of Hepatic Metabolism", F. Lindquist and N. Tygstrup, eds., Academic Press, New York, NY (1974).
10. R.M. Coloso and M.H. Stipanuk, Metabolism of cyst(e)ine in rat enterocytes, *J. Nutr.* 119:1914 (1989).
11. M.H. Stipanuk, R.M. Coloso, R.A.G. Garcia, and M.F. Banks, Cysteine concentration regulates cysteine metabolism to glutathione, sulfate and taurine in rat hepatocytes, *J. Nutr.*, in press. (1991).
12. P.J. Bagley and M.H. Stipanuk, Assay of cysteine dioxygenase activity, *FASEB J.* 4: A805 (1990).
13. J. De La Rosa and M.H. Stipanuk, Evidence for a rate-limiting role of cysteinesulfinate decarboxylase activity in taurine biosynthesis in vivo, *Comp. Biochem. Physiol.* 81B:565 (1985).
14. M.H. Stipanuk, Effect of cysteine on the metabolism of methionine in rats, J. Nutr. 107:1455 (1979).
15. O.H. Lowry, N.J. Rosebrough, A.L. Farr, and R.J. Randall, Protein measurement with the Folin phenol reagent, *J. Biol. Chem.* 193:265 (1951).
16. K. Yamaguchi, Y. Hosokawa, N. Kohashi, Y. Kori, S. Sakakibara, and I. Ueda, Rat liver cysteine dioxygenase (cysteine oxidase). Further purification, characterization, and analysis of the activation and inactivation, *J. Biochem.* 83:479 (1978).
17. M.H. Stipanuk and P.W. Beck, Characterization of the enzymatic capacity for cysteine desulphhydration in liver and kidney of the rat, *Biochem. J.* 206:267 (1982).
18. Y. Hosokawa, S. Niizeki, H. Tojo, I. Sato, and K. Yamaguchi, Hepatic cysteine dioxygenase activity and sulfur amino acid metabolism in rats: possible indicators in the evaluation of protein quality, *J. Nutr.* 118:456 (1988).
19. S.-M. Kuo and M.H. Stipanuk, Changes in cysteine dioxygenase and cysteinesulfinate decarboxylase activities and taurine levels in tissues of pregnant or lactating rat dams and of their fetuses or pups, *Biol. Neonate* 46:237 (1984).
20. N.S. Moise, L.M. Pacioretty, F.A. Kallfelz, M.H. Stipanuk, J.M. King, and R.F. Gilmour, Jr., Dietary taurine deficiency and dilated cardiomyopathy in the fox, *Am. Heart. J.* 121:541 (1991).

RECENT DEVELOPMENTS IN ASSAYS FOR TAURINE, HYPOTAURINE AND SOME METABOLIC PRECURSORS

Z.K. Shihabi[1], H.O. Goodman[2], R.P. Holmes[3]

Departments of Pathology[1], Pediatrics[2], and Urology[3]
Bowman Gray School of Medicine, Wake Forest University
Winston-Salem, NC 27103

The distinguishing chemical characteristics of taurine, hypotaurine and cysteinesulfinic acid that facilitate their separation are their charge and a reactive amine group. Their absorption of light is low and in the low UV range. They are non-fluorescent, they are not volatile, and they are not readily amenable to assay by enzyme-linked reactions. These features limit the detection methods available to the analyst. A brief synopsis of the methods that have been used are outlined below, together with attention to their relative advantages and disadvantages. Prospects for the future are discussed with a particular emphasis on capillary electrophoresis.

PAPER AND THIN LAYER CHROMATOGRAPHY

Paper and thin layer chromatography were the earliest methods used for the separation of taurine and its precursors as well as for amino acids (AA). One- and two-dimensional separation methods have been described (e.g.,[1]). These methods depend on the partitioning of taurine between the stationary phase of paper or silica and the mobile phase. After chromatographic separation spots were visualized with ninhydrin. The disadvantages of paper chromatography are that it requires several hours for development, it gives a limited resolution and is subject to interference by high concentrations of salt or protein. Thin layer chromatography (TLC) gives a better resolution and a faster separation compared to paper chromatography but is still subject to interference unless substantial sample clean-up is utilized. Quantitation of AA with either paper or TLC lacks precision but modern scanning devices may produce adequate quantitation for screening purposes. The principal advantage of TLC and paper chromatography are that multiple samples can be simultaneously run in a one dimension system and that a rapid and adequate separation can be achieved for separating radioisotopes when following metabolic conversions. The introduction of high performance TLC with small size stationary phase particles should enhance this application. Reversed-phase TLC plates would also be suitable for separating taurine after pre-derivatization with a label such as o-phthaldialdehyde (OPA).

Taurine, Edited by J.B. Lombardini *et al.*
Plenum Press, New York, 1992

ION EXCHANGE CHROMATOGRAPHY

The different pK_as of the cationic and anionic groups of these molecules dictate that they can readily be separated by ion exchange chromatography. A variety of approaches has been utilized with this separation mode and are outlined below.

A. Mini-Columns

Taurine is not well retained on either cation or anion columns and can be eluted simply with water. In some methods the effluents from these columns have been used directly (or after evaporation) to quantitate taurine without further purification. In other methods, the eluents have been subjected to further purification, for example, by paper chromatography. Short columns containing 2-5 cm of Dowex 50 W in the H^+ form or Amberlite IR have been successfully used to separate taurine from the other AA in urine or tissues extracts[2,3]. The taurine in eluents may be detected by reaction with ninhydrin, dinitrofluorobenzene or fluorescamine. The flow rate in these columns is very critical. Furthermore, we have shown that the analysis of taurine with these columns is not accurate because other small molecules as well as proteins which react with the detection reagents are also eluted[4]. A mixed cation/anion bed of Dowex AG2-X8 and Dowex W-X8 has been used to decrease coelution[5].

B. Automated Analysis

In 1987 we described a fully automated instrument for analyzing taurine in biological fluids and tissues without sample deproteinization or clean-up[6]. An important feature of the instrument is the removal of proteins and peptides by dialysis through a C-type membrane in a continuous flow system. All AA in the dialysate with the exception of taurine are subsequently retained on a small mixed bed of anion and cation exchange resin. The taurine eluting is reacted with a stream of OPA to produce a fluorescent derivative. The specificity of the autoanalyzer for taurine stems from several factors: (1) Using a mixed resin bed to trap the majority of the AA. (2) Using dialysis to remove proteins and small peptides. (3) Use of acetic acid in the diluent stream to aid in trapping certain AA on the column. (4) Delivering a constant flow rate through the column with a pump.

Distinct advantages of this instrument are that it can be assembled from inexpensive parts, and that it is fully automated requiring no sample preparation. Tissue homogenates, plasma and urine can be analyzed directly without deproteinization. It is fast (20 samples/hr) and is ideally suited for the analysis of a large number of samples. Since the method uses a fluorescent reagent it is very sensitive (down to $5\,\mu M$). It has good between-run precision with a CV of 3.1%. We have improved the preparation of the OPA reagent as follows: 120 mg OPA is dissolved in 50 ml ethanol and mixed with 950 ml of sodium borate buffer (17 mM, pH 9.0). This reagent is stable for at least 6 months refrigerated. An aliquot of this reagent sufficient for 2-3 days of use is removed and mixed with mercaptoethanol (1 ml/L) or N-acetylcysteine (2 g/L). In order to analyze hypotaurine we treat the sample (500 μl) with 50 μl of sodium periodate (10% in 0.2 M acetic acid) for 15 min at room temperature to oxidize the hypotaurine to taurine. Samples treated with and without periodate are measured on the analyzer to determine the hypotaurine content.

To demonstrate the sensitivity of this analyzer we have investigated the uptake and synthesis of taurine by MDCK renal epithelial cells grown in culture. The time course of uptake of $500\,\mu M$ taurine is shown in Figure 1A and taurine content of cells incubated for 3 days with varying concentrations of taurine is shown in Figure 1B. Such data demonstrate that these cells can take up taurine against a concentration gradient and that the intracellular compartment saturates at $\sim 250\,\mu$moles/g protein.

Incubating these cells with 500 umol/L of cysteine, cystine, methionine or hypotaurine did not increase the cell content of taurine, while incubating the cells with cysteine sulfinic acid and cysteic acid increased taurine cell content by 8 and 99%, respectively. Based on these data the MDCK cells lack the key enzymes required to convert cysteine into taurine. However, they have a carboxylase of unknown specificity, which can convert cysteic acid, which is almost absent in renal cells, into taurine (Figure 2). These experiments demonstrate that the analyzer being used was sufficiently sensitive to obviate the need for radioisotopes for studying the metabolic pathways involved in taurine synthesis.

In many instances it is desirable to study also the total concentration of free AA in the sample in addition to taurine. This assay can be performed on this instrument by removing the column[7].

C. Amino Acid Analyzer

These instruments can resolve taurine and sometimes hypotaurine in addition to all the other AA. Both taurine and hypotaurine emerge early in the run. The

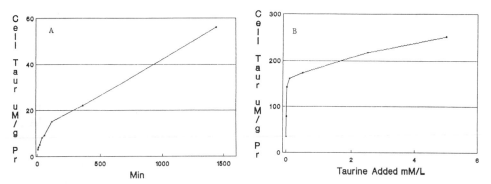

Figure 1. Uptake of taurine by MDCK cells in tissue cultures. MDCK cells were grown in the presence of added taurine, sonicated and the taurine content determined by the autoanalyzer. A - Time dependence (500 μM taurine). B - At different concentrations of added taurine for 3 days.

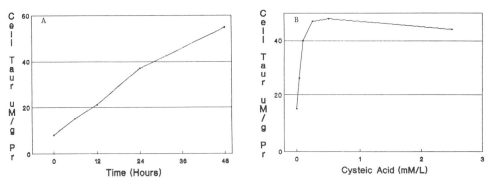

Figure 2. Taurine content of MDCK cells grown with cysteic acid. A - Taurine accumulation in cells grown with 500 μM cysteic acid. B - Taurine content of cells grown for 3 days with varying concentrations of cysteic acid.

425

instruments accurately quantify taurine but are expensive, require a long analysis time (1-6 hours) and require sample deproteinization. Originally the instrument used ninhydrin for the reaction which is not very sensitive; later some instruments were modified to use OPA[8]. The elution gradient can be modified to improve the separation of taurine from other AA and to rapidly wash off other AA. Because of the long analysis time these instruments are not suitable for the analysis of a large number of samples.

D. Ion Exchange HPLC Analysis

Ion exchange HPLC with post-column derivatization has had limited use for analyses of taurine, hypotaurine, and other related compounds[9]. For specialized needs, such as following the catabolism of cysteine, it could be the method of choice because of the diverse column packing and elution conditions that can be used to obtain good separations.

REVERSED PHASE HPLC

Reversed phase HPLC is currently widely used to separate and quantitate taurine, hypotaurine and other AA following derivatization with fluorescent reagents. Fluorescamine, dansyl Cl, 9-fluorenylmethyl chloroformate and OPA can been used as derivatizing reagents. The most popular of these derivatizing agents is OPA. It is inexpensive, easy to prepare, reacts rapidly with the majority of the AA and produces highly fluorescent products. However, OPA does not react with proline and hydroxyproline and gives a low yield with cysteine[10]. The addition of sodium hypochlorite to the reaction mixture in order to oxidize proline, hydroxyproline, cysteine and the use of N-acetylcysteine to increase the fluorescence overcome these two problems[11]. When analyzing for taurine or hypotaurine by HPLC, all the AA emerge from the column, slowing the analysis. To overcome this we modified the gradient originally described by Jones et al.[10]. Stipanuk et al.[12] have described a similar modified gradient. The separation of taurine and hypotaurine under our conditions is illustrated in Figure 3A. These two compounds elute close to each other and to alanine.

The present trend in HPLC is to employ very short columns (3-10 cm in length) packed with small spherical particles, about 3 μm average particle size, which give very fast separations. The use of o-diacetylbenzene[13], column switching and on-column clean-up (by introducing a small Dowex column in the injector) in order to speed and automate the assay, have not yet been explored for taurine analysis.

GAS CHROMATOGRAPHY

Taurine is not volatile and must be derivatized to a volatile product before gas chromatography. This technique has only been sparingly used for the determination of taurine as well as other amino acids[14]. The technique is very cumbersome especially in the derivatization step, is lengthy, and requires more skill than other chromatographic methods.

ELECTROPHORESIS

Since AA are zwitterions, they will migrate in an electrical field. Electrophoresis, on paper, TLC plates or cellulose acetate sheets, was used earlier for analyzing taurine precursors such as cysteine, cysteic acid and hypotaurine[15]. More recently, commercial equipment for capillary electrophoresis has become available. In this method the separation of the sample components is accomplished in capillary tubing 10-100 μm in diameter and 20-50 cm in length at 1-30 KV. This technique is much more flexible than high voltage electrophoresis since the buffers can be modified to accomplish the separation based on factors such as size and hydrophobicity, in addition to charge. The technique is simpler than many of the other chromatographic techniques, is rapid, and is suitable for automation. The dansyl derivatives of several AA have been well resolved by this technique[16]. The elution and separation of the OPA adducts of cysteine, hypotaurine and taurine is illustrated in Figure 3B. The sensitivity was low as absorption at 254 nm was used. With the introduction of fluorescence detectors the sensitivity will be greatly enhanced. Our experience with this type of instrumentation indicates that it might be a potentially useful technique for the analysis of taurine and its precursors.

TAURINE ANTISERA

Taurine has been conjugated to albumin[17,18] through glutaraldehyde and antibodies developed against the conjugate. This antisera has been used primarily to

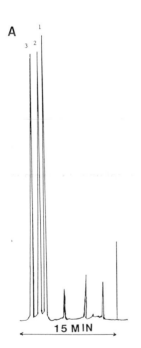

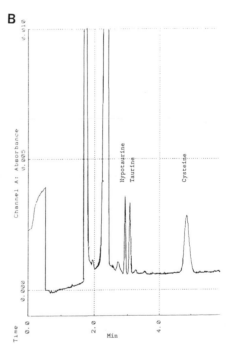

Figure 3A. Chromatogram of taurine (1), hypotaurine (2), and alanine (3) analyzed by HPLC on Novapak. Pre-column derivatization: equal volumes of sample and OPA[10] are mixed together. Solvents A: 80% 40 mM acetate buffer pH 5.9, 1% tetrahydrofuran, 19% methanol. B: 20% 10 mM acetate pH 5.9, 80% methanol. Gradient (2-steps): 1-25% B in A for 5 min at 1 ml/min, 2-40% B in A for 10 min (retention time for alanine 17 min).

Figure 3B. Taurine, hypotaurine and cysteine standards, each 200 μM detected by Beckman capillary electrophoresis. Wavelength 254 nm; voltage 11 KV; capillary 25 cm long, 50 μm in diameter; buffer 180 mM borate, pH 10.3; OPA: 3 mg OPA/ml of borate buffer (20 mmol/L pH 9.0 containing 100 μL ethanol and 10 μL of mercaptoethanolamine).

localize taurine in tissues by cytochemical methods[18,19]. These antisera have low cross reactivity with other AA but react with peptides containing taurine such as gamma glutamyl taurine and glycyl taurine. Monoclonal antibodies also have been prepared against taurine. These antibodies are ideally suited for the automated assay of taurine in fluids and possibly even tissues. The high sensitivity of such methods coupled with the high concentration of taurine in most tissues suggests that in many instances matrix effects could be diluted out.

REFERENCES

1. C.E. Dent, A study of the behavior of some sixty amino acids and other ninhydrin-reacting substances on phenol-"collidine" filter paper chromatograms, *Biochem J.* 43:169 (1948).
2. E.I. Pentz, C.H. Davenport, W. Glover, and D.D. Smith, A test for the determination of taurine in urine, *J. Biol. Chem.* 228:433 (1957).
3. B.A. Sorbo, A method for the determination of taurine in urine, *Clin. Chim. Acta.* 6:87 (1961).
4. S.G. Hartley, H.O. Goodman, and Z.K. Shihabi, Urinary excretion of taurine in epilepsy, *Neurochem. Res.* 14:149 (1989).
5. S.I. Baskin, E.M. Cohn, and J.J. Kocsis, Taurine changes in visual tissues with age, *in*: "Taurine", R. Huxtable and A. Barbeau, eds., pp. 201, Raven Press, New York (1976).
6. H.O. Goodman and Z.K. Shihabi, Automated analysis for taurine in biological fluids and tissues, *Clin. Chem.* 33:835 (1987).
7. Z.K. Shihabi, H.O. Goodman, R.P. Holmes, and M.L. O'Connor, The taurine content of avian erythrocytes and its role in osmoregulation, *Comp. Biochem. Physiol.* 92A:545 (1989).
8. B. Connolly and H.O. Goodman, Potential sources of errors in cation exchange chromatographic measurement of plasma taurine, *Clin. Chem.* 26:508 (1979).
9. K. Kuriyama and Y. Tanaka, Cysteinesulfinic acid and cysteic acid high-performance liquid chromatograph, *Meth. Enzymol.* 148:164 (1987).
10. B.N. Jones, S. Paabo, and S. Stein, Amino acid analysis and enzymatic sequence determination of peptides by an improved o-phthaldialdehyde precolumn labeling procedure, *J. Liq. Chrom.* 4:565 (1981).
11. N. Nimura and T. Kinoshita, o-Phthaldehyde-N-acetyl-L-cysteine as a chiral derivatization reagent for liquid chromatographic optical resolution of amino acid enantiomers and its application to conventional amino acid analyses, *J. Chrom.* 352:169 (1986).
12. M.H. Stipanuk, L.L. Hirschberger, and J. de la Rosa, Cysteinesulfinic acid, hypotaurine, and taurine: reversed-phase high-performance liquid chromatography, *Meth. Enzymol.* 143:155 (1987).
13. J. Fourche, H. Jensen, and E. Neuzil, Fluorescence reactions of aminophosphonic acids, *Anal. Chem.* 48:155 (1976).
14. R.T. Coutts and J.M. Yeung, Gas chromatographic analysis of amino acids, *in*: "Neuromethods", Vol. 3, pp, 29, Humana Press, Clifton, New Jersey (1985).
15. C.J.G. van der Horst and H.J.G. Grooten, The occurrence of hypotaurine and other sulfur-containing amino acids in seminal plasma and spermatozoa of boar, bull and dog, *Biochim. Biophys. Acta.* 117:495 (1966).
16. I.Z. Atamna, C.J. Metral, G.M. Muschik, and H.J. Issaq, Factors that influence mobility, resolution, and selectivity in capillary zone electrophoresis. 3. The role of buffers anion, *J. Liq. Chrom.* 13:3201 (1990).
17. S. Madsen, O.P. Ottersen, and J. Storm-Mathisen, Immunocytochemical visualization of taurine: neuronal localization in the rat cerebellum, *Neurosci. Lett.* 60:255 (1985).
18. S. Ida, K. Kuriyama, Y. Tomida, and H. Kimura, Antisera against taurine: quantitative characterization of the antibody specificity and its application to immunohistochemical study in the rat brain, *J. Neurosci. Res.* 18:626 (1987).
19. S. Madsen, O.P. Ottersen, J. Storm-Mathisen and J.A. Sturman, Immunocytochemical localization of taurine: methodological aspects, *Prog. Clin. Biol. Res.* 351:53 (1989).

ANION-EXCHANGE HPLC OF TAURINE, CYSTEINESULFINATE AND CYSTEIC ACID

Martha H. Stipanuk, Lawrence L. Hirschberger
and Pamela J. Bagley

Division of Nutritional Sciences
Cornell University
Ithaca, New York 14853, U.S.A.

INTRODUCTION

Taurine is synthesized from cysteine by mammalian liver and other tissues. Cysteinesulfinate is believed to be the major intermediate in hypotaurine and taurine synthesis. The oxidation product of cysteinesulfinate, cysteic acid, can also be decarboxylated to taurine; the sulfur is completely oxidized prior to decarboxylation rather than after decarboxylation as is the case for conversion of cysteinesulfinate to hypotaurine and on to taurine. Cysteic acid appears to form spontaneously from cysteine and is present in plasma.

In order to further our understanding of cysteine metabolism and taurine synthesis, an accurate chromatographic method we could use for the determination of radiolabeled taurine, cysteinesulfinate and cysteic acid and their specific activities was required. Although a number of HPLC methods for the determination of taurine and other metabolites of cysteine, including one developed and widely used in our own laboratory[1,2], have been reported, we have had limited success in using any of these methods for the measurement of radiolabeled taurine or cysteinesulfinate. Problems of incomplete and inconsistent recovery of radioactivity in taurine or cysteinesulfinate derivatives appeared to be due to incomplete derivatization, formation of minor or secondary derivatives, instability of the derivatives during chromatography, or poor resolution of radiolabeled taurine and cysteinesulfinate from other radiolabeled compounds. Methods that were satisfactory for analysis of standards often yielded much more variable results when applied to physiological samples.

The method described in this paper was developed specifically for the measurement of radiolabeled taurine, cysteinesulfinate and cysteic acid in physiological samples. Compounds are resolved on a strong anion-exchange column without pre-column derivatization, thus eliminating problems associated with formation and stability of derivatives. With the addition of post-column o-phthalaldehyde (OPA) derivatization of compounds that elute from the column, molar concentrations of taurine, cysteinesulfinate and cysteic acid can be determined as well. If desired, hypotaurine can be measured by its oxidation to taurine and its measurement as such.

Sample Preparation

The samples that we have analyzed consist of tissue preparations that have been incubated with [^{35}S]cysteine or other ^{35}S-labeled precursor for the production of ^{35}S-labeled cysteinesulfinate or taurine. Details have been reported previously[3-5].

Samples were deproteinized by the addition of either sulfosalicylic acid or perchloric acid, followed by centrifugation at 2000 x g for 10 min. The acid supernatant was collected and stored at -20°C until it was chromatographed. Nonenzymatic oxidation of cysteinesulfinate to cysteic acid sometimes occurred during storage; if this is unacceptable, samples should be analyzed immediately.

To remove [^{35}S]cyst(e)ine and other radiolabeled cations prior to HPLC, a 0.5-mL portion of the acid supernatant was applied to a 3 x 0.6 cm column of Dowex-50W (Sigma Chemical Co., St. Louis, MO, prepared in H$^+$-form) in a Pasteur pipet. Taurine, cysteinesulfinate and cysteic acid were then eluted with 2 mL of water; hypotaurine was retained by the Dowex-50W column. The entire 2.5 mL of eluate was routinely collected.

For the analysis of total taurine and hypotaurine, hypotaurine was first oxidized to taurine by neutralizing a 0.5-mL portion of the acid supernatant with KOH, incubation of the neutralized supernatant with 0.05 mL of 30% H$_2$O$_2$ at 37°C for 15 min, and treatment of the incubation mixture with 0.025 mL of a catalase solution (Sigma #C-30, Sigma Chemical Co., solution prepared as 900 Sigma units per 0.025 mL of 1 M potassium phosphate buffer, pH 7) for 15 min at ambient temperature. The sample was then heated in a boiling water bath for 3 min to denature the catalase and centrifuged to remove the precipitated protein. A 0.5-mL aliquot of this supernatant was then applied to a Dowex-50W column and eluted with water as described above.

Chromatography

The chromatographic system consisted of a Waters Model 680 Automated Gradient Controller, two Waters Model 510 pumps, and a WISP 710B automatic sample injector (all from Waters Division of Millipore, Milford, MA). The analytical column was a HEMA-IEC BIO Q analytical column (150 x 4.6 mm, 10 μm particles), which is a strong anion exchange column with quaternary trimethylamino active groups. The analytical column was used in conjunction with a HEMA-IEC BIO DEAE (10 x 4.6 mm, 10 μm particles) guard cartridge, which is a weak anion exchange pre-column with diethylaminoethyl active groups. Both columns were from Alltech Associates, Inc. (Deerfield, IL). It was only this combination of resins that gave acceptable separation for physiological samples; using either DEAE or BIO Q resin in both guard and analytical columns resulted in inadequate retention of taurine.

A binary gradient was used for the mobile phase. Eluant A was water (purified with Milli-Q Plus Ultra Pure Water System, Millipore, Molsheim, France); eluant B was 100 mmol/L boric acid with 250 mmol/L NaCl (adjusted to pH 8.0 with KOH). Eluants were filtered and degassed under vacuum prior to use. The flow rate was 1 mL/min and the injection volume was 100 μL.

Three different gradients were used, depending on the compound(s) of interest, as shown in Table 1. Gradient A resolves all three compounds. Gradient B was used when only taurine was measured, and gradient C was used when only cysteinesulfinate and cysteic acid were measured. Gradients B and C were developed to shorten the analysis time for taurine or cysteinesulfinate plus cysteic acid, respectively, for particular applications. All gradients were run as linear gradients between the times indicated in Table 1. The sample was injected onto a column in 100% eluant A (H$_2$O), and the % of buffer B was increased linearly in one (Gradient B) or two

Table 1. Summary of gradients and elution times for analysis of taurine, cysteinesulfinate and cysteic acid by anion-exchange HPLC

Gradient A		Gradient B		Gradient C	
Time (min)	%B (mL/100mL)	Time (min)	%B (mL/100mL)	Time (min)	%B (mL/100mL)
0	0	0	0	0	0
10	10	10	10	5	15
20	33	13	100	15	33
23	100	18	100	18	100
27	100	21	0	22	100
30	0	33	0	25	0
42	0			37	0

Elution time (min)	Elution time (min)	Elution time (min)
Taurine - 10.1	Taurine - 10.1	Cysteinesulfinate - 9.7
Cysteinesulfinate - 14.0	Cysteic acid - 12.2	Cysteic acid - 18.1

(Gradients A and C) increments. After the desired compounds had eluted from the column, the percentage of buffer B was increased to 100% and held constant at 100% for 4 or 5 min to wash the column. Finally, the percentage of buffer B was decreased to zero and the column was re-equilibrated with eluant A (H_2O) for a 12-min period before the next sample was injected.

The retention times for taurine, cysteinesulfinate and cysteic acid are reported for each gradient. Retention times were stable for subsequent runs, permitting the collection of radioactive fractions by designated elution time. Standard hypotaurine was retained by the analytical column (elution time of approximately 5 min) but was removed by the Dowex-50W clean-up step, which was necessary for physiological samples.

Post-column Derivatization and Detection

If desired to quantitate molar amounts of taurine, cysteinesulfinate or cysteic acid or to define the location of radioactive peaks, post-column derivatization was done using an Eldex Model A-30-S pump (Eldex Laboratories, Inc., San Carlos, CA) connected to the post-column tubing using a tee connector. The derivatizing solution was pumped at a rate of 0.5 mL/min and contained 3 mmol/L OPA, 0.4% methanol (v/v), 0.2% mercaptoethanol (v/v), and 200 mmol/L boric acid, pH 10.4 (adjusted with KOH). A 3-m length of coiled tubing (0.38-mm ID) was inserted between the tee connector and the fluorometer to allow time for the derivatization reaction. Fluorescence was detected using a Model 901 Spectra-Glo Fluorometer (Gilson Medical Electronics, Middleton, WI) as described previously[1] and recorded using a Model 3390A Integrator (Hewlett-Packard Company, Palo Alto, CA). Standard taurine, cysteinesulfinate and cysteic acid (Sigma Chemical Co., St. Louis, MO) were used to generate standard curves, which were acceptable over the range of zero to 10 nmol per 100-μL injection volume. Fluorescence detection limits were 50, 250, and 500 pmol for taurine, cysteinesulfinate and cysteic acid, respectively, with recoveries of internal standards averaging 99, 102 and 104% for these same compounds.

Collection of fractions and measurement of radioactivity

The eluate (after or without post-column derivatization) was collected in fractions corresponding to 1-min or less of run time by a LKB 2211 SuperRac fraction collector (LKB, Bromma, Sweden), which could be programmed to collect only the desired portions of the run based either on elution time or appearance of fluorescent peaks. ^{35}S-Radioactivity was measured in the eluate fractions corresponding to the taurine, cysteinesulfinate and cysteic acid peaks by liquid scintillation spectrometry. Corrections were made for counting efficiency and decay.

Sample chromatograms

Chromatography of standard solutions is illustrated in Figure 1. Applications of this HPLC method to the measurement of taurine plus hypotaurine production by freshly isolated rat hepatocytes incubated with L-[35S]cysteine are shown in Figure 2. Both the fluorescence of the OPA-derivatives and the radioactivity in fractions of eluate are shown.

Application of this method to measurement of cysteinesulfinate production from L-[35S]cysteine during the assay of cysteine dioxygenase activity by a rat liver preparation is illustrated in Figure 3. Little or no radiolabeled cysteic acid was observed if the acid supernatant from the enzyme assay was analyzed immediately. However, when samples were frozen and stored prior to analysis, a portion of the cysteinesulfinate was further oxidized nonenzymatically to cysteic acid. Therefore, it was important to measure both the cysteinesulfinate and cysteic acid peaks to quantitate total cysteinesulfinate formation. Both the fluorescence of the OPA-derivatives as recorded by the integrator and a plot of the radioactivity in 1-min fractions of eluate are shown in Figure 3.

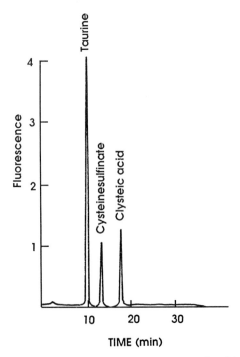

Figure 1. Anion-exchange HPLC of taurine, cysteinesulfinate, and cysteic acid. Detection was by post-column derivatization with OPA and detection with a fluorometer. Fluorescence units are relative units. Gradient A was used. The injection volume contained 200 pmol taurine, 150 pmol cysteinesulfinate and 300 pmol cysteic acid.

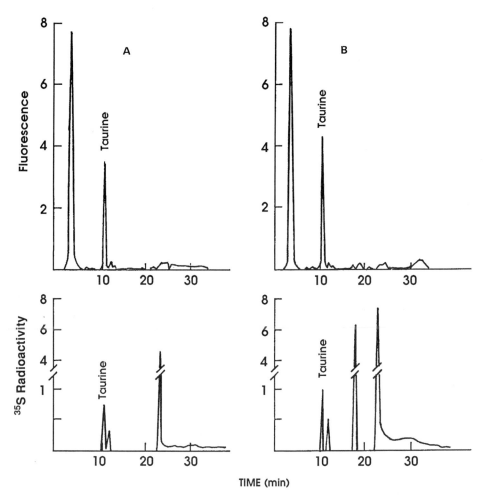

Figure 2. Anion-exchange HPLC of taurine in acid-supernatants from incubations of rat hepatocytes with [^{35}S]cysteine. Panel A was obtain by HPLC of a sample without oxidation of hypotaurine to taurine; panel B is a chromatogram of the same sample analyzed after treatment of the supernatant with H_2O_2 and catalase to convert hypotaurine to taurine. Fluorescence was detected after post-column derivatization of the eluate with OPA; fluorescence units are arbitrary. Radioactivity was counted in fractions of eluate and is plotted below the chromatograms; radioactivity units are also arbitrary. Gradient A was used.

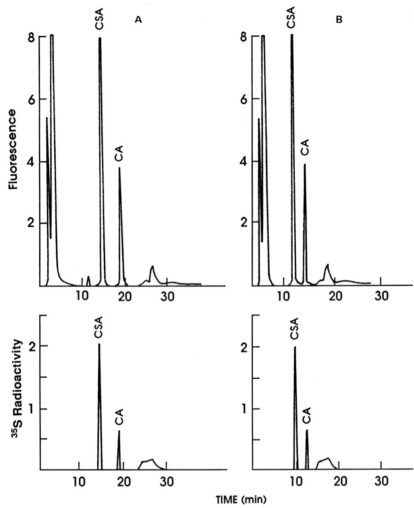

Figure 3. Anion-exchange HPLC of supernatant from assays of cysteine dioxygenase activity in rat liver homogenate. Panel A was obtained using Gradient A; panel B was obtained using Gradient C. Fluorescence was detected after post-column derivatization of the eluate with OPA; fluorescence units are arbitrary. Radioactivity was counted in fractions of eluate and is plotted below the chromatograms, radioactivity units are also arbitrary. CSA = cysteinesulfinate; CA = cysteic acid.

ACKNOWLEDGMENTS

This work was supported by the Cooperative State Research Service, United States Department of Agriculture under Agreement No. GAM900895, by New York State Hatch Project No. 399-492, and by National Institutes of Health Research Grant No. DK-26959.

REFERENCES

1. L.L. Hirschberger, J. De La Rosa, and M.H. Stipanuk, Determination of cysteinesulfinate, hypotaurine and taurine in physiological samples by reversed-phase liquid chromatography, *J. Chromatog.* 343:303 (1985).

2. M.H. Stipanuk, L.L. Hirschberger, and J. De La Rosa, Cysteinesulfinate, hypotaurine and taurine: reversed-phase high-performance liquid chromatography, *Methods Enzymol.* 143:155 (1987).

3. R.M. Coloso, M.R. Drake, and M.H. Stipanuk, Effect of bathocuproine disulfonate, a copper chelator, on cyst(e)ine metabolism by freshly isolated rat hepatocytes, *Am. J. Physiol.* 259:E443 (1990).

4. M.H. Stipanuk, R.M. Coloso, R.A.G. Garcia, and M.F. Banks, Cysteine concentration regulates cysteine metabolism to glutathione, sulfate and taurine in rat hepatocytes, *J. Nutr.* 122: in press (1992).

5. P.J. Bagley and M.H. Stipanuk, Assay of cysteine dioxygenase activity, *FASEB J.* 4:A805 (1990).

HYPOTAURINE IN MALE REPRODUCTION

Ross P. Holmes[1], Harold O. Goodman[2], Chris H. Hurst[1],
Zak K. Shihabi[3], and Jonathan P. Jarow[1]

Departments of Urology[1], Pediatrics[2], and Pathology[3]
Bowman Gray School of Medicine, Wake Forest University
Winston-Salem, NC 27157

INTRODUCTION

Hypotaurine was first identified as a major amino acid of sperm by Horst and Grooten[1]. These studies were extended by Kochakian who demonstrated that hypotaurine was present in most rodent productive tissues and that in some tissues levels were androgen responsive[2]. Within the rodents there was marked species variation in accessory gland hypotaurine content but in reproductive tissues hypotaurine was high in the epididymis and low in the testes. Possible functional roles identified for hypotaurine in sperm include its acting as an antioxidant in preventing sperm lipid peroxidation[3], as a capacitating agent[4], and as a sperm motility factor[5]. The first two roles were also equally well served by taurine, whereas in the latter, only hypotaurine was tested. Thus, from these studies a functional role for hypotaurine, distinct from that of taurine, appears unresolved.

We have recently identified in human sperm an apparent association between sperm hypotaurine content and fertility[6]. this was not shared by taurine which in fact had a negative correlation with fertility. We speculated that the reason for the positive correlation with hypotaurine and the negative correlation with taurine was related to the oxidation of hypotaurine to taurine. Thus, in sperm exposed to oxidant stress, the conversion of hypotaurine to taurine may serve an indicator that the fertilization potential of the sperm has been compromised. Hypotaurine is known to have antioxidant properties[7,8] and this could protect sperm lipids from peroxidative damage. Sperm lipids may be especially vulnerable to oxidation because of their high docosahexenoic acid (22:6) content. This fatty acid has been shown to comprise 47.5% by weight of all the phospholipid fatty acids in human sperm[9]. This indicates that the bulk of the phospholipid molecules have a 22:6 fatty acid at the sn-2 position of the glycerol backbone. Preserving the structural integrity of this fatty acid may be extremely important in maintaining several vital sperm functions.

The association of hypotaurine with male fertility raises several questions that need to be addressed to further understand the physiological function of hypotaurine and how hypotaurine levels are regulated, and ultimately to determine if intervention to modulate sperm hypotaurine levels or in other ways diminish oxidant stress can alleviate infertility in males. Broader questions also addressed in such studies include

Taurine, Edited by J.B. Lombardini *et al.*
Plenum Press, New York, 1992

whether hypotaurine serves an important cellular function distinct from taurine in other tissues and the role of oxidant stress in contributing to male infertility.

The Origin of Sperm Hypotaurine

Sperm hypotaurine could conceivably originate either in the sperm cell via synthesis from cysteine at some stage during spermatogenesis or it could be transported into the cell during its passage along the reproductive tract after its secretion into reproductive fluids. As the circulating concentration of hypotaurine in plasma is ~1 μM[10] and a high-affinity, high-capacity transporter capable of concentrating hypotaurine within cells from this low concentration has not been identified, it is most unlikely that it can be derived from tissue fluid. At present it is not possible to distinguish between these two possibilities but we will examine the available evidence and review some of our preliminary observations.

Although several pathways have been identified for hypotaurine synthesis[11], the major pathway in mammals appears to be from cysteine. Cysteine is first converted to cysteinesulfinic acid (CSA) by cysteine dioxygenase and this is subsequently converted to hypotaurine by CSA decarboxylase. To our knowledge there have been no reports of the activity of these enzymes in reproductive tissues. The first indication as to which tissues are capable of synthesizing hypotaurine is from their hypotaurine content. High concentrations of hypotaurine occur in all sections of the epididymis in rodents[2] and we have confirmed these findings in rats and guinea pigs. The high concentration of hypotaurine in epididymal tissue suggests that it is at least an important site of hypotaurine synthesis. Much lower concentrations occur in the testes and this suggests both that extensive synthesis does not occur there and that the developing spermatozoa have not yet acquired a large amount hypotaurine. However, as immature sperm in the testes comprise only a small proportion of testicular tissue it remains possible that all the hypotaurine and hypotaurine synthesis in testes occurs in immature sperm. This will only be resolved with the isolation and examination of immature sperm from testicular tissue.

When we examined the CSA decarboxylase activity in reproductive tissues of adolescent guinea pigs (450-500 g) we found that this activity paralleled the tissue content: cauda epididymis > caput epididymis > testes in both hypotaurine content and CSA decarboxylase activity (manuscript in preparation). By contrast, epididymal sperm could not synthesize hypotaurine but had a high hypotaurine content. This suggests that the entire epididymis is capable of synthesizing hypotaurine. We have examined epididymal tissues from 4 males (69-80 yrs old) who underwent an orchiectomy for advanced prostate cancer. CSA decarboxylase activity was detected in tissue from two of the patients (480 and 520 nmoles/g/hr) but none in the other two. However, the lack of activity could be attributed to the age of the individual, the disease process, or previous therapy received.

Hypotaurine Secretion into Reproductive Fluids

Johnson et al.[12] examined porcine epididymal fluid and found it to be extremely high in hypotaurine (35 mM) and taurine (10 mM). More recently, Hinton[13] measured the amino acid composition of rat testicular and epididymal fluids. Unfortunately, the amino acid analyzer he used did not separate taurine and hypotaurine adequately but a sample chromatogram shown of caput luminal fluid suggested that the "taurine" peak consisted of 2 components of near equal concentration. His results showed that rat testicular fluid is similar to circulating fluids in taurine/hypotaurine whereas epididymal fluid was high (2-7 mM) in taurine/ hypotaurine. These results suggested that all epididymal segments secreted taurine/hypotaurine into the epididymal lumen. Glutamate, similarly had a high secretion rate into the lumen from the caput and corpus epididymis. However, the secretion of glutamate, by contrast, was low in the caudal epididymis. Hinton proposed that luminal glutathione, the bulk of which is derived from the testes, was the source of the epididymal glutamate[13]. The tripeptide, glutathione, also contains cysteine, the precursor of hypotaurine. In the liver, which is another organ that produces hypotaurine, Higashi et al.[14] suggested that glutathione serves as a reservoir for cysteine. Cysteine may be released from the breakdown of glutathione initiated by the epididymal brush border enzyme, gamma glutamyl-transferase. Since the concentration of cysteine in epididymal fluid is low[13], we propose that the cysteine released from the catabolism of glutathione is taken up by epididymal epithelium and converted to hypotaurine via CSA. The hypotaurine formed is then secreted into the epididymal lumen.

We have calculated that the intracellular concentration of hypotaurine in sperm is ~30 mM in fertile humans. The concentration in the cytoplasm may in fact be much higher as this measurement assumes that hypotaurine is uniformly distributed throughout the cell. The sperm nucleus and acrosome are two specialized sub-structures that occupy a significant portion of the intracellular volume and their hypotaurine content may differ. The analyses of Velazquez et al.[15] on human acrosomes suggested that they contained very little hypotaurine but because they used an amino acid analyzer these results are not definitive. If sperm are able to concentrate hypotaurine during their passage through the epididymis they must possess transport mechanisms that facilitate the process. Most cells possess a ß-amino acid transport system that is capable of sequestering taurine and hypotaurine[11]. This system has a high affinity for taurine with a K_m usually of $<30\,\mu M$ and in most cells has a low capacity. We have measured the ability of sperm to take up $50\,\mu M$ ^{14}C-taurine as an index of their ß-transport system. The competitive effect of hypotaurine on taurine uptake in a number of experimental systems (e.g.) suggests that the transporter recognizes them with near equal affinity. In brain, hypotaurine uptake appeared to be faster than that of taurine[16]. The uptake rates obtained are shown in Table 1. They provide only a crude picture of the capacity of this amino acid transporter. The uptake rate most probably represents the activity of more than one transporter or uptake process and the different components have to be dissected out. Furthermore, as the concentration of taurine used, $50\,\mu M$, is close to the K_m for taurine transport via the ß-amino acid system, the maximal uptake is likely to be twice that listed in the Table. These results indicate that normal human sperm which contain ~500 pmoles hypotaurine/10^6 cells[6] would require ~500 hrs to completely accumulate their required hypotaurine, assuming they started with no hypotaurine and

Table 1. Taurine uptake and hypotaurine content of sperm.

Sperm	Taurine Uptake (pmol/10^6 cells/hr)	Hypotaurine content (pmol/10^6 cells)
Human ejaculate	0.48 ± 0.20	488 ± 299
Rat caput epididymis	1.76 ± 1.15	703 ± 286
Rat cauda epididymis	1.09 ± 0.47	108 ± 59

Sperm were obtained from ejaculates or epididymal segments minced in PBS by centrifugation. They were washed twice in PBS containing 0.1% BSA. Uptake was measured at 37°C for 1 hr in the same buffer containing ^{14}C-taurine. They were washed twice after incubation by centrifugation and the radioactivity measured. Uptake was corrected for non-specific binding by measuring uptake at 0°C and subtracting this value. Results are presented as the means of 3 samples each assayed in duplicate.

took it up at the maximal rate. As human sperm takes ~12 days to move through the epididymis this uptake system could almost account for the hypotaurine uptake required. It is not known whether regulation of the uptake rate occurs as the intracellular taurine compartment becomes saturated and the measured rates may be under-estimated because of down regulation. The uptake rates in rat sperm from the two epididymal segments were similar. This suggests that there is little change in the sperm cell ß-transport system between the time of entry of the sperm into the epididymis and ejaculation. However, sperm from the testes, epididymis and ejaculate of just one species should be studied before final conclusions can be drawn. Substantial transport may also occur via a low-affinity (1-5 mM K_m) and high-capacity amino acid transporting system which has been identified in several cells and tissues[11,17] as the current evidence suggests that epididymal fluid contains mM concentrations of hypotaurine[12,13]. The relative contribution of the two transport processes will only be clarified by further investigation.

Regulation of Reproductive Tissue Hypotaurine

The content of hypotaurine in reproductive tissues and fluids appears to be under hormonal control in both males[2] and females[18]. Kochakian observed in rodents that castration reduced the hypotaurine content of guinea pig seminal vesicles and prostate tissue and that treatment with testosterone restored it. It will be of importance to determine whether the synthesis of hypotaurine in the epididymis is under androgen control. In the ewe, the hypotaurine content changed significantly during the oestrus cycle peaking during ovulation[18]. Studying the regulation of hypotaurine synthesis is important for understanding its role in reproduction and the pathologic process which may result in infertility.

REFERENCES

1. C.J.G. van der Horst and H.J.G. Grooten, The occurrence of hypotaurine and other sulfur-containing amino acids in seminal plasma and spermatozoa of boar, bull and dog, *Biochim. Biophys. Acta* 117:495 (1966).
2. C.D. Kochakian, Sulfur amino acids in the reproductive tract of male guinea pigs, rats, mice and several marine animals, in: "Natural Sulfur Compounds. Novel Biochemical and Structural Aspects", D. Cavallini, G.E. Gaull, and V. Zappia, eds., p. 213, Plenum Press, New York (1980).
3. J.G. Alvarez and B.T. Storey, Taurine, hypotaurine, epinephrine and albumin inhibit lipid peroxidation in rabbit spermatozoa and protect against loss of motility, *Biol. Reprod.* 29:548 (1983).
4. S. Meizel, C.W. Lui, P.K. Working, and R.J. Mrsny, Taurine and hypotaurine: their effects on motility, capacitation and the acrosome reaction of hamster sperm *in vitro* and their presence in sperm and reproductive tract fluids of several mammals, *Develop. Growth Differ.* 22:483 (1980).
5. D.E. Boatman, D.B. Bavister, and E. Cruz, Addition of hypotaurine can reactivate immotile golden hamster spermatozoa, *J. Androl.* 11:66 (1990).
6. R.P. Holmes, H.O. Goodman, Z.K. Shihabi, and J.P. Jarow, The taurine and hypotaurine content of human semen, *J. Androl.* (in press).
7. J.H. Fellman and E.S. Roth, The biological oxidation of hypotaurine to taurine: hypotaurine as an antioxidant, *Prog. Clin. Biol. Res.* 179:71 (1985).
8. O.I. Aruomo, B. Halliwell, B.M. Hoey, and J. Butler, The antioxidant action of taurine, hypotaurine and their metabolic precursors, *Biochem. J.* 256:251 (1988).
9. R. Jones, T. Mann, and R. Sherins, Peroxidative breakdown of phospholipids in human spermatozoa, spermicidal properties of fatty acid peroxides, and protective action of seminal plasma, *Fertil. Steril.* 31:531 (1979).
10. D.B. Learn, V.A. Fried, and E.L. Thomas, Taurine and hypotaurine content of human leukocytes, *J. Leuk. Biol.* 48:174 (1990).
11. C.E. Wright, H.H. Tallan, and Y.Y. Lin, Taurine: biological update, *Ann. Rev. Biochem.* 55:427 (1986).
12. L.A. Johnson, V.G. Pursel, R.J. Gerrits, and C.H. Thomas, Free amino acid composition of porcine seminal epididymal and seminal vesicle fluids, *J. Animal Sci.* 34:430 (1972).

13. B.T. Hinton, The testicular and epididymal luminal amino acid microenvironment in the rat, *J. Androl.* 11:498 (1990).
14. T. Higashi, N. Ateishi, and Y. Sakamoto, Liver glutathione as a reservoir of L-cysteine, *Prog. Clin. Biol. Res.* 125:419 (1983).
15. A. Velazquez, N.M. Delgado, and A. Rosado, Taurine content and amino acid composition of human acrosome, *Life Sci.* 38:991 (1986).
16. P. Kontro and S.S. Oja, Properties of hypotaurine uptake in mouse brain slices, *in*: "The Effects of Taurine on Excitable Tissues", S.W. Schaffer, S.I. Baskin and J.J. Kocsis, eds., p. 49, Spectrum Publications, New York (1981).
17. P. Lahdesmaki and S.S. Oja, On the mechanism of taurine transport at brain cell membranes, *J. Neurochem.* 20:1411 (1973).
18. C.J.G. van der Horst and A. Brand, Occurrence of hypotaurine and inositol in the reproductive tract of the ewe and its regulation by pregnenolone and progesterone, *Nature* 223:67 (1969).

CONTRIBUTORS

J. Azuma
Department of Medicine III
Osaka University Medical School
Osaka, Japan

A. Baba
Department of Pharmacology
Faculty of Pharmaceutical Sciences
Osaka University
1-6 Yamada-Oka
Suita, Osaka 565, Japan

P.J. Bagley
Division of Nutritional Sciences
Cornell University
Ithaca, NY 14853

D.H. Baker
Department of Animal Sciences
  and Division of Nutritional Sciences
University of Illinois
Urbana, IL 61801

M.A. Banks
Division of Ford Chemistry
American Bacteriological and
  Chemical Research Corp.
Gainsville, FL 32608

M.F. Banks
Division of Nutritional Sciences
Cornell University
Ithaca, NY 14853

J. Bao
Department of Physiology and
  Cell Biology
University of Kansas
Lawrence, KS 66045

F. Bennardini
Institute of Biological Chemistry
University of Sassari
Sassari, Italy

L. Beverly
Department of Physiological Sciences
School of Veterinary Medicine
University of California
Davis, CA 95619

A. Bhattacharyya
Department of Physiology and
  Cell Biology
University of Kansas
Lawrence, KS 66045

G. Bkaily
Department of Biophysics
University of Sherbrooke
Sherbrooke, Quebec, Canada

J. Bourguignon
Centre de Neurochimie
Université L. Pasteur
67084 Strasbourg, France

A.M. Budreau
Department of Pediatrics
The University of Tennessee
Memphis College of Medicine
Memphis, TN

V. Castranova
Biochemistry Section, NIOSH
Division of Animal and Veterinary
  Sciences
West Virginia University
Morgantown, WV

M. Cereijido
CINVESTAV
Mexico City, Mexico

R.W. Chesney
Departments of Pediatrics
The University of Tennessee
Memphis College of Medicine
Memphis, TN

R.M. Coloso
Division of Nutritional Sciences
Cornell University
Ithaca, NY 14853

D. Conte Camerino
Pharmacology Unit
Department of Pharmacobiology
University of Bari
Bari, Italy

J. Covarrubias
Nativelle Institute of Pharmacology
Florence, Italy

S. Croswell
Department of Pharmacology
College of Medicine
University of Arizona
Tucson, AZ 85724

G.L. Czarnecki-Maulden
Friskies Research
Nestec Inc.
St. Louis, MO 64503

A. De Luca
Pharmacology Unit
Department of Pharmacobiology
University of Bari
Bari, Italy

B. Drujan
Instituto Venezolano de
  Investigaciones Cientificas
Laboratorio de Neuroquimica
Caracas, Venezuela

J. Duan
Department of Pharmacology
Tianjin Medical College
Tianjin 30070
People's Republic of China

G.R. Dutton
Department of Pharmacology
University of Iowa
  College of Medicine
Iowa City, IA

K.E. Earle
Waltham Centre for Pet Nutrition
Freeby Lane
Waltham-on-the-Wolds
Melton Mowbray
Leicestershire LE14 4RT
United Kingdom

P. Falli
Department of Pharmacology
University of Florence
Florence, Italy

F. Franconi
Institute of Biological Chemistry
Faculty of Pharmacy
University of Sassari
Sassari, Italy

M. Frauli-Meischner
Centre de Neurochimie
67084 Strasbourg, France

A. Gargano
Department of Developmental
  Biochemistry
Institute for Basic Research in
  Developmental Disabilities
Staten Island, NY

A. Giotti
Department of Pharmacology
University of Florence
Florence, Italy

S.N. Giri
Department of Veterinary
  Pharmacology and Toxicology
University of California
Davis, CA 95616

E.N. Glass
Cornell University
School of Veterinary Medicine
Ithaca, NY 14853

H.O. Goodman
Department of Pediatrics
Bowman Gray School of Medicine
Wake Forest University
Winston-Salem, NC 27157

R.E. Gordon
Department of Pathology
Mt. Sinai School of Medicine
New York, NY 10029

K. Greene
School of Veterinary Medicine
University of California
Davis, CA 95619

P. Guerin
Centre de Neurochimie
67084 Strasbourg, France

R. Gupta
Centre de Neurochimie
Université L. Pasteur
67084 Strasbourg, France

T. Hamaguchi
Department of Medicine III
Osaka University Medical School
Osaka, Japan

H. Hara
Department of Medicine III
Osaka University Medical School
Osaka, Japan

H. Harada
Department of Medicine III
Osaka University Medical School
Osaka, Japan

K.C. Hayes
Foster Biomedical Research
   Laboratory
Brandeis University
Waltham, MA 02254

Rachel F. Heller
Department of Pathology
   Box 1194
Mount Sinai School of Medicine
New York, NY 10029

Richard F. Heller
Department of Pathology
   Box 1194
Mount Sinai School of Medicine
New York, NY 10029

M.A. Hickman
Department of Physiological Sciences
School of Veterinary Medicine
University of California
Davis, CA 95619

L.L. Hirschberger
Division of Nutritional Sciences
Cornell University
Ithaca, NY 14853

P.M. Hogan
Department of Companion Animals
Altantic Veterinary College
University of Prince Edward Island
Charlottetown, P.E.I, Canada

R.P. Holmes
Department of Urology
Bowman Gray School of Medicine
Wake Forest University
Winston-Salem, NC 27157

C.H. Hurst
Department of Urology
Bowman Gray School of Medicine
Wake Forest University
Winston-Salem, NC 27157

R.J. Huxtable
Department of Pharmacology
College of Medicine
University of Arizona
Tucson, AZ 85724

Y. Ihara
Department of Medicine III
Osaka University Medical School
Osaka, Japan

H. Imaki
Institute for Basic Research in
   Developmental Disabilities
Staten Island, NY

T. Ishibashi
Department of Pharmacology
Faculty of Pharmaceutical Sciences
Osaka University
1-6 Yamada-Oka
Suita, Osaka, Japan

H. Iwata
Department of Pharmacology
Faculty of Pharmaceutical Sciences
Osaka University
1-6 Yamada-Oka
Suita, Osaka 565, Japan

J.P. Jarow
Department of Urology
Bowman Gray School of Medicine
Wake Forest University
Winston-Salem, NC 27157

D.P. Jones
Department of Pediatrics
The University of Tennessee
Memphis College of Medicine
Memphis, TN

Y. Kang
Department of Pharmacology
Tianjin 30070
People's Republic of China

C. Kirk
Department of Physiological Sciences
School of Veterinary Medicine
University of California
Davis, CA 95619

M.D. Kittleson
Department of Medicine
School of Veterinary Medicine
University of California
Davis, CA 95619

I. Koyama
Research Center
Taisho Pharmaceutical Co., Ltd.
Tokyo, Japan

Y. Koyama
Department of Pharmacology
Faculty of Pharmaceutical Sciences
Osaka University
1-6 Yamada-Oka
Suita, Osaka, Japan

N. Lake
Departments of Physiology and
 Ophthalmology
McGill University
Montreal, Quebec, Canada H3G 1Y6

W.S. Lapp
Department of Physiology
McGill University
Montreal, Quebec, Canada H3G 1Y6

A. Lázaro
CINVESTAV
Mexico City, Mexico

Y.H. Lee
Department of Physiology and
 Cell Biology
University of Kansas
Lawrence, KS 66045

I.H. Lelong
Centre de Neurochimie
67084 Strasbourg, France

J. Lewis
School of Veterinary Medicine
University of California
Davis, CA 95619

P. Li
Department of Pharmacology
Tianjin Medical College
Tianjin 30070
People's Republic of China

L. Lima
Instituto Venezolano de
 Investigaciones Cientificas
Laboratorio de Neuroquimica
Caracas, Venezuela

P.L. Lleu
Department of Pharmacology
College of Medicine
University of Arizona
Tucson, AZ 85724

J.B. Lombardini
Department of Pharmacology
Texas Tech University Health
 Sciences Center
Lubbock, TX 79430

C. Lopez Apreza
Institute of Cellular Physiology
The National University of Mexico
04510 Mexico City, Mexico

P. Lu
Department of Developmental
  Biochemistry
Institute for Basic Research in
  Developmental Disabilities
Staten Island, NY

W.B. Martin
Biochemistry Section, NIOSH
Division of Animal and Veterinary
  Sciences
West Virginia University
Morgantown, WV

A. Mattana
Institute of Biological Chemistry
University of Sassari
Sassari, Italy

P. Matus
Instituto Venezolano de
  Investigaciones Cientificas
Laboratorio de Neuroquimica
Caracas, Venezuela

J.M. Messing
Department of Developmental
  Biochemistry
Institute for Basic Research in
  Developmental Disabilities
Staten Island, NY

M. Miceli
Clinical Laboratory
Annunziata Hospital
Florence, Italy

L.A. Miller
Department of Pediatrics
The University of Tennessee
Memphis College of Medicine
Memphis, TN

L.M. Miller
Altantic Veterinary College
University of Prince Edward Island
Charlottetown, P.E.I, Canada

J. Morán
Institute of Cell Physiology
National University of Mexico
04510 Mexico City, Mexico

J.G. Morris
Department of Physiological Sciences
School of Veterinary Medicine
University of California
Davis, CA 95619

T. Nakamura
Research Center
Taisho Pharmaceutical Co., Ltd.
Tokyo, Japan

B. Nathan
Department of Physiology and
  Cell Biology
University of Kansas
Lawrence, KS 66045

S. Negoro
Department of Medicine III
Osaka University Medical School
Osaka, Japan

M. Nemoto
Research Center
Taisho Pharmaceutical Co., Ltd.
Tokyo, Japan

M. Neuringer
Oregon Regional Primate Research
  Center
Beaverton, OR

M.J. Novotny
Department of Anatomy and Physiology
Altantic Veterinary College
University of Prince Edward Island
Charlottetown, P.E.I, Canada

J. Odle
Department of Animal Sciences and
  Division of Nutritional Sciences
University of Illinois
Urbana, IL 61801

M. Ogasawara
Research Center
Taisho Pharmaceutical Co., Ltd.
Tokyo, Japan

S.S. Oja
Tampere Brain Research Center
Department of Biomedical Sciences
University of Tampere
SF-33101 Tampere, Finland

T. Palackal
Institute for Basic Research in
  Developmental Disabilities
Staten Island, NY

H. Pasantes-Morales
Institute of Cell Physiology
The National University of Mexico
04510 Mexico City, D.F. Mexico

J. Petegnief
Centre de Neurochimie
Université L. Pasteur
67084 Strasbourg, France

S. Pierno
Pharmacology Unit
Department of Pharmacology
University of Bari
Bari, Italy

P.D. Pion
Department of Medicine
School of Veterinary Medicine
University of California
Davis, CA 95619

D.W. Porter
Biochemistry Section, NIOSH
Division of Animal and
  Veterinary Sciences
West Virginia University
Morgantown, WV

S. Punna
Department of Medicine III
Osaka University Medical School
Osaka, Japan

O. Quesada
Department of Developmental
  Biochemistry
Institute for Basic Research in
  Developmental Disabilities
Staten Island, NY

G. Rebel
Centre de Neurochimie
67084 Strasbourg, France

Q.R. Rogers
Department of Physiological Sciences
School of Veterinary Medicine
University of California
Davis, CA 95619

K.L. Rogers
Department of Psychiatry
University of Iowa College of Medicine
Iowa City, IA 52242

R. Sánchez-Olea
Institute of Cell Physiology
National A. University of Mexico
04510 Mexico City, Mexico

P. Saransaari
Tampere Brain Research Center
Department of Biomedical Sciences
University of Tampere
SF-33101 Tampere, Finland

H. Satoh
Department of Physiology
  and Biophysics
University of Cincinnati College of
  Medicine
Cincinnati, OH 45267

A. Sawamura
Department of Medicine III
Osaka University Medical School
Osaka, Japan

P. Scala
Department of Developmental
  Biochemistry
Institute for Basic Research in
  Developmental Disabilities
Staten Island, NY

S. Schaffer
Department of Pharmacology
University of South Alabama School
  of Medicine
Mobile, AL 36688

A. Schousboe
PharmaBiotec Reserch Center
Department of Biological Sciences
The Royal Danish School of Pharmacy
DKSK-2100 Copenhagen, Denmark

G.B. Schuller-Levis
Institute for Basic Research in
  Developmental Disabilities
Staten Island, NY

G. Seghieri
Diabetes Unit
Riuniti Hospital
Pistoia, Italy

Z.K. Shihabi
Department of Pathology
Bowman Gray School of Medicine
Wake Forest University
Winston-Salem, NC 27157

M.L. Skiles
Department of Medicine
School of Veterinary Medicine
University of California
Davis, CA 95616

G.L. Smith
Institute of Physiology
Glasgow University
G12 8QQ Scotland

P.M. Smith
Waltham Centre for Pet Nutrition
Freeby Lane
Waltham-on-the-Wolds
Melton Mowbray
Leicestershire LE14 4RT
United Kingdom

N. Sperelakis
Department of Physiology and
  Biophysics
University of Cincinnati College of
  Medicine
Cincinnati, OH 45267

D.S. Steele
Institute of Physiology
Glasgow University
G12 8QQ Scotland

M.H. Stipanuk
Division of Nutritional Sciences
Cornell University
Ithaca, NY 14853

J.A. Sturman
Department of Developmental
  Biochemistry
Institute for Basic Research in
  Developmental Disabilities
Staten Island, NY

K. Takahashi
Department of Medicine III
Osaka University Medical School
Osaka, Japan

X.W. Tang
Department of Physiology and
  Cell Biology
University of Kansas
Lawrence, KS 66045

E.A. Trautwein
Foster Biomedical Research
  Laboratory
Brandeis University
Waltham, MA 02254

E. Trenkner
Institute for Basic Research in
  Developmental Disabilities
Staten Island, NY

W.H. Tsai
Academia Sinica
Taipei, Taiwan, Republic of China

N.M. van Gelder
Centre de recherche en sciences
  neurologiques
Département de physiologie
Université de Montréal
Montréal, Québec, Canada

G.-X. Wang
Department of Pharmacology
Tianjin Medical College
Tianjin 30070
People's Republic of China

H. Wang
Department of Pharmacology
Tianjin Medical College
Tianjin 30070
People's Republic of China

Q. Wang
Department of Veterinary Pharmacology
  and Toxicology
University of California
Davis, CA 95616

E.D. Wright
Department of Physiology
McGill University
Montreal, Quebec, Canada H3G 1Y6

J.-Y. Wu
Department of Physiology and
  Cell Biology
University of Kansas
Lawrence, KS 66045

M. Yarom
Department of Physiology and
  Cell Biology
University of Kansas
Lawrence, KS 66045

T. Yoshida
Research Center
Taisho Pharmaceutical Co., Ltd.
Tokyo, Japan

S. Zhou
Department of Pharmacology
Tianjin Medical College
Tianjin 30070
People's Republic of China

INDEX

Action potential, myocardial
  effect of taurine on, 131
  effect of taurine depletion on, 174
α-Adrenoceptors myocardial, 110, 115
Aging effect of
  on brain taurine content, 248, 249
  on lymphocyte function, 229-239
  on synaptosomal taurine transport,
    215-220
ß-Alanine
  inhibition of taurine transport by,
    407
  interaction with taurine receptors,
    265
  tissue taurine depletion by
    effect on contraction, 106
    effect on cysteine sulfinic acid
      decarboxylase activity, 11
    effect on immune function, 241-
      243
  transport in glial cells, effect of
    HEPES, 280
Alzheimer's disease, effect on taurine
  content, 248
Amiodarone-induced lung fibrosis,
  329-340
Antiarrhythmic activity, 105, 106,
  187-192
Anticholinergic activity, 191
Antithrombotic activity, 185
Arrhythmias
  effect of taurine depletion on,
    173-179
  effect of taurine on
    adrenalin-induced, 105, 189-191
    ischemia-reperfusion induced,
      174, 175, 188, 189
    strophanthin-induced, 105
    veratrine-induced, 189-191
Aspirin-induced inhibition of platelet
  aggregation, 181-186

Benzodiazepine receptor
  effect of endogenous brain
    modulators, 295-301
  relation to taurine receptor, 264,
    267
Bile acid conjugation
  involvement of taurine in, 45-54
  relation to taurine depletion
    in cats, 39, 40
    in rhesus monkeys, 209, 212
Bleomycin-induced lung fibrosis
  effect of taurine and niacin on,
    329-340
Brain
  alterations of taurine content by
    Alzheimer's disease, 248
    hypertension, 249
    irradiation, 249
    pregnancy, 249
    seizures, 247, 248
  ischemia-induced taurine release
    from, 245-247
  phosphorylation of P_2 synaptosomal
    proteins, 310-314

Caffeine-induced calcium transients,
  163-172
Calcium paradox, 153-161
Calcium transport
  in bleomycin-induced lung fibrosis,
    331-336
  in brain, 254
  in heart, 113-115, 129-143, 149-151,
    191, 193-198
  in lymphocytes, 229-239
  in retinal cells, 291-293
Calmodulin
  interaction with taurine, 115
  involvement in taurine-dependent
    phosphorylation, 310-317
cAMP production, 110, 129, 189-191

451

Taurine (continued)
  depletion
    effect of supplementation
      by amino acids, 15-22
      by cysteic acid, 23-32
    influence on
      axon characteristics, 304
      bile salt turnover, 48-51
      bleomycin or paraquat-induced
        lung injury, 319-328
      calcium homeostasis, 113
      contractile function, 3, 63-81,
        106-108, 173-179
      development, 4
      immune function, 3, 4, 241-243
      ischemia-induced arrhythmias,
        174, 175
      leukocyte function, 83-90
      membrane conductance, 199-
        205
      membrane damage, 355-359
      retinal degeneration, 3, 33, 38,
        207, 211, 212
    effect of high dietary intake, 91-
      98
    efflux
      effect of aging, 217-219
      effect of glutamate-induced
        swelling, 375-380
      effect of ischemia-hypoxia, 245-
        247
      effect of osmolality, 361-374,
        381-384, 388, 391-397
      from cultured cerebral cortex
        neurons, 391-397
      from cultured cerebellar
        neurons, 269-276
      from retina, effect of potassium,
        399-404
    excretion, effect of commercial
      diets, 47-52
    fetal and lactational losses of, 41
    in development
      of cerebral cortex, 215, 216
      of neuronal tissue, 255-260
    lipid peroxidation, 188, 191
    membrane phosphorylation, 309-
      318
    phospholipid methylation, 121-128,
      224-226
    platelet aggregation, 181-186
    positive inotropic effect, 129-131,
      140, 143, 160, 163-172, 193-
      198
    potassium current, 134-142

Taurine (continued)
  protective role of
    in calcium paradox, 153-161
    in lung injury, 319-354
  receptors
    in brain, 263-268
    modulation by endogenous brain
      modulators, 295-301
  sarcoplasmic reticular calcium
  release, 142, 167, 168, 195-197
  seizures, 249, 250
  sodium calcium exchange, 140-143,
    195-197
  sodium current, 129-143
  trophic role, 287-294
  uptake
    by astrocytes, 366, 367, 375-380
    by cerebral cortex slices, 215-220
    by lymphocytes, 231
    by renal epithelial cells, 366, 405-
      411
    by sperm, 439, 440
    effect of aging, 215-220
    effect of HEPES, 277-285
    effect of osmolality, 366, 367, 387,
      388, 407-411
    effect of sodium, 407, 409

Thromboxane A_2, synthesis of, 181,
  185

Veratridine
  effect on taurine release, 246, 249,
    270, 275

BRAD